COOPERATIVE WIRELESS COMMUNICATIONS

WIRELESS NETWORKS AND MOBILE COMMUNICATIONS

Dr. Yan Zhang, Series Editor
Simula Research Laboratory, Norway
E-mail: yanzhang@ieee.org

Cooperative Wireless Communications
Yan Zhang, Hsiao-Hwa Chen and Mohsen Guizani
ISBN: 1-4200-6469-X

Unlicensed Mobile Access Technology: Protocols, Architectures, Security, Standards and Applications
Yan Zhang, Laurence T. Yang and Jianhua Ma
ISBN: 1-4200-5537-2

Wireless Quality-of-Service: Techniques, Standards and Applications
Maode Ma, Mieso K. Denko and Yan Zhang
ISBN: 1-4200-5130-X

Broadband Mobile Multimedia: Techniques and Applications
Yan Zhang, Shiwen Mao, Laurence T. Yang and Thomas M Chen
ISBN: 1-4200-5184-9

The Internet of Things: From RFID to the Next-Generation Pervasive Networked Systems
Lu Yan, Yan Zhang, Laurence T. Yang and Huansheng Ning
ISBN: 1-4200-5281-0

Millimeter Wave Technology in Wireless PAN, LAN and MAN
Shao-Qiu Xiao, Ming-Tuo Zhou and Yan Zhang
ISBN: 0-8493-8227-0

Security in Wireless Mesh Networks
Yan Zhang, Jun Zheng and Honglin Hu
ISBN: 0-8493-8250-5

Resource, Mobility and Security Management in Wireless Networks and Mobile Communications
Yan Zhang, Honglin Hu, and Masayuki Fujise
ISBN: 0-8493-8036-7

Wireless Mesh Networking: Architectures, Protocols and Standards
Yan Zhang, Jijun Luo and Honglin Hu
ISBN: 0-8493-7399-9

Mobile WIMAX: Toward Broadband Wireless Metropolitan Area Networks
Yan Zhang and Hsiao-Hwa Chen
ISBN: 0-8493-2624-9

AUERBACH PUBLICATIONS

www.auerbach-publications.com
To Order Call: 1-800-272-7737 • Fax: 1-800-374-3401
E-mail: orders@crcpress.com

COOPERATIVE WIRELESS COMMUNICATIONS

Edited by

Yan Zhang
Hsiao-Hwa Chen
Mohsen Guizani

CRC Press
Taylor & Francis Group
Boca Raton London New York

CRC Press is an imprint of the
Taylor & Francis Group, an **informa** business

AN AUERBACH BOOK

CRC Press
Taylor & Francis Group
6000 Broken Sound Parkway NW, Suite 300
Boca Raton, FL 33487-2742

First issued in paperback 2019

© 2009 by Taylor & Francis Group, LLC
CRC Press is an imprint of Taylor & Francis Group, an Informa business

No claim to original U.S. Government works

ISBN-13: 978-1-4200-6469-8 (hbk)
ISBN-13: 978-0-367-38590-3 (pbk)

Library of Congress Cataloging-in-Publication Data

Zhang, Yan, 1977-
   Cooperative wireless communications / Yan Zhang, Hsiao-Hwa Chen, and Mohsen Guizani.
     p. cm. -- (Wireless networks and mobile communications)
   Includes bibliographical references and index.
   ISBN 978-1-4200-6469-8 (alk. paper)
   1. Wireless communication systems. I. Chen, Hsiao-Hwa. II. Guizani, Mohsen. III. Title.

TK5103.2.Z525 2009
621.384--dc22   2008037931

Visit the Taylor & Francis Web site at
http://www.taylorandfrancis.com

and the CRC Press Web site at
http://www.crcpress.com

Contents

PART II   Techniques

Editors

Yan Zhang received his BS in communication engineering from Nanjing University of Post and Telecommunications, China; his MS in electrical engineering from Beijing University of Aeronautics and Astronautics, China; and his PhD from the School of Electrical & Electronics Engineering, Nanyang Technological University, Singapore. He is an associate editor on the editorial board of *Wiley Wireless Communications and Mobile Computing* (*WCMC*), *Security and Communication Networks* (Wiley); *International Journal of Network Security*; *International Journal of Ubiquitous Computing*; *Transactions on Internet and Information Systems* (*TIIS*); *International Journal of Autonomous and Adaptive Communications Systems* (*IJAACS*); *International Journal of Ultra Wideband Communications and Systems* (*IJUWBCS*); and *International Journal of Smart Home* (*IJSH*). He is currently serving as the book series editor for *Wireless Networks and Mobile Communications* book series (Auerbach Publications, CRC Press, Taylor & Francis Group). He serves as guest coeditor for the following: *IEEE Intelligent Systems*, special issue on "Context-Aware Middleware and Intelligent Agents for Smart Environments"; *Wiley Security and Communication Networks* special issue on "Secure Multimedia Communication"; *Springer Wireless Personal Communications* special issue on selected papers from ISWCS 2007; *Elsevier Computer Communications* special issue on "Adaptive Multicarrier Communications and Networks"; *Inderscience International Journal of Autonomous and Adaptive Communications Systems* (*IJAACS*) special issue on "Cognitive Radio Systems"; *The Journal of Universal Computer Science* (*JUCS*) special issue on "Multimedia Security in Communication"; *Springer Journal of Cluster Computing* special issue on "Algorithm and Distributed Computing in Wireless Sensor Networks"; *EURASIP Journal on Wireless Communications and Networking* (*JWCN*) special issue on "OFDMA Architectures, Protocols, and Applications"; and *Springer Journal of Wireless Personal Communications* special issue on "Security and Multimodality in Pervasive Environments."

He is serving as coeditor for several books: *Resource, Mobility and Security Management in Wireless Networks and Mobile Communications*; *Wireless Mesh Networking: Architectures, Protocols and Standards*; *Millimeter-Wave Technology in Wireless PAN, LAN and MAN*; *Distributed Antenna Systems: Open Architecture for Future Wireless Communications*; *Security in Wireless Mesh Networks*; *Mobile WiMAX: toward Broadband Wireless Metropolitan Area Networks*; *Wireless Quality-of-Service: Techniques, Standards and Applications*; *Broadband Mobile Multimedia: Techniques and Applications*; *Internet of Things: From RFID to the Next-Generation Pervasive Networked Systems*; *Unlicensed Mobile Access Technology: Protocols, Architectures, Security, Standards and Applications*; *Cooperative Wireless Communications*; *WiMAX Network Planning and Optimization*; *RFID Security: Techniques, Protocols and System-on-Chip Design*; *Autonomic Computing and Networking*; *Security in RFID and Sensor Networks*; *Handbook of Research on Wireless Security*; *Handbook of Research on Secure Multimedia Distribution*; *RFID and Sensor Networks*; *Cognitive Radio Networks*; *Wireless Technologies for Intelligent Transportation Systems*; *Vehicular Networks: Techniques, Standards and Applications*; *Orthogonal Frequency Division Multiple Access (OFDMA)*; *Game Theory for Wireless Communications and Networking*; and *Delay Tolerant Networks: Protocols and Applications*.

Dr. Zhang serves as symposium cochair for ChinaCom 2009; program cochair for BROAD-NETS 2009; program cochair for IWCMC 2009; workshop cochair for ADHOCNETS 2009; general cochair for COGCOM 2009; program cochair for UC-Sec 2009; journal liasion chair for IEEE BWA 2009; track cochair for ITNG 2009; publicity cochair for SMPE 2009; publicity cochair for COMSWARE 2009; publicity cochair for ISA 2009; general cochair for WAMSNet 2008; publicity cochair for TrustCom 2008; general cochair for COGCOM 2008; workshop cochair for IEEE APSCC 2008; general cochair for WITS-08; program cochair for PCAC 2008; general cochair for CONET 2008; workshop chair for SecTech 2008; workshop chair for SEA 2008;

workshop co-organizer for MUSIC'08; workshop co-organizer for 4G-WiMAX 2008; publicity cochair for SMPE-08; international journals coordinating cochair for FGCN-08; publicity cochair for ICCCAS 2008; workshop chair for ISA 2008; symposium cochair for ChinaCom 2008; industrial cochair for MobiHoc 2008; program cochair for UIC-08; general cochair for CoNET 2007; general cochair for WAMSNet 2007; workshop cochair FGCN 2007; program vice-cochair for IEEE ISM 2007; publicity cochair for UIC-07; publication chair for IEEE ISWCS 2007; program cochair for IEEE PCAC'07; special track cochair for "Mobility and Resource Management in Wireless/Mobile Networks" in ITNG 2007; special session co-organizer for "Wireless Mesh Networks" in PDCS 2006; and a member of technical program committee for numerous international conference, including ICC, GLOBECOM, WCNC, PIMRC, VTC, CCNC, AINA, and ISWCS. He received the Best Paper Award at the IEEE 21st International Conference on Advanced Information Networking and Applications (AINA-07).

From August 2006, Dr. Zhang has been working with Simula Research Laboratory, Norway (http://www.simula.no/). His research interests include resource, mobility, spectrum, data, energy, and security management in wireless networks and mobile computing. He is a member of IEEE and IEEE ComSoc.

Hsiao-Hwa Chen is currently a full professor in the Department of Engineering Science, National Cheng Kung University, Taiwan. He received his BSc and MSc with the highest honor from Zhejiang University, China, and his PhD from the University of Oulu, Finland, in 1982, 1985, and 1990, respectively, all in electrical engineering. He worked as a research associate at the Academy of Finland from 1991 to 1993, and as a lecturer and then a senior lecturer at the National University of Singapore from 1992 to 1997. He joined the Department of Electrical Engineering, National Chung Hsing University, Taiwan, as an associate professor in 1997 and was promoted to a full professor in 2000. In 2001, he joined National Sun Yat-Sen University, Taiwan, as the founding chair of the Institute of Communications Engineering of the University. Under his strong leadership the institute was ranked the second in the country in terms of SCI journal publications and National Science Council funding per faculty member in 2004. In particular, National Sun Yat-Sen University was ranked first in the world in terms of the number of SCI journal publications in wireless LANs research papers during 2004 to mid-2005, according to a research report released by the Office of Naval Research, Arlington. He was a visiting professor to the Department of Electrical Engineering, University of Kaiserslautern, Germany in 1999; the Institute of Applied Physics, Tsukuba University, Japan in 2000; the Institute of Experimental Mathematics, University of Essen, Germany in 2002 (under DFG fellowship); the Chinese University of Hong Kong in 2004; and the City University of Hong Kong in 2007.

Dr. Chen's current research interests include wireless networking, MIMO systems, information security, and Beyond 3G wireless communications. He is the inventor of next generation CDMA technologies and is a recipient of numerous research and teaching awards from the National Science Council, the Ministry of Education, and other professional groups in Taiwan. He has authored or coauthored over 200 technical papers in major international journals and conferences, 5 books, and several book chapters in the area of communications, including the book titled *Next Generation Wireless Systems and Networks* (512 pages) and *The Next Generation CDMA Technologies* (468 pages), both by Wiley in 2005 and 2007, respectively. He has been an active volunteer for IEEE's various technical activities for over 15 years. Currently, he is serving as the chair of IEEE Communications Society Radio Communications Committee, and the vice-chair of IEEE Communications Society Communications & Information Security Technical Committee. He served or is serving as symposium chair/cochair of many major IEEE conferences, including IEEE VTC 2003 Fall, IEEE ICC 2004, IEEE Globecom 2004, IEEE ICC 2005, IEEE Globecom 2005, IEEE ICC 2006, IEEE Globecom 2006, IEEE ICC 2007, and IEEE WCNC 2007.

Dr. Chen served or is serving as editorial board member and/or guest editor of *IEEE Communications Letters*; *IEEE Communications Magazine*; *IEEE Wireless Communications Magazine*; *IEEE JSAC*; *IEEE Network Magazine*; *IEEE Transactions on Wireless Communications*; and *IEEE Vehicular Technology Magazine*. He is the editor-in-chief of Wiley's *Security and Communication Networks* journal (www.interscience.wiley.com/journal/security), and the special issue editor-in-chief of *Hindawi Journal of Computer Systems, Networks, and Communications* (http://www.hindawi.com/journals/jcsnc/). He is also serving as the chief editor (Asia and Pacific) for Wiley's *Wireless Communications and Mobile Computing* journal and Wiley's *International Journal of Communication Systems*. His original work in CDMA wireless networks, digital communications, and radar systems has resulted in five U.S. patents, two Finnish patents, three Taiwanese patents, and two Chinese patents, some of which have been licensed to industry for commercial applications. He is an adjunct professor of Zhejiang University, China, and Shanghai Jiao Tong University, China. Professor Chen is a recipient of the Best Paper Award in IEEE WCNC 2008.

Mohsen Guizani is currently a full professor and the chair of the Computer Science Department at Western Michigan University since 2003. Previously, he served as the chair of the Computer Science Department at the University of West Florida from 1999 to 2003. He was an associate professor of electrical and computer engineering and the director of graduate studies at the University of Missouri-Columbia from 1997 to 1999. Prior to joining the University of Missouri, he was a research fellow at the University of Colorado-Boulder. From 1989 to 1996, he held academic positions at the Computer Engineering Department at the University of Petroleum and Minerals, Dhahran, Saudi Arabia. He was also a visiting professor in the Electrical and Computer Engineering Department at Syracuse University, Syracuse, New York during the academic year 1988–1989. He received his BS (with distinction) and MS in electrical engineering; MS and PhD in computer engineering in 1984, 1986, 1987, and 1990, respectively, all from Syracuse University, Syracuse, New York.

Dr. Guizani's research interests include wireless communications and computing, computer networks, design and analysis of computer systems, and optical networking. He served or is serving on the editorial boards of more than 20 national and international journals, such as the *IEEE Transaction on Wireless Communications* (*TWireless*); *IEEE Transaction on Vehicular Technology* (*TVT*); *IEEE Communications Magazine*; the *Journal of Parallel and Distributed Systems and Networks*; and the *International Journal of Computer Research*. He served as a guest editor in the *IEEE Communication Magazine*; *IEEE Journal on Selected Areas in Communications*; *IEEE Network Magazine*; *Journal of Communications and Networks*; *The Simulation Transaction*; *International Journal of Computer Systems and Networks*; *International Journal of Communication Systems*; *International Journal of Computing Research*; and *Journal of Cluster Computing*. Dr. Guizani is the founder and editor-in-chief of the *Wireless Communications and Mobile Computing* journal published by Wiley (http://www.interscience.wiley.com/jpages/1530-8669/). He is also the founder and general chair of the two international conferences: International Wireless Conference on Wireless Communications and Mobile Computing (ACM IWCMC); and Wireless Networks, Communications, and Mobile Computing (IEEE WirelessCom). He is the author/coauthor of six books and more than 180 articles in refereed journals and conferences in the areas of high-speed networking, wireless networking and communications, mobile computing, and optical networking and network security. He has served as a keynote speaker for many international conferences as well as presented a number of tutorials and workshops. He served as the general chair for the Parallel and Distributed Computer Systems (PDCS 2002); IEEE Vehicular Technology Conference 2003 (VTC'03); PDCS 2003; IEEE WirelessCom 2005; ACS/IEEE AICCSA 2006; and IEEE IWCMC 2006. He also served as the program and symposia chair for many conferences and symposia in IEEE Globecom and IEEE ICC.

Dr. Guizani is the chair of the IEEE Communications Society Technical Committee on Transmissions, Access, and Optical Systems (IEEE TAOS), the vice-chair of the IEEE Communications Society of Personal Communications (IEEE TCPC), and a member of other IEEE ComSoc technical committees. He was the IEEE Computer Society Distinguished National Speaker from 2003 to

2005. He is also ABET Accreditation Evaluator for Computer Science and Information Technology Programs.

Dr. Guizani received both the Best Teaching Award and the Excellence in Research Award from the University of Missouri-Columbia in 1999 (a college-wide competition). He won the best Research Award from KFUPM in 1995 (a university-wide competition). He was selected as the Best Teaching Assistant for two consecutive years at Syracuse University in 1988 and 1989.

Dr. Guizani is a senior member of IEEE, and a member of IEEE Communication Society, IEEE Computer Society, ASEE, and ACM.

For more details, please visit: http://www.cs.wmich.edu/ mguizani/

Contributors

Baher Abdulhai
Civil Engineering Department
University of Toronto
Toronto, Ontario, Canada

Raviraj S. Adve
Department of Electrical and Computer
 Engineering
University of Toronto
Toronto, Ontario, Canada

Hazim Ahmed
Software Systems Engineering Department
University of Regina
Regina, Saskatchewan, Canada

Almudena Alcaide
Computer Science Department
Carlos III University of Madrid
Madrid, Spain

Y. Bar-Ness
Center for Wireless Communications
 and Signal Processing Research
ECE Department
New Jersey Institute of Technology
Newark, New Jersey

Elena-Veronica Belmega
Université Paris-Sud XI
SUPELEC, Signals and Systems Laboratory
Gif-sur-Yvette, France

Aggelos Bletsas
Department of Physics
Radio Communications Laboratory
Aristotle University of Thessaloniki
Thessaloniki, Greece

and

Electronic and Computer Engineering
 Department
Technical University of Crete
Crete, Greece

Lin Cai
Department of Electrical and Computer
 Engineering
University of Victoria
Victoria, British Columbia, Canada

Symeon Chatzinotas
Centre for Communication System Research
University of Surrey
Guildford, United Kingdom

Yifan Chen
School of Engineering
University of Greenwich
Greenwich, United Kingdom

Yuanzhu Peter Chen
Department of Computer Science
Memorial University of Newfoundland
Saint John's, Newfoundland, Canada

Josephine P. K. Chu
Department of Electrical and Computer
 Engineering
University of Toronto
Toronto, Ontario, Canada

Zaher Dawy
Electrical and Computer Engineering
 Department
American University of Beirut
Beirut, Lebanon

Mérouane Debbah
SUPELEC, Alcatel-Lucent Chair on
 Flexible Radio
Gif-sur-Yvette, France

Nikos Dimokas
Department of Informatics
Aristotle University of Thessaloniki
Thessaloniki, Greece

Andrew W. Eckford
Department of Computer Science
 and Engineering
York University
Toronto, Ontario, Canada

Mohamed El-Darieby
Software Systems Engineering Department
University of Regina
Regina, Saskatchewan, Canada

David Fusté-Vilella
Computer Architecture Department
Technical University of Catalonia
Barcelona, Spain

Jorge García-Vidal
Computer Architecture Department
Technical University of Catalonia
Barcelona, Spain

Zhu Han
Electrical and Computer Engineering
  Department
University of Houston
Houston, Texas

Julio C. Hernández-Castro
Computer Science Department
Carlos III University of Madrid
Madrid, Spain

Luoquan Hu
Suzhou Testing Center for Information
  Technology Products
Suzhou Entry-Exit Inspection and
  Quarantine Bureau
People's Republic of China
Suzhou, Jiangsu

and

School of Electronics and Information
Soochow University
People's Republic of China
Suzhou, Jiangsu

Muhammad Ali Imran
Centre for Communication System Research
University of Surrey
Guildford, United Kingdom

Alexandros Kaloxylos
Department of Telecommunications
  Science and Technology
University of Peloponnese
Tripoli, Greece

Vasileios Karyotis
School of Electrical and Computer
  Engineering
National Technical University of Athens
Athens, Greece

Dimitrios Katsaros
Department of Computer and
  Communication Engineering
University of Thessaly
Thessaly, Greece

Samson Lasaulce
CNRS
SUPELEC, Signals and Systems Laboratory
Gif-sur-Yvette, France

Jijun Luo
Radio Access LTE System Product
  Management
Nokia Siemens Networks
Munich, Germany

Veluppillai Mahinthan
Centre for Wireless Communications
Department of Electrical and Computer
  Engineering
University of Waterloo
Waterloo, Ontario, Canada

Yannis Manolopoulos
Department of Informatics
Aristotle University of Thessaloniki
Thessaloniki, Greece

Jon W. Mark
Centre for Wireless Communications
Department of Electrical and Computer
  Engineering
University of Waterloo
Waterloo, Ontario, Canada

Albena Mihovska
Center for TeleInfrastruktur
Aalborg University
Aalborg, Denmark

Emilio Mino
Radio Access Networks
Telefonica I+D
Madrid, Spain

Nelson Moniz
Department of Computer Science
 and Engineering
York University
Toronto, Ontario, Canada

Yasser Morgan
Software Systems Engineering Department
University of Regina
Regina, Saskatchewan, Canada

Julián Morillo-Pozo
Computer Architecture Department
Technical University of Catalonia
Barcelona, Spain

Esther Palomar
Computer Science Department
Carlos III University of Madrid
Madrid, Spain

Symeon Papavassiliou
School of Electrical and Computer
 Engineering
National Technical University of Athens
Athens, Greece

Nikos Passas
Department of Informatics
 and Telecommunications
University of Athens
Athens, Greece

Vincent H. Poor
Electrical Engineering Department
Princeton University
Princeton, New Jersey

Michael Portnoy
Department of Computer Science
 and Engineering
York University
Toronto, Ontario, Canada

Predrag Rapajic
School of Engineering
University of Greenwich
Greenwich, United Kingdom

Arturo Ribagorda
Computer Science Department
Carlos III University of Madrid
Madrid, Spain

Erica Cecilia Ruiz-Ibarra
Department of Electrical and Electronics
 Engineering
Instituto Tecnologico de Sonora
Sonora, Mexico

Xuemin (Sherman) Shen
Centre for Wireless Communications
Department of Electrical and Computer
 Engineering
University of Waterloo
Waterloo, Ontario, Canada

O. Simeone
Center for Wireless Communications
 and Signal Processing Research
ECE Department
New Jersey Institute of Technology
Newark, New Jersey

U. Spagnolini
Dipartimento di Elettronica e
 Informazione
Politecnico di Milano
Milano, Italy

Juan M. E. Tapiador
Computer Science Department
Carlos III University of Madrid
Madrid, Spain

Kamel Tourki
Electrical and Computer Engineering
 Department
Texas A&M University at Qatar
Doha, Qatar

Elias Tragos
Computer Networks Laboratory
Institute of Communication and Computer
 Systems (ICCS)
National Technical University of Athens
Athens, Greece

Dionysia Triantafyllopoulou
Department of Informatics and
 Telecommunications
University of Athens
Athens, Greece

Costas Tzaras
Centre for Communication
 System Research
University of Surrey
Guildford, United Kingdom

Luis Armando Villasenor-Gonzalez
Department of Electronics and
 Telecommunications
Centro de Investigacion Cientifica y de
 Educacion Superior de Ensenada
Ensenada, Baja California, Mexico

Natalija Vlajic
Department of Computer Science
 and Engineering
York University
Toronto, Ontario, Canada

Yong Wang
Department of Computer Science
University of Northern British Columbia
Prince George, British Columbia, Canada

Chau Yuen
Modulation and Coding Department
Institute for Infocomm Research
Singapore

Yan Zhang
Networks and Distributed Systems
 Department
Simula Research Laboratory
Oslo, Norway

Zhenrong Zhang
School of Computer, Electronic,
 and Information
Guangxi University
Guangxi, People's Republic of China

Part I

Fundamentals

1 Capacity of Cooperative Channels: Three Terminal Case Study

Elena-Veronica Belmega, Samson Lasaulce, and Mérouane Debbah

CONTENTS

This chapter focuses on three cooperative channels: the relay channel (RC), the broadcast channel (BC) with cooperating receivers, and the multiple access channel (MAC) with cooperating transmitters. The chapter comprises two main parts. The first part is dedicated to discrete channels, their information theoretic performance limits and coding schemes. In particular, special cases where the channel capacity or capacity region is known are reported. The second part is dedicated to Gaussian cooperative channels and is presented from a more pragmatic standpoint. Not only channel capacities and achievable rates are provided but also certain practical technical issues are discussed. Open issues are provided in the conclusion.

1.1 INTRODUCTION

This chapter considers the most elementary cooperative network [1], which comprises three terminals T_1, T_2, T_3 with $T_i \in \{T_x, R_x, \mathcal{R}\}$, where the notation $T_x, R_x, \mathcal{R}$ means that the considered terminal acts as a transmitter, receiver, or a relay. One of the main objectives of this chapter is to give to the reader a recent, precise, and overall picture of what is known about the capacity or capacity region of the channels associated with this system. Currently, the corresponding information is spread over many papers, mainly in the information theory literature, and the connections/differences between the underlying channels are often partially discussed. Each channel corresponds to a given choice of certain degrees of freedom, which are listed here.

A terminal can be a transmitter, receiver, or a relay. We focus on three scenarios (Figure 1.1). Scenario 1: $T_1 \equiv T_x, T_2 \equiv \mathcal{R}, T_3 = R_x$; this corresponds to the relay channel (RC) [2]. Scenario 2: $T_1 \equiv T_x, T_2 \equiv R_x/\mathcal{R}, T_3 = R_x/\mathcal{R}$; this corresponds to the cooperative broadcast channel (CBC),

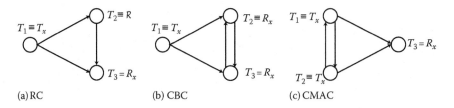

FIGURE 1.1   The three scenarios considered.

introduced by [3]. Scenario 3: $T_1 \equiv T_x/\mathcal{R}, T_2 \equiv T_x/\mathcal{R}, T_3 = R_x$; this corresponds to the cooperative multiple access channel (CMAC), originally studied in Ref. [4].

A link between two terminals can be a discrete or continuous channel. In this chapter particular attention is given to the discrete case, that is the channel where inputs and outputs are in finite alphabets. The main reason for this choice is that the results derived for the general discrete case readily lead to results for the most used channel models such as binary symmetric, erasure, and Gaussian channels. Erasure channels have experienced a recent resurgence of interest because they provide a good channel model from the high layers standpoint (which generally implement retransmission protocols and process packets instead of symbols). Regarding Gaussian channels, despite the fact that they are continuous, the results derived for the discrete case are generally applicable and can even be optimum in the continuous case. A good example illustrating this assertion is the work of Ref. [5] where the author found the capacity of the Gaussian dirty paper channel from the discrete case in Ref. [6] just by making a good choice of the auxiliary random variable used in the capacity formula. Also, a short discussion on the time-varying three-terminal cooperative channels is provided at the end of the chapter.

A terminal can be interested in sending or receiving a private or common information. To clearly illustrate the difference between private and common information, let us consider the case of the downlink in a cellular system. A user who wants to have a phone conversation with his partner corresponds to the case where the receiver has to decode a private message. On the other hand, users who have access to the same bouquet of TV channels correspond to the case of the common message. In this chapter, for each channel, we provide, when it is available, the capacity region for the general case where both private and common messages are sent.

A cooperation link can be unidirectional or bidirectional. In most papers the cooperation between two terminals is unidirectional. In this chapter we also briefly report the main results concerning the case where multiple cooperation rounds are allowed.

A link can be with or without feedback. We briefly mention the case with feedback, because its treatment generally does not lead to new cooperation schemes but can lead to the capacity by inducing physical degradedness of the channel.

For these different situations of interest the most tight inner and outer bounds for the channel capacity or capacity region are provided in this chapter. Of course, in the special cases where it is available, the capacity/capacity region is reported.

The chapter is structured in two main parts and a concluding part. The first part is dedicated to the case of discrete channels. This part is presented from the information-theoretic point of view. In the second part, the channels are always static but have continuous inputs and outputs (Gaussian case). This part is treated in a more pragmatic way to have a complementary approach to the information theoretic approach used in the first part. In the third part a short discussion is made on quasistatic (Rayleigh block fading) cooperative channels and this part is concluded by summarizing certain key points of this chapter and providing open issues related to the different channels presented.

1.2   DISCRETE CHANNELS

1.2.1   RELAY CHANNELS

The RC is a three-terminal channel composed of a source node, a destination node, and one node called the relay, which is neither a source nor a sink (Figure 1.2). The role of the relaying node is to improve the overall performance of the communication between the source and destination. Note that, by definition of the RC [1,2], the relay node does not need to decode the source messages reliably. The most important results concerning the capacity of the discrete RC were given by Cover and El Gamal in Ref. [2]. Since the time Ref. [2] has been published little progress has been made in the determination of the capacity of the general discrete case, which is still unknown.

Mathematically speaking, the RC consists of four finite sets: $\mathcal{X}, \mathcal{X}_{12}, \mathcal{Y}_1$, and $\mathcal{Y}_2$ and a collection of probability distributions $p(.,.|x, x_{12})$ on $(\mathcal{Y}_1, \mathcal{Y}_2)$, for all $(x, x_{12}) \in (\mathcal{X}, \mathcal{X}_{12})$. The channel is supposed to be memoryless and also the relay encoder is supposed to be strictly causal, which means that the relay output at a given moment depends only the past relay observations of the transmitted messages, which is written as $x_{12}(i) = f_i(y_1(1), ..., y_1(i-1))$. At the end of the chapter we mention a broader definition of the RC, which leads to different performance in terms of capacity.

The following theorem gives the best upper bound available, at least in the general case. For instance, for the special case of erasure channels, a tighter upper bound can be derived.

THEOREM 1   ([2], general RC capacity upper bound)

Let R be an achievable transmission rate for the RC. Then, it necessarily verifies the following inequality:

$$R \le \sup_{p(x,x_{12})} \min\{I(X, X_{12}; Y_2), I(X; Y_1, Y_2|X_{12})\}. \tag{1.1}$$

As mentioned in Ref. [7] this bound has an intuitive min-cut max-flow [8] (or cut-set) interpretation. It is a particular case of the elegant upper bound for general multiterminal networks provided by Ref. [7], Chapter 15.

Now let us turn our attention to the best relaying scheme available in the discrete case. It has been stated in Theorem 7 of Ref. [2]. This theorem is as follows.

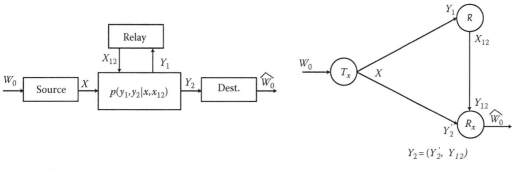

(a) General RC   (b) Orthogonal RC

FIGURE 1.2   The discrete RC.

THEOREM 2   **([2], general RC capacity lower bound)**

Let U, V, and $\hat{Y}_1$ be three auxiliary random variables. For any RC $(\mathcal{X} \times \mathcal{X}_{12}, p(y_1, y_2|x, x_{12}), \mathcal{Y}_1 \times \mathcal{Y}_2)$, the following rate is achievable:

$$R^* = \sup_{p(u,v,x,x_{12},\hat{y}_1)} \min\{R_a, R_b\} \qquad \text{with} \tag{1.2}$$

$$R_a = I(X; \hat{Y}_1, Y_2|X_{12}, U) + I(U; Y_1|X_{12}, V) \tag{1.3}$$

$$R_b = I(X, X_{12}; Y_2) - I(\hat{Y}_1; Y_1|X_{12}, X, U, Y_2) \tag{1.4}$$

where the supremum is taken over all joint probability mass functions of the form
$p(u, v, x, x_{12}, y_1, y_2, \hat{y}_1) = p(v)p(u|v)p(x_{12}|v)p(y_1, y_2|x, x_{12})p(\hat{y}_1|x_{12}, y_1, u)$, *subject to the constraint* $I(\hat{Y}_1; Y_1|Y_2, X_{12}, U) \leq I(X_{12}; Y_2|V)$.

The cooperation scheme used to achieve this rate relies on a combination of the estimate-and-forward (EF) and partial decode-and-forward (DF) protocols. The three auxiliary random variables U, V, and $\hat{Y}_1$ can be chosen arbitrarily provided they meet the Markov chains involved in the decomposition of the joint probability mentioned in the above theorem. For example, if $U \equiv X$, $V \equiv X_{12}$, and $\hat{Y}_1 \equiv \varnothing$,[*] we find the DF protocol, which is capacity-achieving for the physically degraded RC characterized by the Markov chain $X - (X_{12}, Y_1) - Y_2$. In particular, this means that when the source–relay–destination link is better than the source–destination link in the sense that $I(X; Y_1, X_{12}) \geq I(X; Y_2)$, the optimum relaying scheme consists in decoding the source messages and forwarding them to the destination. Assuming physical degradedness of the RC, the channel capacity C_d is expressed as

$$C_d^{(\text{RC})} = \sup_{p(x,x_{12})} \min\{I(X, X_{12}; Y_2), I(X; Y_1|X_{12})\}. \tag{1.5}$$

Unfortunately, the physical degradedness assumption is not met in Gaussian RC. The consequence of this is that the capacity of the Gaussian RC cannot be deduced using the DF protocol and deriving an ad hoc converse proof. The RC differs from the broadcast channel (BC) on this point. Even though the Gaussian BC $[Y_1 = X + Z_1, Y_2 = X + Z_2]$ is not physically degraded, that is, the Markov chain $X - Y_1 - Y_2$ is not verified, it is physically degradable in the sense that there exists a degraded Gaussian channel $[Y_1 = X + Z_1, Y_2 = Y_1 + Z$, where Z is also a gaussian random variable independent of $Z_1]$, which has the same capacity region. This is due to a result of Bergmans [9], which states that the BC capacity region depends only on the marginal probabilities $p(y_1|x)$ and $p(y_2|x)$. No such result exists for the RC.

Now if one makes the choice $U \equiv \varnothing$ and $V \equiv \varnothing$, one obtains the rate achieved by the estimate-and-forward (EF) protocol:

$$R_{\text{ef}}^{(\text{RC})} = \sup_{p(\hat{y}_1, x, x_{12})} \{I(X; Y_2, \hat{Y}_1|X_{12})\} \tag{1.6}$$

subject to the constraint $I(X_{12}; Y_2) \geq I(Y_1; \hat{Y}_1|X_{12}, Y_2)$, which physically means that the relay–destination channel has to be sufficiently good to reliably convey the compressed signal $\hat{Y}_1$. The EF protocol has at least two interesting properties. It does not require the relay to be in a better

---

[*] Note that we used the notation $A \equiv B$, which stands for the random variable. A has the same distribution as the random variable B.

reception condition than the destination and it reaches the max-flow min-cut bound, which coincides with the equivalent two-input single-output channel capacity, when $\hat{Y}_1 \equiv Y_1$, that is, when the relay–destination link is sufficiently good.

We have seen that for the particular case of the degraded RC the capacity is known. It turns out that there are other special cases for which the capacity has been fully determined. The first special case of interest is the RC with feedback, that is, at each time instant i, both the source and relay know the realizations. $(y_1(1), ..., y_1(i-1), y_2(1), ..., y_2(i-1))$. As mentioned in Ref. [2] the feedback changes an arbitrary RC into a degraded RC in which X transmits information X_{12} through (Y_1, Y_2). The channel capacity can be shown to be

---

THEOREM 3   ([2], capacity of the general RC with feedback)

$$C_{\text{fb}}^{(\text{RC})} = \sup_{p(x,x_{12})} \min\{I(X, X_{12}; Y_2), I(X; Y_1, Y_2|X_{12})\}. \tag{1.7}$$

In Ref. [10] the authors have studied a special class of RCs called the semi deterministic RC. In this case the signal received by the relay is a deterministic function of the channel inputs, that is, $Y_1 = f(X, X_{12})$ but no assumptions are made on the signal received by the destination. It turns out that the hybrid cooperation scheme presented in Theorem 4 is capacity-achieving. By choosing $\hat{Y}_1 \equiv \varnothing$, $U \equiv (X_{12}, Y_1)$, and $V \equiv X_{12}$ one can show that the corresponding coding scheme is optimum, which leads to the following coding theorem.

THEOREM 4   ([10], semi deterministic RC capacity)

If Y_1 is a deterministic function of X and X_{12}, then the capacity of the RC is

$$C_{\text{sdet}}^{(\text{RC})} = \max_{p(x,x_{12})} \min\{I(X, X_{12}; Y_2), H(Y_1|X_{12}) + I(X; Y_2|X_{12}, Y_1)\}. \tag{1.8}$$

The capacity was determined also for the deterministic RC; both Y_1 and Y_2 are deterministic functions of the inputs X and X_{12}, as a special case of the semi deterministic channel [10]. Recently, another version of the deterministic RC, where the relay output Y_1 is a deterministic function of the input X and of the receiver output Y_2, was proposed in Ref. [11] and for which the capacity was derived.

Another useful special case of the RC is the case where some links are orthogonal. In Ref. [12] the authors considered the situation where the source transmits a signal with two independent components so that the source signal X can be written as $X = (X_r, X_d)$ with $p(x_r, x_d) = p(x_r)p(x_d)$. For continuous channels this would be implemented by using two nonoverlapping bands of frequency, for instance. Under the aforementioned orthogonality assumption, the channel transition probability can be expressed as $p(y_1, y_2|x, x_{12}) = p(y_2|x_d, x_{12})p(y_1|x_r, x_{12})$ and the capacity is known.

THEOREM 5   **([12], capacity of the orthogonal RC with orthogonality at the source)**

The capacity of the RC with orthogonal components is given by

$$C_{\text{orth},s}^{(\text{RC})} = \max_{p(x_{12})p(x_d|x_{12})p(x_r|x_{12})} \min\{I(X_d, X_{12}; Y_2), I(X_r; Y_1|X_{12}) + I(X_d; Y_2|X)\}. \tag{1.9}$$

We have just analyzed the case where the source–relay and source–destination channel are ortho-gonal. In practice, a more interesting situation, at least for existing applications in wireless networks, is the case where the relay–destination and source–destination (or equivalently the source–relay) channels are orthogonal. This assumption is more realistic than assuming a full duplex RC as we did so far, because it seems to be a hard task to design a relay that can transmit and receive at the same time and in the same band of frequency. This is why the case where the source–relay and relay–destination signals are orthogonal (e.g., in time) is of particular interest. Unfortunately, the capacity is not known for this class of orthogonal discrete channels. Therefore, only lower and upper bounds are available for these channels. To find these bounds one just needs to reuse the bounds provided for the nonorthogonal RC and particularize them by taking into account the Markov chains induced by a proper definition of orthogonality. In this chapter we propose and use the following conditions for the RC with orthogonality at the destination:

1. $p(x, x_{12}) = p(x)p(x_{12})$;
2. $Y_2 = (Y_{12}, Y_2')$ with $p(y_{12}, y_2') = p(y_{12})p(y_2')$;
3. $(Y_1, Y_2') - X - (X_{12}, Y_{12})$; and
4. $Y_{12} - X_{12} - (X, Y_1, Y_2')$.

Now we go a step further in specializing the orthogonal RC, always with orthogonality at the destination, by assuming the different links to be erasure channels. An erasure channel is a discrete channel in which either the transmitted symbols are received errorless at the destination or they are lost with a probability p called the erasure probability [7]. Indeed, the erasure-RC is precisely a discrete RC for which the relay–destination and source–destination channels have to be assumed orthogonal. As mentioned in Section 1.1, erasure channels have experienced a recent resurgence of interest because they can be used in modeling the networks in which there exist a mechanism by which the receiver at the end of a given link can be informed of a packet dropping on its incident erasure link. Usually, this information is transmitted in the packet header. In fact the channel capacity expression is known in the case where the destination knows, in addition to the erasure locations of its incident links, the erasure locations at the source–RC output (side information case) and also in the case where it does not know them. Note that erasure networks without the receiver knowledge of the erasure locations of all the relays are more scalable than networks with this side information because the size of the packet header to acquire this side information would increase drastically. The channel capacities corresponding to these two situations are stated through the following theorems.

THEOREM 6   **([13], capacity of the erasure-RC with side information at the receiver)**

Let p_2, p_1, and p_{12} be the erasure probability over the source–destination, source–relay, and relay–destination links, respectively. The capacity of this channel is given by

$$C_{\text{e,si}}^{(\text{RC})} = \min\{1 - p_2 + 1 - p_{12}, 1 - p_1 p_2\}. \tag{1.10}$$

THEOREM 7   **([14], capacity of the erasure-RC without side information at the receiver)**

The capacity of this channel is given by

$$C_{e,\overline{si}}^{(RC)} = \begin{cases} 1 - p_1 p_2 & \text{if} \quad p_{12} \le p_1 \\ \max\left(R, (1 - p_2)\right) & \text{if} \quad p_{12} > p_1 \end{cases}$$

where $R = \min\{(1 - p_1), (1 - p_2) + (1 - p_{12})\}$.

It can be shown [14] that the knowledge of the erasure locations on the source–relay link at the receiver allows one to reach the max-flow min-cut upper bound. However, when this knowledge is not available the upper bound is not attained. This is why the authors of Ref. [15] derived a tighter upper bound, which relies on the fact that, in erasure-RC, all coding schemes can be interpreted in terms of coding and decoding lists. It turns out that this bound coincides with the lower bound achieved by using maximum distance separable (MDS) codes.*

Therefore, we see that the RC capacity is known in some useful cases. The erasure-RC is of particular interest since a received data packet can be seen as a symbol that is an erasure when the detection error code (e.g., a cyclic redundancy code) declares the packet false or a useful symbol when it is found to be correctly received. The erasure channel is thus well adapted to model a system where the receiver can detect the false packets. In this situation the capacity gives the limit transmission rate for any forward error correcting codes.

1.2.2   COOPERATIVE BROADCAST CHANNELS

The three-terminal CBC consists of a transmitter and two receivers, which can cooperate through a cooperation link to decode their messages (Figure 1.3). Each destination has therefore two roles, the role of a receiver for itself and that of a relay for its partner. The source can broadcast two types of messages: the common message W_0, which is intended for all the receivers and the private messages W_1 and W_2, respectively, intended for users 1 and 2. A satellite and two TV receivers is an example of a system where only a common message is sent by the source. The downlink of a single base-station cell where only phone conversations are exchanged is an example where (almost) only private messages are transmitted. As there are many common points between the cooperative

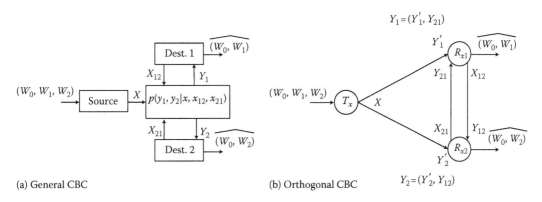

(a) General CBC   (b) Orthogonal CBC

FIGURE 1.3   The discrete CBC.

---

* Recall that these codes attain the Singleton bound, i.e., their minimum distance is such that $d_{\min} = n - k + 1$, where k is the size of the input words and n the length of the codewords.

broadcast and RC, we do not provide, in this section, the lower and upper bounds for the CBC capacity region [16,17], which can be quite naturally derived from the RC analysis. Rather, we provide only coding theorems, that is, analyze the cases where the capacity or capacity region is fully determined.

The CBC has been introduced by the authors of Ref. [3]. The authors focused on the single common message case. In the problem formulation proposed in Ref. [18] the decoders can cooperate through an arbitrary number of conference links, that is, noiseless links with finite capacity. Although assuming a certain form of channel degradedness, the authors have not determined the channel capacity. One can note that the authors used the concept of conference links introduced in Ref. [4], which simplifies the problem but leads to coding schemes applicable to classical noisy cooperation links. The first special case of the CBC, with a common message for which the channel capacity is known, is when the cooperation channel is unidirectional ($X_{21} \equiv \varnothing$).

Even though the capacity is not known for the general (discrete) RC, the capacity of the channel under consideration can be found both in the discrete and Gaussian cases. In the case under investigation receiver 1 not only acts like a simple relay node but has also to decode a message, which is a different situation from that of the RC. Technically speaking, the capacity of the general RC is defined from only one decoding constraint $\Pr[g_2(\underline{Y_2}) \neq W_0)] \to 0$ while the CBC with a single common message is defined from two decoding constraints ($\Pr[g_1(\underline{Y_1}) \neq W_0] \to 0$ and $\Pr[g_2(\underline{Y_2}) \neq W_0] \to 0$). Here we denote the two decoding functions at the destinations by $g_1(.)$ and $g_2(.)$. Exploiting this observation one can easily prove, along the lines of Ref. [2], that the DF protocol is optimum without assuming any explicit form of degradedness [2]. For the general (discrete) CBC we have the following theorem.

THEOREM 8 **([19], capacity of the general discrete CBC with a single common message)**

The capacity of the CBC with a unidirectional cooperation channel is given by

$$C_1^{(CBC)} = \sup_{p(x,x_{12})} \min\{I(X;Y_1|X_{12}), I(X,X_{12};Y_2)\}. \tag{1.11}$$

This expression is the same as that for the physically degraded RC, even though here we have made no degradedness assumption (we did not assume that $p(y_1,y_2|x,x_{12}) = p(y_1|x,x_{12})p(y_2|x,y_1)$). This is useful because the Gaussian counterpart of this channel is not physically degraded. Therefore, applying this theorem in the Gaussian case leads to the channel capacity instead of an achievable rate.

A second case where the capacity is known is when the channel output (Y_1, Y_2) is assumed to be a deterministic function of the channel input X: $Y_1 = f_1(X)$ and $Y_2 = f_2(X)$.

THEOREM 9 **([20], capacity of the deterministic CBC with a single common message)**

The capacity of this channel when conference links with capacities C_{12} and C_{21} are assumed for cooperation is given by

$$C_{det}^{(CBC)} = \min\{H(Y_2) + C_{12}, H(Y_1) + C_{21}, H(Y_1, Y_2)\}. \tag{1.12}$$

This result was initially proved in Ref. [20]. The interesting point is that the max-flow min-cut bound is reached by using a hybrid coding scheme using the EF idea and that of the DF protocol (see

Ref. [16] for more details on this hybrid scheme specific to the discrete CBC with a single common message).

As a last case for which the capacity of the CBC with a single common message is known, we mention the erasure CBC, which is a special case of orthogonal discrete CBCs. In this respect the authors of Ref. [13] have studied networks without cycles and have given the capacity for the CBC with a single common message, unidirectional cooperation (no cycle), and receiver side information on the erasure locations of the source–relay link.

THEOREM 10 **([13], capacity of the unidirectional erasure CBC with a single common message and side information)**

Assume the erasure locations of the source–relay link known at the destination. Then the capacity is

$$C_{1,e,si}^{(CBC)} = \min\{1 - p_1, 1 - p_2 + 1 - p_{12}\}. \tag{1.13}$$

The authors of this chapter have proven that this result extends to the bidirectional CBC, which leads to the following theorem. Before stating the respective theorem, we first give the definition of the orthogonality in CBC channels with a bidirectional cooperation link with orthogonality at the two destinations. The definition we used for the orthogonality in the RC, which is the same definition that can be used for CBC channels with only a unidirectional cooperation channel, can be easily extended to the case of the CBC channels with two or more cooperation links:

1. $p(x, x_{12}, x_{21}) = p(x)p(x_{12})p(x_{21})$;
2. $Y_1 = (Y_{21}, Y_1')$ with $p(y_{21}, y_1') = p(y_{21})p(y_1')$;
3. $Y_2 = (Y_{12}, Y_2')$ with $p(y_{12}, y_2') = p(y_{12})p(y_2')$;
4. $(Y_1', Y_2') - X - (X_{12}, Y_{12}, X_{21}, Y_{21})$;
5. $Y_{12} - X_{12} - (X, X_{21}, Y_1', Y_2', Y_{21})$;
6. $Y_{21} - X_{21} - (X, X_{12}, Y_1', Y_2', Y_{12})$.

THEOREM 11 **(Capacity of the bidirectional erasure CBC with a single common message and side information)**

Based on the same assumption as in the previous theorem, the channel capacity is expressed as

$$C_{2,e,si}^{(CBC)} = \min\{1 - p_1 p_2, 1 - p_1 + 1 - p_{21}, 1 - p_2 + 1 - p_{12}\}. \tag{1.14}$$

If the knowledge of the source–relay link erasures are not available at the destination, the problem is more difficult. The authors of Ref. [21] have tried to determine the channel capacity without this side information by allowing the receivers to do an infinite number of cooperation rounds (as in Ref. [3]). It turns out that, even though the capacity is known for the erasure-RC without side information, the capacity has not been determined for the erasure CBC with a common message. This problem, which seems to be doable, remains therefore an open problem.

Now let us allow the source to send not only a common message but also private messages. As for the discrete RC, the optimum performance of the CBC is known when the channel is physically degraded and also when each terminal has a feedback from the output of the channels it transmits over. This is stated in the following coding theorems.

THEOREM 12 **([17], capacity region of the physically degraded CBC)**

If one assumes that one of the two Markov chains, $X - (X_{12}, X_{21}, Y_1) - Y_2$ or $X - (X_{12}, X_{21}, Y_1) - Y_2$, is met, then the channel capacity region is expressed as

$$
\mathcal{C}_{\text{d}}^{\text{(CBC)}} = \bigcup_{p(u,x_1,x_2)p(x|u)p(y_1,y_2|x,x_1,x_2)} \left\{ \begin{array}{rcl} & & (R_0, R_1, R_2): \\ R_0 + R_2 & \leq & \min\{I(U, X_1; Y_2|X_2), I(U; Y_1|X_1, X_2)\} \\ R_1 & \leq & I(X; Y_1|U, X_{12}, X_{21}) \end{array} \right\} \quad (1.15)
$$

where there is no loss of optimality by choosing an auxiliary random variable U in a set $\mathcal{U}$ such that $|\mathcal{U}| \leq |\mathcal{X}||\mathcal{X}_{12}||\mathcal{X}_{21}| + 2$.

This theorem has also been proven independently in Ref. [16] in a more specific case where there is no common message and the receivers are connected through conference links instead of classical noisy point-to-point channels. In both cases [16,17], the proposed capacity-achieving coding scheme consists in mixing the superposition coding idea introduced in Ref. [22] and exposed clearly in Ref. [23] for the (noncooperative) BC and the DF protocol introduced in Ref. [2].

If one assumes the presence of a feedback mechanism over all the links, the results of Theorem 3 of Ref. [2] for the general RC extend to the general BC with cooperative receivers. It is also shown in Ref. [17] that assuming the presence of perfect feedback links implies the physical degradedness of the channel. Therefore, the capacity region can be fully determined as described below.

THEOREM 13   ([17], capacity region of the CBC with feedback)

Assume that, at each time instant i, each terminal is informed with the outputs $Y_j(1), ..., Y_j(i-1)$, $j \in \{1, 2\}$, of the channel over which it transmits. Then the channel capacity region is as follows:

$$
\mathcal{C}_{\text{fb}}^{\text{(CBC)}} = \bigcup_{p(u,x_1,x_2)p(x|u)p(y_1,y_2|x,x_1,x_2)} \left\{ \begin{array}{rcl} & & (R_0, R_1, R_2): \\ R_0 + R_2 & \leq & \min\{I(U, X_{12}; Y_2|X_{21}), I(U; Y_1, Y_2|X_{12}, X_{21})\}, \\ R_1 & \leq & I(X; Y_1, Y_2|U, X_{12}, X_{21}) \end{array} \right\}
$$
$$(1.16)$$

where U is bounded in cardinality by $|\mathcal{U}| \leq |\mathcal{X}||\mathcal{X}_{12}||\mathcal{X}_{21}| + 2$.

One can mention [17] that this capacity region can be attained by removing the feedback mechanism from certain links. Indeed, there is no loss of optimality by removing the feedback links from the receivers to the main source. Additionally, the cooperation link from the bad receiver to the good one can also be removed because the feedback link provides enough information to the best receiver to ensure the existence of a capacity-achieving cooperation scheme.

To conclude the section dedicated to the CBC we mention two other cases where the capacity region has been derived: the semideterministic CBC with unidirectional cooperation and the determinist CBC with bidirectional cooperation.

THEOREM 14   ([24], capacity region of the semideterministic CBC with unidirectional cooperation)

Assume $Y_1 = f_1(X, X_{12})$ and $X_{21} \equiv \varnothing$. Then we have

$$
\mathcal{C}_{1,\text{sdet}}^{(\text{CBC})} = \bigcup_{p(t,u,x_{12},x)p(y_1,y_2|x,x_{12})} \left\{ \begin{array}{rcl} & & (R_0, R_1, R_2): \\ R_0 & \leq & \min\{I(T;Y_1|X_{12}), I(T,X_{12};Y_2)\} \\ R_0 + R_1 & \leq & H(Y_1|X_{12}) \\ R_0 + R_2 & \leq & I(T,U,X_{12};Y_2) \\ R_0 + R_1 + R_2 & \leq & I(T;Y_1|X_{12}) + H(Y_1|T,U,X_{12}) + I(U;Y_2|T,X_{12}) \\ R_0 + R_1 + R_2 & \leq & H(Y_1|T,U,X_{12}) + I(T,X_{12},U;Y_2) \end{array} \right\}
$$

(1.17)

As the optimum rate region for the semideterministic CBC with bidirectional cooperation has not been determined yet, we provide that of the deterministic CBC and also assume no common message.

THEOREM 15 ([20,25], **capacity region of the deterministic CBC with bidirectional cooperation and private messages**)

Assume that the receivers are connected through conference links with finite capacities C_{12} and C_{21}. Additionally assume that $Y_1 = f_1(X)$ and $Y_2 = f_2(X)$. Then,

$$
\mathcal{C}_{2,\text{det}}^{(\text{CBC})} = \bigcup_{p(t,u,x_1,x)p(y_1,y_2|x,x_1)} \left\{ \begin{array}{rcl} & & (R_1, R_2): \\ R_1 & \leq & H(Y_1) + \min\{C_{21}, H(Y_2|Y_1)\} \\ R_2 & \leq & H(Y_2) + \min\{C_{12}, H(Y_1|Y_2)\} \\ R_1 + R_2 & \leq & H(Y_1, Y_2) \end{array} \right\}
$$

(1.18)

Two capacity-achieving coding schemes have been found to derive this rate region. In Ref. [25] the coding scheme is based on the Slepian–Wolf coding whereas the authors of Ref. [20] rederived this region independently by using a simpler coding scheme, which consists in partitioning the messages (before encoding them) at the source. This essentially shows that there exist situations where projecting information symbols onto two orthogonal directions can lead to capacity-achieving cooperation schemes.

1.2.3 COOPERATIVE MULTIPLE ACCESS CHANNELS

The three-terminal CMAC corresponds to the dual situation of the cooperative BC: there are two sources that can cooperate and one destination. This channel has been defined properly for the first time by Willems [4]. In his formulation Willems implicitly assumed the cooperation channels to be orthogonal to the downlink channels (Figure 1.4). More precisely, the two sources are assumed to be connected through an arbitrary number of conference (i.e., noiseless) links with total finite capacities C_{12} and C_{21}. We report here the result found in Ref. [4].

THEOREM 16 ([4], **capacity region of the discrete CMAC with conference links and private messages**)

Let U be an auxiliary random variable. For the discrete memoryless multiple access channel (MAC) $(\mathcal{X}_1 \times \mathcal{X}_2, p(y|x_1, x_2), \mathcal{Y})$ with encoders connected by communication links with total finite capacities C_{12} et C_{21}, the capacity region $\mathcal{C}_{\text{conf}}^{(\text{cmac})}$ is the set of all positive pairs (R_1, R_2) such that

$$
\left\{ \begin{array}{rcl} R_1 & \leq & I(X_1;Y|X_2,U) + C_{12}, \\ R_2 & \leq & I(X_2;Y|X_1,U) + C_{21}, \\ R_1 + R_2 & \leq & \min\{I(X_1,X_2;Y|U) + C_{12} + C_{21}, I(X_1,X_2;Y)\} \end{array} \right.
$$

(1.19)

with $p(u,x_1,x_2,y) = p(u)p(x_1|u)p(x_2|u)p(y|x_1,x_2)$.

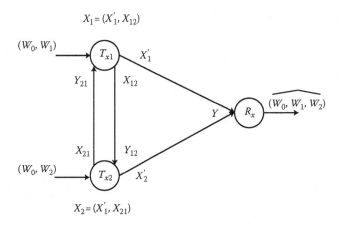

FIGURE 1.4   The discrete orthogonal CMAC.

The authors have proved that this result easily generalizes to the case where the conference links are replaced with noisy point-to-point channels and a common information is sent by the sources. The latter feature is particulary simple to integrate because the two sources cooperate only for increasing the rates associated with the private messages and not that of the common message, since the two sources know this message perfectly without any cooperation. The corresponding result is as follows.

THEOREM 17   (Capacity region of the discrete CMAC with common and private messages)

The capacity region for the discrete memoryless MAC with K bidirectional orthogonal cooperation links is the closure of the convex hull $\cup \mathcal{R}(p)$, where the union is taken over the joint probability distribution $p(u,x_1,x_2,x_{12},x_{21},y_{12},y_{21},y) = p(u)p(x_1|u)p(x_2|s)p(x_{12})p(y_{12}|x_{12})p(y_{12}|x_{12})p(y|x_1,x_2)$. The set $\mathcal{R}(p)$ is the set of tuples $R = (R_0, R_1, R_2)$ such that

$$
\begin{cases}
R_1 & \leq \ I(X_1; Y|X_2, U) + I(X_{12}; Y_{12}) \\
R_2 & \leq \ I(X_2; Y|X_1, U) + I(X_{21}; Y_{21}) \\
R_1 + R_2 & \leq \ I(X_1, X_2; Y|U) + I(X_{12}; Y_{12}) + I(X_{21}; Y_{21}) \\
R_0 + R_1 + R_2 & \leq \ I(X_1, X_2; Y).
\end{cases}
\tag{1.20}
$$

To prove the achievability part of this theorem we used the coding technique introduced by Slepian and Wolf in Ref. [26] and proved, as [4] for the CMAC with conference links, that a two-round cooperation is optimal, that is nothing is gained for the general CMAC by implementing a multiple-round cooperation. This effect is due to the fact that the total capacity in a given cooperation direction is fixed, namely C_{12} for the direction source $1 \rightarrow$ source 2 and C_{21} for the direction source $2 \rightarrow$ source 1. This result differs from those obtained in Refs. [3,21] where the authors observed that the rate of the cooperative BC with a common message increases with the number of cooperation exchanges (or decoding rounds). This is due to the fact that these works did not take into account the finite resource aspect for cooperation. In general, when the discrete channels are analyzed, we do not know how the spectral resources of the system are to be considered. For more information about the multiple-round cooperation see also Refs. [27,28]. In Ref. [27] the authors analyzed the AF-based Gaussian CBC by discussing the influence of the spectral resources on performance improvement with respect to the number of cooperation rounds, which leads to conclusions different from [3,21].

The best performance obtained w.r.t. the achievable rates or different bit error rate (BER) based system criteria is generally achieved for one or two cooperation rounds and thus shows that the performance does not increase with the number of cooperation rounds.

The capacity region of the erasure CMAC without side information at the destination can be easily derived as a special case of the discrete CMAC capacity region. However, when side information is assumed, the problem of finding the optimum region when the cooperation is bidirectional is still open. Only the unidirectional case has been solved so far, as stated below.

For the cooperative MAC the capacity region was found in the case where the receiver has the side information w.r.t. the erasure locations and only one link of cooperation between the encoders.

THEOREM 18   **([13], capacity region of the erasure CMAC with unidirectional cooperation)**

The capacity region for the erasure CMAC with one cooperation link and side information is given by

$$\begin{cases} R_1 & \leq & 1 - p_1 p_{12} \\ R_2 & \leq & 1 - p_2 \\ R_1 + R_2 & \leq & 1 - p_1 + 1 - p_2. \end{cases} \qquad (1.21)$$

We observe that for the erasure CMAC channel, the min-cut max-flow upper bound is also achievable, provided the receiver disposes of the side information w.r.t. the erasure locations.

1.3   CONTINUOUS CHANNELS: THE CASE OF AWGN CHANNELS

In this section we analyze three-terminal cooperation channels with continuous inputs and outputs. We focus on the most used model for these channels, which is the additive white Gaussian noise (AWGN) model. All the coding and relaying schemes derived for the general discrete three-terminal cooperative channels can be directly reused for their Gaussian counterparts. This is why we do not provide the corresponding inner bounds, outer bounds, or capacity expressions. In fact, the achievable or optimum information rates or rate regions can be found essentially by calculating entropies of (possibly vector) Gaussian random variables. One important point to mention is that an optimal coding scheme derived for the discrete case is not necessarily optimal in the Gaussian case. For this reason a specific converse or a more sophisticated scheme might be needed. For sake of completeness, we mention a few references where this kind of extension has been conducted.

- Nonorthogonal Gaussian degraded RC [2]. As the conventional Gaussian RC [$Y_1 = X + Z_1$, $Y_2 = X + X_{12} + Z_2$] is not physically degraded, the authors of Ref. [2] introduced a Gaussian degraded RC for which the received signals can be expressed as $Y_1 = X + Z_2$ and $Y_2 = Y_1 + X_{12} + Z$.
- Orthogonal Gaussian RC [18].
- Gaussian RC [29].
- Orthogonal Gaussian CBC with a single common message [19].
- Nonorthogonal CBC with private and common messages [17].
- Nonorthogonal CMAC [30].

Although the capacity of the Gaussian RC is still unknown, numerical results provided in the above references show that the receiver performance achieved by using the existing relaying schemes is not so far from the available capacity outer bounds (see e.g., [18] for the RC and [19] for the cooperative

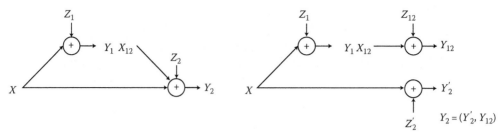

(a) Nonorthogonal AWGN CBC   (b) Orthogonal AWGN CBC

FIGURE 1.5 Gaussian CBC with a unidirectional cooperation link.

BC with a common message). It is thus not clear if the determination of a capacity-achieving relaying scheme will provide a real breakthrough in terms of relaying protocols or if it will just be the solution to a theoretical challenge.

In this section we therefore focus on coding and relaying schemes that are more inherent to the Gaussian model, especially the AF protocol and its modified versions. Indeed, this protocol cannot be implemented in discrete cooperative channels by its definition. Additionally, we restrict our attention to the case of orthogonal Gaussian cooperative channels (Figure 1.5), which is the most useful assumption in practice, since it is a very hard task to design a relay device that can transmit and receive at the same time in the same band of frequency. It also turns out that the performance loss due to the RC orthogonalization can be small in situations of interest. As mentioned by [19] this loss can be evaluated exactly in the case of the Gaussian CBC with a unidirectional cooperation channel. Indeed, in this case the DF protocol is optimum without assuming physical degradedness, which means that the capacity is known for the nonorthogonal and orthogonal cases.

THEOREM 19 ([19], capacity of the nonorthogonal Gaussian CBC with a unidirectional cooperation link [$Y_1 = X + Z_1, Y_2 = X + X_{12} + Z_2$], the single common message case)

Denote by P the source transmit power, P_{12} the user-relay transmit power, B the total system bandwidth, and for $i \in \{1,2\}$, n_i the noise spectral density on the source–destination i channel. The capacity is expressed as

$$C_{1,\text{orth}}^{(\text{grc})} = \max_{r \in [0,1]} \min \left\{ C\left(\frac{rP}{n_1 B}\right), C\left(\frac{P + P_{12} + 2\sqrt{\bar{r}PP_{12}}}{n_2 B}\right) \right\} \tag{1.22}$$

where $C(x) = B\log_2(1+x)$ and $\bar{r} = 1 - r$.

If $n_2 \le n_1$ the capacity is simply $C(P/n_1 B)$. On the other hand, if $n_2 > n_1$ there are two possible working regimes [2]. If $P_{12} \ge P(n_2 - n_1/n_1)$ then $r^* = 1$ and $C_{\text{orth}} = C(P/n_1 B)$. Now if $P_{12} < P(n_2 - n_1/n_1)$ the best r is given by $\bar{r}^* = \left(a_1^2 + a_3^2 \pm 2\sqrt{a_1^2 a_3^2}\right)/a_2^2$ where $r^* \in [0, 1]$, $a_0 = P + P_{12}$, $a_1 = \sqrt{PP_{12}}$, $a_2 = Pn_2/n_1$, and $a_3^2 = a_1^2 + a_2^2 - a_0 a_2$. A sufficient condition for $a_1^2 + a_2^2 - a_0 a_2$ being nonnegative is precisely $n_2 \geq n_1$. An important point to notice here is that the saturation regime is reached for a finite cooperation power $P_{12}^* = P(n_2 - n_1/n_2)$.

THEOREM 20 ([19], capacity of the orthogonal Gaussian CBC with a unidirectional cooperation link [$Y_1 = X + Z_1$, $Y_2' = X + Z_2'$, $Y_{12} = X_{12} + Z_{12}$, $Y_2 = (Y_2', Y_{12})$], the single common message case)

Denote by B the total system bandwidth, $B_0 = \alpha_0 B$ the downlink channel bandwidth, and $B_{12} = \overline{\alpha}_0 B$ the cooperation channel bandwidth with $B_0 + B_{12} = B$. The capacity is written as

$$C_{1,\text{orth}}^{(\text{grc})} = \max_{\alpha_0} \min \{R_1(\alpha_0), R_2(\alpha_0)\} \qquad with \tag{1.23}$$

$$R_1(\alpha_0) = \alpha_0 C\left(\frac{\rho_1}{\alpha_0}\right) \tag{1.24}$$

$$R_2(\alpha_0) = \alpha_0 C\left(\frac{\rho_2'}{\alpha_0}\right) + \overline{\alpha}_0 C\left(\frac{\rho_{12}}{\overline{\alpha}_0}\right) \tag{1.25}$$

where $\rho_1 = P/(n_1 B)$, $\rho_2' = P/(n_2' B)$, and $\rho_{12} = P/(n_{12} B)$.

As can be seen we considered in Equation 1.25 that when the parameter α_0 is not fixed, the capacity is obtained by optimizing the bandwidth allocation. Indeed, it can be shown that there is a unique $\alpha_0^{(m)} \in [0, 1]$ maximizing the second term of the min function ($R_2(.)$). Depending on the channel parameters the optimum bandwidth allocation α_0^* is either given by $\alpha_0^{(m)}$ or the intersection of $R_1(.)$ and $R_2(.)$. Without loss of generality assume now that $n_1 < n_2'$. A natural question is then to ask what is the cooperation power needed for being in the saturation regime, that is $C_{\text{orth}} = C(\rho_1)$. For a given α_0, $R_2(\alpha_0) \geq R_1(\alpha_0)$ is equivalent to $\rho_{12} \geq \overline{\alpha}_0 \left[\left((1 + \rho_1)/(1 + \rho_2')\right)^{\alpha/\overline{\alpha}_0} - 1 \right]$. For a fixed $\alpha_0 \neq 0$ the required cooperation power is clearly finite. On the other hand if we are interested in optimizing bandwidth allocation, α_0 can take all the values between 0 and 1. When $\alpha_0 \to 1$ the saturation condition becomes $\rho_{12} \geq \exp\left\{1/\overline{\alpha}_0 \left[\ln\left((1 + \rho_1)/(1 + \rho_2')\right) + \overline{\alpha}_0 \ln \overline{\alpha}_0\right]\right\}$. Since $\rho_1 > \rho_2'$, it is clear that the cooperation power has to be infinite for $\overline{\alpha}_0 \to 0$. Contrary to the nonorthogonal case, one cannot reach the saturation regime for finite cooperation powers and there will always be a performance loss due to orthogonalizing the interuser channel. Let us consider a numerical example. We chose $n_1 B = 1$, $n_2' B = n_{12} B = 4$. For three different values of the transmit power $P \in \{1, 10, 100\}$, Figure 1.6 represents the relative capacity loss due to orthogonalization as a function of P_{12}: $\Delta C[\%] \triangleq 100(C_{\overline{\text{orth}}} - C_{\text{orth}})/C_{\overline{\text{orth}}}$. The performance loss is clearly driven by the ratio P_{12}/P. If this ratio is greater than 20 dB the relative capacity loss is less than 10 percent for the considered range of transmit powers. In real contexts, such a situation can appear when the link budget corresponding to the cooperation channel is much better than the budget of the downlink channels, which is in fact a very common scenario in a cellular networks (e.g., two users in the same room or building). On the other hand if the available cooperation power is limited, the nonorthogonal solution performs much better than its orthogonal counterpart. In practice, complexity and feasibility issues have also to be accounted for.

Now that we have also justified the orthogonality assumption in terms of performance, we focus on Gaussian cooperative channels using AF-type protocols. Consider a Gaussian RC where the source–relay and relay–destination channels are assumed to be orthogonal in the frequency domain (without loss of generality). First, recall the original version of the amplify-and-forward protocol introduced by Ref. [31] and analyzed in more detail in Ref. [32]. The cooperation signal transmitted by the relay is written at time i as $X_{12}(i) = \sum_{j=1}^{i-1} a_{i,j} Y_1(j)$, which can also be rewritten in a vector

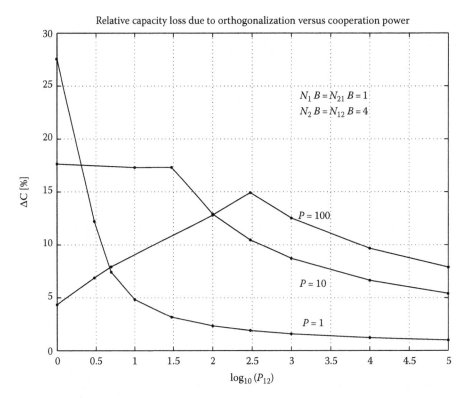

FIGURE 1.6 Relative capacity loss versus $\log_{10}(P_{12})$. (From Lasaulce, S. and Klein, A.G., *IEEE Proceedings of International Conference on Acoustic, Speech and Signal Processing (ICASSP)*, 2006. With permission.)

form $\underline{X}_{12} = \mathbf{A}\underline{Y}_1$, where $\underline{X}_{12} = (X_{12}(1), ..., X_{12}(n))^T$, $\underline{Y}_1 = (Y_1(1), ..., Y_1(n))^T$, and the matrix $\mathbf{A}$ is a strictly lower triangular matrix that has to meet the transmit power constraint at the relay. For this particular protocol one can evaluate the maximum transmission rates such that the destination can decode reliably the source messages: this is the concept of channel capacity for a fixed relaying protocol used in Ref. [31]. This capacity is given by the following theorem.

THEOREM 21 ([32], **capacity of the frequency division AWGN RC with linear relaying functions** $[Y_1 = aX + Z_1, Y'_2 = X + Z'_2, Y_{12} = bX_{12} + Z_{12},$ **and** $Y_2 = (Y'_2, Y_{12})]$)

Denote by P the source transmit power, P_{12} the relay transmit power, $N'_2 = B_0 n'_2$ the variance of the source–destination channel noise, $N_1 = B_0 n_1$ the variance of the source–RC noise, and $N_{12} = B_{12} n_{12}$ the variance of the relay–destination channel noise and also a, the source–RC gain and b the relay–destination channel gain (the gain of the source–destination channel is supposed to be equal to 1). Supposing that the noise variances of the links are all equal $N_1 = N'_2 = N_{12} = N$, the channel capacity is then

$$C_{\text{lin}}(P) = \max_{\underline{\delta}, \underline{\theta}, \underline{\eta}} \delta_0 C\left[\frac{\theta_0 P}{\delta_0 N}\right] + \sum_{j=1}^{4} \delta_j C\left[\frac{\theta_j P}{\delta_j N}\left(1 + \frac{a^2 b^2 \eta_j}{1 + b^2 \eta_j}\right)\right] \qquad (1.26)$$

where $\underline{\delta} = [\delta_0, ..., \delta_4]^T$, $\underline{\theta} = [\theta_0, ..., \theta_4]^T$, and $\underline{\eta} = [\eta_0, ..., \eta_4]^T$, subject to $\delta_j \geq 0$, $\theta_j \geq 0$, and $\eta_j > 0$,

$$\sum_{j=0}^{4} \delta_j = \sum_{j=0}^{4} \theta_j = 1$$

and

$$\sum_{j=1}^{4} \eta_j (a^2 \theta_j P + N \delta_j) = P_{12}.$$

This theorem gives the capacity for the vector AF protocol. In the literature many authors consider a simplified version of this protocol, which is the scalar AF protocol. The main motivation for this choice is the simplicity of the scalar version of the AF (SAF) protocol both in terms of implementation and simplification of the related theoretical performance analyses. In this special but useful case the capacity is given by

$$C_{\text{SAF}} = C \left[\frac{P}{N} \left(1 + \frac{a^2 b^2 P_{12}}{N + a^2 P + b^2 P_{12}} \right) \right]. \tag{1.27}$$

We see that in the high cooperation regime, that is $P_{12} \to \infty$, C_{SAF} tends to the equivalent 2×1 single-input multiple-output (SIMO) system capacity, which is also the channel capacity:

$$\lim_{P_{12} \to \infty} C_{\text{SAF}} = C_{\text{SIMO}} = C \left(\frac{P}{N} (1 + a^2) \right). \tag{1.28}$$

Even though the scalar AF protocol is a suboptimum version of the vector AF presented here, it still tends to achieve the channel capacity when the relay–destination channel quality increases. However, in general the SAF protocol is not capacity-achieving and its performance can be improved. In fact, if one restricts one's attention to scalar relaying schemes, it can be shown that there exist nonlinear functions that perform better than the simple linear multiplication by a scalar. For example, the authors of Ref. [33] have introduced the clipped AF cooperation strategy for the Gaussian RC. The relaying function in this case is the linear threshold function. The rationale for the proposed function is that it preserves the important soft information but does not needlessly expend power-relaying large noise samples.

$$f_\beta(y_1) = \begin{cases} y_1, & |y_1| \le \beta \\ \beta * sgn(y_1), & |y_1| > \beta. \end{cases} \tag{1.29}$$

We see that the AF protocol is a particular case of the clipped AF, obtained for $\beta \gg 1$ large. Also, for $\beta = 0$ we obtain a scalar DF for a binary phase-shift keying (BPSK) modulation. More sophisticated nonlinear and scalar-relaying functions have been proposed in Refs. [34–37]. In Refs. [34,35] the authors derived the optimum relaying function in the sense of the mutual information when no direct link is assumed. The proposed solution bridges the gap between the scalar AF, scalar DF, and scalar EF protocols. In Ref. [37] the authors have also studied the optimal relay function when no direct link is assumed and for a BPSK input signal modulation. They found that the relaying function that minimizes the raw BER is a Lambert function. Also, they observed that in the high signal-to-noise ratio (SNR) case the relaying function resembles a hard limiter, the DF scheme. In the low SNR case, the function resembles a linear AF relaying function (Figure 1.7).

So far, we have seen that the scalar AF protocol and its modified versions have a strong advantage in terms of implementation because they are simple. However, this argument implicitly relies on the relaying device. Indeed, the scalar AF is an analog protocol since it takes a signal in $\mathbb{R}$ or $\mathbb{C}$ and produces a new signal also in $\mathbb{R}$ or $\mathbb{C}$. If recent trends in radio design are any indication, it is almost certain that many relays would be digital devices to maintain cost-effectiveness and to

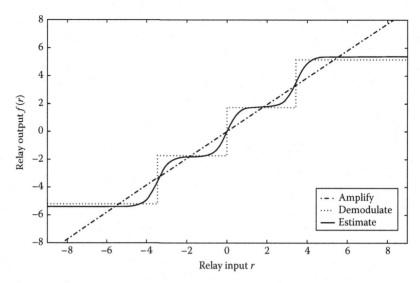

FIGURE 1.7 Relay functions for a 4-PAM modulation. (From Gomadam, K.S. and Jafar, S.A., Optimizing soft information in relay networks, *Proceedings of Asilomar Conference on Signals, Systems and Computers*, Oct.–Nov. 2006. With permission.)

permit the relays to communicate with other nodes in the system to exchange channel information, acquire frame/packet synchronization, enter into relaying agreements, and so on. Therefore, digital counterparts of the AF protocol have to be proposed. Solving this issue was one of the motivations of the work presented in Ref. [33]. The authors of Ref. [33] proposed a scalar quantize-and-forward (QF) protocol based on the joint source–channel coding originally introduced in Ref. [38] and studied more deeply by Farvardin et al. (see e.g., [39]). The proposed protocol is a (generally nonuniform) quantizer, which optimizes the end-to-end distortion by exploiting the knowledge of the SNRs of source–relay and relay–destination channels. Therefore, by increasing the number of bits, the QF protocol behaves more like the scalar AF protocol while, for a low numbers of quantization bits, it behaves more like the scalar DF protocol. The proposed solution can also be seen as a way of implementing a channel optimized AF-type protocol in a digital relay transceiver. Obviously, the equivalent analog-relaying function is not linear in general, which makes a connection with the works on the optimization of the relaying function cited previously.

1.4 CONCLUSION AND OPEN ISSUES

In this chapter we treated in detail the case of discrete three-terminal cooperative channels, in particular, by providing the Shannon capacity or capacity region for different special cases. One of the useful interesting special cases is the case of erasure channels, since it provides a good model of communication networks from the higher layers point of view. From the physical layer standpoint, the most used channel models are the Gaussian model for static channels and Rayleigh fading model for time-varying channels. In this respect, only the static case has been addressed in this chapter (Section 1.3, dedicated to continuous inputs continuous outputs channels). In order to adopt a complementary approach to the first part and avoid redundancy, we did not provide all the Gaussian counterparts of the discrete cases but rather focused on more practical issues inherent to Gaussian RC. One has to keep in mind that the coding and relaying schemes derived in the discrete case can be readily applied to the Gaussian case. Concerning the Rayleigh fading cooperation channels, no special part has been dedicated to them. This is why we mention here a few important issues concerning these channels before providing a list of open problems related to this chapter.

First of all one has to distinguish fast fading from slow fading (or quasistatic) channels. When fast fading is assumed, that is, when the channel gains vary from symbol to symbol, the ergodic capacity coincides with the Shannon capacity. The Shannon capacity is a good performance criterion since it effectively provides the limit information rates possible over a given and arbitrary block of data. The coding and relaying schemes derived for the Gaussian (and therefore static) case can be applied to the fast fading case and leads to the knowledge of the limit performance of the considered cooperative channel. A good example of this type of work is the work of Host Madsen on multiple-input and multiple-output (MIMO) RC [29]. An interesting result found in Ref. [29] is that capacity lower and upper bounds, not necessarily tight in the Gaussian case, become tight in the fast fading case provided some regularity conditions are assumed. When slow fading cooperative channels are considered, the Shannon capacity is no longer a suitable performance criterion since it represents only an upper bound of the transmission rates achieved over all the blocks of data. This is why Ref. [40] introduced the concept of outage probability. But outage probability is more difficult to evaluate. For example, for a single-user channel, if the mutual information between the channel input and output can be modeled as a Gaussian random variable, the ergodic rate is the first order of the mutual information whereas the outage probability is related to the second-order moment of it. This is one of the reasons why the authors of Ref. [41] assume the high SNR regime. Indeed, in this regime, the outage probability analysis is equivalent to the diversity-multiplexing trade-off (DMT) analysis as shown by Zheng and Tse in Ref. [41]. In Ref. [42] the authors assumed fixed relaying protocols and derived the corresponding DMTs for the half duplex (i.e., orthogonal) RC, the half duplex cooperative BC with a single common message and unidirectional cooperation, and the nonorthogonal AF protocol-based cooperative MAC with private messages. The authors proved that without the half duplex constraint the optimal trade-off is attained by using the AF-protocol. On the other hand, when half-duplex channels are assumed, it is generally necessary to construct more sophisticated strategies aiming at transmitting independent symbols as frequently as possible. In the same spirit the authors of [43] derived DMTs for MIMO AF-based cooperative channels and also provide space-time codes that can reach the derived limits of performance.

We conclude this chapter by mentioning a few open issues related to the channels investigated in this chapter:

1. Discrete RC capacity. The most powerful lower bound has been provided in Theorem 7 of Ref. [2]. Since then, no significant progress has been made at least not in the general case.
2. Erasure CBC with a single common message, bidirectional cooperation, and without side information. Even though the capacity of the erasure RC has been determined, the capacity of the CBC remains an open issue. This is quite surprising if we take into account the fact that the best relaying scheme is known.
3. Semi deterministic CBC capacity region with common and private messages and bidirectional cooperation. Currently, the capacity region is available only in the unidirectional case [24].
4. New definitions for the relay node. The authors of Refs. [44–46] recently showed that depending on the delay tolerated at the relay, the limit information rates for the RC can change significantly. Of course, this would also modify the results for all the channels when relaying-type operations are used (RC, CBC, CMAC, etc.).
5. Cooperative networks are more exposed to security issues than their noncooperative counterparts since the relaying nodes can try to decode a private message that is not intended for them. This leads to the secrecy capacity of the capacity region of cooperative channels [47,48].

6. Note that the three-terminal cooperative channels studied in this chapter can be further complicated by allowing bidirectional cooperation between all pairs of terminals, as suggested by Ref. [1].

7. Also, a more unifying approach has still to be developed to study the limit performance of cooperative networks. The three channels presented here can also be seen as special cases of a general interference channel, for which the capacity determination is an even more challenging open problem.

8. More connections between dirty paper coding and network coding for relay channels could be established. Recent contributions [49] show that by bringing network coding at the physical layer (also known as analog network coding) combined with dirty paper precoding, time is saved compared with classical DF protocols; interference resulting from nonorthogonality is mitigated, leading to a better use of resources and improved spectral efficiency.

REFERENCES

[1] E. C. van der Meulen, Three-terminal communication channels, *Adv. Appl. Prob.*, 3, 120–154, 1971.

[2] T. M. Cover and A. A. El Gamal, Capacity theorems for the relay channel, *IEEE Trans. Inform. Theory*, 52(5), 572–584, Sept. 1979.

[3] S. C. Draper, B. J. Frey, and F. R. Kschischang, Interactive decoding of a broadcast message, *Conference on Communication, Control and Computing*, Allerton, Oct. 2003.

[4] F. M. J. Willems, The discrete memoryless multiple access channel with partially cooperating encoders, *IEEE Trans. Inform. Theory*, IT-29(3), 441–445, May 1983.

[5] M. H. M. Costa, Writting on dirty paper, *IEEE Trans. Inform. Theory*, IT-29(3), 439–441, May 1983.

[6] S. I. Gel'fand and M. S. Pinsker, Coding for channel with random parameters, *Problems of Control and Inform. Theory*, 9(1), 19–31, 1980.

[7] T. M. Cover and J. A. Thomas, *Elements of Information Theory*. 2nd edn., Wiley-Interscience, Hoboken, NJ, 2006.

[8] L. R. Ford and D. R. Fulkerson, *Flows in Networks*, Princeton University Press, Princeton, NJ, 1962.

[9] P. Bergmans, Random coding theorem for broadcast channels with degraded components, *IEEE Trans. Inform. Theory*, 19(2), 197–207, March 1973.

[10] A. El Gamal and M. Aref, The capacity of the semideterministic relay channel, *IEEE Trans. Inform. Theory*, IT-28(3), 536, May 1982.

[11] T. M. Cover and Y.-H. Kim, Capacity of a class of deterministic relay channels, *IEEE Internat. Symposium on Information Theory (ISIT)*, Nice, France, 591–595, June 2007.

[12] A. El Gamal and S. Zahedi, Capacity of a class of relay channels with orthogonal components, *IEEE Trans. on Inform. Theory*, 51(5), 1815–1817, May 2005.

[13] R. Gowaikar, A. F. Dana, R. Palanki, B. Hassibi, and M. Effros, Capacity of wireless erasure networks, *IEEE Trans. Inform. Theory*, 52(3), 789–804, March 2006.

[14] R. Khalili and K. Salamatian, An information theory for erasure channels, *Conference on Communication, Control and Computing*, Allerton, Sept. 2005.

[15] R. Khalili and K. Salamatian, On the achievability of cut-set bound for a class of erasure-relay channels, *Workshop on Wireless Ad-Hoc Networks*, IWWAN, May 2004.

[16] R. Dabora and S. D. Servetto, Broadcast channels with cooperating decoders, *IEEE Trans. Inform. Theory*, 52(12), 5438–5454, Dec. 2006.

[17] Y. Liang and V. V. Veeravalli, Cooperative relay broadcast channels, *IEEE Trans. Inform. Theory*, 53(3), 900–928, March 2007.

[18] Y. Liang and V. V. Veeravalli, Gaussian orthogonal relay channels: Optimal resource allocation and capacity, *IEEE Trans. Inform. Theory*, 51(9), 3284–3289, Sept. 2005.

[19] S. Lasaulce and A. G. Klein, Gaussian broadcast channels with cooperating receivers: The single common message case, *IEEE Proceedings of International Conference on Acoustic, Speech and Signal Processing (ICASSP)*, Toulouse, France, May 2006.

[20] T. D. Nguyen and S. Lasaulce, Capacity region of the deterministic broadcast channel with cooperative decoders, 2005, Oct., LSS technical report, CNRS, France.

[21] R. Khalili, S. Lasaulce, and P. Duhamel, Broadcast cooperative erasure channels, *Conference on Communication, Control and Computing*, Allerton, Sept. 2006.

[22] T. M. Cover, Broadcast channels, *IEEE Trans. Inform. Theory*, IT-18(1), 2–14, Jan. 1972.

[23] T. M. Cover, Comments on broadcast channels, *IEEE Trans. Inform. Theory*, 44(6), 2524–2530, Oct. 1998.

[24] Y. Liang and G. Kramer, Rate regions for relay broadcast channels, *IEEE Trans. Inform. Theory*, 53(10), 3517–3535, Oct. 2007.

[25] S. C. Draper, B. J. Frey, and F. R. Kschischang, On interacting encoders and decoders in multiuser settings, *International Symposium on Information Theory*, Chicago, IL, June–July 2004.

[26] D. Slepian and J. K. Wolf, Coding theorem for multiple access channels with correlated sources, *Bell Syst. Tech. J.*, 52(7), 1037–1076, Sept. 1973.

[27] E. V. Belmega, B. Djeumou, and S. Lasaulce, Performance analysis for the AF-based frequency division cooperative broadcast channel, *IEEE Proceeding of Signal Processing Advances in Wireless Communications Conference (SPAWC)*, Helsinki, Finland, June 2007.

[28] C. T. K. Ng, I. Maric, A. J. Goldsmith, S. Shamai, and R.D. Yates, Iterative and one-shot conferencing in relay channel, *Proceedings IEEE ITW*, Punta del Este, Uruguay, March 2006.

[29] B. Wang, J. Zhang, and A. Host-Madsen, On the capacity of ergodic MIMO relay channels, *IEEE Trans. Inform. Theory*, 51(1), 29–43, Jan. 2005.

[30] A. Sendonaris, E. Erkip, and B. Aazhang, User cooperation diversity—part I: System description, *IEEE Trans. Commun.*, 51(11), 1927–1938, Nov. 2003.

[31] S. Zahedi, M. Moheni, and A. El Gamal, On the capacity of AWGN relay channels with linear relaying functions, *IEEE International Symposium on Information Theory (ISIT)*, Chicago, IL, June–July 2004.

[32] A. A. El Gamal, M. Mohseni, and S. Zahedi, Bounds on capacity and minimum energy-per-bit for AWGN relay channels, *IEEE Trans. Inform. Theory*, 52(4), 1545–1561, April 2006.

[33] B. Djeumou, S. Lasaulce, and A. G. Klein, Practical quantize-and-forward schemes for the frequency division relay channel, *EURASIP J. Wireless Commun. Netw.*, 2007, 1–11, Nov. 2007.

[34] K. S. Gomadam and S. A. Jafar, On the capacity of memoryless relay networks, *Proceedings of IEEE International Conference on Communications*, Istanbul, Turkey, June 2006.

[35] K. S. Gomadam and S. A. Jafar, Optimizing soft information in relay networks, *Proceedings of Asilomar Conference on Signals, Systems and Computers*, Oct.–Nov. 2006.

[36] S. A. Jafar, C. Huang, and K. S. Gomadam, Beyond links—soft information optimization for memoryless relay networks, *Proceedings of UCSD Workshop on Information Theory and its Applications*, Feb. 2006.

[37] I. Abou-Faycal and M. Medard, Optimal uncoded regeneration for binary antipodal signaling, *IEEE Proceedings of International Conference on Communications*, Paris, France, June 2004.

[38] A. Kurtenbach and P. Wintz, Quantizing for noisy channels, *IEEE Trans. Commun.*, 17, 291–302, April 1969.

[39] N. Farvardin and V. Vaishampayan, Optimal quantizer design for noisy channels: An approach to combine source–channel coding *IEEE Trans. Inform. Theory*, 33(6), 827–837, Nov. 1987.

[40] L. H. Ozarow, S. Shamai, and A. D. Wyner, Information theoretic considerations for cellular mobile radio, *IEEE Trans. Vehicular Technology*, 43(2), 359–378, May 1994.

[41] L. Zheng and D. Tse, Diversity-multiplexing: A fundamental tradeoff in multiple antenna channels, *IEEE Trans. Inform. Theory*, 49(5), 1073–1096, May 2003.

[42] K. Azarian, H. El Gamal, and P. Schniter, On the achievable diversity-multiplexing tradeoff in half-duplex cooperative channels, *IEEE Trans. Inform. Theory*, 51(12), 4152–4172, Dec. 2005.

[43] S. Yang and J.-C. Belfiore, Optimal space-time codes for the MIMO amplify-and-forward cooperative channel, *IEEE Trans. Inform. Theory*, 53(2), 647–663, Feb. 2007.

[44] A. El Gamal and N. Hassanpour, Relay-without-delay, *IEEE Proceedings of International Symposium on Information Theory (ISIT)*, pp. 1078–1080, Adelaide, Australia, Sept. 2005.

[45] A. El Gamal, N. Hassanpour, and J. Mammen, Relay networks with delays, *IEEE Trans. Inform. Theory*, 53(10), 3413–3431, Oct. 2007.

[46] E. C. van der Meulen and P. Vanroose, The capacity of a relay channel, both with and without delay, *IEEE Trans. Inform. Theory*, 53(10), 3774–3776, Oct. 2007.

[47] R. Tannious and A. Nosratinia, Relay channel with private messages, *IEEE Trans. Inform. Theory*, 53(10), 3777–3785, Oct. 2007.

[48] M. Yuksel and E. Erkip, The relay channel with a wire-tapper, *Conference on Information Sciences and Systems*, Baltimore, MD, March 2007.

[49] N. Fawaz, D. Gesbert, and M. Debbah, When network coding and dirty paper coding meet in a cooperative ad-hoc network, *IEEE Transactions on Wireless Communications*, 7(5), 1862–1867, May 2008.

2 Capacity Limits in Cooperative Cellular Systems

Symeon Chatzinotas, Muhammad Ali Imran, and Costas Tzaras

CONTENTS

During the last years, cooperative cellular systems have emerged as the next advancement toward the implementation of broadband wireless networks. However, the term "cooperation" can have diverse meanings in the context of wireless networks. Therefore, the purpose of this chapter is to identify the major cooperative strategies and to investigate the advantages of the cooperative cellular systems, which employ multicell joint processing with comparison to the conventional interference-limited cellular systems. More specifically, the architecture of cooperative cellular systems is described and analyzed and the channel coding schemes which fully exploit the cooperative architecture and achieve the optimal capacity are identified. In this context, the capacity performances of cooperative and conventional cellular systems are compared and the involved practical limitations are investigated.

2.1   INTRODUCTION

During the last decades, wireless cellular systems have gone through an intensive evolutionary process, moving from analog voice streams to digital data services. As the demand for cost-efficient high-rate wireless services increases, the wireless network operators have to employ new wireless architectures to achieve higher data-rates. However, in spite of the evolution of the wireless cellular technologies, the increase in the system complexity becomes disproportional with respect to the provided capacity gain. Therefore, the research community as well as the industrial actors have begun a quest for alternative cellular architectures that have the ability to provide high spectral efficiencies. In this direction, cooperative wireless cellular architectures are gaining momentum as a dominant candidate for an alternative approach in wireless cellular networks. Cooperation in wireless networks can take many forms, such as user terminal (UT) cooperation, base station (BS) cooperation and relaying. UT cooperation is theoretically possible but practically it involves many complications, because the UTs have to communicate either on a separate wireless frequency band or through the BS to exchange cooperative information. This fact results in a waste of spectrum and energy, which is very important in terms of battery life in mobile devices. Relaying can be beneficial, but it either consumes the resources of relaying UTs or requires the installation of additional transponders by the network operator.

Based on the previous discussion, the approach of BS cooperation is analyzed and compared to the conventional cellular systems. In general, BS cooperation assumes that all the BSs or clusters of BSs are connected through high-capacity error-free channels to a central processor that jointly encodes or decodes the UT signals. This multicell joint processing has the ability to transform the cellular network to a wide-area Multiple Input Multiple Output (MIMO) system or distributed antenna system (DAS). The benefit of BS cooperation is a high capacity gain compared with conventional cellular systems, because intercell interference is no longer harmful for the capacity performance of the cellular system. This is due to the fact that in the uplink, the received power transmitted by the UT can be received by more than one adjacent BSs and it can be transmitted to the central processor (hyper-receiver) for joint detection [1–4]. Similarly, in the downlink the central processor (hyper-transmitter) jointly computes the UT signals before transmission and all the BSs form an antenna array to optimally transmit these signals to the UTs [5–8].

The rest of this chapter is structured as follows. Section 2.1 introduces the main concepts of cooperation in wireless cellular systems. Section 2.2 focuses more on distinguishing among UT cooperation, relaying, and BS cooperation. Furthermore, in this section the feasibility of the cooperative architectures is discussed. The remaining sections of this chapter focus on a cooperative cellular system that employs multicell joint processing. Section 2.3 provides a thorough investigation of the channel coding schemes, which achieve the capacity limit in the context of the considered cooperative cellular system. More specifically, the process of Successive Interference Cancellation (SIC) is analyzed for the uplink cellular channel, whereas for the downlink cellular channel the Dirty Paper Coding (DPC) scheme (coding with known interference cancellation) is described. Subsequently, the capacity performance of a multicell joint processing system that employs the aforementioned coding schemes is evaluated and compared. Section 2.4 is dedicated to presenting the differences between the design objectives of the cooperative and conventional cellular systems. More specifically, this section describes how the conventional cellular systems attempt to mitigate interference to maximize the single channel capacity of each UT. This approach is compared to the cooperative scheme, according to which the interference can be exploited to maximize the network sum-rate capacity. In this context, a typical cellular scenario is presented and analyzed to evaluate the spectral efficiency gap between the two approaches. Section 2.5 is focused on identifying the practical limitations of the considered cooperative architecture, taking into account the currently available technological resources. Finally, Section 2.6 offers a set of conclusive observations based on the previous comparative analysis and discusses the open issues in the area of multicell joint processing.

2.2 COOPERATIVE CELLULAR SYSTEMS

Over the last few decades, there has been rapid growth in personal (digital and voice) data communication. In this direction, the paradigm of point-to-point private communication links has shifted toward the shared use of common communication channels by multiple users. Just as in any other sharing scenario, this necessitates a "cooperative" approach that optimally utilizes the shared medium to jointly benefit all the users. In a cellular system, three types of cooperation scenarios emerge: cooperation between the UTs, cooperation between the geographically spaced BSs, and relaying over intermediate transceivers (relay stations [RSs]) (Figure 2.1).

2.2.1 UT COOPERATION

According to this approach, the UTs of the system can cooperate to increase the throughput or the fairness of the cellular system. The UT cooperation can appear in diverse forms, such as scheduling, power control, and virtual antenna diversity. More specifically, scheduling and power control are used to promote fair distribution of the available capacity by regulating the transmission strategy of each UT is terms of time sharing and interference mitigation, respectively. On the other hand, virtual antenna diversity can be used to achieve larger channel capacity by allowing adjacent UTs to form a multiple-antenna transceiver [9]. However, this capacity gain comes with a price, because the

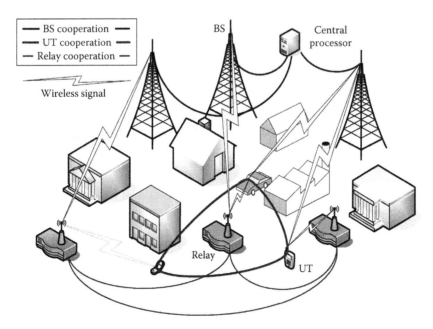

FIGURE 2.1 Types of cooperation in a cellular system.

UTs have to exchange synchronization data and channel state information (CSI) to collaboratively communicate with the adjacent BSs and this exchange results in a significant signaling overhead.

2.2.2 RELAYING

This approach refers to multi-hop cellular networks [10], where the communication link between BS and UT can comprise multiple hops, which are implemented through relaying elements (RS). The basic distinction in multi-hop cellular networks is whether fixed wireless transponders are employed for relaying or the UTs can act as relays under certain conditions. In the first case, the positioning of the transponders can be engineered for optimal coverage, but there is an additional cost of deploying and maintaining active network elements. In the second case, the extension of infrastructure is redundant, but the battery life of UTs can decrease substantially. The networks that adopt the second approach are also known as hybrid cellular networks [11,12], because they combine the principles of cellular and ad hoc networks. Another point of distinction in multi-hop networks is whether the relays can cooperate while transmitting data. This scenario seems more feasible in the case of fixed wireless transponders, because they can be wirelessly interconnected to cooperatively communicate data from/to the desired destination. Furthermore, another important characteristic of multi-hop cellular networks is the employed relaying strategy, which can be decode-and-forward (DF), compress-and-forward (CF), or amplify-and-forward (AF) [13–16]. In the first case, the signal is first decoded to remove the noise and then reencoded for forwarding. This strategy is more appropriate for dedicated relays, which can have higher power and processing capabilities. In the second case, a quantized and compressed observation of the received signal is relayed, whereas in the third case both signal and noise are amplified before forwarding. The later strategies are more appropriate for UT relays due to their reduced complexity. Finally, the relays can be categorized to static and adaptive according to their triggering mechanisms. More specifically, the static relays always forward signals, whereas the adaptive relays incorporate algorithms to evaluate if the forwarding is going to be beneficial for the system or not.

2.2.3 BS COOPERATION

The signal that is transmitted over the wireless medium, connecting the mobile users and the fixed BSs, suffers attenuation with distance traveled. Hence, a single BS is unable to establish a strong wireless connection with all the users, which are distributed over a sufficiently large geographical area. To overcome this problem, the cellular architecture was devised, where multiple BSs are deployed to efficiently cover the system area.

In traditional cellular systems, each UT associates with the BS that has the strongest channel to the specific terminal. While communicating data from the UT to the associated BS, the communication between a user and its associated BS can be overheard by neighbouring BSs. This acts as interference in the detection of the other signals at the neighboring BSs. If the BSs cooperate with one another, this unwanted overheard signal can be dealt with appropriately, noting that this interfering signal is still a desired information-bearing signal for the others. The idea of cooperation between multiple receivers to jointly decode multiuser signals is called multiuser detection [17,18]. This joint multiuser detection has the potential of turning the unwanted received signal into useful received signal by means of cooperation between the receivers. This concept of multicell joint decoding at the uplink of a cellular system was first introduced by the authors of Ref. [3] and also independently proposed by the seminal work of the authors of Ref. [1], further extended in Refs. [2,19] and discussed in Ref. [20]. This ensued the use of the term "Hyper receiver" for such a system wide joint decoder. The principles of joint decoding have been extended to the downlink channel, where the cooperating BSs can precancel the interference at the cooperating transmitters end (a hyper-transmitter), using a precoding technique termed as DPC [21] (see Section 2.3.2).

The geographical separation between the BSs is a hindering factor toward their cooperation. One way is to connect all BSs with high-speed and unlimited bandwidth links (a close physical

TABLE 2.1
Feasibility of Various Cooperation Scenarios

Cooperation Type	Pros	Cons
UT	No need for additional infrastructure	Increased UT complexity, battery life penalty and signaling overhead
RS (fixed transponders)	No UT cooperation needed Battery life saving	New sites needed Increased complexity for BS and RS signal processing
BS	No UT cooperation needed Use of the existing infrastructure	BS interconnection needed (cabling) Increased complexity for BS signal processing

reality is the optic fiber links) to allow them to cooperate without any rate and delay limits. However, this approach has vast economic implications and many will argue that it would not be financially viable to connect all the BSs in a geographically large system. Do we need to jointly process "all" the received signals? The answer to this question is hinged on another phenomenon. The spacing between the adjacent BSs defines how many "tiers of surrounding BSs" will overhear any transmitted signal with sufficient strength. Based on this fact the authors in Refs. [22–25] propose that clusters of cooperating BSs can achieve results close to the hyper-receiver decoding. A commonly used BS cooperation strategy in current wireless cellular systems (based on Code Division Multiple Access [CDMA] scheme) is termed as soft-handover [26–28]. In this scheme diversity gain is achieved by connecting a mobile user to two BS antennas (or two sector antennas of same BS in softer handover scenario) simultaneously. For a mobile user whose channel is varying rapidly, this increases the instantaneous chance of having a strong channel to at least one of the two antennas supporting the connection. The capacity of the cellular system with an extension to Wyner's model that closely represents this soft handover scenario is investigated in Refs. [29,30].

A detailed evaluation of the major cooperation strategies in terms of practical implementation can be found in Table 2.1. As can be seen, BS cooperation can take advantage of the existing infrastructure, because the processing complexity is transferred to the hyper-receiver without affecting the UTs or requiring new transceiving sites. A more detailed discussion about the practical limitations of multicell joint processing is given in Section 2.5. It should be noted that multiple cooperation strategies can be deployed at the same time, although initial results have shown that the capacity gain derived from this combination is not considerable. More specifically, it was shown [31–33] that relaying does not result in a significant capacity gain, when BS cooperation is in place. On the other hand, according to recent findings [34,35] multicell joint processing can increase the order of magnitude of the spectral efficiency in a typical macrocellular scenario. Based on the above discussion, the rest of this chapter is dedicated to investigating the benefits and limitations of BS cooperation in a cellular system.

2.3   COOPERATIVE CAPACITY LIMITS

In the cellular context, both uplink (UTs to BSs) and downlink (BSs to UTs) communication channels can be classified as Gaussian interference channels. The capacity region of these channels is not known in general [36]. However, the concept of BS cooperation in a cellular system can convert the uplink and the downlink interference channels into a MIMO Multiple Access Channel (MAC) and MIMO Broadcast Channel (BC), respectively. The capacity regions of these MIMO channels have been found and they can be interpreted accordingly in the context of cooperative cellular systems. It is depicted in Figure 2.2 that a MIMO multiple access channel can be obtained from

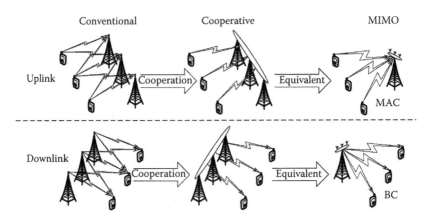

FIGURE 2.2 BS cooperation converts the uplink and downlink cellular channel to a wide-area MIMO MAC and BC channel, respectively.

a general uplink interference channel, introducing cooperation among BSs. Similarly, the lower part of Figure 2.2 shows that a MIMO broadcast channel can be formed from a general downlink interference channel by allowing the BS cooperation. The main objective of this BS cooperation is to convert the unwanted interference power into useful signal power by cooperating in signal encoding (downlink) or decoding (uplink). In the information-theoretic literature, multiuser channels such as cellular channels are represented as linear vector baseband channels of the form

$$\mathbf{y}[i] = \mathbf{H}[i]\mathbf{x}[i] + \mathbf{z}[i] \tag{2.1}$$

where
  $\mathbf{y}$ is the output vector
  $\mathbf{H}$ is the channel matrix
  $\mathbf{x}$ is the input vector
  $\mathbf{z}$ is the Additive White Gaussian Noise (AWGN) vector
  i is the time index

The channel is said to be *memoryless* if the output depends only on the current input and it is conditionally independent from the previous input or output values. On the grounds of the memoryless channel, the time index can be dropped, that is,

$$\mathbf{y} = \mathbf{H}\mathbf{x} + \mathbf{z} \tag{2.2}$$

The AWGN vector comprises independent identically distributed (i.i.d.) complex Gaussian noise variables with zero mean and variance normalized to unity, which are symbolized as $\mathcal{CN}(0, 1)$. In this context, the covariance matrix of $\mathbf{z}$ is $\mathbb{E}\left[\mathbf{zz}^{\dagger}\right] = \mathbf{I}$. According to Ref. [37], the per-cell sum-rate capacity of this cellular system is given by

$$C_{\mathrm{opt}}(\gamma) = \frac{1}{N}\mathbb{E}\left[\log \det \left(\mathbf{I}_N + \gamma\mathbf{HH}^{\dagger}\right)\right] \tag{2.3}$$

where
  $\gamma = P/\sigma^2$ is the transmit power over noise ratio
  N is the number of BSs in the system

As can be seen, the capacity depends on γ and on the eigenvalue distribution of the random matrix $\mathbf{HH}^{\dagger}$. In the following sections, we discuss in more detail the implications of this cooperation in theory and in practice.

2.3.1 CELLULAR UPLINK CHANNEL

As discussed earlier, the multiple BS receivers in a cellular uplink channel, cooperate in decoding the received signals. All received signals are transported (assuming an unconstrained, delayless link) to a central processor where they are jointly decoded. In this context, these geographically dispersed BS receivers act as multiple antennas of a single receiver [1–3].

2.3.1.1 Successive Interference Cancellation

An intuitive approach to decode these signals jointly is to start with the signal that can be decoded with highest certainty. All subsequent decodings subtract this known signal from the received aggregate signal before decoding other signals. This cancels the effect of the first signal as interference from all subsequent decodings. After the next signal is decoded, its effect as interference can also be cancelled in subsequent decodings, and the process continues. This decoding scheme is known as Successive Interference Cancellation (SIC) and combined with Minimum Mean Square Error (MMSE) filtering it can achieve the performance of optimum joint decoding [38, Chap. 8]. The block diagram of the SIC process is depicted in Figure 2.3.

Several techniques are used to cancel interference and optimize the capacity in the cellular uplink channel. In some scenarios, parallel interference cancellation is used instead of successive cancellation. This requires higher computational power but reduces the latency involved in the process [39]. A trade-off between the computational power and latency can be achieved using multistage interference cancellation techniques [40,41], where groups of users are detected in parallel and interference cancellation performed successively to improve the detection performance. To improve performance in practical systems, error control coding and interference cancellation can be done iteratively by exchanging the soft decisions in these subsequent stages of signal detection [42,43].

2.3.2 CELLULAR DOWNLINK CHANNEL

In the cellular downlink channel, multicell joint processing is employed by the transmitting BSs to cooperatively calculate and encode all the signals destined for the system UTs. In this context, the cellular downlink channel can be viewed as a MIMO broadcast channel with spatially distributed antennas.

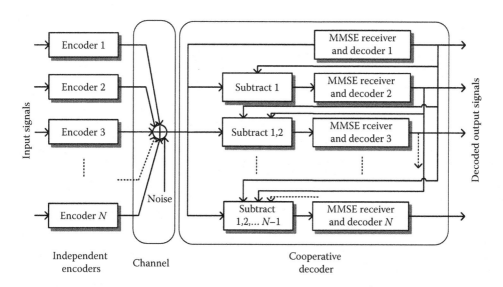

FIGURE 2.3 SIC block diagram.

2.3.2.1   Dirty Paper Coding

Celebrated results in the information-theoretic literature [6,44–46] have shown that the optimal capacity region of the cellular downlink channel can be achieved by DPC [21]. According to the single-channel DPC, interference does not compromise the channel capacity as long as this interference is known a priori to the transmitter [47]. DPC applies even though the receiver is not aware of the interference. This fact inspired the title of DPC as an analogy to the problem of writing on dirty paper, where the reader is not able distinguish between dirt and written symbols.

In the multiuser case, DPC can be extended by considering that the interference is due to the signals transmitted to multiple UTs over a shared medium. These signals are constructed by the transmitter and hence they can be considered to be known a priori. By establishing a serial encoding order for the system UTs, DPC can be used to minimize interference. More specifically, the UT signal that is encoded first is bound to receive interference from all the other signals, because none of them is known to the transmitter at the time of the encoding. Similarly, the UT signal that is encoded last does not receive any interference, because all of the other UT signals have been constructed and are known to the receiver. In general, a UT signal that is encoded at a random order is not affected by the preceding UT signals, but it receives interference from the following UT signals. The block diagram of the DPC process is depicted in Figure 2.4. The DPC principles have been applied to practical coding schemes, such as Tomlinson–Harashima precoding [48,49] and the vector perturbation technique [50].

2.3.3   Cooperative Channel Capacity

This section is dedicated to analyzing the uplink and downlink channel models of a cellular system in the presence of flat fading and power-law path loss. Furthermore, the uplink and downlink spectral efficiency is evaluated under the assumption of multicell joint processing.

2.3.3.1   Cellular Uplink Channel

Wyner's model [1] assumes that all the UTs in the cell of interest have equal channel gains, which are normalized to 1. It considers interference only from the UTs of the two neighboring cells, which

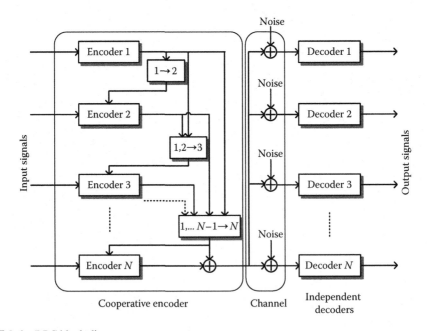

FIGURE 2.4 DPC block diagram.

are all assumed to have a fixed channel gain, also known as interference factor α, which ranges in $[0, 1]$. According to Wyner's model, the received signal at cell n is

$$y^n = \sum_{k=1}^{K} x_k^n + \alpha \left(\sum_{k=1}^{K} x_k^{n-1} + \sum_{k=1}^{K} x_k^{n+1} \right) + z^n \tag{2.4}$$

where

x_k^n is the transmitted complex signal from UT k in cell n

z^n is an i.i.d. complex circularly symmetric (c.c.s.) random variable representing AWGN with mean $\mathbb{E}[z^n] = 0$ and variance $\mathbb{E}[z^n z^{n*}] = \sigma^2$

All UTs are subject to an average power constraint, that is, $\mathbb{E}[x_k^n x_k^{n*}] \leq P$ for all (n, k). The parameter $\gamma = P/\sigma^2$ is defined as the transmit power over noise ratio. Assuming that there is a power-law path loss model that affects the channel gain, Wyner has modeled the case where the UTs of each cell are collocated with the cell's BS, because no distance-dependent degradation of the channel gain is considered. The same assumption is made by Somekh–Shamai [19], who have extended Wyner's model for flat fading environment. According to this model, the received signal at cell n and discrete time i is

$$y^n[i] = \sum_{k=1}^{K} b_k[i]^n x_k^n[i] + \alpha \left(\sum_{k=1}^{K} c_k[i]^n x_k^{n-1}[i] + \sum_{k=1}^{K} d_k[i]^n x_k^{n+1}[i] \right) + z^n[i] \tag{2.5}$$

where

x_{ki}^n is the ith complex channel symbol of the kth UT in the nth cell

$\{b_k^n\}, \{c_k^n\}, \{d_k^n\}$ are independent, strictly stationary, and ergodic complex random processes, respectively, in the time index i, which represent the flat fading processes experienced by the UTs

The fading coefficients are normalized to unit power, that is, $\mathbb{E}[b_k^n[i] b_k^n[i]^*] = \mathbb{E}[c_k[i]^n c_k^n[i]^*] = \mathbb{E}[d_k[i]^n d_k^n[i]^*] = 1$, and all UTs are subject to an average power constraint, that is, $\mathbb{E}[x_k^n x_k^{n*}] \leq P$.

In both Refs. [1,19], a single interference factor α is used to model both the cell density and the path loss. The interference factor α ranges in $[0, 1]$, where $\alpha = 0$ represents the case of perfect isolation among the cells and $\alpha = 1$ represents the case of BSs' collocation, namely a MIMO MAC channel. Based on the results in Ref. [19] for the model of Equation 2.5, Figure 2.5 depicts the per-cell sum-rate capacity for decreasing α in a Rayleigh flat fading environment, assuming that intercell interference is received only from the first tier of adjacent cells.

The model in Ref. [4] differs from the aforementioned models in the sense that it considers interference from all UTs of the cellular system. For the UTs of each cell, an interference coefficient is defined w.r.t. each BS, which depends on the power-law path loss model. Although the author in Ref. [4] takes into account the path loss effect, the UTs of each cell have still equal channel gain, and this refers to the case where the UTs of each cell are collocated with the cell's BS. However, this model is more detailed than the previously described models, because it decomposes the interference factor α, so that the cell density Π and the path loss exponent η can be modeled and studied separately. According to this model, the received signal at cell n for the flat fading case is

$$y^n[i] = \sum_{k=1}^{K} b_k[i]^n x_k^n[i] + \sum_{j=1}^{N/2} \alpha_j \left(\sum_{k=1}^{K} c_{kj}^n[i] x_{kj}^{n-j}[i] + \sum_{k=1}^{K} d_{kj}^n[i] x_{kj}^{n+j}[i] \right) + z^n[i] \tag{2.6}$$

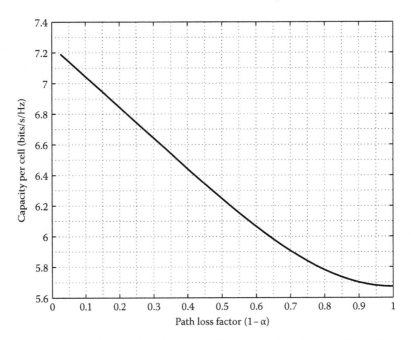

FIGURE 2.5 Capacity per cell C (bits/s/Hz) vs. $1 - \alpha$. Parameter values $K = 5$ and $\gamma = 10$.

where the interference factors α_j of the $n - j$ and $n + j$ cells are calculated according to the "modified" power-law path loss model [52]:

$$\alpha_j = (1 + j/\Pi)^{-\eta/2} \tag{2.7}$$

Based on the model of Equation 2.6, Figure 2.6 (solid lines) depicts the high and low-γ per-cell capacity for varying cell densities. The dashed lines of Figure 2.6 represent the high and low-γ per-cell capacity, if the model of Equation 2.6 is modified to comply with the main assumption of Wyner-like models, according to which intercell interference is received only by the first tier of adjacent cells. As expected, for high cell densities (small cellular system range D) this assumption does not hold and the capacity gap between the two models increases. For low-γ regime, the capacity gap becomes proportionally even larger, as it can be seen in Figure 2.6. However, for low cell densities (large cellular system range D) the two models converge, because the bulk of the interference comes from the first tier of neighboring cells. In other words, interference factors α_j for $j > 1$ become negligible and can be ignored without having an effect on the sum-rate capacity.

According to the model used herein, K UTs are distributed in each cell of a cellular system comprising N BSs. The evolution of the aforementioned information-theoretic models can be seen graphically in Figure 2.7. Initially, Wyner introduced the concept of interference factor a, which quantifies the amount of intercell interference with the first interfering tier. Subsequently, Somekh and Shamai introduced the i.i.d. fading coefficients b and c. Multiple tiers of interference were introduced by Letzepis, who has considered variable interference factors depending on the distance from the interfering tier and the power-law path loss. Finally, Chatzinotas et al. [34] have alleviated the assumption of collocated UTs by introducing user distribution. Figure 2.7 also depicts a comparison of the interference factors used in each model. In the current model, by assuming power-law path loss, flat fading, and uniformly distributed users, the received signal at cell n, at time index i, is given by

$$y^n[i] = \sum_{m=1}^{N} \sum_{k=1}^{K} \varsigma_k^{nm} g_k^{nm}[i] x_k^m[i] + z^n[i] \tag{2.8}$$

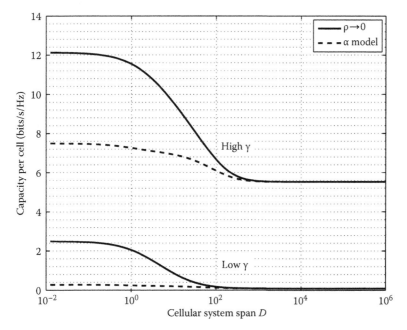

FIGURE 2.6   Capacity per cell C (bits/s/Hz) vs. cellular system range D for Letzepis's model (solid lines) and Wyner-like models (dashed lines). Parameter values $N = 100$, $\eta = 2$, $K = 5$, low $\gamma = .01$, and high $\gamma = 10$.

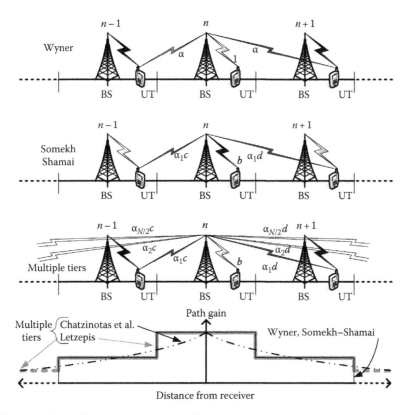

FIGURE 2.7   Evolution of information-theoretic cellular models.

where

$x_k^m[i]$ is the ith complex channel symbol transmitted by the kth UT of the mth cell

$\{g_k^{nm}\}$ are independent, strictly stationary, and ergodic complex random processes in the time index i, which represent the flat fading processes experienced in the transmission path between the nth BS and the kth UT in the mth cell

The fading coefficients are assumed to have unit power, that is, $\mathbb{E}[g_k^{nm}[i]g_k^{nm}[i]^*] = 1$ for all (n, m, k) and all UTs are subject to an average power constraint, that is, $\mathbb{E}[x_k^m[i]x_k^m[i]^*] \leq P$ for each (m, k). The interference factors ς_k^{nm} in the transmission path between the mth BS and the kth UT in the nth cell are calculated according to the "modified" power-law path loss model [4,52]:

$$\varsigma_k^{nm} = \left(1 + d_k^{nm}\right)^{-\eta/2} \tag{2.9}$$

Dropping the time index i, the aforementioned model can be more compactly expressed as a vector memoryless channel of the form

$$\mathbf{y} = \mathbf{H}_{UL}\mathbf{x} + \mathbf{z} \tag{2.10}$$

The channel matrix $\mathbf{H}_{UL}$ can be written as

$$\mathbf{H}_{UL} = \boldsymbol{\Sigma} \odot \mathbf{G} \tag{2.11}$$

where

$\boldsymbol{\Sigma}$ is an $N \times KN$ deterministic matrix

$\mathbf{G}$ is a complex Gaussian $N \times KN$ matrix with elements of variance 1, comprising the corresponding Rayleigh fading coefficients

The entries of the $\boldsymbol{\Sigma}$ matrix are defined by the variance profile function:

$$\varsigma(u, v) = \left[1 + d\,(u, v)\right]^{-\eta/2} \tag{2.12}$$

where $u \in [0, 1]$ and $v \in [0, K]$ are the normalized indexes for the BSs and the UTs, respectively, and $d\,(u, v)$ is the normalized distance between BS u and user v. According to Ref. [53], the per-cell capacity if $\mathbf{H} = \mathbf{G}$ is given by

$$\begin{aligned}
\lim_{N \to \infty} C_{opt}(\gamma) &= \lim_{N \to \infty} \frac{1}{N} \mathcal{I}\,(\mathbf{x}; \mathbf{y} \mid \mathbf{H}\,) \\
&= \lim_{N \to \infty} \frac{1}{N} \mathbb{E}\left[\log \det \left(\mathbf{I}_N + \gamma \mathbf{H}\mathbf{H}^\dagger\right)\right] \\
&= \lim_{N \to \infty} \mathbb{E}\left[\frac{1}{N} \sum_{i=1}^{N} \log\left(1 + \gamma \lambda_i\left(\mathbf{H}\mathbf{H}^\dagger\right)\right)\right] \\
&= \int_0^\infty \log(1 + \gamma x)\mathrm{d}F_{\mathbf{H}\mathbf{H}^\dagger}(x) = \mathcal{V}_{\mathbf{H}\mathbf{H}^\dagger}(\gamma) = K\mathcal{V}_{\mathbf{H}^\dagger\mathbf{H}}(\gamma)
\end{aligned} \tag{2.13}$$

where

$\gamma = P/\sigma^2$ is the transmit power over noise ratio

$\lambda_i\left(\mathbf{H}\mathbf{H}^\dagger\right)$ are the eigenvalues of matrix $\mathbf{H}\mathbf{H}^\dagger$

$$\mathcal{V}_{\mathbf{X}}(\gamma) \triangleq \mathbb{E}[\log(1 + \gamma \lambda_i\,(\mathbf{X}))] = \int_0^\infty \log(1 + \gamma \lambda_i\,(\mathbf{X})\,\mathrm{d}F_{\mathbf{X}}(x) \tag{2.14}$$

is the Shannon transform [53] of a random square Hermitian matrix $\mathbf{X}$, whose limiting eigenvalue distribution has a cumulative function denoted by $F_{\mathbf{X}}(x)$.

Based on the MP-law approximation [34], the per-cell capacity for $\mathbf{H} = \Sigma \odot \sqrt{N}\mathbf{G}$ is given by

$$C_{\text{opt}}(\gamma) = K\mathcal{V}_{f_K}\left(q_K(\Sigma)\frac{\tilde{\gamma}}{K}\right) \tag{2.15}$$

where $\tilde{\gamma} = KN\gamma$ is the system transmit power over noise ratio. In addition, $\mathcal{V}_{f_K}$ is the Shannon transform of the limiting eigenvalue distribution of the Marčenko–Pastur law:

$$\mathcal{V}_{f_K}(\gamma) = log\left[1 + \gamma - \frac{1}{4}\phi(\gamma, K)\right] + \frac{1}{K}log\left[1 + \gamma K - \frac{1}{4}\phi(\gamma, K)\right] - \frac{1}{4K\gamma}\phi(\gamma, K) \tag{2.16}$$

$$\phi(\gamma, K) = \left[\sqrt{\gamma\left(1 + \sqrt{K}\right)^2 + 1} - \sqrt{\gamma\left(1 - \sqrt{K}\right)^2 + 1}\right]^2 \tag{2.17}$$

Finally, considering that Σ is row-regular [54] due to the symmetry of the system, $q_K(\Sigma)$ is given by

$$\lim_{N \to \infty} q_K(\Sigma) = \frac{1}{K}\int_0^K \varsigma^2(v)dv, \quad \forall u \in [0, 1] \tag{2.18}$$

For a planar cellular system with uniformly distributed UTs, it has been shown in Ref. [34] that

$$\varsigma(v) = \varsigma_{\text{CO}}(v) \sqcap \left(\frac{Nt - \frac{K}{2}}{K}\right)$$

$$+ \sum_{m=1}^{M} \varsigma_{\text{I}}^m \left(Nt - K - \sum_{c=1}^{m-1} 6cK\right) \sqcap \left(\frac{Nt - K - \left(\sum_{c=1}^{m-1} 6cK + 3mK\right)}{6mK}\right) \tag{2.19}$$

$$\varsigma_{\text{CO}}(v) = \left[1 + d_{\text{CO}}(v)\right]^{-\eta/2} \quad \text{and} \quad \varsigma_{\text{I}}^m(v) = \left[1 + d_{\text{I}}^m(v)\right]^{-\eta/2}$$

where
M represents the number of interfering tiers of the planar cellular system
$\sqcap(t/T)$ is the rect function, where T is the width of the pulse

2.3.3.2 Cellular Downlink Channel

The cellular downlink channel is defined in accordance with the uplink channel, as described in Section 2.3.3.1. In this direction, let us assume that K users are uniformly distributed in each cell of a cellular system comprising N BSs. Each UT and BS is equipped with one omnidirectional antenna. Assuming flat fading, the received signal at UT $k = 1 \ldots K$ of cell $n = 1 \ldots N$, at time index i, will be given by

$$y_k^n[i] = \sum_{j=1}^{N} \varsigma_k^{nj} g_k^{nj}[i]x^j[i] + z_k^n[i] \tag{2.20}$$

where
$x^j[i]$ is the complex signal transmitted by BS $j = 1 \ldots N$ at the ith time instant
$\{g_k^{nj}\}$ are independent, strictly stationary, and ergodic complex random processes in the time index i, which represent the flat fading processes experienced in the transmission path between the jth BS and the kth UT in the nth cell

The fading coefficients are assumed to have unit power, that is, $\mathbb{E}[g_k^{nj}[i]g_k^{nj}[i]^*] = 1$ for all (n,j,k), and all BSs are subject to a sum power constraint, that is, $\sum_{j=1}^{N} \mathbb{E}[x^j[i]x^j[i]^*] \leq P_{\text{tot}}$. The interference factors ς_k^{nj} in the transmission path between the jth BS and the kth UT in the nth cell are calculated according to the modified power-law path loss model [4,52]:

$$\varsigma_k^{nj} = \left(1 + d_k^{nj}\right)^{-\eta/2} \tag{2.21}$$

Dropping the time index i, the aforementioned model can be more compactly expressed as a vector memoryless channel of the form

$$\mathbf{y} = \mathbf{H}_{\text{DL}}\mathbf{x} + \mathbf{z} \tag{2.22}$$

where
the vector $\mathbf{y} = [y_1^1 \ldots y_K^N]^T$ represents received signals by all the UTs of the cellular system
the vector $\mathbf{x} = [x^1 \ldots x^N]^T$ represents transmit signals by the BSs
the components of vector $\mathbf{z} = [z_1^1 \ldots z_K^N]^T$ are i.i.d c.c.s. random variables representing AWGN
with $\mathbb{E}[z_k^n] = 0$, $\mathbb{E}[z_k^n z_k^{n*}] = \sigma^2$. The channel matrix $\mathbf{H}_{\text{DL}}$ can be written as

$$\mathbf{H}_{\text{DL}} = \Sigma^\dagger \odot \mathbf{G}^\dagger = \mathbf{H}_{\text{UL}}^\dagger \tag{2.23}$$

where
$\Sigma^\dagger$ is a $KN \times N$ deterministic matrix
$\mathbf{G}^\dagger$ is a standard complex Gaussian $KN \times N$ matrix with variance 1, comprising the corresponding fading coefficients

The entries of the Σ matrix are defined by the variance profile function

$$\varsigma(u,v) = \left[1 + d(u,v)\right]^{-\eta/2} \tag{2.24}$$

where $u \in [0,K]$ and $v \in [0,1]$ are the normalized user indexes for the UTs and the BSs, respectively, and $d(u,v)$ is the normalized distance between BS u and user v. To evaluate the optimal downlink sum-rate capacity of a cellular system, the problem of power allocation has to be solved. Power allocation determines how the available power is distributed among the UTs. The aim is to find the power allocation strategy that maximizes the sum-rate capacity of the linear cellular system at each time instance. For the downlink cellular channel, this objective can be translated to the following maximization problem:

$$\text{maximize } C_{\text{DL}}(\mathbf{H}_{\text{DL}}, P_{\text{tot}}) = \sum_{i=1}^{KN} \log \frac{\det\left[1 + \mathbf{h}_i\left(\sum_{j=1}^{i}\Gamma_j\right)\mathbf{h}_i^\dagger\right]}{\det\left[1 + \mathbf{h}_i\left(\sum_{j=1}^{i-1}\Gamma_j\right)\mathbf{h}_i^\dagger\right]} \tag{2.25}$$

$$\text{subject to } \sum_{i=1}^{KN} \text{Tr}(\Gamma_i) \leq P_{\text{tot}} \quad \text{and} \quad \Gamma_i \geq 0 \tag{2.26}$$

where
$\Gamma_i \in \mathbb{C}^{N \times N}$ are the downlink input covariance matrices
$\mathbf{h}_i \in \mathbb{C}^{1 \times N}$ are the row vectors of the channel matrix $\mathbf{H}_{\text{DL}}$, namely the channel gains of UT i w.r.t. N BSs

Equation 2.25 is neither a concave nor a convex function of Γ_i, and thus the entire space of covariance matrices should be examined to meet the objective [6,55]. However, this obstacle can be overcome by using the principles of duality. More specifically, instead of solving Equation 2.25, the following dual maximization problem can be considered:

$$\text{maximize} \quad C_{\text{UL}}\left(\mathbf{H}_{\text{DL}}, P_{\text{tot}}\right) = \log \det \left(\mathbf{I} + \sum_{i=1}^{KN} \mathbf{h}_i^{\dagger} \mathbf{q}_i \mathbf{h}_i\right) \tag{2.27}$$

$$\text{subject to} \quad \sum_{i=1}^{KN} \text{Tr}\left(\mathbf{q}_i\right) \leq P_{\text{tot}} \quad \text{and} \quad \mathbf{q}_i \geq 0 \tag{2.28}$$

where $\mathbf{q}_i \in \mathbb{C}$ are the transmit powers of each single antenna UT. Equation 2.27 is a convex function of $\mathbf{q}_i$ [56]. According to Ref. [35], the sum-rate capacity bounds under per-BS and per-system power constraint converge in high-SNR cellular systems, which are adequately populated with BS, so that the available power budget is adequate to serve a high number of UTs per cell at each time instant.

2.3.3.3   Uplink vs. Downlink Spectral Efficiency

Figure 2.8 depicts the per-cell sum-rate capacity for the uplink (dashed line) and the downlink linear array (solid line) for $N = 10$ BSs and $K = 10$ UTs per cell in the high-γ regime. The UT power constraint is $P = 10$, whereas the BS power constraint is $P_{\text{BS}} = KP$. Perfect channel reciprocity is assumed in the uplink and downlink and therefore $\mathbf{H}_{\text{DL}} = \mathbf{H}_{\text{UL}}^{\dagger}$. Furthermore, the AWGN thermal noise at BS and UT receivers is assumed to be equal to σ^2 and a path-loss exponent of $\eta = 2$ is considered. The sum-rate capacities of the uplink and downlink channel are evaluated based on Refs. [34,35], respectively. As can be seen, the downlink capacity is always higher than the uplink capacity, which is to be expected because the power constraint (per-BS) in the downlink scenario is

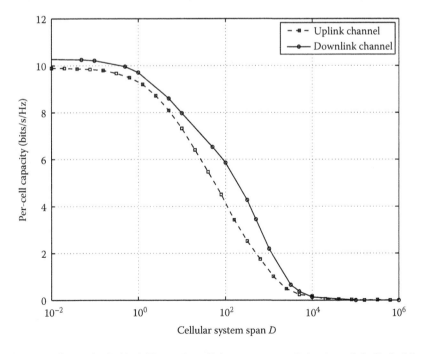

FIGURE 2.8   Per-cell capacity in bits/s/Hz vs. the cellular system span D for the uplink (dashed line) and the downlink linear array (solid line).

laxer than the uplink scenario (per-UT). The largest value of the capacity gap is observed for medium cell densities ($D = 10^2$).

2.4 COOPERATIVE VS. CONVENTIONAL CELLULAR SYSTEMS

This section focuses on discussing the different treatments of interference in cooperative and conventional systems and evaluating the spectral efficiency gap produced by this treatment.

2.4.1 CONVENTIONAL CELLULAR SYSTEMS

In conventional cellular systems, the UTs measure and compare the channel gain to several BSs in their range. Each UT associates with the BS that has the highest channel gain.

2.4.1.1 Single-Channel Capacity

Considering the example of the uplink channel, each BS attempts to decode the signal of each individual UT associated with it. In this attempt, all signals from the other UTs associated with the same BS are treated as intracell interference and the signals from UTs associated with other BSs are called intercell interference. In the context of a conventional cellular system, these interfering signals are detrimental to the successful decoding of the intended signal. The capacity of this link between the UT and its associated BS is given by the single link capacity formula of Shannon [57]:

$$C = \log(1 + \text{SNIR}) \tag{2.29}$$

where SNIR is the ratio of the power of the intended signal to the power of intercell interference, intracell interference, and AWGN, combined. Similarly on the downlink, any independent message transmitted to one UT acts as interference to all other UTs in the system, limiting the achievable capacity. For the uplink case, the problem is aggravated due to different levels of received signal strengths for the UTs that are at a short distance from the BS in comparison with the UTs that are at a long distance from the BS. This problem is known as near-far problem and tight power control is used to minimize the effects of this phenomenon.

2.4.1.2 Interference Mitigation

Several techniques are used in conventional cellular systems to keep the interference in control. In a CDMA-based cellular system the main approaches are to control the power [58] or to design the CDMA codes in such a way that the interference is minimized [59,60]. In the uplink case, the interference effect can be minimized by using the linear multiuser detection [61,62]. On the downlink, "blind" multiuser detection techniques can be used to reduce the effect of interference [42,63,64] at the receiving end.

2.4.2 COOPERATIVE CELLULAR SYSTEMS

In cooperative cellular systems, the notion of cell becomes of less importance, because a single UT can communicate with multiple adjacent BSs. More specifically, the number of BSs that can communicate with a single UT depends on the path loss impairment of the wireless channel and is proportional to the cell density of the system.

2.4.2.1 DAS and MIMO Systems

A cellular system that employs multicell joint processing can be represented by a MIMO channel. More specifically, in the uplink case, all the BSs jointly decode the received signal, and thus they can be considered as a large MIMO receiver with spatially distributed antennas. Similarly, in the downlink case, all the BSs jointly encode the transmitted signal, and thus they can be considered as a large MIMO transmitter with spatially distributed antennas. Another term, which is sometimes used for multicell joint processing cellular systems, is DAS [65]. The main difference between the DAS and the MIMO systems is that the antennas in the DAS channel are adequately spatially separated, so that the channel coefficients among the multiple antennas can be considered uncorrelated, in contrast with the coefficients of the MIMO channel. This representation of the cellular system is very useful, because a number of theorems from the MIMO information-theoretic literature can be applied in the calculation of the cellular system capacity.

2.4.2.2   Interference Exploitation

In the practical engineering design of cellular systems, the main figure of merit that determines the capacity rate of a UT is the SINR $= P_R/(I + \sigma_R^2)$, where P_R is the received power at the BS of interest, σ_R^2 is the thermal AWGN at the receiving BS, and I is the intercell and intracell interference received from other UTs of the system. However, in the information-theoretic analysis of hyper-receiver cellular systems, the main figure of merit that determines the per-cell capacity is $\gamma = P/\sigma_{HR}^2$, where P is the transmit power of the UT and σ_{HR}^2 is the AWGN thermal noise at the hyper-receiver. The main reason that SINR does not constitute an appropriate figure of merit for multicell joint processing analysis is that intercell interference is not harmful and thus the term I can be transferred to the nominator. Because there is no harmful interference, there is no need for power control, and thus the UTs constantly transmit with the maximum available power P [66]. In this context, the transmit power P remains fixed for all the UTs, whereas the received power and as a result the produced rate differs for each UT. Because the UTs achieve different rates, a more appropriate objective function for multicell processing would be the per-cell sum-rate capacity. In this context, the variable affecting the value of this function should have a value that is UT independent and encompasses all the system parameters. Taking this into account, an appropriate variable would be the rise over thermal (RoT). In hyper-receiver cellular networks, the RoT of the system is defined as the ratio of the total signal power received from all UTs of the system at the BS end to the thermal AWGN. More specifically, assuming distributed UTs, RoT is given by

$$\text{RoT} = \gamma \sum_{i=1}^{KN} (\varsigma_i)^2 \tag{2.30}$$

where $\sum_{i=1}^{KN} (\varsigma_i)^2$ is the sum of the path loss coefficients of all UTs w.r.t. a single BS. Figure 2.9 depicts the per-cell capacity vs. RoT for a typical macrocellular scenario, which is characterized by the parameters of Table 2.2. More specifically, the parameter L_0 is used to denormalize the path loss coefficients according to the following equation:

$$\varsigma(d(v)) = \sqrt{L_0 \left[1 + \hat{d}(v)/d_0 \right]^{-\eta}} \tag{2.31}$$

where $\hat{d}(v)$ is the normalized distance. The RoT curves (thick lines) have been drawn on top of the $\log(1 + x)$ curve (thin line). It can be easily seen that hyper-receiver cellular systems are not interference-limited, because the per-cell capacity increases monotonically with RoT.

2.4.3   SPECTRAL EFFICIENCY ANALYSIS

In this section, a simplified comparative model is used to highlight the conceptual difference in terms of capacity between a conventional cellular system and a multicell joint processing system. It should be noted that in the context of this mathematical model, some simplifying assumptions are made that do not strictly hold in a realistic system. Assuming the uplink channel of a linear cellular system,

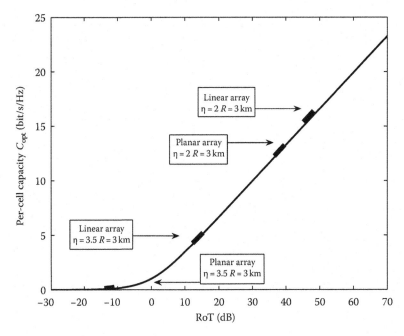

FIGURE 2.9 Per-cell capacity (bit/s/Hz) vs RoT (dB) for the linear cellular system. Parameters: $\eta = 2, 3.5$ and $P_T = 100$–200 mW.

TABLE 2.2

Value/Range of Parameters Used for a Typical Macrocellular Scenario

Parameter	Symbol	Value/Range (Units)
Cell radius	R	0.1–3 km
Reference distance	d_0	1 m
Path loss at ref. distance	L_0	-38 dB
Path loss exponent	η	2 usual values {2, 3.5}
UTs per cell	K	20
UT transmit power	P_T	200 mW
Thermal noise density	N_0	-169 dBm/Hz
Channel bandwidth	B	5 MHz

BSs and UTs are located on a straight line, as shown in Figure 2.10. For instance, this model can closely fit a cellular system designed to provide coverage on a motorway. This linear cellular array comprises N symmetric cells deployed over a distance D, and thus each cell has a span of D/N. Furthermore, it is assumed that K UTs are uniformly distributed on each cell's coverage span. Let each BS be indexed as $n = 1 \ldots N$ and each user within a cell be indexed by $k = 1 \ldots K$. Without loss of generality, the edge effects of the cellular system can be ignored to preserve cell symmetry. The central cell's BS is located in the middle of the linear array (origin), and the variable distance of any UT from the origin is represented as r. Due to the symmetry of the system, we can focus on the UTs positioned on the right side of the central BS and find similar expressions for the UTs on the left side.

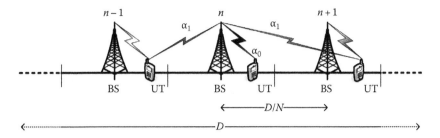

FIGURE 2.10 Illustration of a linear cellular array. Users transmit to the BS of their cell and their signal interferes with the other BSs. Assuming the uplink communication, the receiver of the central cell is receiving signal from the UTs within the cell as well as the interference signals from two neighbour cells.

First, the performance of a conventional system is considered by focusing on the decoder of the central BS receiver. The objective of the receiver is to decode the signals transmitted by all K UTs located in the central cell. In the decoding process, the transmissions in the (at least two immediate) neighbor cells are treated as unwanted interfering signals, also known as the intercell interference, I^o. While decoding the signal for any of the K desired UTs in cell n, $K - 1$ signals of the remaining desired UTs also act as interfering signals termed as intracell interference, I_n^i. Assuming that the AWGN power at the receiver of the central cell is σ_n^2, the received SNIR for each user, k, is denoted by $\gamma_{k,n}$ and is given as

$$\gamma_{k,n} = \frac{h_{k,n}p_{k,n}}{I_n^o + I_n^i + \sigma_n^2} \tag{2.32}$$

where the intercell interference is the aggregate received signals from all UTs located in adjacent (at least two immediate neighboring) cells. To calculate the intercell interference, assume that each transmitted signal from a user $\acute{k}$ in another cell $\acute{n}$ experiences a distance-dependent power gain given as

$$h_{\acute{k},\acute{n}} = \frac{1}{(1 + r_{\acute{k},\acute{n}})^\eta} \tag{2.33}$$

with $r_{\acute{k},\acute{n}}$ being the distance between the user $\acute{k}$ in the adjacent cell $\acute{n}$ and the BS of interest n. In this direction, intercell interference is given as [67]

$$I_n^i = \sum_{\substack{\forall \acute{k},\acute{n} \\ \acute{n} \neq n}} h_{\acute{k},\acute{n}}p_{\acute{k},\acute{n}} \tag{2.34}$$

As this is a function of total power transmitted in each cell and the distance-dependent path gains of all users in the cell, appealing to the symmetry of the cellular system and the uniform distribution of the users, we can model the intercell interference as in Ref. [1] using the intercell interference factor:

$$I_n^o = \alpha P_T \tag{2.35}$$

where P_T is the sum power transmitted in each cell (same for all cells, due to symmetry). Similarly, the intracell interference for user k can be modeled as

$$I_n^o = \sum_{\acute{k} \neq k} h_{\acute{k},n}p_{\acute{k},n} \tag{2.36}$$

To promote rate fairness in the cellular system, all UTs are power-controlled to achieve equal received power [67] and hence equal SNIR at the BS receiver. For the studied model, this power control requires that the UT at the edge of the cell (having the largest distance from the receiver) should transmit maximum power, P and all other UTs positioned closer to the receiver scale down their transmissions to equalize their received power to the received power of the cell edge UT. In this direction, the transmitted power of the kth UT in the nth cell is given by

$$p_{k,n} = \frac{P}{h_e} \cdot \frac{1}{h_{k,n}} = \frac{P}{\left(1 + \frac{D}{2N}\right)^\eta} \cdot \frac{1}{h_{k,n}} \tag{2.37}$$

where h_e is the channel gain for the cell-edge UT. Hence, all UTs except for the cell-edge UTs transmit with a lower power than the maximum allowed p_{max}. Using this power control, we can approximate I_n^o and I_n^i, for any cell as

$$I_n^o = \alpha K h_e P \quad \text{and} \quad I_n^i = (K-1)h_e P \tag{2.38}$$

It follows that all UTs can achieve an equal SNIR, which can be expressed as (see Ref. [67], voice activity factor of unity is assumed here for simplified notation)

$$\gamma_k = \frac{h_e P}{[(1+\alpha)K - 1]h_e P + \sigma^2} = \frac{h_e \gamma}{[(1+\alpha)K - 1]h_e \gamma + 1} \tag{2.39}$$

The conventional system is designed such that the equalized γ_k is adequate to support a reliable communication link. Communication is established by transmitting symbols from a finite set over the noisy channel and then using a detection scheme at the receiver side. This scheme results in a finite probability of detection error, which can be arbitrarily decreased by transmitting more power to make the symbols more distinct and capable of combating the noise. In Figure 2.11, we depict how

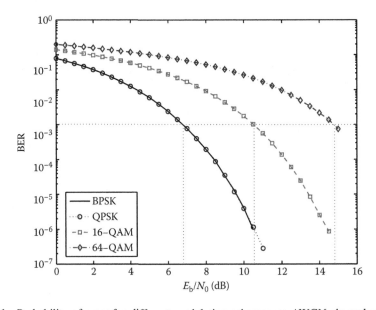

FIGURE 2.11 Probability of error for different modulation schemes on AWGN channel without channel coding.

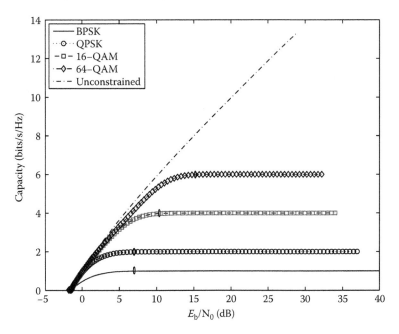

FIGURE 2.12 Average mutual information for some modulation constrained communication links and capacity upper bound for unconstrained AWGN transmission (Shannon limit). The oval mark on each curve shows the point of operation for an error rate of 10^{-3} on an uncoded communication link (cf. Figure 2.11).

this error rate decreases when more energy is used for each bit of transmitted information (higher E_b/N_0, energy per bit to noise spectral density, implies higher SNIR target considering identical coding and modulation schemes). If the criterion for reliable communication is to achieve an error rate less than one error in a thousand bits (indicated with a dotted line at the error rate of 10^{-3}), we must maintain an E_b/N_0 of 7 dB for BPSK and QPSK, 10.5 dB for 16-QAM, and 15 dB for 64-QAM. The benefit of employing higher-order modulation schemes can be observed by examining the achieved mutual information depicted in Figure 2.12. It can be observed that the mutual information for any modulation scheme gets saturated after a certain increase in transmitted power. To increase the capacity at high E_b/N_0, higher-order modulation schemes have to be employed, so that each symbol maps a larger number of information bits.

Subsequently, an identical cellular model is considered, but this time the performance of multicell joint processing is investigated. In this context, we assume unconstrained modulation, that is, the input is Gaussian distributed and is transmitted over a Gaussian channel. The mutual information, per cell, is given by Equation 2.3. It should be noted that in the context of multicell processing all UTs transmit at their maximum available power P without any power control, because interference is no longer harmful.

We consider a specific scenario for comparison. There are $N = 10$ cells with $K = 10$ users in each cell. The length of each cell is one distance unit. The path loss exponent is $\eta = 2$ and noise power is normalized to unity. We find the capacity using the equalized SNIR for the conventional system for $\alpha = 0.01$ changing the transmit power constraint P and compare it with the information-theoretic capacity calculated using Equation 2.3. As can be seen (Figure 2.13), conventional cellular systems are interference-limited, because increasing the transmit power does not provide a substantial capacity gain. On the other hand, multicell processing systems can provide a capacity gain proportional to the transmit power.

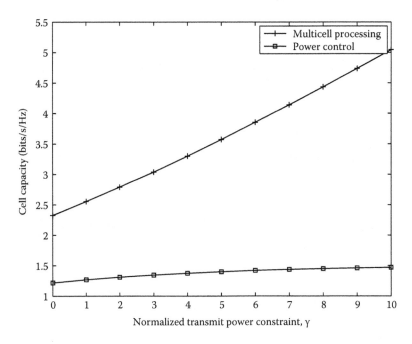

FIGURE 2.13   Comparison of multicell processing and power control approach.

2.5   PRACTICAL LIMITATIONS

This section is dedicated to discussing the practical issues that can limit the capacity performance of multicell joint processing systems.

2.5.1   CHANNEL-STATE INFORMATION

To achieve the optimal capacity region, the employed coding schemes (SIC and DPC) require perfect Channel State Information (CSI). In general, the channel state is estimated at the receivers by measuring the received power of pilot signals originating at the transmitters. More specifically, each receiver quantizes the pilot signal of each transmitter and subsequently it estimates its channel vector to the closest of a quantization vector set. The accuracy of the estimation is proportional to the size of the quantization vector set. However, the practical channel estimation in a cellular system differs between the uplink and the downlink channel. More specifically, the pilot signals are always broadcast by the BSs and measured at the UTs. Therefore, the UTs are aware of the downlink channel gains and a reverse channel is needed to feedback the channel estimates to the BSs. In this direction, the required rate of the reverse channel is proportional to the quantization vector set. This fact gives rise to an accuracy-throughput trade-off, because by increasing the accuracy of the channel estimation, the throughput of the reverse uplink channel is decreased due to the CSI data and vice versa. An efficient practice of mitigating this problem is to allow only a subset of UTs to feedback channel estimates at each time instant. In this context, the receiver selection algorithm can be channel-dependent or -independent. In channel-dependent algorithms the receivers that have strong channel gains or small quantization error are allowed to feedback, resulting in higher sum-rate capacities. In channel-independent algorithms, the receiver selection is random to promote fairness and facilitate the transmission of delay-sensitive traffic.

Another problem in the downlink case is that there is a time delay of the channel estimation, which corresponds to the time needed for the pilot signal to reach the UT, to be measured and quantized, and to be fed back to the BS. Hence, when the UTs are highly mobile, the time coherence of the channel is very short and this process can introduce estimation errors. This problem is tackled by transmitting pilot signals more frequently, but this results once again in an accuracy-throughput trade-off, because by increasing the frequency of the pilot signal transmission, the throughput of the downlink channel is decreased and vice versa. In the uplink channel, the BSs require the CSI of the reverse channel to jointly decode the received signals. However, this information is not available and therefore the downlink channel gains are used instead. This discrepancy is of minor importance if the uplink–downlink channel reciprocity applies. In Frequency Division Multiplexing (FDM) systems, this means that there is no frequency-selective fading and thus the channel gains are not frequency-dependent. In Time Division Multiplexing (TDM) systems, it means that there is no time selectivity and thus the channel gains are not time-dependent. Nevertheless, these ideal assumptions rarely apply and hence channel estimation errors are inevitable. In general, the level of channel reciprocity is higher in TDM than in FDM systems. The degradation of the information-theoretic capacity due to imperfect channel estimation is studied in Refs. [68–74].

2.5.2 CHANNEL CODING SCHEMES

A concise review of interference cancellation for cellular systems is presented in Ref. [39], and the practical limitations of the known schemes are discussed. Perfect channel knowledge is assumed for optimum interference cancellation. However, it is not always feasible in practice to estimate the channel perfectly. Any imperfect estimation of the channel makes the interference cancellation imperfect (even if the signal is detected error free, it needs to be reconstructed perfectly to be cancelled from aggregate signal and this requires perfect channel estimate). The signals that are decoded in later stages of SIC are particularly affected by the residual interference and errors propagate, resulting in higher error rates for the users decoded later. Various approaches are suggested to address this problem. One approach is to perform the interference cancellation in several stages [75]. Another approach is to perform the channel estimation and signal detection iteratively [43,75]. Optimal power control can also be used, so that it accounts for the statistics of channel estimation error [76,77] and allocates higher power to signals that are decoded in the presence of residual errors.

2.5.3 COOPERATIVE INFRASTRUCTURE

The global cooperation at the BS end as envisaged by Refs. [1,3] has two practical limitations. Connecting all BSs, geographically dispersed over a very wide area, with delayless and infinite bandwidth channels is very resource intensive and economically expensive. This motivates the search for suboptimal and localized joint decoding arrangements. As the distance-dependent signal attenuation ensures that any signal is physically localized to a limited area, joint decoding of all BSs that receive this signal with significant strength should suffice to achieve high capacity. Following this idea, several schemes have been developed for joint decoding of cell clusters [22,24,25,78]. Limitations of the backhaul system that interconnects the various BSs should also be taken into account. Authors in Ref. [79] describe how capacity can be achieved in a system with constrained backhaul. The recent survey in Ref. [80] summarizes the effects of limited-capacity backhaul for nonfading uplink and downlink channels for simplified cellular system models.

2.5.4 PROCESSING LOAD

A global joint decoder will be extremely demanding in terms of computational load because the complexity of symbol-by-symbol multiuser detection increases exponentially as the number of users to be detected in the system increases [18]. However, for a coded system MMSE in combination with SIC yields linear complexity in the number of users or at least polynomial if one considers that the computation of the MMSE filters, matrix-vector multiplications, and subtraction are quadratic or cubic in the number of users [38, Chap. 8]. Another approach to keep the processing load within constraints is to use local joint decoding techniques suggested in Refs. [22,24,25,78]. Here, it should be noted that the processing load, is a relatively trivial issue as this processing is required at the fixed-position BSs, which have laxer constraints on size and available processing power.

2.6 SUMMARY AND OPEN ISSUES

With the advent of wireless data networks, the increasing demand of high-rate wireless communication from/to mobile users has promoted research toward cooperative strategies, which can be employed to achieve higher data-rates. Cooperation in cellular networks can take many forms, such as UT cooperation, relaying, and BS cooperation. BS cooperation, also known as multicell joint processing, is a very promising strategy in terms of spectral efficiency as well as in terms of feasibility. In this chapter, the performance of multicell joint processing systems has been evaluated and compared with legacy cellular systems. Furthermore, the optimal coding schemes and the practical limitations of multicell joint processing systems have been discussed and analyzed. However, to approach the practical implementation of multicell joint processing systems, further research is required in the following directions:

- Extension of information-theoretic models based on realistic assumptions, for example, distributed UTs, multiple tier interference, and per-antenna power constraints.
- Evaluation of performance degradation due to clustered joint processing. Optimal resource allocation strategies among clusters. Effect of limited-rate backhaul interconnection.
- Practical capacity-achieving coding schemes that are characterized by low encoding/decoding complexity and resilience to error propagation. Evaluation of coding performance under realistic channel estimation conditions.

REFERENCES

[1] A. Wyner, Shannon-theoretic approach to a Gaussian cellular multiple-access channel, *IEEE Transactions on Information Theory*, 40, 1713–1727, November 1994.

[2] O. Somekh and S. Shamai, A Shannon-theoretic view of Wyner's multiple-access cellular channel model in the presence of fading, in *IEEE International Symposium on Information Theory–1998*, p. 393, Cambridge, MA, August 1998.

[3] S. Hanley and P. Whiting, Information-theoretic capacity of multi-receiver networks, *Telecommunications Systems*, 1, 1–42, 1993.

[4] N. Letzepis, Gaussian cellular multiple access channels, PhD thesis, Institute for Telecommunications Research, University of South Australia, December 2005.

[5] T. Lan and W. Yu, Input optimization for multi-antenna broadcast channels with per-antenna power constraints, in *IEEE Global Telecommunications Conference–2004 (GLOBECOM '04)*, 1, 420–424, Dallas, TX, 2004.

[6] S. Vishwanath, N. Jindal, and A. Goldsmith, Duality, achievable rates, and sum-rate capacity of Gaussian MIMO broadcast channels, *IEEE Transactions on Information Theory*, 49(10), 2658–2668, 2003.

[7] H. Viswanathan, S. Venkatesan, and H. Huang, Downlink capacity evaluation of cellular networks with known-interference cancellation, *IEEE Journal on Selected Areas in Communications*, 21(5), 802–811, 2003.

[8] S. Jing, D. Tse, J. Soriaga, J. Hou, J. Smee, and R. Padovani, Multicell downlink capacity with coordinated processing, in *EURASIP Journal on Wireless Communications and Networking*, Article ID 586878, 2008.

[9] A. Ozgur, O. Leveque, and D. Tse, Hierarchical cooperation achieves optimal capacity scaling in ad hoc networks, *IEEE Transactions on Information Theory*, 53, 3549–3572, October 2007.

[10] Y. Lin and Y. Hsu, Multi-hop cellular: A new architecture for wireless communications, in *Nineteenth Annual Joint Conference of the IEEE Computer and Communications Societies INFOCOM 2000*, pp. 1273–1282, Tel Aviv, Israel, 2000.

[11] B. Liu, Z. Liu, and D. Towsley, On the capacity of hybrid wireless networks, in *Twenty-Second Annual Joint Conference of the IEEE Computer and Communications Societies (INFOCOM 2003)*, 2, 1543–1552, San Francisco, CA, 2003.

[12] A. Zemlianov and G. de Veciana, Capacity of ad hoc wireless networks with infrastructure support, *IEEE Journal on Selected Areas in Communications*, 23, 657–667, March 2005.

[13] J. Laneman, D. Tse, and G. Wornell, Cooperative diversity in wireless networks: Efficient protocols and outage behavior, *IEEE Transactions on Information Theory*, 50(12), 3062–3080, 2004.

[14] G. Kramer, M. Gastpar, and P. Gupta, Cooperative strategies and capacity theorems for relay networks, *IEEE Transactions on Information Theory*, 51(9), 3037–3063, 2005.

[15] T. Cover and A. El Gamal, Capacity theorems for the relay channel, *IEEE Transactions on Information Theory*, 25(5), 572–584, 1979.

[16] S. Borade, L. Zheng, and R. Gallager, Amplify-and-forward in wireless relay networks: Rate, diversity, and network size, *IEEE Transactions on Information Theory*, 23, 3302–3318, October 2007.

[17] S. Verdú, Minimum probability of error for asynchronous Gaussian multiple-access channels, *IEEE Transactions on Information Theory*, 32, 85–96, January 1986.

[18] S. Verdú, *Multiuser Detection*. University Press, Cambridge, U.K., 1998.

[19] O. Somekh and S. Shamai, Shannon-theoretic approach to a Gaussian cellular multiple-access channel with fading, *IEEE Transactions on Information Theory*, 46, 1401–1425, July 2000.

[20] S. Shamai (Shitz) and A. Wyner, Information theoretic considerations for symmetric, cellular multiple access fading channels part I, and part II, *IEEE Transactions on Information Theory*, 43, 1877–1911, November 1997.

[21] M. Costa, Writing on dirty paper, *IEEE Transactions on Information Theory*, 29(3), 439–441, 1983.

[22] M. Bacha, J. Evans, and S. Hanly, On the capacity of MIMO cellular networks with macrodiversity, in *Seventh Australian Communications Theory Workshop–2006*, pp. 105–109, Perth, Australia, February 2006.

[23] B. Ng, J. Evans, S. Hanly, and A. Grant, Distributed linear multiuser detection in cellular networks, in *Fifth Australian Communications Theory Workshop*, pp. 127–132, Newcastle, Australia, February 2004.

[24] E. Aktas, J. Evans, and S. Hanly, Distributed decoding in a cellular multiple-access channel, in *IEEE International Symposium on Information Theory*, p. 484, Chicago, IL, June 2004.

[25] A. Grant, S. Hanly, J. Evans, and R. Muller, Distributed decoding for Wyner cellular systems, in *Fifth Australian Communications Theory Workshop*, pp. 77–81, Newcastle, Australia, February 2004.

[26] W. Lee, *Mobile Cellular Telecommunications Systems*. McGraw-Hill, Inc., New York, 1990.

[27] A. Viterbi, A. Viterbi, K. Gilhousen, and E. Zehavi, Soft handoff extends CDMA cell coverage and increases reverse link capacity, *IEEE Journal on Selected Areas in Communications*, 12, 1281–1288, October 1994.

[28] D. Wong and T. Lim, Soft handoffs in CDMA mobile systems, *IEEE Wireless Communications*, 4, 6–17, December 1997.

[29] Y. Liang, T. Yoo, and A. Goldsmith, Coverage spectral efficiency of cellular systems with cooperative base stations, in *IEEE Global Telecommunications Conference–2006 (GLOBECOM '06)*, pp. 1–5, San Francisco, CA, November 2006.

[30] O. Somekh, B. Zaidel, and S. Shamai, Sum rate characterization of joint multiple cell-site processing, *IEEE Transactions on Information Theory*, 53(12), 4473–4497, December 2007.

[31] O. Simeone, O. Somekh, Y. Bar-Ness, and U. Spagnolini, Uplink throughput of TDMA cellular systems with multicell processing and amplify-and-forward cooperation between mobiles, *IEEE Transactions on Wireless Communications*, 6, 2942–2951, August 2007.

[32] O. Somekh, O. Simeone, H. V. Poor, and S. Shamai, Cellular systems with full-duplex amplify-and-forward relaying and cooperative base-stations, in *IEEE International Symposium on Information Theory (ISIT07)*, pp. 16–20, Nice, France, June 2007.

[33] O. Simeone, O. Somekh, Y. Bar-Ness, and U. Spagnolini, Low-SNR analysis of cellular systems with cooperative base stations and mobiles, in *Fortieth Asilomar Conference on Signals, Systems and Computers–2006 (ACSSC '06)*, pp. 626–630, Pacific Grove, CA, 2006.

[34] S. Chatzinotas, M. A. Imran, and C. Tzaras, Optimal information theoretic capacity of the planar cellular uplink channel, in *9th IEEE International Workshop on Signal Processing Advances in Wireless Communications-2008*, pp. 196–200, Recife, Pernambuco, Brazil, July 2008.

[35] S. Chatzinotas, M. A. Imran, and C. Tzaras, Spectral efficiency of variable density cellular systems, in *IEEE International Symposium on Personal, Indoor and Mobile Radio Communications, International Workshop on Efficiency-2008*, Cannes, France, September 2008.

[36] T. Cover and J. Thomas, *Elements of Information Theory*, 2nd ed. John Wiley & Sons, New York, 2006.

[37] I. Telatar, Capacity of multi-antenna Gaussian channels, *European Transactions on Telecommunications*, 10, 585–595, November 1999.

[38] D. Tse and P. Viswanath, *Fundamentals of Wireless Communications*. University Press, Cambridge, U.K., 2005.

[39] J. Andrews, Interference cancellation for cellular systems: A contemporary overview, *IEEE Transactions on Wireless Communications*, 12, 19–29, April 2005.

[40] M. Varanasi, Group detection for synchronous Gaussian code-division multiple-access channels, *IEEE Transactions on Information Theory*, 41, 1083–1096, July 1995.

[41] F. Wijk, G. Janssen, and R. Prasad, Groupwise successive interference cancellation in a DS/CDMA system, in *IEEE International Symposium on Personal, Indoor and Mobile Radio Communications, PIMRC*, Toronto, Canada, Vol. 2, 742–746, 1995.

[42] X. Wang and A. Host-Madsen, Group-blind multiuser detection for uplink CDMA, *IEEE Journal on Selected Areas in Communications*, 17, 1971–1984, November 1999.

[43] P. Alexander and A. Grant, Iterative channel and information sequence estimation in CDMA, in *IEEE Sixth International Symposium on Spread Spectrum Techniques and Applications, 2000*, Vol. 2, 593–597, Parsippany, NJ, 2000.

[44] H. Weingarten, Y. Steinberg, and S. Shamai, The capacity region of the Gaussian multiple-input multiple-output broadcast channel, *IEEE Transactions on Information Theory*, 52(9), 3936–3964, 2006.

[45] P. Viswanath and D. Tse, Sum capacity of the vector Gaussian broadcast channel and uplink-downlink duality, *IEEE Transactions on Information Theory*, 49(8), 1912–1921, 2003.

[46] W. Yu, Uplink-downlink duality via minimax duality, *IEEE Transactions on Information Theory*, 52(2), 361–374, 2006.

[47] C. Peel, On dirty-paper coding, *IEEE Signal Processing Magazine*, 20, 112–113, May 2003.

[48] H. Harashima and H. Miyakawa, Matched-transmission technique for channels with intersymbol interference, *IEEE Transactions on Communications*, 20, 774–780, August 1972.

[49] M. Tomlinson, New automatic equalizer employing modulo arithmetic, *Electronics Letters*, 7, 138–139, March 1971.

[50] B. Hochwald, C. Peel, and A. Swindlehurst, A vector-perturbation technique for near-capacity multi-antenna multiuser communication—Part II: Perturbation, *IEEE Transactions on Communications*, 53, 537–544, March 2005.

[51] T. Rappaport, *Wireless Communications: Principles and Practice*. Prentice Hall PTR, Upper Saddle River, NJ, 2001.

[52] L. Ong and M. Motani, On the capacity of the single source multiple relay single destination mesh network, *Ad Hoc Networks*, 5(6), 786–800, 2007.

[53] A. Tulino and S. Verdú, Random matrix theory and wireless communications, *Foundations and Trends in Communication and Information Theory*, 1(1), 1–182, 2004.

[54] A. Tulino, A. Lozano, and S. Verdu, Impact of antenna correlation on the capacity of multiantenna channels, *IEEE Transactions on Information Theory*, 51(7), 2491–2509, July 2005.

[55] N. Jindal, W. Rhee, S. Vishwanath, S. Jafar, and A. Goldsmith, Sum power iterative water-filling for multi-antenna Gaussian broadcast channels, *IEEE Transactions on Information Theory*, 51, 1570–1580, April 2005.

[56] W. Yu, W. Rhee, S. Boyd, and J. Cioffi, Iterative water-filling for Gaussian vector multiple-access channels, *IEEE Transactions on Information Theory*, 50(1), 145–152, 2004.

[57] C. Shannon, A mathematical theory of communication, *Bell Systems Technical Journal*, 27, 379–423, 1948.

[58] J. Zander, Performance of optimum transmitter power control in cellular radio systems, *IEEE Transactions on Vehicular Technology*, 41, 57–62, February 1992.

[59] C. Rose, S. Ulukus, and R. Yates, Wireless systems and interference avoidance, *IEEE Transactions on Wireless Communications*, 1, 415–428, July 2002.

[60] S. Ulukus and R. Yates, Iterative construction of optimum signature sequence sets in synchronous CDMA systems, *IEEE Transactions on Information Theory*, 47, 1989–1998, July 2001.

[61] U. Madhow and M. Honig, MMSE interference suppression for direct-sequence spread-spectrum CDMA, *IEEE Transactions on Communications*, 42, 3178–3188, December 1994.

[62] S. Ulukus and R. Yates, Adaptive power control with MMSE multiuser detectors, in *IEEE International Conference on Communications 'Towards the Knowledge Millennium' ICC, 97*, Montreal, Canada 1, 361–365, June 1997.

[63] X. Wang and H. Poor, Blind multiuser detection: A subspace approach, *IEEE Transactions on Information Theory*, 44, 677–690, March 1998.

[64] M. Honig, U. Madhow, and S. Verdú, Blind adaptive multiuser detection, *IEEE Transactions on Information Theory*, 41, 944–960, July 1995.

[65] H. Hu, Y. Zhang, and J. Luo, eds., *Distributed Antenna Systems: Open Architecture for Future Wireless Communications*, 1st edn. Auerbach Publications, Taylor & Francis Group, New York, 2007.

[66] A. Feiten, R. Mathar, and S. Hanly, Eigenvalue-based optimum-power allocation for gaussian vector channels, *IEEE Transactions on Information Theory*, 53(6), 2304–2309, 2007.

[67] K. Gilhousen, I. Jacobs, R. Padovani, A. Viterbi, L. Weaver, Jr., and C. Wheatley, III, On the capacity of a cellular CDMA system, *IEEE Transactions on Vehicular Technology*, 40, 303–312, May 1991.

[68] M. Sharif and B. Hassibi, On the capacity of MIMO broadcast channels with partial side information, *IEEE Transactions on Information Theory*, 51, 506–522, February 2005.

[69] P. Ding, D. Love, and M. Zoltowski, Multiple antenna broadcast channels with shape feedback and limited feedback, *IEEE Transactions on Signal Processing*, 55, 3417–3428, July 2007.

[70] N. Jindal, MIMO broadcast channels with finite-rate feedback, *IEEE Transactions on Information Theory*, 52, 5045–5060, November 2006.

[71] A. Dana, M. Sharif, and B. Hassibi, On the capacity region of multi-antenna Gaussian broadcast channels with estimation error, in *IEEE International Symposium on Information Theory (ISIT'06)*, pp. 1851–1855, Seattle, WA, July 2006.

[72] B. Hassibi and M. Sharif, Fundamental limits in MIMO broadcast channels, *IEEE Journal on Selected Areas in Communications*, 25, 1333–1344, September 2007.

[73] A. Lapidoth, S. Shamai, and M. Wigger, On the capacity of fading MIMO broadcast channels with imperfect transmitter side-information, in *Allerton Conference on Communication, Control and Computing*, Monticello, IL.

[74] J. Roh and B. Rao, Multiple antenna channels with partial feedback, in *IEEE International Conference on Communications–2003 (ICC '03)*, 5, 3195–3199, Anchorage, Alaska, May 2003.

[75] M. Kopbayashi, J. Boutros, and G. Caire, Successive interference cancellation with siso decoding and EM channel estimation, *IEEE Journal on Selected Areas in Communications*, 19, 1450–1460, August 2001.

[76] R. Buehrer, Equal BER performance in linear successive interference cancellation for CDMA systems, *IEEE Transactions on Communications*, 49, 1250–1258, July 2001.

[77] J. Andrews and T. Meng, Optimum power control for successive interference cancellation with imperfect channel estimation, *IEEE Transactions on Wireless Communications*, 2, 375–383, March 2003.

[78] D. Aktas, M. Bacha, J. Evans, and S. Hanly, Scaling results on the sum capacity of cellular networks with MIMO links, *IEEE Transactions on Information Theory*, 52, 3264–3274, July 2006.

[79] P. Marsch and G. Fettweis, A decentralized optimization approach to backhaul-constrained distributed antenna systems, in *Sixteenth IST Mobile and Wireless Communications Summit, 2007*, pp. 1–5, Budapest, Hungary, July 2007.

[80] S. Shamai, O. Somekh, O. Simeone, A. Sanderovich, B. Zaidel, and H. Poor, Cooperative multi-cell networks: Impact of limited-capacity backhaul and inter-users links, in *Joint Workshop on Coding & Communications*, Durnstein, Austria, October 2007.

3 Low-Complexity Strategies for Cooperative Communications

Andrew W. Eckford, Josephine P. K. Chu, and Raviraj S. Adve

CONTENTS

The challenging nature of the wireless channel poses problems for wireless networks, in that devices within the network must ensure reliable communication in the presence of effects such as fading. To mitigate these challenges, a useful technique is to make use of spatial diversity, which exploits the large number of antennas available in the network. Because the antennas are likely to be far enough apart to experience independent fading environments, at least one of the antennas is likely has a good link to the data sink. Because these methods of obtaining diversity require nodes throughout a wireless network to work together, systems that use them are said to attain cooperative diversity.

The strategies of cooperative diversity are of interest in ad hoc and sensor networks (e.g., Ref. [1]), which are currently of intense interest to researchers. Such networks are intended to be inexpensive and easily deployable, and thus they generally admit nodes with stringent constraints on their computational and energy resources. As a result, ensuring reliable communication requires the use of low-complexity and energy-efficient algorithms. In this chapter, we explore the current state of the art in low-complexity strategies for cooperative communication.

3.1 INTRODUCTION

It is well known that wireless links suffer from multipath fading, in which reflected versions of the transmitted signal can randomly combine and destructively interfere at the receiver. A common strategy to avoid the effects of fading is to use the channel at points widely separated in time, frequency, or space, to obtain many independent realizations of the fading random process. The strategy of seeking many independent realizations of the fading process is known as increasing the system's diversity.

In the presence of nonergodic Rayleigh fading, a key measure of the performance of a wireless device is its diversity order. Let γ represent the random signal-to-noise ratio (SNR) of a link experiencing Rayleigh fading, and let $\bar{\gamma}$ represent the average SNR on the link. The probability that γ is less than or equal to some minimum, $\gamma_{\min}$, is given by

$$P(\gamma \leq \gamma_{\min}) = 1 - e^{-\alpha\gamma_{\min}/\bar{\gamma}}, \tag{3.1}$$

where α is a positive constant. One may think of $\gamma_{\min}$ as the minimum SNR to guarantee reliable communication, so that below this SNR, an "outage" occurs. Using the Θ order notation, it is easy to show that

$$1 - e^{-\alpha\gamma_{\min}/\bar{\gamma}} = \Theta(\bar{\gamma}^{-1}). \tag{3.2}$$

From Equations 3.1 and 3.2, given d independent links, each with the same $\bar{\gamma}$, the probability that every link has an SNR less than $\gamma_{\min}$ is $\Theta(\bar{\gamma}^{-d})$. Because this outage effect dominates the probability of error in any wireless system, a system whose probability of error is of order $\Theta(\bar{\gamma}^{-d})$ is said to have a diversity order of d.

We can see from Equations 3.1 and 3.2 why the diversity order is increased with the number of independent channels from source to destination. Of particular interest to this chapter is the use of spatial diversity, where multiple independent paths from the transmitter to receiver are used, separated widely in space. For instance, this can be accomplished by using multiple antennas at both the transmitter and receiver [2]. However, this approach is not practical for small, inexpensive, and low-complexity hardware. Instead, in a wireless network, we may use relaying to achieve a similar effect. In a three-terminal wireless system, if two terminals wish to communicate, but the link between them is too weak to support reliable communication, the third terminal might be able to act as a relay between them. This arrangement is known as the relay channel, and has been studied formally by researchers since the work of van der Muelen in 1971 [3]. Furthermore, in a large wireless network, it is easy to see that multiple relays may be used, each of which provides a new path to increase spatial diversity. More complicated techniques that use this idea can also be used. For instance, instead of only the source node transmitting information of its own, nodes organize themselves into groups and assist each other to improve the error performance. Such techniques are said to achieve cooperative diversity [5,6].

As in any noisy channel, complicated signal processing, including the use of error-correcting codes, is required to achieve optimal performance in the relay channel. The seminal work of Cover and El Gamal [4] was the first to examine the relay channel from an information-theoretic perspective and found achievable rates for two useful special cases. There has been much recent work on adapting powerful error-correcting codes, such as turbo and turbo-like codes, to achieve these bounds. Meanwhile, it has been recognized that error-control coding can significantly improve the performance of a system employing cooperative diversity.

Because both sensor networking and cooperative communication are nascent fields, there has been little study thus far of the sorts of low-complexity cooperative solutions that sensor networks would require. In particular, many of the methods that have been proposed for the relay channel, or for coded cooperative diversity, assume that the relay is capable of decoding the source's transmitted codeword. Because the decoding algorithms for the most powerful error-correcting codes (such as turbo codes) are relatively complex, decoding these codes is likely beyond the abilities of most contemporary sensor networking hardware. Even encoding such an error-correcting code may be a complex task (depending on the code chosen), while to maximize versatility and adaptability, a sensor networking device should be able to operate efficiently in many potential channel conditions, regardless of the strength of its link with its neighbors.

We have performed a considerable amount of work to investigate the problem of low-complexity coded cooperation (e.g., [8–10,35,44,46,47]). In this chapter, we review some useful low-complexity strategies, both from our own work and elsewhere, that can be employed using contemporary hardware. These techniques make use of codes that are easy to encode and that have adaptable rates, without requiring complex operations such as decoding at the relay. We also discuss fractional cooperation, which is an efficient and low-complexity method for cooperative diversity in large networks.

The remainder of this chapter is organized as follows. In Section 3.2, we give our system model and notation, which is used throughout the remainder of the chapter. In Section 3.3, we give a comprehensive review of the related literature. In Section 3.4, we introduce several low-complexity relaying techniques, especially coded demodulate-and-forward, which satisfies the requirements that we mentioned earlier. In Section 3.5, we discuss fractional cooperation, which is a robust and useful technique to increase the diversity order in low-complexity cooperative wireless networks. Finally, in Section 3.6, we give some open problems for future research.

3.2 MODEL AND DEFINITIONS

Here we state our models and definitions, which are both useful in describing our proposed methods as well as in discussing the related literature. These models closely resemble those in Refs. [8–10] and are quite similar to models of comparable wireless relaying systems throughout the literature.

The simplest form of the relay channel is shown in Figure 3.1. In this system, the task of the source, marked S in the figure, is to convey a message to the destination, also called the data sink, marked D in the figure. The task of the relay, marked R in the figure, is to assist the source in transmitting its message to the destination. The system may be generalized to several relays, as shown in Figure 3.2. In this case, there is an index set $\mathcal{I} = \{1, 2, \ldots, |\mathcal{I}|\}$ of relays, all of which assist the source to transmit its information to the destination. From now on, we use $r = |\mathcal{I}|$ to represent the number of relays in the system.

We give some general assumptions concerning the relay systems that are discussed in this chapter. We assume that all devices in the system are half-duplex, that is, they may not transmit and receive at the same time. This assumption was previously made for "cheap" hardware in Ref. [11], and is a typical feature of currently available sensor networking devices. We assume that channel-state

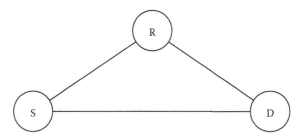

FIGURE 3.1   Relay network with a single relay. Node S is the source, node R is the relay, and node D is the destination.

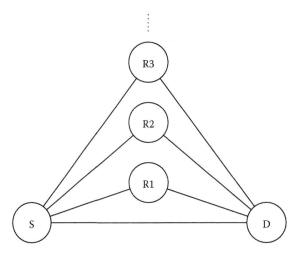

FIGURE 3.2 Relay network with multiple relays, where the relays are numbered R1, R2, R3, and so on.

information is known throughout the network, and we assume that all nodes throughout the network are aware of each other's relaying strategies. Generally, in this chapter we disregard any issue that may be thought of as a "protocol" issue, such as selecting relays, coordinating transmissions, or allocating resources; although these problems are important, the solutions to them are not specific to our setup, so they are beyond the scope of this chapter.

3.2.1 SINGLE-RELAY MODEL

For illustration, we introduce our notation and definitions for the single-relay system, which will be generalized to multiple relays in the next section. With a single relay, there are three radio links: the source-to-relay (SR) link, the relay-to-destination (RD) link, and the source-to-destination (SD) link. For convenience, throughout the remainder of the chapter we assume that all nodes in the network use binary phase shift keying (BPSK) modulation, so the transmission alphabet is given by $\mathcal{X} = \{+1, -1\}$. However, nothing fundamental changes if any binary modulation scheme is used.

Communication in a relay network occurs in two phases: first, the source transmits its information, observed by both the relay and the destination; and second, the relay forms a sequence to assist the source and transmits that sequence to the destination. In the first phase, the source encodes its information sequence using an error correcting code, forming the length-n_S sequence $\mathbf{x}_S \in \mathcal{X}^{n_S}$. As a result of this transmission, the relay and destination observe $\mathbf{y}_R$ and $\mathbf{y}_{SD}$, respectively, where

$$\mathbf{y}_R = a_{SR}\mathbf{x}_S + \mathbf{n}_{SR}, \tag{3.3}$$

with a_{SR} and $\mathbf{n}_{SR}$ representing the amplitude and noise on the SR link, respectively; and

$$\mathbf{y}_{SD} = a_{SD}\mathbf{x}_S + \mathbf{n}_{SD}, \tag{3.4}$$

with a_{SD} and $\mathbf{n}_{SD}$ representing the amplitude and noise on the SD link, respectively. The vectors $\mathbf{y}_R$ and $\mathbf{y}_D$ are both defined on $\mathbb{R}^{n_S}$, where $\mathbb{R}$ is the set of real numbers. In the second phase, the relay forms a length-n_R sequence $\mathbf{x}_R = \phi_R(\mathbf{y}_R)$, where $\phi_R : \mathbb{R}^{n_S} \rightarrow \mathcal{X}^{n_R}$ represents the relay processing function, which we discuss further in Section 3.4.2. The relay transmits its sequence, and the destination observes $\mathbf{y}_{RD}$, where

$$\mathbf{y}_{RD} = a_{RD}\mathbf{x}_R + \mathbf{n}_{RD}, \tag{3.5}$$

with a_{RD} and $\mathbf{n}_{RD}$ representing the amplitude and noise on the SD link, respectively.

We assume that the noise is zero-mean Gaussian noise, so the relay system is completely parametrized by the triple of amplitudes (a_{SR}, a_{SD}, a_{RD}). Without loss of generality, we assume that the noise has unit variance. In a fading environment, we typically assume that the amplitudes (a_{SR}, a_{SD}, a_{RD}) are realizations of independent Rayleigh random variables, which (as we stated in Section 3.1) is a necessary assumption to obtain diversity gain. For a Rayleigh-distributed random variable A, we use the second moment $\bar{\gamma} = E[A^2]$ as a parameter. Thus, the random amplitudes are completely parametrized by $(\bar{\gamma}_{SR}, \bar{\gamma}_{SD}, \bar{\gamma}_{RD})$. Because the noise has unit variance (by assumption), these quantities also represent the average SNRs on each link.

3.2.2   MULTIPLE-RELAY MODEL

The model in the previous section can be straightforwardly extended to multiple relays, and here we present the necessary notation and definitions to do so. For each $i \in \mathcal{I}$ (recalling that $\mathcal{I}$ is the index set corresponding to the relays), the ith relay observes

$$\mathbf{y}_{R,i} = a_{SR,i}\mathbf{x}_S + \mathbf{n}_{SR,i}, \tag{3.6}$$

analogous to Equation 3.3. The transmission from source to destination $\mathbf{y}_{SD}$ is the same as in Equation 3.4, because there remains only one source and only one destination. Concerning the RD links, for each $i \in \mathcal{I}$, the destination observes

$$\mathbf{y}_{RD,i} = a_{RD,i}\mathbf{x}_{R,i} + \mathbf{n}_{SR,i}, \tag{3.7}$$

where $\mathbf{x}_{R,i}$ is the signal transmitted by the ith relay. Furthermore, for each $i \in \mathcal{I}$, we assume that $\mathbf{x}_{R,i} = \phi_{R,i}(\mathbf{y}_{R,i})$, where $\phi_{R,i} : \mathbb{R}^{n_S} \to \mathcal{X}^{n_{R,i}}$ is the relay processing function available at the ith relay. Note that these functions are possibly different from relay to relay, and that each relay may form a sequence $\mathbf{x}_{R,i}$ of a length $n_{R,i}$, which also may differ from relay to relay.

From Equations 3.6 and 3.7, the parameter set of the multiple relay case is augmented by the amplitudes $a_{SR,i}$ and $a_{RD,i}$ for each $i \in \mathcal{I}$. Letting

$$\mathbf{a}_{SR} = [a_{SR,1}\ a_{SR,2},\ \ldots,\ a_{SR,r}],$$

and

$$\mathbf{a}_{RD} = [a_{RD,1}\ a_{RD,2},\ \ldots,\ a_{RD,r}],$$

and assuming that the noise is unit-variance zero-mean additive Gaussian noise, the multiple-relay system is completely parametrized by $(\mathbf{a}_{SR}, a_{SD}, \mathbf{a}_{RD})$, noting that a_{SD} is still a scalar. Where these amplitudes are realizations of independent Rayleigh fading processes, the system is parametrized by $(\bar{\boldsymbol{\gamma}}_{SR}, \bar{\gamma}_{SD}, \bar{\boldsymbol{\gamma}}_{RD})$, where $\bar{\boldsymbol{\gamma}}_{SR}$ and $\bar{\boldsymbol{\gamma}}_{RD}$ are defined analogously to $\mathbf{a}_{SR}$ and $\mathbf{a}_{SD}$.

3.3   REVIEW OF RELATED LITERATURE

Here we present a comprehensive review of the literature related to our work. Generally, the work we outline here concerns communication techniques in general relay networks, without considering the constraints of practical hardware. However, this literature provides the theoretical and practical context for our work.

One of the earliest and most pivotal articles on this topic is Ref. [4], where the capacity of relay networks, one of the simplest form of cooperative networks, was analyzed. In the article, upper bounds were presented for relay networks. Two different schemes, decode-and-forward (DF) and estimate-and-forward (EF) (also known compress-and-forward or quantize-and-forward), were introduced, and their associated achievable rates were found. In DF, the relay decodes the source codeword and transmits data to assist in the decoding of the codeword in the destination. In EF, instead of decoding the source codeword, the relay estimates and compresses the received signal and sends data based on the estimated data to assist the decoding at the destination. It was shown that in degraded channels, where the quality of the SR channel is better than that of the SD channel, DF achieves capacity. The concept of using other nodes' antennas to provide diversity improvement has been extended to scenarios where instead of only the source node transmitting information of its own, nodes organize themselves into groups and assist each other to improve the error performance, which (as mentioned earlier) is known as cooperative diversity [5,6].

Recently, many articles have been published to provide us with further insights into relay channels. In Ref. [7], the cooperative diversity protocols amplify-and-forward (AF) and incremental relaying were introduced. With AF, no coding or decoding is performed at the relay node, and the relay simply scales the received signal to satisfy power constraints. The incremental relaying protocol requires the relay to perform AF only when the transmission from source to destination does not allow successful decoding and feedback from the destination is required. In the paper, the multiplex-diversity trade-off for DF, AF, and incremental relaying were analyzed for networks with half-duplex channels. Another article that presented results on the diversity-multiplex trade-off is Ref. [12], where the diversity-multiplex trade-off for AF, DF, and cooperative multiple-access (CMA) was presented. Under CMA, each node transmits a summation of its own data and a scaled version of its received signal. The analysis done in this paper is different from that of Ref. [7] as the definition of the half-duplex channel is slightly different. In this case both the source and relay transmit in the second phase, instead of just the relay transmitting.

In Ref. [13], the ergodic and outage capacity for three different time-division multiple-access-based cooperative protocols with the use of AF or DF are presented, and space-time code design for relay channels using AF are shown. In Ref. [14], achievable rates for networks with multiple source, relay, and destination nodes for full-duplex channels with the use of both DF and EF were found. Mutual information for wireless channels where phase fading is present was also shown. In Ref. [15], ergodic and outage capacity for wireless relay channels with one relay were found for full-duplex and half-duplex channels, and the effects of power allocation were studied. In that article, correlation between the source and relay codewords are taken into account in the capacity derivations. The achievable rates for half-duplex additive white Gaussian noise (AWGN) channels using DF are presented in Ref. [16], where binary and Gaussian input are used. The achievable rates with the use of EF under full-duplex channels were studied in detail in Ref. [17]. The achievable rates with the use of different methods of quantization at the relay were presented, and the channel conditions where DF and the different EF schemes are optimal were identified.

Coding schemes have been introduced to implement cooperation in relay networks. AF and DF using repetition codes and a received signal combiner at the destination node were presented in Ref. [18]. In Ref. [19], a memoryless EF scheme was proposed and was shown to be optimal for a single-relay system where only memoryless relay functions are allowed. Another well-known scheme, which is also one of the earliest cooperative coding schemes, is coded cooperative diversity [20], where the partnering node transmits only additional parity bits in the second phase when the cyclic redundancy check of the received codeword is satisfied. Otherwise, the node discards the received signal and transmits additional parity bits for its own data. In Ref. [21], a distributed turbo code was introduced, where the source node transmits use a rate-1/2 turbo code, and after decoding the source codeword, the relay interleaves the data and forms a new codeword based on the interleaved bits and transmits it to the destination. Different encoding and decoding schemes with the use of turbo codes at the source and relay were introduced in Ref. [22], and their performance was illustrated. In Ref. [23], a turbo code codeword was partitioned such that after receiving the source transmission, each relay, upon successful decoding, transmits the part of the codeword for which it is responsible, or upon decoding failure, stays quiet and allows the source to transmit the codeword chunk. In Ref. [24], rateless codes were used by both the source and relay nodes. In that scheme, the relay receives signals from the source until it can decode correctly, and then assists the source to transmit to the destination until it sends an acknowledgment to let both source and relay know that successful decoding has been achieved. This is similar to the incremental relaying scheme suggested in Ref. [7].

As stated in Ref. [4], for DF, it is necessary for the relay to decode the source codeword correctly to cooperate in the transmission. With this principle in mind, several articles have been published to design optimized codes for the relay channel. In Ref. [25], a cooperative diversity coding scheme using a convolutional code was introduced, where the source transmits a punctured version of the codeword, and upon correct decoding, the relay transmits the rest of the codeword to the destination, emphasizing the importance of a good codeword over the SR channel. Several papers are available to provide suggestions on how to build good low-density parity-check (LDPC) codes for the relay channel. In Ref. [26], a coding scheme was introduced, where transmission of the source codeword is divided between the source and relay. As the channel model assumes half-duplex channels, the time-division between the receiving and transmitting mode of the relay was optimized, and in the case where the transmission of the systematic bits of the codeword is split between both time slots, parity bit assignment for the systematic bits was optimized as well. In Ref. [27], the optimal relationship between the SR and SD LDPC codewords' variable and check degree profiles was provided. Differential evolution [28] was used in Ref. [29] to find optimal edge distributions to achieve maximum transmission rate while assuring successful decoding at both relay and destination. In Ref. [30], bilayer LDPC codes were devised to provide degree distributions for codes that are optimal given the required code rates over the SR and SD channels.

Even though DF provides excellent performance when the SR channel is good, its performance is limited by the quality of the SR channel as successful decoding is required at the relay. When the SR channel is less than ideal, EF provides good performance, as successful decoding is not required at the relay. A practical EF implementation following the steps provided in Ref. [4] can be found in Ref. [31], where nested scalar quantization and systematic irregular repeat-accumulate codes are used for Wyner-Ziv coding at the relay. Cooperation with the use of Slepian-Wolf and Wyner-Ziv coding were presented in Refs. [32,33], respectively; we discuss these in some more detail below. When using either type of coding, the relay can choose to use either DF or EF, or provide no cooperation, depending on the given SR channel. In Ref. [34], achievable rates for Gaussian channels are presented, and code design decisions such as quantization scheme and power allocation are provided.

3.4 CODED COOPERATION FOR LOW-COMPLEXITY DEVICES

In this section, we discuss techniques for coded cooperation using low-complexity devices, including the AF and demodulate-and-forward strategies. Throughout, we bear in mind the constraints that we mentioned in Section 3.1, namely the use of codes that are easy to encode and that have adaptable rates, without requiring complex operations such as decoding at the relay. We discuss some low-complexity coding techniques that can be adapted to demodulate-and-forward. These techniques are generally presented in terms of a single-relay system for the purpose of illustration and can be easily generalized to the multiple-relay case. Finally, we discuss how to adapt coded demodulate-and-forward to achieve diversity gains from relay selection.

3.4.1 AMPLIFY-AND-FORWARD

In AF, the relay simply acts as a "pipe" for the received signal, performing no processing other than amplification. Thus, the relay processing function $\phi_R(\cdot)$ multiplies its argument by a scalar s_R (i.e., $\phi_R(\mathbf{y}_{SR}) = s_R \mathbf{y}_{SR}$). The RD signal in the single-relay context is then given by

$$
\begin{aligned}
\mathbf{y}_{RD} &= a_{RD}\mathbf{x}_R + \mathbf{n}_{RD} \\
&= a_{RD}s_R\mathbf{y}_{SR} + \mathbf{n}_{RD} \\
&= a_{RD}s_R a_{SR}\mathbf{x}_S + a_{RD}s_R\mathbf{n}_{SR} + \mathbf{n}_{RD},
\end{aligned}
\tag{3.8}
$$

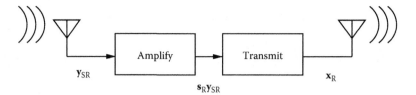

FIGURE 3.3   Block diagram depicting the operations in AF at the relay.

with obvious generalizations to the multiple-relay case. Given the lack of processing, AF is a feasible technique for low-complexity wireless devices. As we noted in Section 3.3, much work has been done concerning the capacity of AF. Coding can be easily accomplished, wherein $\mathbf{x}_S$ is the codeword of a powerful error-correcting code. From Equations 3.8 and 3.4, the destination observes copies of $\mathbf{x}_S$ with different amplitudes, and through different independent Gaussian noise channels, which can be straightforwardly handled by the decoder. A block diagram of AF is given in Figure 3.3.

Although AF is theoretically simple from the perspective of signal processing, there are issues that may make it impractical in certain contexts, most important of which is our assumption of "cheap" hardware of a half-duplex constraint [11]. Such a constraint could take two forms. If time division duplexing (TDD) is used, the relay would need to somehow delay the analog signal $s_R \mathbf{y}_{SR}$ until it was clear to transmit. This would require high-resolution sampling of the signal and storage of the sampled signal in memory, at significant expense in terms of computational resources. If frequency division duplexing is used, the relay could simply shift the received signal into another frequency band. However, this strategy does not easily scale to a large network, because the relaying would occur simultaneously, and each relay would need to be allocated an independent frequency band to avoid interference. Apart from being an inefficient use of resources, a large enough network would eventually run out of bandwidth. Aside from these issues, a system designer might prefer the flexibility of digital transmission and might want to avoid having both a digital radio and an analog radio on the same device.

Due to its conceptual simplicity and due to these implementational issues, we do not discuss AF in any further detail in this chapter.

3.4.2   DEMODULATE-AND-FORWARD

We have assumed that a low-complexity wireless device is incapable of decoding a powerful error-correcting code. In Section 3.4.1 we pointed out some problems in dealing with a strictly analog signal. Because it would be desirable to continue to deal with digital signals without decoding, it is natural to consider a case where the relay demodulates the source's digital transmission, but does not decode it—a scenario known as demodulate-and-forward. Demodulate-and-forward has been proposed as a technique to achieve cooperative diversity gains in sensor networks in Refs. [18,35]. In Ref. [19], demodulate-and-forward was used to discuss the capacity of relay systems.

We can define the relay processing function $\phi_R(\cdot)$ under demodulate-and-forward in a similar manner to Refs. [8–10]. Similarly to the AF case, for simplicity we state the function for the single-relay case and briefly discuss how to generalize the function to multiple relays. Because this function is difficult to state in closed form, we state the operations that are required to calculate it.

Assuming BPSK modulation, let $\sigma : \mathbb{R} \to \{0, 1\}$ represent the demodulation function mapping the real-value received signal to a binary alphabet, where

$$\sigma(y) := \begin{cases} 0, & y \geq 0, \\ 1, & y < 0. \end{cases} \tag{3.9}$$

Also let $m : \{0, 1\} \rightarrow \{-1, +1\}$ represent the modulation function, mapping the binary symbols to a signed BPSK modulation value. With a slight abuse of notation, if the argument of either of these functions is a vector, the operation is applied to every element of the vector.

The relay first obtains the demodulated symbol sequence $\mathbf{b}$, calculating

$$\mathbf{b} = \sigma(\mathbf{y}_{SR}). \tag{3.10}$$

In some published descriptions of demodulate-and-forward, such as Ref. [18], the demodulated sequence is immediately transmitted by forming $\mathbf{x}_R = m(\mathbf{b})$. (From Equation 3.10, this simply means that $\mathbf{x}_R = \text{sign}(\mathbf{y}_{SR})$.) However, we can provide a description of coded demodulate-and-forward by encoding the sequence $\mathbf{b}$ using an additional error-correcting code. In the remainder of the chapter we consider cases where only part of $\mathbf{b}$ is retransmitted. To be as general as possible, suppose k of the n_S symbols in $\mathbf{b}$ are to form part of the relay's transmission, and let $\mathbf{w}$ represent a length-k sequence containing the indices of the elements of $\mathbf{b}$ that are to be relayed (e.g., if every second symbol is to be relayed, then $\mathbf{w} = [1, 3, 5, \ldots]$). Letting $\mathbf{b}'$ represent the vector of symbols to be relayed, where

$$\mathbf{b}' = [b_{w_1}, b_{w_2}, \ldots, b_{w_k}], \tag{3.11}$$

the elements of $\mathbf{b}'$ are then encoded as a codeword $\mathbf{c}$ of a new error-correcting code, and we finally form $\mathbf{x}_R = m(\mathbf{c})$. Codes to accomplish demodulate-and-forward in a low-complexity manner are discussed in Section 3.4.3. A block diagram of coded demodulate-and-forward is given in Figure 3.4.

We make two remarks on generalizing this scheme. First, the scheme can be easily generalized to the case of multiple relays by defining a mapping $\mathbf{w}$ for each relay; furthermore, k could be different from relay to relay, and each relay could employ a different error-correcting code. Second, it is straightforward to generalize this to any binary modulation other than BPSK; this is done through appropriate modifications to $\sigma(\cdot)$ and $m(\cdot)$.

Demodulate-and-forward may be thought of as a type of EF, as noted in Ref. [19]. In particular, the demodulated sequence represents the relay's estimate of the value of each received symbol, which is then forwarded to the destination (without decoding). Bounds on achievable rates of these schemes were given for full-duplex channels in Ref. [17] and for half-duplex channels in Ref. [34]. Such a scheme can be straightforwardly generalized in a number of ways, such as by allowing different quantization schemes, apart from a simple hard decision. Indeed, with this in mind, one may think of the TDD version of AF as also being a special case of EF (because, as we argued in Section 3.4.1, the time-shifting the signal requires quantization and storage), where the estimation is done with very high fidelity.

3.4.3 CODED DEMODULATE-AND-FORWARD: ENCODING

In Section 3.4.2, we allowed the relay to reencode the demodulated and punctured string $\mathbf{b}'$ using a second error-correcting code, thus forming the string $\mathbf{c}$. Such a method improves the performance of demodulate-and-forward by mitigating the effect of the noise on the RD link. Here, we present details on some coded demodulate-and-forward schemes in the literature, bearing in mind the complexity constraints at the relay nodes.

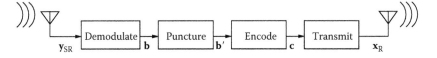

FIGURE 3.4   Block diagram depicting the operations in demodulate-and-forward at the relay.

The operation of coded demodulate-and-forward is suggested in a very simple scheme described in Ref. [35]. Using this scheme, all symbols in **b** are involved in relaying, so $\mathbf{w} = [1, 2, \ldots, n_S]$. The encoded word **c** is formed by performing a modulo-2 "accumulate" operation on adjacent bits in **b**. In other words,

$$c_i = \begin{cases} b_i, & i = 1, \\ b_i \oplus b_{i-1}, & 2 \le i \le n_S, \end{cases} \tag{3.12}$$

where $\oplus$ represents modulo-2 addition. This simple technique is clearly appropriate for reduced-complexity implementations. Furthermore, it was shown in Ref. [35] that, combined with a message-passing decoding algorithm at the destination, full diversity order is achieved in a one-relay system, with better performance than maximal ratio combining. As a very simple type of recursive convolutional code, the code in Equation 3.12 is not especially powerful, so better performance could be expected by using the method in Ref. [35] along with a more powerful code. This approach to coded demodulate-and-forward was taken in Refs. [8,9], which use low-density generator matrix (LDGM) codes and punctured systematic repeat-accumulate (PSRA) codes, respectively.

We first discuss LDGM-coded demodulate-and-forward. In an LDGM code, the codeword is created by forming parity checks of randomly selected information bits [37,38]; as such, they are similar to LT codes [39]. Let h represent the degree of a parity check, that is, the number of information bits that participate in it. Given the length-k vector $\mathbf{b}'$ from Equation 3.11, and letting **p** represent a length-k vector with exactly h entries equal to one, and the rest equal to zero (i.e., with Hamming weight h), one such parity check c is given by

$$c = \mathbf{b}'\mathbf{p}^\mathrm{T},$$

where the superscript T represents transposition, and all operations are modulo-2. A codeword is formed by collecting several parity checks together. For instance, let $\mathbf{p}_i$, $i = \{1, 2, \ldots, n_P\}$, represent a sequence of parity check vectors, where h_i is the degree of the ith parity check; furthermore, for each i, the vector $\mathbf{p}_i$ is selected at random, uniformly from the set of all possible vectors with exactly h_i ones. As a result, we can collect the parity checks together in a matrix **P**, defined as

$$\mathbf{P} = \begin{bmatrix} \mathbf{p}_1 \\ \mathbf{p}_2 \\ \vdots \\ \mathbf{p}_{n_P} \end{bmatrix}.$$

In Ref. [8], systematic LDGM codes were used (in which the information bits are transmitted as part of the codeword), so the codeword **c** is formed by

$$c = \mathbf{b}'[\mathbf{I} \ \mathbf{P}^\mathrm{T}] = \mathbf{b}\mathbf{G},$$

where **I** is an identity matrix, and $\mathbf{G} = [\mathbf{I} \ \mathbf{P}^\mathrm{T}]$ is then the generator matrix of the code. If $h_i \ll k$ for all i, then **G** is sparse, hence the name of the code. Again, for the single-relay case, the results in Ref. [8] indicate diversity gains and good performance using this code.

Although LDGM codes are relatively powerful, it is known that they suffer from error floors, in which the probability of error decreases only gradually as the channel strength increases [38]. In Ref. [9], this concern was addressed by using PSRA codes instead of LDGM codes. A PSRA code contains a recursive modulo-2 "accumulator" as in Equation 3.12, but with a more complicated code structure [41,42]. In particular, the information bits $\mathbf{b}'$ are first repeated an integer number of times ℓ; for instance, letting $\ell = 3$, the length-ℓk vector $[\mathbf{b}' \ \mathbf{b}' \ \mathbf{b}']$ would be formed. This vector would then be subject to a random permutation, selected uniformly at random from the set of all possible permutations on ℓk letters. Finally, the codeword **c** is formed by applying Equation 3.12, with the permuted vector in place of **b**. Again, in Ref. [9], a systematic form of this code was used, so **b′** is transmitted in addition to the permuted and accumulated symbols. Finally, the rate of the code can be increased from the nominal rate of $1/(1 + \ell)$ by puncturing, that is, by not transmitting certain preselected bits (the bits to be "punctured" can be selected, e.g., at random, although the systematic bits are never punctured). Results in Ref. [9] indicated diversity gains in the single-relay case.

It is possible to use virtually any other error-correcting code to implement coded cooperation, but the LDGM and PSRA codes were selected because of the low complexity of their encoding operations. In particular, the LDGM code can be implemented with only a pseudorandom number generator (to select elements of **b′** for parity checking) and a modulo-2 adder (to calculate the parity checks). Similarly, the PSRA code can be implemented with only a permuter and a modulo-2 adder; furthermore, given a pseudorandom number generator, there exist fast and simple algorithms for permuting a vector [40]. Furthermore, LDGM and PSRA codes both have the desirable feature that their rates can be changed dynamically. For the LDGM code, note that it is straightforward to increase or decrease the number of transmitted symbols by generating more parity checks, each of which are independent of the others. For PSRA codes, the rate can be varied through the amount of puncturing, but (unlike the LDGM codes) the PSRA codes have a fundamental minimum rate of $1/(1 + \ell)$, corresponding to no puncturing.

As mentioned in Section 3.3, practical work to implement such schemes has also been done as an extension of the Slepian-Wolf problem [32] and the Wyner-Ziv problem [31,33], which is closely related to coded demodulate-and-forward. In particular, the Slepian-Wolf problem considers the rate at which two correlated sources must transmit to accurately transmit their data to a destination. Focusing particularly on Slepian-Wolf cooperation, the source transmits a word $\mathbf{x}_S$, which is immediately demodulated and relayed by the relay, so that $\mathbf{x}_R = \text{sign}(\mathbf{y}_{SR})$. Now, the destination has two "correlated sources," $\mathbf{x}_R$ and $\mathbf{x}_S$. Somewhat similarly to the case of LDGM-coded cooperation, the source then transmits some parity checks to help the source recover $\mathbf{x}_S$ given $\mathbf{x}_R$, as in the Slepian-Wolf coding scheme from Ref. [43], although these parity bits are not retransmitted by the relay. Unlike the usual Slepian-Wolf setup, the destination has observed $\mathbf{y}_{SD}$ and thus has some information already about $\mathbf{x}_S$.

3.4.4 CODED DEMODULATE-AND-FORWARD: DECODING

Here we briefly discuss the decoding of a coded cooperation scheme using demodulate-and-forward, which takes place entirely at the destination. The decoding of these schemes can be accomplished using message-passing decoding techniques [36], as pointed out in Ref. [10]. As an example, a factor graph representing the LDGM-coded demodulate-and-forward scheme is given in Figure 3.5. There are two strategies that the destination may employ to decode. First, the destination could decode the relay's codeword (or in a multiple relay setting, each relay's codeword) in isolation and use the results to decode the source's codeword, known as serial decoding. Second, the destination could decode all the codewords simultaneously on the joint factor graph in Figure 3.5, using the sum-product algorithm throughout the entire graph, known as parallel decoding. Parallel decoding is expected to have better performance in general, but serial decoding is simpler to analyze.

Under serial decoding in a two-relay system, the destination first decodes $\mathbf{y}_{RD}$, thereby recovering **b′**, as given in Equation 3.11. Suppose for convenience that $\mathbf{b′} = \mathbf{b}$, that is, every symbol is relayed. If the decoding of the relay's code is successful, the destination recovers **b′**, which contains demodulated symbols from $\mathbf{y}_{SR}$. In particular, the symbols contained in **b′** are formed by taking hard decisions on $\mathbf{y}_{SR}$ and are thus equivalent to observing $\mathbf{x}_S$ through a binary symmetric channel. Thus, the task of the decoder is to recover $\mathbf{x}_S$ given the noisy sequences $\mathbf{y}_{SR}$ and **b′**, which can be easily accomplished using the sum-product algorithm. Furthermore, if not every symbol is relayed, then only those symbols that are relayed have an additional noisy observation from **b′**, which is also straightforward to accommodate in the sum-product algorithm.

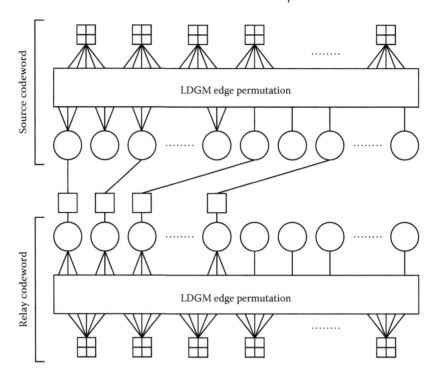

FIGURE 3.5 Factor graph depicting LDGM codes in use at the source and the relay. Note that in this example, not all the source's symbols are relayed.

The parallel case is slightly more complicated to describe. Considering the factor graph representing each codeword in the system (i.e., representing $\mathbf{x}_S$ and $\mathbf{x}_R$), the factor graphs are joined by nodes and edges connecting those elements in $\mathbf{x}_S$ that are relayed in each $\mathbf{x}_R$, as is shown in Figure 3.5. For example, supposing $x_{S,u}$ is chosen and relayed as $x_{R,v}$ (noting that the indices u and v would be different in general), and noting that the demodulation operation at the relay implies that $x_{R,v}$ is an observation of $x_{S,u}$ through a binary symmetric channel (with, say, inversion probability p), then the factor connecting the two variables is given by

$$p(x_{R,v}|x_{S,u}) = \begin{cases} 1-p, & x_{R,v} = x_{S,u}, \\ p, & x_{R,v} \neq x_{S,u}. \end{cases}$$

Using this factor as a connector between codes, messages are passed throughout the entire factor graph in accordance with the sum-product algorithm [36]. The reader is directed to Ref. [10] for further details.

Although these decoding operations take place at the destination (where, by assumption, computational complexity is not an issue), we can show that the computational burden at the destination is reasonable. To see this, note that the destination must decode the transmissions of r relays, for which the ith relay is of length $n_{R,i}$, although the $n_{R,i}$ are proportional in size to n_S; furthermore, the destination must use the results of these decoding operations (either in serial or in parallel) to assist in decoding the length-n_S sequence $\mathbf{x}_S$. The order of the computational complexity is thus $\Theta(rn_S)$.

3.4.5 CODED DEMODULATE-AND-FORWARD: RELAY SELECTION

To achieve a diversity order of d, it is generally necessary to have $d-1$ relays in operation, as these relays (plus the direct SD link) provide d independent paths from source to destination. However, it can be shown that all d relays need not be active to achieve diversity order d; only the "best" relay needs to be used.

Using AF, the "best" relay is the one with the best compound SR and RD channel [44]; the equivalent SNR on this compound channel can be easily evaluated because the relay processing is simple. Using DF, the "best" relay is the one with the best RD channel, subject to the constraint that the SR channel can successfully decode the source's code [45]; this is intuitive because a good SR channel is of no further use to the destination once the relay has decoded the code. Considering relay selection for coded demodulate-and-forward, unlike AF the relay processing function is complicated, and unlike DF there is no separation between the SR and RD links. Thus, a different approach, taking into account the error-correcting codes in use on each link, is required.

As noted in Ref. [46], optimal relay selection would involve calculating the probability of error associated with each point in the single-relay parameter space (a_{SR}, a_{SD}, a_{RD}) and using the relay with the lowest probability of error. The single-relay parameter space is used because, after the selection is carried out, only one relay (the selected one) assists the source. The probabilities of error could be calculated, e.g., by simulation, and stored in a lookup table at each relay. However, because the lookup table would need to be three-dimensional, this is not a desirable solution. However, choosing a heuristic for relay selection is a nontrivial problem, as failing to properly choose the best relay can lead to a loss of diversity order.

In Ref. [46], a simple relay selection method was proposed for coded demodulate-and-forward, specifically when using PSRA codes. Because PSRA codes are good error-correcting codes that have performance close to the Shannon limit, it was proposed that the "best" relay be defined as triple (a_{SR}, a_{SD}, a_{RD}) with the largest mutual information. In particular, because the destination observes $\mathbf{y}_{SD}$ and $\mathbf{y}_{SR}$ and because the channel noise is AWGN, we need to find the largest $I(X_S; Y_{SD}, Y_{SR})$, where

$$
\begin{aligned}
I(X_S; Y_{SD}, Y_{SR}) &= H(Y_{SD}, Y_{SR}) - H(Y_{SD}, Y_{SR} \mid X_S) \\
&= H(Y_{SD}, Y_{SR}) - H(Y_{SD} \mid X_S) - H(Y_{SD} \mid X_S),
\end{aligned}
$$

where the last line follows from the fact that $\mathbf{y}_{SD}$ and $\mathbf{y}_{SR}$ are conditionally independent given $\mathbf{x}_S$. These quantities can be straightforwardly calculated. It was shown in the article that full diversity gain is achieved by this method.

There are difficulties with this scheme, the most important of which is that mutual information measures the minimum rate at which arbitrarily low probability of error is achievable, while the performance metric of the relaying system is the probability of error at a given rate. Furthermore, the scheme in Ref. [46] is a poor heuristic for LDGM codes, because those codes do not approach the Shannon limit as closely as PSRA codes. An alternative heuristic for relay selection is a promising topic for future work.

3.4.6 CODED DEMODULATE-AND-FORWARD: EXAMPLES

In Figure 3.6, we give results of an example coded demodulate-and-forward implementation, compared to implementations using AF and DF, similar to results given in Ref. [10]. These example results are plotted with respect to E_b/N_0, where E_b represents the energy per information bit and N_0 represents the amplitude of the noise power spectral density. Furthermore, in each case, the average SNR on each link is the same. In AF, no code is used anywhere in the system, whereas for both demodulate-and-forward and DF, both the source and the relay use rate-1/2 PSRA codes. Furthermore, for comparison, we include results in the absence of relaying with rate 1/2 and 1/4. In each case, 2000 information bits were used. Not surprisingly, DF is the best system, but not much better than coded demodulate-and-forward, which is better than all the other systems at useful frame error rates. The results in Figure 3.6 clearly demonstrate that demodulate-and-forward is a viable cooperation scheme, in spite of its low complexity.

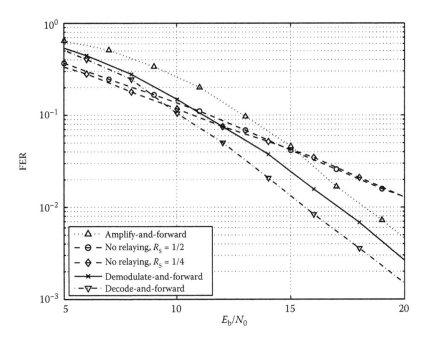

FIGURE 3.6 Plot comparing the performance of amplify-and-forward, demodulate-and-forward, and decode-and-forward, similar to a plot from Ref. [10]. (From Eckford, A.W., Chu, J.P.K., and Adve, R.S., *IEEE Trans. Wireless Comm.*, p. 1926, 2008. With permission.)

3.5 FRACTIONAL COOPERATION FOR LOW-COMPLEXITY DEVICES

Fractional cooperation was introduced in Refs. [10,47] as a means for nodes employing coded demodulate-and-forward to achieve energy-efficient cooperative diversity gains. Furthermore, it is ideal in cooperative settings, as it allows nodes to devote a fraction of their resources to the relaying task. In this section, we discuss fractional cooperation as a low-complexity technique for achieving cooperative diversity gains.

3.5.1 FRACTIONAL COOPERATION: SINGLE-RELAY CASE

We begin our discussion of fractional cooperation by focusing on the single-relay case [47], which provides a good example to illustrate the main idea. As before, the relay observes $\mathbf{y}_{SR}$ and forms the demodulated sequence $\mathbf{b}'$. In the model, we specified that $\mathbf{b}'$ could represent the demodulation of only some of the received symbols in $\mathbf{y}_{SR}$, and here we make use of that specification.

Recall that $\mathbf{w}$ represented the vector of indices from $\mathbf{y}_{SR}$ that appear in $\mathbf{b}'$, which was made explicit in Equation 3.11. In fractional cooperation, the relay has a parameter v, $0 < v \leq 1$, such that $\mathbf{w}$ contains vn_S unique indices, chosen arbitrarily. In other words, v is the fraction of $\mathbf{y}_{SR}$ that the relay is willing to retransmit. This means of cooperation is clearly interesting to low-complexity networked devices. The arbitrary selection of symbols for relaying implies that a low-complexity means for selecting these symbols can be implemented (e.g., random selection), and centralized control is not required under the definition; the relay is assisting "as much as it can," making its own decision as to how much to assist.

Given the definition of fractional cooperation, we may ask how the cooperation task is affected by selecting only a fraction of the available symbols. The following theoretical result answers this question (simplified from Ref. [47]):

THEOREM 1

Let R_S represent the rate of the source's codeword. Then if $R_S > v$, the diversity order in the frame error rate is 1, and if $R_S \leq v$, the diversity order is either 1 or 2.

Proof: We provide a sketch of the proof, and the reader is directed to Ref. [47] for the details. Say a_{SR} is large enough that the SR link is error-free (which is the best possible case); then $m(\sigma(\mathbf{y}_{SR})) = \mathbf{x}_S$. Selecting an arbitrary fraction v of $\mathbf{x}_S$ is equivalent to passing $\mathbf{x}_S$ through an erasure channel with erasure probability $1 - v$. It is known that the capacity of such a channel is v, so if $R_S > v$, the relay link operates at a rate higher than capacity. Because the relay link with $R_S > v$ is not capable of reliable communication on its own, then in order for successful communication to occur, a necessary condition is that $a_{SD} > 0$. Thus, the diversity order must be 1. Otherwise, because there is only one relay, the diversity order is at most 2. ∎

It can be shown that this theorem also applies if the relay uses fractional cooperation along with DF.

It is straightforward to see that, if $R_S = v$, then vn_S is equal to the number of information bits in $\mathbf{x}_S$. Thus, to achieve a diversity gain, it is necessary to relay at least as many bits as were in the original information sequence. However, it is straightforward to show that a fractional cooperation system with any $v > 0$ will have a lower frame error rate than a system with no relaying, even if gains in diversity order are not achieved.

3.5.2 Fractional Cooperation: Multiple-Relay Case

In Section 3.5.1, we saw for the single-relay case that a relatively large amount of cooperation was required to achieve diversity gains. However, this is not true in the multiple-relay case, which extends fractional cooperation in a way that is both theoretically and practically interesting. In this case, each relay $i \in \mathcal{I}$ observes $\mathbf{y}_{SR,i}$ and selects a fraction v of that observation for demodulation and relaying. (For convenience, we assume that the fraction v is the same for every relay, but this assumption can be relaxed.) Furthermore, each relay $i \in \mathcal{I}$ again selects its symbols from $\mathbf{y}_{SR,i}$ arbitrarily, with the condition that there exists a sufficiently large (but finite) number of relays r so that the probability that a particular index is not selected by any relay is less than ϵ, where $\epsilon \rightarrow 0$ as $r \rightarrow \infty$ (e.g., this condition is satisfied if symbols are selected independently at random by each relay). Again, such a system can be easily implemented in a low-complexity and distributed manner, with no need for centralized coordination.

There exists a minimum number r_c of relays in order for diversity gains to be achieved, where r_c is equal to the minimum number of relays so that successful communication is possible when the $a_{SD} = 0$, that is, the direct link is absent. Given this definition, the following result was obtained by the authors of Ref. [10]:

THEOREM 2

For fractional cooperation system with r relays, and with r_c defined as above, the diversity order d of the system is

$$d = \begin{cases} 1, & r < r_c, \\ r - r_c + 2, & r \geq r_c. \end{cases}$$

Proof: The proof is somewhat involved, and the reader is directed to Ref. [10] for the details; here we sketch the proof. By the definition of r_c, if $r < r_c$, then successful communication is not possible when $a_{SD} = 0$. Thus, it must be true that $a_{SD} > 0$, implying a diversity order of 1. If $r = r_c$, then there exist two redundant paths from source to destination (one being the SD link and the other being the set of all relay links), so the diversity order is 2. Finally, consider the case where $r > r_c$. In this case, successful communication on the relay link requires that r_c out of r relays have some minimum signal strength a_{min} on both the SR and RD links. Letting p_a represent the probability that $a_{SR,i} > a_{min}$ or $a_{RD,i} > a_{min}$, the probability of transmission failure using only the relay link is given by

$$P(\text{failure}) = \sum_{j=0}^{r_c-1} \binom{r}{j} p_a^j (1 - p_a)^{r-j}$$

$$= \binom{r}{r_c - 1} p_a^{r_c-1} (1 - p_a)^{r-r_c+1} + \cdots .$$

Since $p_a = \Theta(1)$, $(1 - p_a) = \Theta(\bar{\gamma}^{-1})$, and the neglected terms have order smaller than $\Theta(\bar{\gamma}^{-(r-r_c+1)})$, the outage probability for the relay link alone is $\Theta(\bar{\gamma}^{-(r-r_c+1)})$. Including the direct link, we have a diversity order of $r - r_c + 2$.   ∎

Note that Theorem 1 has a particular meaning in terms of Theorem 2, in particular, that a necessary condition for $r_c = 1$ is $v \geq R_S$. In the single-relay case addressed by Theorem 2, if $r_c = 1$, the diversity order is 2, while if $r_c > 1$, the diversity order is 1. Again, it can be shown that this result also applies to the DF case.

This result implies that, as long as the number of relays is greater than some minimum r_c, adding one more relay always increases the diversity order by 1. This is true regardless of the size of v; in other words, an order-1 increase in diversity can be obtained with arbitrarily little additional assistance from the new relay. Furthermore, these gains are achieved with very limited coordination among the relays and with only low-complexity techniques, as mentioned in Section 3.5.1. Thus, each relay needs to devote only a small amount of its energy and computational resources to relaying, using the rest to transmit its own information. Fractional cooperation is therefore a very useful technique in low-complexity cooperative networks.

The proof of Theorem 2 in Ref. [10] makes no assumptions about the power of any error-correcting codes that may be used in the system. In particular, if no error-correcting codes are used, a consequence of Ref. [10, Lemma 1] is that the value of r_c is enormous. However, it was also shown in Ref. [10] that r_c is lower-bounded by $\log(1 - R_S)/\log(1 - v)$, which is roughly equal to $-\log(1 - R_S)/v$ for small values of v. Furthermore, an error-correcting code that approached the Shannon limit could come close to achieving this bound. In Ref. [10], it was found that PSRA codes could achieve $r_c \simeq 1/v$ at rate $R_S = 1/4$, which is reasonably close to the bound. However, there is room for optimization.

3.5.3   FRACTIONAL COOPERATION: EXAMPLES

In Figure 3.7, we give results for some example systems using fractional coded demodulate-and-forward, which illustrate our claims in Theorem 2; these results are similar to those given in Ref. [10], though with different parameters. In this figure, 2000 information bits are encoded and sent by the source, and the average SNR on each link is the same. For the case where $v = 0.15$, we see that the diversity order is roughly 4 for 9 relays, 5 for 10 relays, and 6 for 11 relays, implying that the value of r_c is 7, which is close to $1/v$. Plots for other values of v are given for comparison. For instance, we see that 11 relays with $v = 0.15$ have the same diversity order as 10 relays with $v = 0.175$, which is remarkable because the 11 relay system relays a total of 6600 bits, while the 10 relay system relays a total of 7000 bits. As a result, we see that using an increased number of relays at smaller v leads to energy-efficient diversity gain.

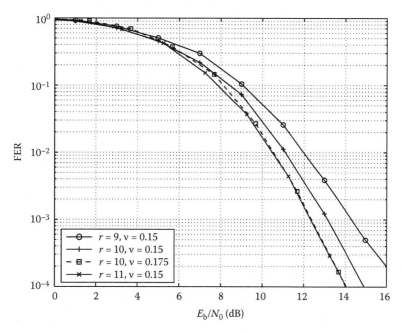

FIGURE 3.7 Plot illustrating the performance of fractional coded demodulate-and-forward for various numbers of relays and various values of ν.

3.6 OPEN PROBLEMS

Our proposed techniques leave many interesting research problems to be addressed. In this section, we give some questions that may be considered in future research.

1. Irregular fractional cooperation. Above, we gave conditions by which diversity gains could be achieved when each relay uses the same ν. However, there are many interesting scenarios in which ν might differ from relay to relay. For example, a relay under heavy load or with a nearly depleted battery might wish to use a smaller ν than another node without these restrictions. Rather than using νn_S as a "threshold" for participation (i.e., requiring that any node participating in cooperation be able to contribute ν bits), it would be preferable to allow each node to truly contribute as much as it can and make that decision locally. A direction of future research is to give the conditions under which fractional diversity gains are achieved under such irregular cooperation schemes.

2. Relay selection for fractional cooperation. We saw in Section 3.4.5 that relay selection could be used to increase the diversity order of a system. However, this usage of relay selection required that the selected relay devote its entire resources to relaying. An interesting direction of future research would be to explore how the gains obtained through fractional cooperation would extend to a relay selection scenario. Although it is likely that several relays would have to be selected (a consequence of Theorem 1), each of those relays would only be required to devote a small fraction of their resources to the relaying task.

3. Alternative codes and code optimization. We previously considered LDGM and PSRA codes as low-complexity but powerful error-correcting codes. Other, possibly more powerful, alternative codes exist; for example, turbo codes can be used. Furthermore, each of these code types can be optimized to improve their performance. Throughout this work we were not particularly careful about designing the best possible codes, and although good performance was observed with our methods, there remains room for improvement. This is

also true of fractional cooperation, where improved codes might lead to smaller values of r_c and therefore increased diversity orders.

4. Low-complexity protocols for cooperation. The focus of this chapter has been on find-ing methods that obtain diversity gains with low complexity. We have generally avoided any "implementation" issue, such as coordinating relay activities, obtaining channel-state information, choosing ν in a fractional cooperation system, and performing relay selection. Thus, there is much work to be done to obtain low-complexity protocols to complement the low-complexity methods that have been presented in this chapter.

In our own work, we are focusing on solutions to the first two problems that we list above. However, all of these open problems represent potentially rich areas of future research.

3.7   CONCLUSION

In this chapter, we have explored low-complexity techniques for performing cooperative communica-tion, specifically focusing on coded demodulate-and-forward. As we have illustrated in this chapter, coded demodulate-and-forward is more flexible than AF, while its performance is comparable to DF, which is a technique with much higher complexity. Paired with simple, but powerful, error-correcting codes such as LDGM and PSRA codes, coded demodulate-and-forward can be easily implemented in contemporary sensor networking hardware. Furthermore, the technique of fractional cooperation, which is paired naturally with coded demodulate-and-forward, provides a low-complexity, robust, and distributed way to achieve diversity gains in a low-complexity network. There remains much interesting research work to be done in this important and emerging field.

REFERENCES

[1] C.-Y. Chong and S. P. Kumar, Sensor networks: Evolution, opportunities and challenges, *Proceedings of the IEEE*, 91, 1247–1256, Aug. 2003.

[2] J. G. Proakis, *Digital Communications* 4th edn. New York: McGraw-Hill, 2001.

[3] E. C. van der Muelen, Three-terminal communication channels, *Advances in Applied Probability*, 3, 120–154, 1971.

[4] T. M. Cover and A. A. El Gamal, Capacity theorems for the relay channel, *IEEE Transactions on Information Theory*, IT-25, 572–584, Sept. 1979.

[5] A. Sendonaris, E. Erkip, and B. Aazhang, User cooperation diversity—Part I: System description, *IEEE Transactions on Communications*, 51, 1927–1938, Nov. 2003.

[6] A. Sendonaris, E. Erkip, and B. Aazhang, User cooperation diversity—Part II: Implementation aspects and performance analysis, *IEEE Transactions on Communications*, 51, 1939–1948, Nov. 2003.

[7] J. N. Laneman, D. N. C. Tse, and G. W. Wornell, Cooperative diversity in wireless networks: Efficient protocols and outage behavior, *IEEE Transactions on Information Theory*, 50, 3062–3080, Dec. 2004.

[8] A. W. Eckford, J. P. K. Chu, and R. S. Adve, Low-complexity cooperative coding for sensor networks using rateless and LDGM codes, in *Proceedings of IEEE International Conference on Communications (ICC)*, Istanbul, Turkey, pp. 1537–1542, 2006.

[9] A. W. Eckford and R. S. Adve, A practical scheme for relaying in sensor networks using repeat-accumulate codes, in *Proceedings of 40th Annual Conference on Information Sciences and Systems*, Princeton, NJ, pp. 386–391, 2006.

[10] A. W. Eckford, J. P. K. Chu, and R. S. Adve, Low complexity and fractional coded cooperation for wireless networks, *IEEE Transactions on Wireless Communications*, 7(5), 1917–1929, May 2008.

[11] M. A. Khojastepour, A. Sabharwal, and B. Aazhang, On the capacity of Gaussian 'cheap' relay channel, in *Proceedings of the IEEE Global Telecommunications Conference (Globecom)*, San Francisco, CA, pp. 1776–1780, 2003.

[12] P. S. K. Azarian, H. El Gamal, and P. Schniter, On the achievable diversity-multiplexing tradeoff in half-duplex cooperative channels, *IEEE Transactions on Information Theory*, 51, 4152–4172, Dec. 2005.

[13] R. U. Nabar, H. Bölcskei, and F. W. Kneubühler, Fading relay channels: Performance limits and space-time signal design, *IEEE Journal on selected Areas in Communications*, 22, 1099–1109, Aug. 2004.

[14] G. Kramer, M. Gastpar, and P. Gupta, Cooperative strategies and capacity theorems for relay networks, *IEEE Transactions on Information Theory*, 51, 3037–3063, Sept. 2005.

[15] A. Høst-Madsen and J. Zhang, Capacity bounds and power allocation for wireless relay channels, *IEEE Transactions on Information Theory*, 51, 2020–2040, June 2005.

[16] A. Chakrabarti, A. Sabharwal, and B. Aazhang, Sensitivity of achievable rates for half-duplex relay channel, in *Proceedings of IEEE Workshop on Signal Processing Advances in Wireless Communications*, New York, pp. 970–974, 2005.

[17] R. Dabora and S. D. Servetto, On the role of estimate-and-forward with time-sharing in cooperative communication, submitted to *IEEE Transactions on Information Theory*.

[18] D. Chen and J. N. Laneman, Modulation and demodulation for cooperative diversity in wireless systems, *IEEE Transactions Wireless Communications*, 5(7), 1785–1794, Jul. 2006.

[19] K. S. Gomadam and S. A. Jafar, On the capacity of memoryless relay networks, in *Proceedings of IEEE International Conference on Communications*, Istanbul, Turkey, pp. 1580–1585, 2006.

[20] T. E. Hunter and A. Nosratinia, Performance analysis of coded cooperation diversity, in *Proceedings of IEEE International Conference on Communications*, Seattle, WA, pp. 2688–2692, 2003.

[21] B. Zhao and M. C. Valenti, Distributed turbo coded diversity for relay channel, *Electronics Letters*, 39, 786–787, May 15, 2003.

[22] Z. Zhang and T. M. Duman, Capacity-approaching turbo coding and iterative decoding for relay channels, *IEEE Transactions on Communications*, 53, 1895–1905, Nov. 2005.

[23] R. Liu, P. Spasojevic, and E. Soljanin, Cooperative diversity with incremental redundancy turbo coding for quasi-static wireless networks, in *Proceedings of IEEE Workshop on Signal Processing Advances in Wireless Communications*, New York, pp. 791–795, June 2005.

[24] J. Castura and Y. Mao, Rateless coding for wireless relay channels, *IEEE Transactions on Wireless Communications*, 6(5), 1638–1642, May 2007.

[25] A. Stefanov and E. Erkip, Cooperative coding for wireless networks, *IEEE Transactions on Communications*, 52, 1470–1476, Sept. 2004.

[26] J. Hu and T. M. Duman, Low density parity check codes over half-duplex relay channels, in *Proceedings of IEEE International Symposium on Information Theory*, Seattle, WA, pp. 972–976, 2006.

[27] A. Chakrabarti, A. de Baynast, A. Sabharwal, and B. Aazhang, Low density parity check codes for the relay channel, *IEEE Journal on selected Areas in Communications*, 25, 1–7, Feb. 2007.

[28] T. J. Richardson, M. A. Shokrollahi, and R. L. Urbanke, Design of capacity-approaching irregular low-density parity-check codes for modulation and detection, *IEEE Transactions on Information Theory*, 47, 619–637, Feb. 2001.

[29] C. Li, M. Khojastepour, G. Yue, X. Wang, and M. Madihian, LDPC code design for half-duplex relay networks, in *Proceedings of IEEE International Conference on Acoustics, Speech and Signal Processing*, Honolulu, HI, pp. 873–876, 2007.

[30] P. Razaghi and W. Yu, Bilayer low-density parity-check codes for decode-and-forward in relay channels, *IEEE Transaction on Information Theory*, 53(10), 3723–3739, Oct. 2007.

[31] Z. Liu, V. Stankovic, and Z. Xiong, Practical compress-and-forward code design for the half-duplex relay channel, in *Proceedings of 39th Annual Conference on Information Sciences and Systems*, Baltimore, MD, 2005.

[32] J. Li and R. Hu, Slepian-Wolf cooperation: A practical and efficient compress-and-forward relay scheme, in *Proceedings of 43rd Annual Allerton Conference on Communication, Control, and Computing*, Monticello, IL, 2005.

[33] R. Hu and J. Li, Practical compress-forward in user cooperation: Wyner-Ziv cooperation, in *Proceedings of IEEE International Symposium on Information Theory*, Seattle, WA, pp. 489–493, 2006.

[34] A. Chakrabarti, A. de Baynast, A. Sabharwal, and B. Aazhang, Half-duplex estimate-and-forward relaying: Bounds and code design, in *Proceedings of IEEE International Symposium on Information Theory*, Seattle, WA, pp. 1239–1243, 2006.

[35] J. P. K. Chu and R. S. Adve, Implementation of co-operative diversity using message-passing in wireless sensor networks, in *Proceedings of IEEE Global Communications Conference*, St. Louis, MO, pp. 1167–1171, 2005.

[36] F. R. Kschischang, B. J. Frey, and H.-A. Loeliger, Factor graphs and the sum-product algorithm, *IEEE Transactions Information Theory*, 47, 498–519, Feb. 2001.

[37] T. R. Oenning and J. Moon, A low density generator matrix interpretation of parallel concatenated single bit parity codes, *IEEE Transactions on Magnetics*, 37, 737–741, Mar. 2001.

[38] J. Garcia-Frias and W. Zhong, Approaching Shannon performance by iterative decoding of linear codes with low-density generator matrix, *IEEE Communications Letters*, 7, 266–268, Jun. 2003.

[39] M. Luby, LT codes, *Proceedings of 43rd Annual IEEE Symposium on Foundations of Computer Science*, Vancouver, British Columbia, Canada, pp. 271–280, 2002.

[40] D. E. Knuth, *The Art of Computer Programming*, Vol. 2, Addison-Wesley, Reading, MA, 1997.

[41] D. Divsalar, H. Jin, and R. J. McEliece, Coding theorems for 'Turbo-like' codes, in *Proceedings of 36th Annual Allerton Conference on Communications, Control, and Computing,* Monticello, IL, pp. 201–210, 1998.

[42] A. Abbasfar, D. Divsalar, and K. Yao, Accumulate repeat accumulate codes, in *Proceedings of IEEE International Symposium on Information Theory,* Chicago, IL, p. 505, 2004.

[43] S. S. Pradhan and K. Ramchandran, Distributed source coding using syndromes (DISCUS): Design and construction, *IEEE Transactions on Information Theory*, 49(3), 626–643, Mar. 2003.

[44] Y. Zhao, R. S. Adve, and T. J. Lim, Improving amplify-and-forward relay networks: Optimal power allocation versus selection, in *Proceedings of IEEE International Symposium on Information Theory*, Seattle, WA, pp. 1234–1238, 2006.

[45] E. Beres and R. S. Adve, On selection cooperation in distributed networks, in *Proceedings of 40th Annual Conference on Information Sciences and Systems*, Princeton, NJ, pp. 1056–1061, 2006.

[46] J. P. K. Chu, R. S. Adve, and A. W. Eckford, Relay selection for low-complexity coded cooperation, in *Proceedings of IEEE Global Communications Conference*, Washington, DC, pp. 3932–3936, 2007.

[47] J. P. K. Chu, R. S. Adve, and A. W. Eckford, Fractional cooperation using coded demodulate-and-forward, in *Proceedings of IEEE Global Communications Conference*, Washington, DC, pp. 4323–4327, 2007.

4 Orthogonal Opportunistic Relaying for Cooperative Wireless Communications

Aggelos Bletsas

CONTENTS

Contrary to mainstream cooperative wireless research which focuses on multiple in-band transmissions from several relay nodes, orthogonal opportunistic relaying (OREL) is based instead on cooperative listening and single relay retransmission. This approach simplifies receiver design and allows for immediate implementation with low-cost RF front-ends. The main idea behind OREL is that relays are useful even when they do not transmit but instead, they are utilized as distributed sensors of the wireless environment, requiring:

1. Distributed algorithms that need limited network channel state information (NCSI) and reduced network coordination (NC)
2. Dynamic algorithms that adapt to the time-varying wireless conditions and changing network topologies
3. Efficient algorithms that harvest the maximum cooperative diversity benefits, improving reliability without sacrificing throughput, compared to traditional, noncooperative techniques

Fulfilling all the above requirements becomes challenging. This chapter reviews state-of-art progress on OREL under the perspective of all the above requirements, incorporating both theoretical as well as experimental results from implementation in radio network test beds. Hopefully, the intriguing theoretical properties, practical significance, and importance of OREL discussed in this chapter, will spark interest among academic researchers as well as engineers, wireless technology decision makers, and practitioners.

4.1 INTRODUCTION

4.1.1 MOTIVATIONS

Recent advances on multi-antenna transceivers, both in terms of theoretical foundations as well as practical prototypes/products, have emphasized the importance of space as another useful dimension for wireless communications, apart from time, frequency, and code. In such systems, the statistical richness of the wireless channel, between multiple transmit–receive antenna pairs, provides an additional mode of operation for wireless systems and can be exploited, under certain conditions, for improved reliability or spectral efficiency.

Reliability is provided through diversity, when the same symbol is transmitted across different transmit–receive antenna pairs. Spectral efficiency is provided through multiplexing, when different symbols are appropriately coded and simultaneously transmitted, at the same frequency band (in-band), across different transmit–receive antenna pairs. Both cases are based on the assumption that intricate wireless propagation phenomena such as reflection, scattering, and diffraction, traditionally viewed as undesired in conventional single-antenna communication, could provide for both spatially, statistically "separated," and thus very useful, wireless paths between multiple transmit–receive antennas. In short, multi-input multi-output (MIMO) transceivers exploit, instead of attempting to overcome, the vagaries of the wireless channel, providing significant improvements in terms of reliability or throughput speed [1–4].

Nevertheless, there are physical limitations on the number of antennas per transceiver that can support such operation mode. Statistical richness of multiple antenna links is compromised with increased number of antennas, especially in transceivers with limited physical size (e.g., in handheld terminals). Furthermore, issues of hardware complexity arise since modifications of the conventional RF front-end chain are needed to support multiple antennas, while new algorithms for transmission or reception of simultaneous, in-band signals increase software complexity. As a result, issues of cost are not negligible with multi-antenna transceivers.

While MIMO technologies exploit the statistical richness of the wireless channel, peer-to-peer (or multi-hop) communication principles exploit the increased nonlinearity of wireless propagation: breaking a long-distance, high-power transmission into several local, low-power transmissions through intermediate peer terminals improves reliability; this is due to the fact that effective receive-SNR at neighboring terminals increases, even though transmission power decreases. At the same time, interference to other neighboring transmissions also decreases due to reduced transmission power. The net effect is a communication network that scales graciously with improved number of participating peers [5]. In short, multiple terminals willing to relay information are also useful.

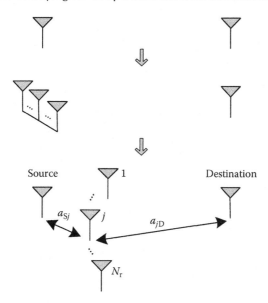

FIGURE 4.1 The motivation of this work: could low-complexity, single-antenna cooperating terminals form a virtual antenna array and provide (some) benefits of classic (colocated), multi-antenna (MIMO) links using existing radio hardware?

Orthogonal opportunistic relaying (OREL) aims to combine the usefulness of multiple antennas in MIMO setups with the benefits of peer-to-peer networks. In other words, OREL attempts to exploit the statistical richness of wireless channels as well as the advanced nonlinearity of wireless propagation. The goal is to enjoy the improved reliability or throughput speed found in multi-antenna systems as well as power (energy) saving benefits of relay networks, with single-antenna relays (Figure 4.1). Apart from the above theoretical objectives, OREL is motivated by the abundant, low-complexity and cost, single-antenna RF transmit and receive architectures found on the market, following an almost 100 year engineering experience: could single-antenna transceivers intelligently cooperate to relay information while providing (some) benefits of MIMO transmissions? Could such schemes be implemented with existing low-cost single-antenna transceivers? If yes, what would be the theoretical performance and how far away from the optimal (according to some generic criterion)?

4.1.2 OREL Approach and Challenges

OREL attempts to form an efficient virtual antenna array at the vicinity of communication between two nodes, the "source" and the "destination" (Figure 4.1). Instead of employing a typical MIMO approach where multiple terminals transmit to the destination using common time and frequency channels (i.e., transmit at the same time and in-band through the use of an appropriate code), OREL employs a single transmission from an appropriately selected relay terminal, which provides an "efficient" path toward the destination, depending on the wireless channel conditions. In other words, OREL exploits cooperating relays as distributed sensors of the wireless environment instead of active information retransmitters and proposes ways for distributed, adaptive, and low-complexity discovery of the most "appropriate" terminal for information retransmission.

OREL applies to local communication between neighboring terminals and resembles intelligent routing in a two-hop setup, between a source–destination pair and several intermediate relays (Figure 4.1), and despite its conceptual simplicity, aims to provide diverse benefits of multi-antenna (MIMO) schemes. Antenna selection is a well-understood technique in classic multi-antenna transceivers, where all used antennas are colocated (i.e., belong to the same terminal). In relay environments, the problem of antenna-selection among single-antenna terminals is more challenging since

1. The number of available relays is unknown and time-varying. For example, intermediate relay nodes might have entered "sleep" mode, depending on their internal state.
2. The number of useful relays is also unknown and time-varying. The usefulness of each available relay depends on the channel conditions, which are in general variable in time/space, and "unknown".
3. Even if a useful relay is found, information needs to be conveyed over noisy (nonperfect) relay links. Thus, careful processing and reliable reception of information at the relay and the final destination are both desired.
4. Energy at intermediate relays is spent not only to retransmit but also to receive information. Reception energy is comparable to transmission energy in modern radios and thus, relaying algorithms might be energy-expensive, even when a single relay transmits (but multiple relays listen)!
5. The distributed nature of cooperative communications requires coordination among different terminals, and thus, appropriate protocols are required to traffic when and which terminal transmits. Such distributed control requires nonnegligible network resources.

These observations underline the fundamental difference of any type of cooperative relaying compared to classic MIMO communication and suggest a new "holistic" research agenda that seriously takes into account all the above inherent challenges of cooperative communications. From this perspective, distributed, adaptive, and efficient algorithms are needed, vastly differentiating OREL from existing approaches of antenna selection. Classic approaches for antenna selection include selection diversity or switched diversity based on signal strength or bit error rate (BER), all examined in classic, colocated multi-antenna (MIMO) systems (see for example the relevant discussion in Ref. [6], chapter 13, pp. 249–257).

Finally, as mentioned before, theoretical analysis should be escorted by implementation today, using commodity and low-cost hardware. This is, perhaps, the most intriguing OREL challenge. It was not surprising to find that many of the existing transceivers found on the market do not support simultaneous, in-band reception of signals transmitted by multiple transmitters. Instead, many transceivers adhere to the principle of orthogonal transmissions (POT) among multiple terminals, exemplifying the practical value of OREL, where only a single terminal (source or relay) transmits at a given frequency band and time instant.

Opportunistic relaying with orthogonal use of time/frequency was first introduced in Refs. [7,8]. The term "opportunistic" describes the dynamic exploitation of wireless channel changes, both in time (due to the nonconstant wireless channel) as well as space (due to several potential relay paths). The term ("opportunistic") has been extensively adopted in the literature to describe exploitation of a time-varying quantity, for example, the wireless channel as in Ref. [9], the retransmitted signal strength as in Ref. [10], or even packet traffic as in Ref. [11].

The following section describes the basic assumptions behind OREL and its analysis.

4.1.3 Main Assumptions

Before proceeding to a discussion of OREL algorithms, the main assumptions followed in this work are emphasized. The received signal follows the discrete baseband model:

$$y_j = a_{ij} x_i + n_j, \tag{4.1}$$

where the received complex signal at node j is a function of the transmitted symbol from node i, scaled by the complex fading coefficient plus additive complex Gaussian noise. It is noted that the fading coefficient remains constant for several consecutive symbols, corresponding to slow fading. Assuming that the original source is unaware of the forward channel toward destination (Figure 4.1), slow fading leads to the most difficult communication problem: there is no transmission rate that guarantees arbitrarily small probability of error, resulting in zero ergodic (Shannon) capacity.

This sharply contradicts the case where fast (ergodic) fading could provide for reliable communication, since time instants (and symbols) under intense fading are compensated by time instants when the channel conditions are beneficial and the received signal strength is adequately strong. Slow fading and absence of channel-state information (CSI) at the original source are assumed throughout this work. Absence of CSI at the source is a valid assumption in cases where no direct link between source and destination exists, or in cases where the source cannot exploit such information (e.g., there is no or limited capability for rate adaptation).

Furthermore, partial CSI at each relay j regarding its own channel strength $|a_{Sj}|^2$, $|a_{jD}|^2$ toward source and destination, respectively, will be discussed. Assumptions about network (global) channel-state information (NCSI) regarding all involved links will not be made, given the distributed character of the relay channel. Section 4.3.4 explains why the assumption of NCSI availability in relay systems is, in general, a bad idea.

Additionally, this work omits any assumptions about availability of beamforming capabilities at the (distributed) transceivers:

$$|a_{1j}\,x + a_{2j}\,x| \neq |a_{1j}\,x| + |a_{2j}\,x|, \tag{4.2}$$

which means that when distributed transmitters coordinate their transmissions and transmit the same symbol, their signals do not necessarily add constructively at the receiver. This is due to the fact that distributed transmitters operate with different oscillators, and therefore, their passband carriers are not necessarily in phase, and as such, the above baseband model does not directly apply. Additional relevant discussion is provided in Section 4.4.1. Carrier synchronization of distributed transmitters vastly increases complexity and cost, and thus, deviates from our initial motivation to use commodity radios.

For the same reason, the common assumption found in the literature about a Gaussian relay channel $|a_{1j} + a_{2j}| = |a_{1j}| + |a_{2j}|$ is not adopted. Rayleigh fading is assumed in most cases, unless generalized fading models are highlighted. Finally, given that all terminals are located in the same "neighborhood," average, thermal noise, power spectral density across all terminals will be assumed constant ($E\left[|n_j|^2\right] = \mathcal{N}_0$).

4.2 OREL STRATEGIES: PART I

4.2.1 STRATEGIES THAT ADAPT TO WIRELESS CHANNEL CONDITIONS: INSTANTANEOUS VS. AVERAGE SIGNAL STRENGTH METRICS

Under the assumptions stated above, an interesting problem arises: what kind of signal metrics should be used in OREL to select the most "appropriate" relay? One could argue that such selection could be performed among the relays that are conveniently located in the vicinity of source and destination. Topology information regarding the location of each relay compared to source or destination could be used. This idea requires some means of distance measurement either through specialized hardware (e.g., global positioning system [GPS] receiver module) or by measuring average signal strength, as the latter is in general distance-dependent:

$$E\left[|a_{ij}|^2\right] \propto \frac{1}{d_{ij}^\nu}, \tag{4.3}$$

where d_{ij} is the distance between nodes i and j and $2 \leq \nu \leq 5$, depending on the type of the environment (free space, indoor, outdoor, urban, etc.). Examples of relay selection based on average signal strength can be found in Refs. [12,13] while examples based on distance can be found in Refs. [14,15]. Both approaches attempt to take advantage of the wireless propagation advanced nonlinearity. However, specialized hardware such as GPS may not be readily available, while estimating the average signal strength (and consecutively distance) in slow-fading channels requires several blocks of symbols; the (slow-fading) channel remains constant for a block of symbols and thus, several consecutive blocks are needed to estimate the statistical average.

On the other hand, exploiting instantaneous signal strength measurements is much faster, given that in slow-fading environments one value is (in principle) adequate to characterize the channel for a block of symbols, during which the channel remains constant. In that way, appropriate relay selection criteria could maximize the received signal strength "seen" at the receiver structure, instead of engineering a "convenient" network topology. Such an approach exploits the opportunities of the wireless channel, as some relay paths might be very "strong" at specific time intervals, while taking into account the advanced nonlinearity of wireless propagation. A natural problem emerges: what kind of end-to-end, instantaneous signal strength metrics could be used?

Obviously, a relay j is useful if both links toward source and destination (a_{Sj}, a_{jD}, respectively) are strong enough. Otherwise, end-to-end communication will be compromised if either relay link is in deep fade. Naturally, a signal strength metric is needed that rewards relay paths with strong channel conditions toward both source and destination, while penalizing relay paths with either link (toward source or destination) in deep fade. Consequently, a potential instantaneous signal strength metric function for relay selection becomes

$$g_j = g\left(a_{Sj}, a_{jD}\right) = \min\left\{|a_{Sj}|^2, |a_{jD}|^2\right\}, \quad g_b = \max_j\left\{g_j\right\}, \quad j \in \{1, 2, \ldots, N_r\}, \qquad (4.4)$$

with b the "best" (selected) relay and N_r the total number of relay terminals.

Section 4.2.3 discusses distributed ways to perform such selection, without the need for global (network) CSI. The above formula suggests that under the assumptions followed in this work (slow fading, equal noise power density across relays), relay selection can be performed proactively, before the initial source submits any information symbol. An alternative approach could be used to perform relay selection reactively, after the source has submitted its message, among the relays that have successfully received it:

$$g_j = |a_{jD}|^2, \quad g_b = \max_j\left\{g_j\right\}, \quad j \in \mathcal{S}_D \subseteq \{1, 2, \ldots, N_r\}, \qquad (4.5)$$

where $\mathcal{S}_D$ is the subset of the relays that have successfully received the message. Notice that selection according to the relay that has successfully received information and is physically located closer to the destination amounts to

$$g_j = E\left[|a_{jD}|^2\right], \quad g_b = \max_j\left\{g_j\right\}, \quad j \in \mathcal{S}_D \subseteq \{1, 2, \ldots, N_r\}. \qquad (4.6)$$

The discontinuous nature of the signal strength metric function $g(.,.) = \min(.,.)$ in Equation 4.4 implies a similarly discontinuous processing approach at each relay. Indeed, such selection metric is appropriate among regenerative relays, that is, relays that decode the received information and regenerate the information signal before retransmission, as discussed in detail in Section 4.3. Additionally, the reactive approach described above is tightly connected to regenerative (decode-and-forward) relays. For less-discontinuous processing at the assisting relays, smoother metric functions for equivalent, end-to-end relay path signal strength could be used. Variations of the metric function below can be used with amplify-and-forward relays (transponders), where received signal plus inherent additive noise are amplified and retransmitted, without any hard decision on the content of the received information (more details are provided in Section 4.3):

$$g_j = g\left(a_{Sj}, a_{jD}\right) = \frac{1}{\frac{1}{|a_{Sj}|^2} + \frac{1}{|a_{jD}|^2}} \equiv \frac{|a_{Sj}|^2 |a_{jD}|^2}{|a_{Sj}|^2 + |a_{jD}|^2}, \quad g_b = \max_j\left\{g_j\right\}, \quad j \in \{1, 2, \ldots, N_r\}. \qquad (4.7)$$

Regardless of the relay processing technique, the metric function, or whether instantaneous versus average signal strength values are used, the wireless channel needs to be sampled by the intermediate relays. The sampling rate of such network functionality depends on the wireless channel coherence time; the latter describes the approximate time interval when the wireless channel can be assumed constant and depends on Doppler shift (which in turn depends on carrier wavelength and mobility speed). Therefore, OREL exploits relays as sensors of the wireless environment and heavily depends relay selection strategies on such functionality. In Section 4.2.3, one approach for such distributed operation is explained.

Finally, for radios where there is no immediate way to measure signal strength, instantaneous or average, an alternative approach based on measured BER can be used to evaluate potential relay paths. Such an approach requires the submission of a known string of bits and the measurement of the BER at the receiver, revealing indirectly a "good" or "bad" channel quality. This method assumes (a) a sufficiently large number of bits used, so that BER can be estimated with reasonable accuracy and (b) a limited BER measurement time or equivalently fast radios, so that the channel quality measured at a specific instant is valid and not stale for the immediate future. Depending on the number of bits used and the radio transmission speed, BER measurements as channel quality assessment could correspond to instantaneous signal strength (when small overall time is needed for BER measurement compared to the channel coherence time) or average signal strength (large overall time for BER) or measurements in between.

4.2.2 STRATEGIES THAT REDUCE NETWORK (TOTAL) RECEPTION ENERGY: PROACTIVE VS. REACTIVE SELECTION

Figure 4.2 depicts proactive relay selection, before any type of information is submitted toward the final destination. In this way, all relays that are not selected to forward information during the second phase can avoid listening to the transmitted information, since they will not forward toward destination. Thus, only the selected "best" relay spends energy for message reception, and total (network) reception energy does not scale with increased number of relays. It is noted that such

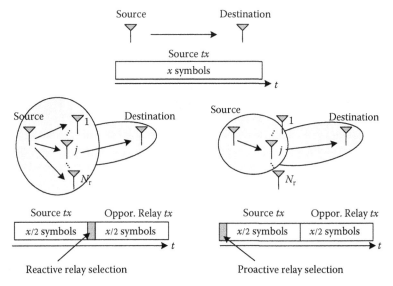

FIGURE 4.2 Opportunistic reactive vs. proactive selection. A relay can be selected before the source transmits the message (proactive selection) or after (reactive selection). In proactive selection, relays not selected can save reception energy during the source transmission.

proactive schemes can be facilitated with either regenerative (decode-and-forward) or transponder (amplify-and-forward) relays.

In contrast, reactive relaying schemes assume that all relays listen to the transmitted information during the first phase, and thus, network reception energy scales with the number of participating relays, even though a single, opportunistically selected relay retransmits. In modern radios, energy used for information reception has become comparable to energy used for transmission (see e.g., [16] for a relevant discussion). This is due to the complexity of forward error correction (FEC) algorithms and the signal conditioning techniques of modern protocol waveforms. Despite the improved transmission energy savings offered by multi-hop relaying, reception energy expenditures may become an overkill for cooperative communications, especially in battery-constrained networks (e.g., sensor networks) and deserve additional research attention.

In short, the issue of transmission energy consumption, especially when relays spend transmission energy to forward information originated from other nodes, has been extensively studied under the prism of increased reliability offered by cooperative communications. It is now time for cooperative communications research to carefully study network (i.e., total) reception energy, which is crucial in practical scenarios. OREL schemes, especially the proactive ones, seem to effectively address reception energy consumption.

4.2.3   STRATEGIES THAT REDUCE NETWORK COORDINATION OVERHEAD

In Section 4.2.1, the presented OREL selection strategies use the maximum of some relay function of end-to-end signal strength metric. From that perspective, one could assume that such selection, at least in principle, requires estimation of CSI at all involved links, propagation of that information to a central controller, decision about which relay maximizes that strength metric, and finally, propagation of that information to all involved cooperating nodes. However, such assumption would not be precise. Discovering the maximum (or minimum) among the elements of a set does not require knowledge of the individual values of the constituent elements. Instead, a comparison logic is sufficient to discover the maximum (or minimum) of a set. For example, finding out the tallest student in a classroom does not require precise measurement of the height of each individual student: instead, all students are asked to stand up and thus, the teacher is able to easily distinguish the tallest, or even better, the tallest student observes the class and waves her presence.

Therefore, selection based on the maximum of some relay signal strength metric can be performed without the need of network CSI at a central controller. This is further assisted by the fact that all kinds of OREL signal strength metric are local, that is, depend on the individual signal strengths of each relay toward source and destination (proactive case) or toward destination (reactive case) and do not incorporate any type of information regarding other relays. The challenge here is to perform such selection in a distributed fashion, with as small an overhead as possible. That overhead includes total time (delay) spent for relay selection as well as additional cost in terms of bandwidth and power (energy), communication packets exchanged among the participating nodes to coordinate and perform such selection (network coordination overhead). One strategy to perform OREL selection in a distributed fashion uses timers, commonly used in networking medium access control (MAC) protocols. In such methods, the tricky part is to minimize packet exchange among all participating nodes.

It is assumed that the source (Figure 4.1) initiates communication, with the transmission of a ready-to-send (RTS) packet toward the relays and the destination. Such transmission can be used from each relay j to estimate its own signal strength $|a_{sj}|^2$ toward the source. The same packet can be directly received from the destination (if such direct link exists) or indirectly, through an intermediate relay (if direct communication link between source and destination does not exist). The destination responds with a clear-to-send (CTS) packet, transmitted at the same (frequency) channel used for communication between relays and destination. That packet is used from each receiving relay to estimate the signal strength $|a_{Dj}|^2$ from destination to relay, which is equivalent to the signal strength $|a_{jD}|^2$ from relay j to destination, due to the reciprocity theorem.

The same CTS packet may be received from the source via the intermediate relay nodes at the worst-case scenario, when no direct link exists between source and destination. When such link exists, the CTS packet is not exploited at the source for CSI estimation and throughput adaptation, simply because its low-complexity radio does not support rate adaption, or equivalently, the channel conditions of the direct link are in deep fade so that only the minimum supported rate can be used.

Upon reception of the CTS packet and estimation of the forward relay signal strength, each relay j initiates a timer T_j with a speed value inversely proportional to the "instantaneous" signal strength metric g_j:

$$T_j \propto \frac{1}{g_j}, \tag{4.8}$$

where g_j was defined according to the proactive strategy described in Sections 4.2.1 and 4.2.2. The timer T_b of the relay with the maximum g_j expires first and that relay waives its presence with the transmission of a "flag" packet. The rest of the relays (as well as source and destination) discover that the "best" relay has been found and back off from the relay selection procedure. To accommodate the case of "hidden" relays, where relays are able to "listen" to transmission from the source and destination, but not transmission from each other, the received flag packet from the source (or destination) could be retransmitted from the source (or destination) toward the relays, to notify all relays about relay selection completion. Therefore, the total number of packets needed for such selection is three (RTS, CTS, flag), for the case of no-hidden relays, and four (RTS, CTS, flag, flag) for the case of hidden relays. Given that RTS/CTS is already incorporated in most MAC protocols and needed anyway, even for noncooperative communication, the number of required packets for OREL selection is maximum two. Figure 4.3 summarizes the above procedure for relay operation.

The above procedure does not require a specific time synchronization protocol, since CTS transmitted from the destination is used to synchronize all relay timers. However, an explicit synchronization protocol could further simplify the whole procedure. It is interesting to calculate the probability of flag packet collision, that is, the probability of worst-case scenario when more than one relays transmit a flag packet, which means that more than one relay competes for selection and information forwarding. Such probability can be upper bounded by the following expression, which essentially describes the probability that two or more relay timers expire within an uncertainty time interval Δ:

$$\mathrm{Pr}_{\mathrm{collision}} = \mathrm{Pr}\left\{T_j \leq T_b + \Delta\right\}, \quad \text{any} \quad j \in \{1, 2, \ldots, N_r\} \quad \text{and} \quad j \neq b, \tag{4.9}$$

where Δ is a time interval that incorporates radio switch time from receive to transmit mode, propagation time differences (due to destination–relay or relay–relay distance differences among several relays), and duration of the flag packet [8,17]. The radio receive-to-transmit switch time and the duration of the flag packet dominate this value, which can be engineered to get values in the microsecond regime.

The above probability can be further simplified if ordered timer values are considered ($T_b \leq T_{(2)} \leq T_{(3)} \cdots \leq T_{(N_r)}$):

$$\mathrm{Pr}_{\mathrm{collision}} \equiv \mathrm{Pr}\left\{T_{(2)} \leq T_b + \Delta\right\}. \tag{4.10}$$

From the above equation, it can be seen that the above probability can be reduced when timer duration T_i, $i \in \{1, 2, \ldots, N_r\}$, is significantly larger than Δ. On the other hand, the average timer duration cannot be large, since in that case the total overhead (delay) for relay selection will increase. Therefore, there is a flexible trade-off regarding how much time is spent for relay selection and how efficiently such selection can be performed.

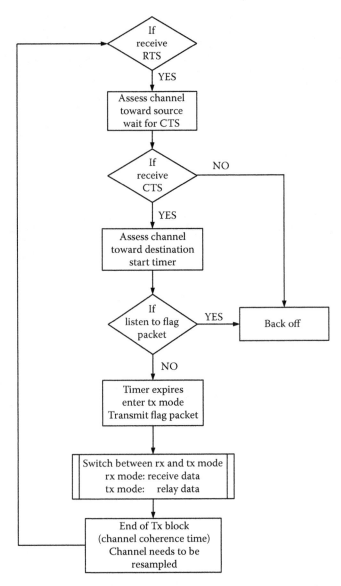

FIGURE 4.3 A flowchart example of an opportunistic relay.

Using order statistics, the above probability has been calculated for any kind of wireless channel statistics in Refs. [8,17]. For the specific cases of Rayleigh or Ricean fading, it was shown that spending a time interval two orders of magnitude greater than Δ (e.g., a few hundred microseconds for $\Delta \approx$ a few microseconds) drops the above probability below 0.6%, for various network topologies. For reasonable mobility speeds up to 3 km/hour (e.g., walking people), the Doppler shift for 900 Mhz carrier frequency becomes ≈ 2.5 Hz, corresponding to channel coherence time approximately equal to 200 ms. Therefore, spending less than 0.25 percent of the channel coherence time provides distributed relay selection with reasonable efficiency, even though the above calculation assumes worst-case scenario conditions (hidden relay terminals, nonoptimized radio switch time, mobile relays, etc.). Using more sophisticated radios, the necessary time for relay selection can be dropped to three orders of magnitude smaller delay than the channel coherence time (e.g., 0.025 percent) for the same collision probability. Allowing for additional time (delay) for relay selection can provide smaller collision probability.

The above example shows that total delay spent for distributed relay selection can be a limited fraction of the channel coherence time, which in turn is a function of terminal mobility. Therefore, delay for relay selection is acceptable (as in the above example), provided that the fraction of coherence time spent for relay selection is kept small. A similar algorithm can be invented for the reactive case, where a subset of the relays participates, with the signal strength metric depending only on the forward link (between relay and destination).

In short, opportunistic relay selection in slow-fading environments can be performed with reasonable delay and in a distributed fashion, without the need for network (global) CSI anywhere in the network. Other more involved approaches can be considered and one of the goals of this work is to motivate further research on the problem of distributed and efficient network coordination and MAC for cooperative communications.

4.3   OREL STRATEGIES: PART II

In this section, the performance of OREL is discussed when regenerative (decode-and-forward) or transponder (amplify-and-forward) relays are used. Analysis will be based on outage probability, that is, the probability that the mutual information between source and destination drops below the target rate:

$$\mathrm{Pr}_{\mathrm{outage}} \stackrel{\triangle}{=} \mathrm{Pr}\left\{ \mathcal{I}_{\mathrm{SD}}(\mathrm{SNR}) < \mathcal{R} \right\}. \qquad (4.11)$$

When such event happens, the probability of erroneous communication becomes one, regardless of the modulation scheme or the error correction techniques used; under the main assumption of zero CSI at the source, as well as slow fading (random, but constant channel throughout the transmitted block of symbols), there is no way to invent a communication scheme that guarantees arbitrarily small probability of error for any desired rate. From that perspective, the ergodic capacity is zero and the above metric of outage probability is used to characterize performance. Note that for ergodic (fast-fading) environments, the receiver can exploit symbols where the channel is strong, compensating for symbols when the channel is poor and thus, capacity is obviously nonzero, at the absence of CSI at the source. It is also important to notice that the assumptions of this work constitute a difficult communication problem, which is addressed with the help of cooperative relays.

For historic reasons, an asymptotic analysis is first presented. Asymptotic analysis was used in the landmark study by Laneman et al. [18] to characterize cooperative relaying in three-node setups (one source–destination pair assisted by a single relay). Similar analysis was also used in the equally important study by Laneman and Wornell [19], to characterize nonopportunistic cooperation of multiple relays that transmit in-band with each other, using an ideal distributed space-time code, with the purpose of assisting a single source–destination pair.

Both studies above employed diversity-multiplexing gain trade-off (DMT), originally invented for classic multi-antenna systems, where antennas are colocated at the source or the destination [20]. Such a tool provides an indication of the inherent trade-off (in any communication system) between reliability, indicated by diversity order, and transmission speed, indicated by multiplexing gain. The DMT tool assumes that signal-to-noise ratio (SNR) is increased to infinity AND simultaneously, target rate R is also increased to infinity, proportionally to $\log_2(\mathrm{SNR})$ ($R \equiv R(\mathrm{SNR}) \propto \log_2(\mathrm{SNR})$). Accordingly, diversity order d and multiplexing gain r for probability of error Pr_e are defined as

$$r \stackrel{\triangle}{=} \lim_{\mathrm{SNR}\to\infty} \frac{R(\mathrm{SNR})}{\log_2 \mathrm{SNR}}, \quad d \stackrel{\triangle}{=} - \lim_{\mathrm{SNR}\to\infty} \frac{\log_2 \mathrm{Pr}_e(R)}{\log_2 \mathrm{SNR}}. \qquad (4.12)$$

Laneman and colleagues did not provide a justification regarding the availability of the above tool for cooperative relaying schemes, as their major motivation and approach were suggesting that cooperative relays could be viewed as virtual antenna arrays, and thus, tools for classic MIMO systems could be readily incorporated. A justification of the above tool validity for distributed cooperative communication is attempted therein: by assuming SNR $\rightarrow \infty$, the benefits of advanced nonlinearity of wireless propagation are bypassed and the DMT result is not specific to a particular topology or wireless propagation coefficient v (Equation 4.3). Therefore, any observed benefits from the DMT analysis will not be the result of a conveniently engineered topology (e.g., a relay located between source and destination indoors, with $v = 5$) but instead, the result of (potential) exploitation of several statistically independent paths through the relays, toward the destination. Furthermore, the balanced increase (in both SNR and target rate R) suggests that results are practically meaningful.

On the other hand, such asymptotic analysis precludes the examination of total transmit power, transmit power allocation, and their impact on cooperative communication performance, especially for comparison studies among different cooperative or noncooperative schemes. Furthermore, schemes with the same DMT performance might perform differently in finite SNR and finite target rate conditions. Therefore, finite-SNR analysis and closed-form calculation of outage probability in Equation 4.11 are also desired. Under such analysis, total transmission power $\mathcal{P}_{tot}$ used and how such power is allocated to source transmission $\mathcal{P}_{source}$ and total power transmitted by all cooperative relay(s) $\mathcal{P}_{rel}$ (as well as how $\mathcal{P}_{rel}$ is allocated among the relays at the case where more than one relays retransmit) are research problems of great interest:

$$\mathcal{P}_{source} = \zeta \mathcal{P}_{tot}, \quad \mathcal{P}_{rel} = (1 - \zeta)\mathcal{P}_{tot}. \tag{4.13}$$

Total transmission power is always bounded by legislation, even for the case of distributed radiators. For example, part 15 of Federal Communication Commission (FCC) for industrial, scientific, and medicine (ISM) bands requires measurement of total power when multiple radiators transmit at the same frequency band and time.[*] Additionally, monitoring total transmission power is essential for comparison studies among different cooperative communication schemes or between cooperative and noncooperative communication.

OREL results for both analysis approaches (asymptotic or not) are summarized in the following sections.

4.3.1 DECODE-AND-FORWARD OREL: ASYMPTOTIC ANALYSIS

It was shown in Ref. [17] that the relay selection rule of Equation 4.4, under a proactive protocol where relay is selected before source transmission (Figure 4.2), provides DMT performance:

$$d(r) = (1 + N_r)(1 - 2r), \tag{4.14}$$

with $d \geq 0$ and $0 \leq r \leq \frac{1}{2}$. Exactly the same result holds for reactive relay selection according to Equation 4.5, when the relay is selected after source transmission, among the relays that have successfully decoded the source message (Figure 4.2). In either case, the diversity orders scales with the number of cooperating nodes (N_r relays plus the source, or just N_r when direct link between source and destination does not exist), even though a single relay transmits. This is due to careful selection of Equation 4.4 (or Equation 4.5) that takes into account both links toward source and destination for each relay, suggesting that any relay selection rule that neglects either link might be DMT-suboptimal. The maximum multiplexing gain of $\frac{1}{2}$ is due to the fact that the communication protocol requires two phases for each symbol, that is, symbol transmitted from source to destination and relays during the first phase and symbol retransmitted from selected relay to destination, during the second phase (Figure 4.2). Therefore, the channel is used twice for each symbol, resulting in $r \leq \frac{1}{2}$.

---

[*] See measuring procedures in Section 15.31, paragraph (h) of part 15, updated recently and found at http://www.fcc.gov/oet/info/rules/part15/part15-9-20-07.pdf

An alternative approach that increases r would require the source to transmit a new symbol to the destination, while the relay(s) retransmits (to the destination) the previous symbol. However, such approach implies: (a) that a direct link between source and destination exists, (b) the destination RF front-end is able to receive signals transmitted from (at least) two (nominally equal but practically different) distributed and unsynchronized carriers, and (c) the receiver signal processing algorithm must be able to decode (at least) two (instead of one) different messages. All the above are feasible but increase overall complexity and cost and practically demand nontrivial engineering. In the following section, a different approach based on feedback is presented that improves multiplexing gain, while low-complexity ("cheap") radios and protocols are used.

Interestingly, the above DMT curve of OREL is achieved with strictly one node (source or selected relay) transmitting at any given time. Even though strictly orthogonal transmissions are used, the DMT curve is exactly the same as that of the protocol suggested by Laneman and Wornell [19], where during the second phase, the relays that have successfully decoded the message employ an optimal (ideal) distributed space-time code and transmit in-band (at the same frequency channel) to the destination. This is the idea of a virtual antenna array formed by a subset of the relays, assuming (a) no CSI at each relay, regarding the forward link toward destination, (b) availability of an ideal distributed space-time code for more than two relays, and (c) a destination RF front-end able to receive signals from more than one (nominally equal but practically different) distributed and unsynchronized carriers. The latter, as mentioned before, increases the complexity and cost of transceiver design.

On the other hand, OREL can be used with many existing and low-cost radios that operate with strictly orthogonal transmissions, requiring however, forward CSI at each relay and distributed selection with minimal overhead, as described before. Additionally, proactive OREL with decode-and-forward relays provides reception energy savings (explained in Section 4.2.2), compared to the distributed space-time coding protocol in Ref. [19].

4.3.2 AMPLIFY-AND-FORWARD OREL: ASYMPTOTIC ANALYSIS

For the case of transponder (amplify-and-forward) relays, the proactive relay selection rule of Equation 4.4 provides the same DMT performance as above [17]:

$$d(r) = (1 + N_r)(1 - 2r). \qquad (4.15)$$

This result suggests that OREL can be very useful (at least in terms of DMT performance), even when "dumb" relays are utilized, that is, relays that only amplify and retransmit the incoming signal without any detection attempt.

The proactive relay selection rule of Equation 4.7 (as well as Equation 4.4) was mainly studied in Ref. [21] in terms of DMT performance, under generalized fading (where Rayleigh is a special case). The objective there was to adhere to strictly orthogonal transmissions (only one terminal transmits at any given time, at a given frequency band) while improving the multiplexing gain (and consequently communication speed). In that way, reliable and fast communication could be realized, even with the help of existing, low-complexity radios, built according to conventional technology.

Feedback from the destination was used and the protocol was slightly modified: the selected relay (discovered in a proactive manner) was used only if direct communication from source to destination failed. In that way, relays would come to the rescue only when they were needed, as opposed to the previous protocols described above, where relay(s) were always used during the second phase. The feedback idea above, also discussed in Ref. [18] for three-node setups (one source, one relay, and one destination and thus, no option for relay selection), improves multiplexing gain and overall DMT performance as it combines the benefits of OREL relay selection (producing cooperative diversity) with the benefits of feedback (saving "wasted" channel usage, and thus improving multiplexing gain). Additionally, such improved DMT performance is feasible with reduced total (network) reception energy (due to the proactive nature of selection), while using "dumb" transponder relays. If both source and relay transmissions fail, then the source could transmit in a second round, assisted by the selected relay, provided that feedback from destination signals such necessity. Additional rounds of feedback may be used, and DMT performance of OREL with feedback can incorporate the number of feedback rounds allowed in the protocol [21].

Feedback schemes are in general very useful when direct communication between source and destination exists, and thus, limited duration (small overhead) packets can be reliably exchanged between source, destination, and relays in between. In cases where such link is not present, simple schemes can be devised that exploit communication between source and destination through the participating relays. Therefore, in such distributed setups, efficient network coordination algorithms become essential and their overhead should be extensively studied, in conjunction with DMT performance (or other performance metrics) analysis.

In the following section, important OREL analytical results are summarized, regarding outage probability as a function of (finite) SNR.

4.3.3 DECODE-AND-FORWARD OREL: FINITE-SNR ANALYSIS

Given that finite-SNR analysis is the objective, the proactive selection rule of Equation 4.4 needs to take into account power allocation ζ between source and selected relay:

$$g_j = g\left(a_{Sj}, a_{jD}\right) = \min\left\{\zeta|a_{Sj}|^2, (1 - \zeta)|a_{jD}|^2\right\}. \tag{4.16}$$

Under such selection rule and assuming no direct path between source and destination, the outage probability at the destination becomes [22–24]

$$\Pr_{DF}^{proactive} = \prod_{j=1}^{N_r}\left[1 - \exp\left\{-\frac{2^{2R} - 1}{\mathcal{P}_{tot}/\mathcal{N}_0}\left(\frac{1}{\zeta\,E[|a_{Sj}|^2]} + \frac{1}{(1 - \zeta)\,E[|a_{jD}|^2]}\right)\right\}\right]. \tag{4.17}$$

Furthermore, the reactive selection rule of Equation 4.5 provides exactly the same outage probability:

$$\Pr_{DF}^{reactive} = \Pr_{DF}^{proactive}. \tag{4.18}$$

It was further shown that reactive relay selection and thus, allocation of total relay power $\mathcal{P}_{rel}$ on a single relay outperform strategies that split the power among relays that retransmit using an ideal, distributed space-time code [22–24].

These findings suggest that the distributed selection mechanism of OREL not only provides DMT performance as good as that of more complex schemes (such as those based on distributed space-time processing) but also provides an efficient power allocation mechanism that achieves good outage-probability performance as well, even though global (network) CSI is not needed anywhere in the network. Reception energy savings are feasible (for the proactive selection) while low-cost radios that strictly operate with orthogonal transmissions can be used.

In short, regenerative OREL can maximize the benefits of cooperation, without the need for nonorthogonal transmissions and distributed space-time codes, under the aforementioned assumptions (slow fading, no direct link between source and destination, efficient relay selection, etc.).

4.3.4 AMPLIFY-AND-FORWARD OREL: FINITE-SNR ANALYSIS

For the transponder case, a similar finite-SNR analysis has been performed and the single-relay selection rule that minimizes outage probability among all single-relay selection strategies has been found [22,24]. The relay selection rule was similar to that in Equation 4.7 (but not exactly the same).

It is noted that expressions of outage probability and relevant results depend on the specific amplify-and-forward strategy followed, which in turn, depends on the used amplification power mechanism and constraints [25]. Under a specific amplification mechanism, outage probability of transponder OREL was calculated [22,24].

Furthermore, interesting observations among total relay power allocation were made. Under the main assumption of OREL, according to which each relay has availability of each own channel conditions toward source and destination and no central controller has network (global) CSI availability, an alternative approach to OREL would be splitting power among multiple transmitting amplify-and-forward relays according to criteria that are not time-dependent and thus, do not depend on instantaneous relay channel fluctuations. This is the case of fixed power allocation among relays (potentially useful in immobile networks):

$$\mathcal{P}_{rel} = \lambda_1 \mathcal{P}_{rel} + \lambda_2 \mathcal{P}_{rel} + \cdots + \lambda_{N_r} \mathcal{P}_{rel} \rightarrow \lambda = [\lambda_1 \ \lambda_2, \ldots, \lambda_{N_r}]^T = \text{const}, \qquad (4.19)$$

where λ_i is the proportion of total relay transmit power $\mathcal{P}_{rel}$ allocated to transponder relay i.

It was shown that if total relay power $\mathcal{P}_{rel}$ is split among a set of transponder relays and kept constant (given that each relay does not have CSI availability in the channel conditions of other relays), OREL amplify-and-forward relaying outperforms multiple relay transmissions, in terms of outage probability [26]:

$$\text{Pr}_{AF}^{OREL} < \text{Pr}_{AF}^{\lambda}, \quad \forall \lambda = \text{const}, \quad \text{NO Network CSI availability.} \qquad (4.20)$$

This is true for any (constant) power allocation λ and from this perspective, OREL is globally optimum, compared to any constant power allocation scheme, under the main assumption behind OREL, that is, network (global) CSI is not available anywhere in the network.

However, if the constraint about global CSI availability is relaxed and each relay (or equivalently a central controller) is knowledgeable about instantaneous CSI of all relays, then power allocation among multiple simultaneously transmitting relays can be time-varying and not constant:

$$\lambda = \lambda(t) = [\lambda_1(t) \ \lambda_2(t), \ldots, \lambda_{N_r}(t)]^T \neq \text{const}, \quad \text{Network CSI availability.} \qquad (4.21)$$

In that case, it can be shown that careful allocation of power among the relays could provide for smaller outage probability than transponder OREL:

$$\text{Pr}_{AF}^{OREL} > \text{Pr}_{AF}^{\lambda(t)}, \quad \lambda = \lambda(t) \neq \text{const}, \quad \text{Network CSI availability.} \qquad (4.22)$$

Nevertheless, simple simulations show that performance benefits of relay power allocation based on network CSI, compared to OREL (which is distributed and thus, not based on network CSI) are marginal when a small number of relays is used. Therefore, the extensive overhead required for network CSI acquisition is not justified for a limited number of relays.

For an extended number of relays, the assumption of global instantaneous CSI availability everywhere in the network is not realistic or practical for the following reason: a relay cannot estimate the channel conditions of another relay toward source and destination and thus, such estimation results need to be performed by each relay (or the source, or the destination) and communicated toward all participating nodes (or a central controller). This assumes not only perfect estimation, but also perfect communication among cooperating terminals, while the objective is to enhance (another) communication between source and destination. Obviously, "the chicken or the egg" (causality) dilemmas arise.

One can counterargue that assumptions regarding perfect CSI are common in wireless research, especially in cellular environments (communication between a base station and multiple users), and thus, extensions to the relay channel should be welcomed. There are two important observations to be made here: (a) in cellular environments, the base station can directly estimate the channel toward each user with the help of pilot tones, without the intervention of a central controller or another terminal. From that perspective, channel needs to be only estimated, and not communicated as well, to another part of the network (at least in principle), and (b) CSI estimation is not cost-free in all practical setups, and thus, performance benefits due to global CSI availability should well justify the respective and nontrivial overhead for network CSI acquisition. Relay selection benefits and relay power allocation for transponder relays, under network CSI availability, are discussed in Ref. [27].

In short, transponder (amplify-and-forward) OREL provides an efficient solution for information forwarding in terms of outage probability and transmit power allocation, with partial CSI at each relay and no global CSI anywhere in the network.

4.4 IMPLEMENTATION AND EXTENSIONS

4.4.1 IMPLEMENTATION ISSUES

One of the common assumptions in cooperative communications research is the assumption of equivalence between passband and baseband signal representation. Even though such assumption is reasonable in classic (colocated) MIMO systems, additional discussion and attention are needed in distributed transmitter schemes. Assuming that distributed transmitters A and B cooperate and transmit information messages x_A, x_B respectively, at the same frequency band using carrier frequency f_c, the aggregate transmitted signal from the two transmitters becomes

$$x_A \cos(2\pi f_c t + \phi_A) + x_B \cos(2\pi f_c t + \phi_B) \neq (x_A + x_B) \cos(2\pi f_c t + \phi). \tag{4.23}$$

Obviously, this is due to the fact that the two carriers are not synchronized (i.e., "not the same"), since they belong to different terminals. Lack of carrier synchronization is mainly due to the different phase ϕ between the two carriers and also due to different oscillators used at the two terminals, resulting in carrier frequencies f_c^A, f_c^B nominally equal but practically different from f_c ($f_c^A \neq f_c$, $f_c^B \neq f_c$). Therefore, the aggregate transmitted signal does not guarantee algebraic addition $x_A + x_B$ of the transmitted messages x_A, x_B at baseband, in sharp contrast to the case of classic multi-antenna transmitters where a common oscillator is used for all antenna branches.

Even if a common frequency reference is established among distributed transmitters (i.e., $f_c^A = f_c^B = f_c$, $\phi_A = \phi_B = \phi$), the aggregate received signal at a destination C will be in principle closer to the left-hand side of Equation 4.23, instead of right-hand side, due to the fact that the two signals travel paths of different length, and thus difference in distance covered is translated to phase difference $\phi_A \neq \phi_B$ between the two useful signals at the receiver antenna. Such phase difference does not necessarily add constructively the two signals, as in the right-hand side of Equation 4.23. In slow-fading environments, as in this work, this phenomenon makes communication difficult, since this phase difference does not change often and thus, time intervals of destructive addition cannot be compensated by time intervals of constructive addition.

Therefore, special attention is needed when implementation of cooperative communication is the main objective and more research is required to realize the performance benefits promised by cooperative communication theory, under the basic assumption of equivalence between baseband and passband signal representations. Is carrier synchronization trivial, especially at the high-frequency regimes modern radios operate? Practice suggests that such synchronization is not trivial at all, as can be seen by the abundant transceivers found in the market that do not support such operation.

Are there any physical limitations on the ability to synchronize distributed carriers in the GHz regime, especially for narrow-band communication where cooperative principles could be of great assistance? That remains to be seen. On the other hand, it should be noted that cooperative communication principles at the physical layer have become considerably trendy quite recently (mainly due to the eloquently written papers by Nick Laneman and colleagues [18,19,28]); therefore, it is natural to expect that more time, effort, and research are needed so that radio modules can be engineered to accommodate in practice relevant cooperative modes, including distributed carrier synchronization.

For the case of OREL, a simple demonstration with low-cost ISM radio modules regime has been implemented and described in Refs. [8,29]. The main objective there was to demonstrate that benefits of cooperation at the local level can be realized, even with neighboring transceivers that strictly adhere to the POT. This was made possible by (a) employing the relays as sensors of the wireless environment, instead of active retransmitters, and (b) implementing an intelligent routing decision (layer 3), based on dynamic and distributed discovery of the most appropriate relay links toward source and destination. Such dynamic and distributed discovery was facilitated by a simple MAC scheme (layer 2) that gave access to the appropriately selected ("best") relay, according to instantaneous end-to-end signal strength (and alongside the distributed and adaptive techniques of Section 4.2.3).

Custom, low-cost transceivers were built using ISM radio modules (at 900 MHz) and 8-bit micro-controllers, while a room-based demonstration aimed to show even to the nonspecialist, the dynamic selection of appropriate relay paths that improved end-to-end communication, when people were moving inside the room [8,29].

In short, cooperative communication benefits can be realized with existing, low-cost transceivers, provided that relay selection is performed according to the distributed (i.e., local CSI, no central network controller), adaptive (i.e., instantaneous signal strength, proactive or reactive selection), and efficient (i.e., limited or controlled overhead for selection) techniques described above. Hopefully, the OREL importance will spark interest among the research and engineering community to bring these ideas to the next level and perhaps, incorporate them in local area network (LAN) protocols.

4.4.2 Extensions to Multi-Hop Scenarios

The basic motivation behind this work was to introduce spatial diversity benefits of multi-antenna communication among local, single-antenna, and low-complexity transceivers, with limited use of the channel (and thus, reduced sacrifice of end-to-end throughput) while trying to minimize all required overhead. Therefore, theoretical analysis focused on a local, two-hop setup and OREL was studied in an exhaustive manner, incorporating both asymptotic (e.g., DMT) as well as finite-SNR analysis (outage probability for constant total transmit power). Additionally, relevant overhead required for network coordination among the cooperating terminals was also studied and finally, a simple indoor demonstration was built with custom, low-cost radios.

An obvious question arises: how could such OREL principles based on local communication, as described in this work, get extended to the multi-hop case? Given that OREL is based on local two-hop communication, extension to the multi-hop case can be realized by employing OREL principles every two hops, over the course of a multi-hop route. In this way, local opportunities of the wireless channel can be exploited in the routing layer, so that information can be efficiently forwarded across several hops. However, one could argue that such operation, where information from layer 2 (datalink layer) is exploited in layer 3 (routing layer), violates the layering principle of the open systems interconnect (OSI); the latter has largely assisted in the development and smooth functioning of the wireless communication industry. Thus, such layering violation might raise adoption and interoperability concerns. There are two different (and debatable) answers to such concerns:

1. It is well known that the OSI model was inspired during the technological era of "wired" networks. That is why a lot of recent research has focused on cross-layer optimization, especially suited for wireless networks.
2. The gains of opportunistic techniques, that is, techniques that exploit the vagaries of the wireless channel, are too significant to bypass.

Therefore, it is both valuable and acceptable to come up with techniques that leverage OREL (local communication) to assist routing operations (macroscopic communication).

Interestingly, parallel and independent research on OREL was conducted by Sunjit Biswas and colleagues [30,31], under the same basic idea: the wireless channel and its idiosyncracies "should not be hidden but exploited". The main goal in Refs. [30,31] was to improve routing between rooftop-based 802.11b transceivers that networked a broad neighborhood, taking into account opportunities of the wireless channel (and hence their term "opportunistic routing"). Despite the same basic idea ("exploit the channel") and the common term ("opportunistic"), the similarities between OREL (opportunistic relaying) and opportunistic routing stop there. Their important differences are summarized below:

1. Opportunistic routing aims to assist macroscopic communication between two distant terminals, and as such, it is not solely based on local information. On the contrary, opportunistic relaying aims to assist local communication only (and hence the term "relaying" versus the term "routing").

2. Opportunistic routing is closer to the reactive case where relays that successfully received the message retransmit according to the forward channel quality and according to some general knowledge about the location of the final destination. In contrast, opportunistic relaying has incorporated proactive relay selection modes as well, allowing for local route (relay-path) selection before message transmission. In that way, valuable network (total) reception energy can be saved, since relays that are not selected, will not listen to the source transmission.

3. Opportunistic routing assesses the quality of the wireless channel by measuring BER, instead of directly estimating instantaneous signal strength, as in opportunistic relaying. This was due to implementation details of 802.11b transceivers. As explained in Section 4.2.1, assessing link quality based on BER might be equivalent to using average signal strength metrics (with suboptimal outage performance), depending on the total number of bits used and the channel coherence time.

4. Opportunistic routing results are based on experimental validation with 802.11b radios, while opportunistic relaying results are grounded in theoretical analysis as well as experimental demonstration with custom-built radios.

Despite their different aims, both approaches reveal significant cooperative communication benefits with existing radios, compared to noncooperative principles and could be viewed as complementary to each other. There is great space for further improvements of both approaches, especially toward the direction of minimizing the overhead required for network coordination and cooperation among the participating nodes.

In any case, extensions of OREL to the multi-hop case are clearly feasible and relevant examples already exist.

4.5 OPEN ISSUES

"To relay or not to relay"? This chapter attempted to highlight the importance of orthogonal opportunistic relaying, where deciding not to relay (but instead cooperatively listen) is also (or in some cases more) beneficial. More importantly, this chapter attempted to address most issues relevant to cooperative wireless communication, which is an inherently network (i.e., distributed) problem. One could safely argue that the required complexity for a specific cooperative scheme might overcome the advertised cooperation benefits and thus, such operation will not be useful. Therefore, it is important for cooperative communication research to incorporate holistic approaches that address (or at least, attempt to address) the following sources of overhead or complexity:

1. Network (global) CSI (NCSI) availability, especially when channel information between two terminals is needed at a third terminal (and thus, cannot be estimated but rather communicated). Discussion regarding NCSI availability should quantify the amount of resources needed for such information acquisition and measure their impact on overall performance.
2. Network coordination overhead (NCO), in terms of the packets that need to be exchanged and the distributed control algorithms that traffic when and which terminals transmit.
3. Network reception energy consumption, in addition to transmission energy, (NEC), since modern radios spent considerable energy to receive information, and thus, cooperation benefits might not scale in battery-constrained networks.
4. Hardware complexity (HC), given that beam-forming capabilities or even simultaneous (in time), in-band (in frequency) transmissions require nontrivial engineering in the transceiver architecture.

Analysis of the above sources of overhead should accompany any proposed technology for cooperative communications, in parallel with the analysis (under idealized conditions) of achieved throughput and error probability, under cooperation.

OREL, in particular, could be improved with new algorithms that attempt to minimize NCO, especially needed for distributed, adaptive, and dynamic relay selection, according to the time-varying channel conditions. Opportunistic extensions for the fast-fading wireless environments (e.g., moving cars) are also important. Finally, incorporation of the OREL ideas into LAN standards, such as 802.11 or TETRA, is a fertile area for investigation due to the promised theoretical benefits of OREL, without major modifications of existing transceiver hardware.

ACKNOWLEDGMENTS

The author is grateful to a number of friends and colleagues for invaluable assistance: the MIT Media Lab Viral Communications Group and its directors, Andy Lippman and David Reed, J. Paradiso and J. Lifton from MIT Media Lab, Ashish Khisti from MIT RLE, Hyundong Shin from Department of Electronics and Radio Engineering, Kyung Hee University, Moe Win from MIT LIDS, and J. N. Sahalos from AUTH RCL.

REFERENCES

[1] J. H. Winters, On the capacity of radio communication systems with diversity in Rayleigh fading environment, *IEEE J. Select. Areas Commun.*, 5(5), 871–878, June 1987.
[2] G. J. Foschini, Layered space-time architecture for wireless communication in a fading environment when using multi-element antennas, *Bell Labs Technical Journal*, 1(2), pp. 41–59, Oct. 1996.
[3] G. J. Foschini and M. J. Gans, On limits of wireless communications in a fading environment when using multiple antennas, *Wireless Personal Commun.*, 6(3), 311–335, Mar. 1998.
[4] G. J. Foschini, G. D. Golden, R. A. Valenzuela, and P. W. Wolniansky, Simplified processing for high spectral efficiency wireless communication employing multi-element arrays, *IEEE J. Select. Areas Commun.*, 17(11), 1841–1852, Nov. 1999.
[5] P. Gupta and P. R. Kumar, The capacity of wireless networks, *IEEE Trans. Inform. Theory*, 46(2), 388–404, Mar. 2000.
[6] A. F. Molisch, *Wireless Communications*. IEEE Press and John Wiley & Sons, The Atrium, Southern Gate, Chichester, West Sunex, PO19 8SQ, U.K., 2007.
[7] A. Bletsas, A. Lippman, and D. P. Reed, A simple distributed method for relay selection in cooperative diversity wireless networks, based on reciprocity and channel measurements, in *Proceedings of IEEE 61st VTC*, Stockholm, Sweden, May 2005.
[8] A. Bletsas, Intelligent antenna sharing in cooperative diversity wireless networks, PhD dissertation, Massachusetts Institute of Technology, Cambridge, MA, Sept. 2005.
[9] P. Viswanath, D. N. C. Tse, and R. Laroia, Opportunistic beamforming using dumb antennas, *IEEE Trans. Inform. Theory*, 48(6), 1277–1294, June 2002.

[10] A. Scaglione and Y.-W. Hong, Opportunistic large arrays: Cooperative transmission in wireless multihop ad hoc networks to reach far distances, *IEEE Trans. Signal Proces.*, 51(8), 2082–2092, Aug. 2003.

[11] C. Cetinkaya and E. Knightly, Opportunistic traffic scheduling over multiple network paths, in *Proceedings of IEEE Infocom*, Hong Kong, Sept. 2004.

[12] J. Luo, R. S. Blum, L. J. Cimini, L. J. Greenstein, and A. M. Haimovich, Link-failure probabilities for practical cooperative relay networks, in *Proceedings of IEEE VTC*, Stockholm, June 2005.

[13] J. Luo, R. S. Blum, L. Cimini, L. Greenstein, and A. Haimovich, Power allocation in a transmit diversity system with mean channel gain information, *IEEE Commun. Lett.*, 9(7), 2415–2525, July 2005.

[14] V. Sreng, H. Yanikomeroglu, and D. Falconer, Relayer selection strategies in cellular networks with peer-to-peer relaying, in *Proceedings of IEEE VTC*, Orlando, FL, Oct. 2003.

[15] B. Zhao and M. C. Valenti, Practical relay networks: A generalization of hybrid-arq, *IEEE J. Select. Areas Commun., (Special Issue on Wireless Ad Hoc Networks)*, 23(1), 7–18, Jan. 2005.

[16] R. Min, M. Bhardwaj, C. Seong-Hwan, N. Ickes, E. Shih, A. Sinha, A. Wang, and A. Chandrakasan, Energy-centric enabling technologies for wireless sensor networks, *IEEE Wireless Commun. Mag.*, 9(4), 28–39, Aug. 2002.

[17] A. Bletsas, A. Khisti, D. P. Reed, and A. Lippman, A simple cooperative diversity method based on network path selection, *IEEE J. Select. Areas Commun. (Special Issue on 4G Wireless Systems)*, 24(9), 659–672, Mar. 2006.

[18] J. N. Laneman, D. N. C. Tse, and G. W. Wornell, Cooperative diversity in wireless networks: Efficient protocols and outage behavior, *IEEE Trans. Inform. Theory*, 50(12), 3062–3080, Dec. 2004.

[19] J. N. Laneman and G. W. Wornell, Distributed space-time coded protocols for exploiting cooperative diversity in wireless networks, *IEEE Trans. Inform. Theory*, 59, 2415–2525, Oct. 2003.

[20] L. Zheng and D. N. C. Tse, Diversity and multiplexing: A fundamental tradeoff in multiple antenna channels, *IEEE Trans. Inform. Theory*, 49, 1073–1096, May 2003.

[21] A. Bletsas, A. Khisti, and M. Z. Win, Opportunistic relaying with feedback and cheap radios, *IEEE Trans. Wireless Commun.*, 7(5), 1823–1827, May 2008.

[22] A. Bletsas, H. Shin, M. Z. Win, and A. Lippman, Cooperative diversity with opportunistic relaying, in *Proc. IEEE Wireless Commun. Networking Conf.*, Las Vegas, NV, April 2006.

[23] A. Bletsas, H. Shin, and M. Z. Win, Outage-optimal cooperative communications with regenerative relays, in *Proc. Conf. Inform. Sci. Sys. (CISS'06)*, Princeton, NJ, March 2006, invited paper.

[24] A. Bletsas, H. Shin, and M. Z. Win, Cooperative communications with outage-optimal opportunistic relaying, *IEEE Trans. Wireless Commun.*, 6(9), 3450–3460, Sept. 2007.

[25] A. Bletsas, H. Shin, and M. Z. Win, Outage analysis for cooperative communication with multiple amplify-and-forward relays, *Electron. Lett.*, 43(6), 51–52, Mar. 2007.

[26] A. Bletsas, H. Shin, and M. Z. Win, Outage optimality of amplify-and-forward opportunistic relaying, *IEEE Commun. Lett.*, 11(3), 261–263, Mar. 2007.

[27] Y. Zhao, R. Adve, and T. J. Lim, Improving amplify-and-forward relay networks: Optimal power allocation versus selection, *IEEE Trans. Wireless Commun.*, 6(8), 3114–3123, Aug. 2007.

[28] N. Laneman and G. Wornell, Energy-efficient antenna sharing and relaying for wireless networks, in *Proc. IEEE Wireless Commun. Networking Conf.*, Chicago, IL, Sept. 2000.

[29] A. Bletsas and A. Lippman, Implementing cooperative diversity antenna arrays with commodity hardware, *IEEE Commun. Mag.*, 44(12), 33–40, Dec. 2006.

[30] S. Z. Biswas, Opportunistic routing in multi-hop wireless networks, MS thesis, Cambridge, MA, Mar. 2005.

[31] S. Biswas and R. Morris, Opportunistic routing in multi-hop wireless networks, in *SIGCOMM*, Philadelphia, PA, Aug. 2005.

5 Cross-Layer Design for Cooperative Wireless Communication

Dionysia Triantafyllopoulou, Nikos Passas,
and Alexandros Kaloxylos

CONTENTS

Modern wireless communication systems provide a variety of adaptation capabilities at different layers of the classical protocol stack. To maximize the overall performance of these systems, there is the need for a scheme that allows the efficient cooperation between protocols that operate in different layers. A number of proposals can be found in the recent literature that try to provide this cooperation and are usually referred to as cross-layer cooperative designs. This chapter provides a categorization of these proposals based on their target objective and discusses their main characteristics. To emphasize on the performance improvement that such a solution can provide to a legacy system, a cross-layer cooperation mechanism for WiMAX networks is described in detail.

5.1   INTRODUCTION

Legacy wireless communication systems provide few or no adaptation capabilities, as a result of the limited transmission requirements they were asked to fulfill. Nowadays, the efficient support of a wide set of network applications under a variety of conditions is considered as one of the most important requirements for modern wireless communication systems. To address this complex issue, different performance improvement mechanisms are introduced in almost all layers of a typical wireless protocol stack. Such mechanisms include the multi-encoding rate functionality at the Application layer, the adaptive automatic repeat request (ARQ) mechanisms at the Transport layer, the adaptive routing at the Network layer, the adaptive forward error correction (FEC) and appropriate resource allocation at the Medium Access Control (MAC) layer, and the adaptive modulation, coding and power control mechanisms at the Physical (PHY) layer. Although these mechanisms have been initially designed to work independently, there is an increasing need for the development of a scheme that will allow for their efficient cooperation toward optimized system performance. This need emerges from the fact that their independent parallel operation may result in inefficiencies caused mainly by possible conflicting actions.

A typical example of inefficient operation is the case of a Transmission Control Protocol (TCP) connection in a path that contains one or more wireless links. In this case, TCP may misinterpret packet errors on the wireless links as indications of congestion and reduce the transmission rate in an attempt to solve the problem. However, in this case, a data rate reduction is not considered as the most appropriate action as it reduces the throughput without facing the actual problem of errors on the wireless link. To overcome this situation, a cross-layer cooperative mechanism between the MAC and the Transport layers could provide the TCP with the necessary information regarding the nature of the packet losses as either transmission errors on the wireless link or packet losses due to congestion. This way, unnecessary transmission rate reductions could be avoided.

The rest of this chapter is organized as follows. We start by defining what constitutes a cross-layer cooperative design. Then, we describe a number of cross-layer solutions categorized based on their primary goal. Finally, we present a case study for Worldwide Interoperability for Microwave Access (WiMAX) networks that combines the operation of adaptive modulation and transmission power control at the Physical layer with a multi-encoding rate scheme at the Application layer.

5.2 CROSS-LAYER COOPERATIVE DESIGN: DEFINITION AND SIGNIFICANCE

According to Ref. [1], "a protocol design that violates a reference layered communication architecture is a cross-layer design with respect to the particular layered architecture." The definition of the cross-layer design principle can be extended to include not only the designs of protocols but also of algorithms and architectures that exploit or provide a set of interlayer interactions [2]. For example, consider how information related to the physical channel can be used from protocols in all layers of the protocol stack. The wireless channel is characterized by a time- and space-varying nature. The channel quality may be affected by environmental factors that cause phenomena such as path loss, shadowing, and fading and deteriorate the quality of the transmitted signals. Additionally, in cases such as ad hoc networks, the topology may be constantly changing due to the mobility of the wireless nodes. The wireless channel quality variations may be taking place on either a small scale of some milliseconds, as in the case of fading, or on a larger scale that depends on the nodes' mobility and the variations in their surrounding environment that influence the propagation of the radio signals. This information can be taken into account by higher layer protocols to adapt their operation accordingly and improve the overall performance of the system.

In a wireless multi-hop network the broadcast nature of the wireless channel can be exploited for the cooperation of multiple intermediate nodes. These nodes may have the ability to relay messages they receive from a sending node, without being their original receivers. Thus, they can assist the communication between sending and receiving nodes and improve the network throughput. The varying nature of the physical medium can also be exploited by the opportunistic (channel-aware) scheduling algorithms. These incorporate Physical layer channel information into the scheduling decision to achieve higher network performance and increase the system throughput [3]. Moreover, routing algorithms in wireless multi-hop networks could also be enhanced by channel state awareness.

In such cases, the routing algorithms could construct paths that include the wireless links experiencing the highest possible channel quality, while avoiding links suffering from deep fades. At the Transport layer, the TCP could defer packet retransmissions when the wireless medium is in bad shape and reschedule them when the channel quality improves. Additionally, the Application layer could take advantage of the fluctuations of the channel quality and adapt its media encoding rate accordingly to avoid overloading the system.

The aforementioned examples give a first insight into the benefits of cross-layer cooperative techniques. However, special attention should be paid to the fact that the uncontrolled use of cross-layer cooperative designs could possibly have a negative effect on long-term system performance [4]. The continuous violations of the principles of traditional layered architecture may result in the loss of the system's modularity, which is the factor that facilitates efficient protocol design and enables its quick and massive proliferation. Hence, the designers of a cross-layer interaction scheme should always consider the consequences that their design choices may have on the overall system stability. Additionally, they should aim at proposing solutions that do not seriously influence the long-term system performance, do not negatively affect the operation of other parts of the wireless network stack, and have the ability to efficiently interoperate with other proposed cross-layer schemes.

5.3 CROSS-LAYER COOPERATIVE DESIGN IN THE RECENT BIBLIOGRAPHY

In the recent bibliography we can find a great variety of proposals on cross-layer cooperative schemes designed for numerous kinds of wireless networks and applications. In this section, we provide a classification of these proposals, based on their operation and target objective. This classification includes schemes focusing on (a) Quality of Service (QoS) and throughput improvement, (b) Radio Resource Management (RRM), (c) mobility support, and (d) security. These general categories apply to both single-hop and multi-hop networks. We also discuss mechanisms designed solely for multi-hop networks and more specifically for (a) cooperative wireless communication environments, (b) energy efficiency in Wireless Sensor Networks (WSNs), and (c) routing.

5.3.1 QUALITY OF SERVICE AND THROUGHPUT IMPROVEMENT

A large number of recent proposals on cross-layer cooperative design aim at improving the QoS provision and the overall performance of the system they are applied to. Such proposals include rate control algorithms based on the experienced channel conditions that aim to normalize the unpleasant variations of the wireless medium quality. Other mechanisms introduce efficient schemes to avoid unnecessary retransmissions and improve the system throughput. Adaptive coding schemes that take into account both the QoS requirements of the applications and the conditions of the wireless medium are also suggested.

More specifically, the authors in Ref. [5] present a general cross-layer feedback architecture for mobile wireless environments. This design uses a "tuning layer" for every layer of the wireless protocol stack. Each tuning layer provides an interface to the data structures that determine the operation of its corresponding protocol layer. The tuning layers are used by "protocol optimizers" that contain the cross-layer optimization algorithms and comprise the "optimization subsystem" that operates concurrently with the wireless protocol stack. Although the authors suggest the introduction of this functionality in all layers, the overall scheme's purpose is the improvement of the TCP feedback mechanism's throughput. This is achieved by (a) making it aware of the channel quality and rescheduling the retransmissions until the channel conditions improve and (b) adjusting the connections' receiver window according to the applications' requirements.

Reference [6] presents two cross-layer algorithms, a dual-based and a penalty-based one, that address the problem of rate control in a multi-hop random access network. These algorithms are implemented in a distributed manner and work both at the Data-Link layer to adjust transmission attempt probabilities and at the Transport layer to adjust session rates. The first algorithm updates the transmission attempt probabilities at the Data-Link layer, based on information regarding the mobile node's neighborhood, whereas the Transport layer adjusts the session rates based on the link rates computed by the Data-Link layer. In the second algorithm, each session computes its rate based on its contribution to the overall traffic of its path whereas the Data-Link layer computes the transmission attempt probabilities taking into account its effect on the neighboring nodes. The main objective of both algorithms is the provision of proportional fairness among end-to-end sessions.

In Ref. [7], the authors study the problem of rate optimization for multicast communications at the MAC layer. The proposed design performs joint optimization of the MAC multicast threshold, that is, the minimum number of existing ready receivers before a sender decides to transmit packets and the transmission rate to maximize the overall throughput. Additionally, the Transport layer erasure coding is introduced which provides reliable transmissions and avoids retransmissions at the MAC layer. Its use achieves throughput improvement compared to a typical MAC layer mechanism that retransmits lost packets.

5.3.1.1 QoS Support of Multimedia Applications

A subgroup of this category deals with the support of demanding multimedia applications. The reason for the growing interest in this area is the fact that modern multimedia applications are characterized by strict QoS requirements especially regarding bandwidth, packet loss, and latency and require special handling by packet-based wireless communication networks that were traditionally designed to support best effort traffic. Thus, cross-layer designs in this area include (a) encoding and rate adaptation algorithms for the support of real-time voice and video applications over wireless packet networks, (b) adaptive packetization and scheduling algorithms that provide the best possible QoS to multimedia applications given specific wireless channel conditions, and (c) algorithms for the efficient support of wireless multicast streaming applications.

For example, Ref. [8] deploys an Application layer adaptive packetization, a prioritized scheduler, and a MAC layer retransmission scheme for the support of delay-sensitive multimedia over wireless data networks. The aim is to fulfill the rate and delay constraints of the applications and provide resilience against distortion. This is achieved by providing the optimal packetization and retransmission scheme that takes into account information from the Application, MAC, and Physical layers. This information includes content characteristics (e.g., signal variance), video encoder features (i.e., compression levels), applications' delay characteristics, and channel conditions. The described mechanism achieves performance improvement equal to 2 dB or more in terms of Peak Signal-to-Noise Ratio (PSNR) in many video sequences under a variety of transmission conditions.

In Ref. [9], information from the Application, MAC, and Physical layers is taken into account to propose a cross-layer architecture for adaptive video multicast streaming over multi-rate Wireless Local Area Networks (WLANs). An automatic rate selection mechanism, employed by the access point, is introduced and its mission is the selection of the most appropriate multicast data rate at the Physical and MAC layers, based on the channel conditions perceived by the mobile terminals. Hierarchical video encoders are used at the Application layer, with the ability to adapt the video transmission to the varying channel conditions. Thus, mobile terminals experiencing low channel quality receive only the base layer of the encoded video, while others experiencing favorable conditions receive video enhancement layers. The use of this mechanism achieves performance improvement in terms of video quality, delay, throughput, and signaling overhead.

In Ref. [10], an architecture for video streaming over IEEE 802.11 (Wi-Fi) networks is presented. The rate adaptation is performed in the Physical layer, the MAC layer, and the video encoder residing at the Application layer. More specifically, the transmission rate of the MAC layer and the video coding rate at the Application layer can be dynamically adjusted, using channel state and medium sharing information. An entity called channel state predictor performs an estimation of the channel state, while a medium sharing predictor performs link throughput estimation based on the throughput observed during previous transmissions. The transmission rate selected is the one that maximizes the user throughput given specific channel conditions. The use of the described architecture achieves performance improvement in terms of PSNR and packet loss rate.

Reference [11] introduces a cross-layer mechanism for real-time traffic over IEEE 802.16 (WiMAX) networks. This mechanism uses information from the Physical and MAC layers, namely the packet error, the packet timeout rates, and the mean delay of real-time applications. It coordinates the adaptive modulation capability at the Physical layer and multi-rate data-encoding capability of modern real-time applications. Its aim is to assist IEEE 802.16 systems to adapt to frequent channel and traffic changes and results in reduced packet loss rates. More details about this mechanism are provided in the case-study of Section 5.4.

Finally, Ref. [12] proposes a cross-layer architecture that performs joint optimization of the Application, Data-Link, and Physical layers for the efficient support of wireless video streaming. More specifically, the Physical and Data-Link layers adapt the transmission to the wireless quality variations, while the Application layer takes into consideration the video distortion information and adapts its encoding rate accordingly. More details about the general architectural framework proposed in this chapter will be provided in Section 5.4.

5.3.2 RADIO RESOURCE MANAGEMENT (RRM)

RRM is an issue of great importance in wireless communication networks. The air interface and hardware resources need to be efficiently used to achieve increased throughput and channel capacity and avoid unpleasant phenomena such as interference. Several proposals on cross-layer cooperative designs regarding RRM can be found in the recent bibliography. The majority of them introduce efficient schemes for power control, resource allocation, admission control, and packet scheduling.

In Ref. [13] the Data-Link layer performs resource allocation taking into account its impact on the Transport layer. The importance of such a design emerges from the fact that the resource allocation strategy directly influences the performance of TCP and User Datagram Protocol (UDP), as it affects the connections' delay and loss rates. Additionally, at the Data-Link layer, the fairness weights for the data traffic are calculated based on cross-layer information from higher layers when scheduling integrated voice/data traffic. The proposed design manages to achieve satisfactory QoS provision and effective MAC scheduling. However, it assumes perfect power control in uplink transmissions, which may be unfeasible in actual cellular networks.

Reference [14] tackles the problem of connection admission control in code division multiple access networks for multi service packet traffic. Packet traffic is modeled as a Markov modulated Poisson process and connection admission control is performed, taking into account the QoS requirements regarding the Physical layer Signal-to-Interference Ratio (SIR) and the Network layer blocking probability. Four cross-layer schemes are proposed, each one providing different levels of QoS guarantees at the Physical and Network layers, network utilization for packet traffic, and computational complexity. More specifically, the first algorithm is a linear programming-based scheme that maximizes network utilization subject to both SIR and blocking probability constraints. The second algorithm aims at guaranteeing the SIR outage probability. Due to the fact that these algorithms are computationally demanding, two more algorithms are proposed. These algorithms are sub optimal but computationally efficient, and they differ mainly in their capability to guarantee SIR outage probability.

Reference [15] proposes a joint MAC–Physical layer resource management algorithm that performs packet scheduling, subcarrier allocation, and power control for wireless Orthogonal Frequency Division Multiplexing (OFDM) networks. The above operations are performed taking into account the interdependencies between the MAC and Physical layers, for example, the impact of channel state information on the performance of packet scheduling and power allocation. Its aim is the maximization of the system power efficiency and overall performance, guaranteeing in parallel QoS provision and fairness. According to simulation results the described scheme achieves the improved QoS and fairness in a degree similar to the one expected by known wired fair queuing systems, while guaranteeing improved power and spectrum efficiency. However, its performance depends highly on the accuracy of the channel state estimations, and QoS provision is negatively affected in cases of imperfect channel state information.

In Ref. [16], a cross-layer rate, scheduling, and power control scheme for multi-hop wireless networks is considered. The joint design of these operations takes into account information such as the channel contention, physical interference, and link capacities. This scheme decomposes the overall problem to the three sub problems of congestion control, scheduling design, and power control. The main goal is to improve system performance as well as the transparent interaction of TCP congestion control algorithms, link scheduling, and transmission power allocation.

5.3.3 MOBILITY SUPPORT

The requirement of continuous support of mobile subscribers combined with the deployment of a variety of access technologies has created the need for efficient mobility management mechanisms that support handovers either within the same (intrasystem) or between different access technologies (intersystem). Numerous proposals on cross-layer cooperative designs regarding mobility management can be found in the recent bibliography. The main target of such designs is the support of seamless handovers, the reduction of the inevitable latency induced by the signaling exchange during a handover procedure, the elimination of the call dropping rate, and the guaranteeing of QoS maintenance during and after a successful handover. These proposals mainly include cross-layer management schemes between the Data-Link and the Network layers.

Reference [17] introduces a handover management protocol that aims to enhance the performance of Hierarchical Mobile IP (HMIP) handover procedure by taking into account information from the Data-Link and Network layers. This information takes into consideration the speed of the mobile terminals and the handover signaling delay. The proposed scheme consists of the "neighbor discovery" and "delay estimation" units that reside at the Network layer and the "speed estimation" and "received signal strength (RSS) measurement" units at the Data-Link layer. These modules discover neighboring base stations (BSs) and estimate handover delay and mobile terminal speed. Additionally, it contains the "handover trigger unit" that collects the necessary information by the previous modules and determines the most appropriate time to initiate a handover procedure and the "handover execution unit" that starts the HMIP registration process at the time indicated by the handover trigger unit.

Reference [18] describes a design that reconstructs the handover procedures of IEEE 802.16e and Mobile IPv6 (MIPv6) by the interleaving of the handover procedures of layers 2 and 3. This is performed through the integration of the correlated messages of the IEEE 802.16e and MIPv6 and the minimization of the flow control. The modified messages refer to the procedures of binding update, handover information, neighbor advertisement, and binding acknowledgment. Its aim is to reduce the number of control messages exchanged during a handover and eliminate the latency and connection dropping rate induced by the handover process.

Reference [19] introduces a cross-layer design that addresses the problem of Wireless Profiled TCP (WP-TCP) premature timeouts in cases of vertical handovers between WLANs and cellular networks. This problem may be caused by the rapid increase in Round-Trip Time (RTT) and false fast retransmit due to packet reordering during the vertical handover. Packet reordering may occur in cases of downward vertical handovers when new data packets transmitted through the WLAN arrive at the WP-TCP receiver before the packets that were previously transmitted by the cellular network. The proposed design uses two proactive schemes that prevent false fast retransmit by equalizing the round-trip delay experienced by all packets and suppress premature timeouts by carefully increasing the retransmission timeout time.

5.3.4 SECURITY

Security is an essential issue in modern wireless communication networks, especially when sensitive data are transferred. Many proposals for cross-layer cooperative designs have been presented aiming to avoid the various Denial-of-Service (DoS) attacks and tracing potential adversary nodes in a wireless network.

For example, Ref. [20] proposes an algorithm with self-adapting nature to avoid DoS attacks in WSNs. This algorithm takes into account performance parameters that could be affected by a DoS attack. Such performance parameters are the packet loss, successful packet delivery to destination, SNR, Bit Error Rate (BER), energy consumption, and distance. According to these parameters, reliable route selection is performed to avoid various kinds of DoS attacks, such as jamming, collision, exhaustion, misdirection, sybil, and worm-hole attacks, that may take place either in the Physical, Data-Link, or Network layer.

Reference [21] deploys a set of security schemes for WSNs, each of which aims at tackling specific security issues. For example, in the case of energy-efficient security provisioning, transmission power can be automatically tuned at the Physical layer according to the interference strength, which reduces energy consumption and avoids congestion attacks. At the MAC layer, the number of retransmissions can be limited to prevent exhaustion attack and save energy. Finally, at the Network layer, multi path routing can be adopted to avoid routing black-hole and reduce the energy consumption due to congestion.

Reference [22] introduces a design for the tracing of malicious relays that try to corrupt the communications by transmitting garbled signals in cooperative wireless communication environments. This scheme combines the operation of adaptive signal detection at the Physical layer of the receiving side and the insertion of pseudo random tracing markers in the transmitted symbol stream at the Application layer of the sender side. Its aim is to efficiently perform the tracing of the malicious relays and avoid the high error rates or the inability to distinguish between the adversarial relays and the legitimate ones. The described scheme manages to trace malicious nodes with a very low missing probability. However, it requires some additional bandwidth for the transmission of tracing markers, and it induces some extra computational complexity for the detection of garbled signals.

5.3.5 COOPERATIVE WIRELESS COMMUNICATIONS

In cooperative wireless communication environments, the wireless nodes take advantage of the broadcast nature of the wireless medium, to enable their cooperation by fully or partly forwarding information from source to destination. Such systems achieve higher spatial diversity, better frequency reuse, and higher throughput. In this area, cross-layer designs introduce efficient channel-aware relaying and diversity reception algorithms, throughput enhancement, as well as improved bandwidth and power allocation schemes.

Reference [23] uses a cross-layer cooperative design between the Physical and the MAC layers for IEEE 802.11 WLANs. In brief, if a source node has a packet to send, it will send it either directly to the destination or to an intermediate node (relay), based on the required transmission time. The transmission time is estimated based on the feasible data rates between the nodes and depends on the channel quality. In some cases, the receiver can hear the signals transmitted by both the source and the intermediate node and can combine them taking advantage of the cooperative diversity. This design exploits the broadcast nature of the wireless channel and the cooperative diversity to achieve performance improvement in terms of throughput, delay, coverage, and interference. However, to achieve this cooperation, it has to maintain up-to-date information regarding the availability of potential relays.

In Ref. [24], a cooperative multiple access protocol that jointly addresses the Physical and MAC layers is proposed. In this protocol, a relay listens to the wireless channel and uses the empty time slots in the time division multiple access frame to forward unsuccessfully transmitted packets. Thus, when the relay receives a negative acknowledgment indicating that the transmission of a packet was unsuccessful, it will store the packet in its queue waiting for a retransmission. Then, at the beginning of each time slot, it will sense the wireless channel and if it does not detect any transmission, it will forward the packet that resides at the head of its queue. The use of this protocol achieves higher system throughput combined with energy efficiency as it reduces the number of transmissions per successfully transmitted packet.

Another example is Ref. [25], which proposes a cross-layer design between the Physical and the Application layers. The proposed design aims at enabling the BS of a cooperative cellular network to perform optimal power and bandwidth allocation, as well as best relay node and relay selection strategy for each source–destination pair. This is achieved by taking into account both the user traffic demands and the Physical layer parameters (e.g., channel state information and power control parameters). The cross-layer problem is decomposed into two sub problems: a utility maximization problem in the Application layer and a joint relay-node selection, power, and bandwidth allocation problem in the Physical layer. The Physical layer problem may be solved using another set of dual variables that accounts for the cost of power expended at each node. These two dual decomposition steps solve the overall utility maximization problem with a complexity that is linear in the number of frequency tones and the number of relay strategies and quadratic in the number of relays.

5.3.6 ENERGY EFFICIENCY IN WIRELESS SENSOR NETWORKS

An issue of great importance in wireless communication networks and especially in wireless ad hoc and sensor networks is energy consumption. In WSNs energy efficiency is one of the most vital issues as the network lifetime depends mainly on the power consumption of the network nodes. Energy consumption is affected by many factors, including processing needs of the mobile nodes, the channel conditions (affecting the required coding and the number of retransmissions to guarantee reliable communication), and the routing protocol employed. Most of the cross-layer cooperative designs in this area attempt to achieve reduced power consumption by implementing efficient power control algorithms, energy-aware routing algorithms, and awake/sleep scheduling schemes.

Reference [26] proposes an approach that employs power control at the Physical layer, retransmission control at the MAC layer, and a new routing protocol at the Network layer. When multiple paths exist between the transmitting and the receiving nodes, the routing algorithm takes into account the energy required for a reliable transmission over each path and performs packet routing in such a way that energy consumption is balanced among multiple paths. At the Physical and MAC layers, the power and ARQ control algorithms take into account the channel conditions and calculate the per-hop success probability of each link. By combining these functionalities, this solution manages to increase the network lifetime and maintain a trade-off between network lifetime maximization and reliability constraint.

Reference [27] introduces a cross-layered architecture based on the notion of initiative determination according to which, each node will have the freedom to decide its participation in a communication. In this architecture, the entire protocol stack is replaced by a unique protocol that performs the operations of all the layers. Based on this concept, a wireless node will decide to participate in a communication if the channel conditions can guarantee the construction of reliable links, the relayed traffic does not create congestion, the wireless node does not experience any buffer overflow, and its remaining energy exceeds a specific minimum value. The aim of this proposal is to achieve reliable communication with minimal energy consumption, adaptive communication decisions, and local congestion avoidance. However, although the replacement of the entire traditional protocol stack by a single module may result in improved performance, such a design choice may result in inability of the system to extend its functionality or efficiently interoperate with other networks.

In Ref. [28], a cross-layer design combines the use of FEC coding and the determination of the awake/sleep periods for narrowband WSNs. This design takes into account the characteristics of the Physical and the MAC layers and aims at obtaining an energy-efficient operation. Due to the fact that the number of users contending for the wireless channel directly affects the multi user diversity at the Physical layer, the scheduling of the awake/sleep periods at the MAC layer should be jointly designed with the Physical layer. Thus, the level of coding required to alleviate unfavorable channel conditions must be set so that the messages can be transmitted and received successfully and energy efficiently during the awake periods. The final goal is the scheduling of awake periods that are as short as possible, so that the energy-saving sleep periods are relatively long.

5.3.7 New Routing Schemes

The deployment of efficient routing algorithms is a dominant factor in the performance of multi-hop wireless networks. The challenges of such algorithms include the mitigation of the unpleasant effects of the wireless transmission and the constraints on throughput capacity and energy consumption per node. In the recent bibliography we can find a variety of proposals on cross-layer cooperative designs that attempt to introduce novel routing schemes for wireless multi-hop communication networks.

In Ref. [29], the contention relations between links are modeled as a conflict graph that indicates the mutual interferences between groups of links. Then, joint routing, congestion control, and scheduling are performed with the source nodes adjusting their sending rate according to the local congestion and the scheduling and routing based on the conditions of the neighboring nodes.

Reference [30] proposes a cross-layer scheme that performs routing and power optimization in wireless mesh networks. This design is based on the parameters of the Physical retransmission rate, interference, and Packet Error Rate (PER) and aims at increasing the average throughput capacity and at improving the power control policy that reduces interference in a wireless mesh network. The transmission power of a node toward each neighbor is optimized by determining the best trade-off between interference and PER, preserving the maximum transmission rate available. When a packet is sent, the transmission power is dynamically changed to the optimal level calculated for the intended next hop. The aim is the maximization of the average throughput routed by the wireless mesh network and the reduction of the average transmitted power that directly affects the interference.

Reference [31] proposes a QoS-aware routing protocol that incorporates an admission control scheme and a scheme that provides feedback for the available bandwidth to the applications. Bandwidth estimation can be achieved by having the mobile nodes to either (a) listen to the wireless channel and estimate the available bandwidth (based on the ratio between free and busy times), or (b) disseminate information regarding their currently used bandwidth in the "Hello" messages of the routing protocol. The routing protocol supports two kinds of applications; one that indicates the minimal bandwidth that must be guaranteed and another that can adjust its coding rate according to the feedback received by the network. The aim of this protocol is to support real-time audio and video applications by the provision of guaranteed services, such as bandwidth, delay, jitter, and packet delivery rate.

5.3.8 Discussion

In the previous sections we have classified a variety of proposals on cross-layer cooperative schemes designed for the performance improvement of several kinds of wireless networks and applications.

In the first category proposals have as a common goal the improvement of throughput and QoS provision in a wireless communication network. This can be achieved mainly by introducing QoS and channel-state awareness to the operations of transmission rate control, retransmission policy adaptation, and Application layer encoding rate adaptation. The schemes that perform cross-layer RRM aim at an efficient management of the radio resources by taking into account the capabilities of the Physical layer as well as the overall network conditions and the applications' requirements. In the area of mobility support, proposals aim at enabling seamless handovers with reduced latency and guaranteed QoS provision. To achieve this goal the handover procedure performed in layers 2 and 3 can be combined and even enhanced with information from other layers. As far as security is concerned, many cross-layer schemes detect malicious nodes in multi-hop networks and eliminate them from path selection procedures. Moreover, they can adapt transmission power and retransmission attempts to combat a variety of DoS attacks.

In multi-hop networks a variety of proposals on cross-layer cooperative designs can be found. In the area of cooperative wireless communications the main goal is to increase the spatial diversity of the network and achieve higher resource utilization. This can be achieved by enabling intermediate nodes in a multi-hop network to cooperate by forwarding information from source to destination taking into consideration channel-state information. An important issue in the performance of WSNs is energy efficiency. The proposals in this area aim at performing energy-aware operations to reduce the power consumption and prolong the network lifetime. This can be achieved by the implementation of energy-aware power control and routing algorithms as well as channel-aware ARQ and FEC schemes. Finally, new cross-layer routing schemes have been proposed that aim at performing efficient path selection in multi-hop networks, by taking into account the unfavorable conditions of the wireless medium, the applications' QoS requirements, and energy and capacity constraints.

The following table includes a summary of the above classification, describing in brief the main objective of each category along with the means it uses to achieve its goals.

Category	References	Main Objective	Means to Achieve Goal
QoS and throughput improvement	[5–12]	To improve the QoS provision to applications and increase the network throughput	The Application layer collects information from the Transport (e.g., time-outs, RTT), MAC (scheduling info), and Physical layers (e.g., SNR) to select the most appropriate video–audio encoding rate. Transmission rate and retransmission policy are performed taking into consideration Physical layer information
RRM	[13–16]	To perform efficient resource management, taking into consideration not only the capabilities of the Physical layer but also the requirements of the applications and the network conditions regarding bandwidth availability	QoS-aware power and resource allocation, connection admission control, and packet scheduling
Mobility support	[17–19]	To efficiently support seamless handovers by reducing the handover latency and guaranteeing QoS provision	The handover decision is enhanced with information from the Data-Link and Network layers. The handover procedures of layers 2 and 3 are incorporated into a unified handover scheme
Security	[20–22]	To protect sensitive data transmission by avoiding DoS attacks across all layers of the protocol stack	Avoid routing traffic through potential malicious nodes in a WSN, tune transmission power to avoid congestion attacks, and eliminate retransmissions to prevent exhaustion
Cooperative wireless communications	[23–25]	To enable efficient channel-aware relaying and achieve spatial diversity, improved frequency reuse, and higher throughput	Intermediate nodes in a multi-hop network forward information from source to destination taking into consideration channel-state information
Energy efficiency in WSNs	[26–28]	To reduce energy consumption by the network nodes and prolong the network lifetime	Implementation of efficient power control schemes, energy aware routing algorithms, and channel-aware ARQ and FEC schemes
New routing algorithms	[29–31]	To perform efficient routing in multi-hop networks, alleviating the unfavorable conditions of the wireless medium and taking into account energy and capacity constraints	Routing algorithms take into account information regarding the channel conditions, energy consumption, interference, congestion, and QoS requirements

In the following section we use the standard IEEE 802.16(e) [32,33] as a case study to describe a specific cross-layer cooperation mechanism for real-time traffic support that interacts with and coordinates the Physical, MAC, and Application layers.

5.4 CASE STUDY: CROSS-LAYERING IN WiMAX NETWORKS

Two very promising technologies in the field of wireless broadband communication systems are the IEEE 802.16 [32] and 802.16e [33] standards that are used as a basis for the popular WiMAX networks. To enhance the system's performance, the IEEE 802.16e standard has adopted the Orthogonal Frequency Division Multiple Access (OFDMA) scheme that provides multiplexing in both uplink and downlink directions by subdividing the available bandwidth into multiple frequency subcarriers. The user data streams are also divided into several parallel substreams, of increased symbol duration, each of which is modulated and transmitted on a separate orthogonal subcarrier. Thus, the OFDMA scheme manages to offer multi user diversity with the use of efficient subcarrier, bit, and power allocation algorithms. Additionally, it provides increased robustness against frequency-selective fading and Inter Symbol Interference (ISI).

As a case study we present an extension of the heuristic cross-layer mechanism's functionality proposed in Ref. [11] to support and cooperate with systems employing the OFDMA access scheme. More specifically, we describe an extension of the heuristic decision algorithm's operation introduced in Ref. [11] that can also adapt its operation to cooperate with an OFDMA scheduler and control the signal power to improve QoS. Its aim is the improvement of the overall system's performance by introducing an efficient adaptive modulation, power control, and multi-encoding rate scheme of real-time applications in IEEE 802.16e networks.

5.4.1 ARCHITECTURE

The cross-layer cooperation mechanism for real-time traffic over IEEE 802.16e networks described in this chapter is based on the architectural framework introduced in Ref. [12], which consists of N layers and a cross-layer optimizer (Figure 5.1).

According to this framework, the optimization process is performed in three steps:

1. Layer abstraction: Computes an abstraction of layer-specific parameters that are processed by the optimizer. This process aims at reducing the overall data processing and communication overhead while maintaining consistency.

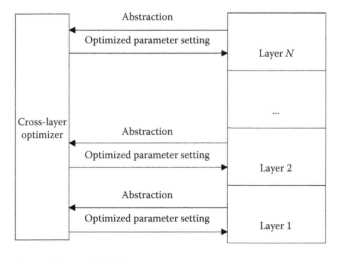

FIGURE 5.1 General cross-layer architecture.

2. Optimization: Determines the values of layer parameters that optimize a specific objective function.
3. Layer reconfiguration: Distributes the optimal values of the abstracted parameters to the corresponding layers that in turn translate them back into layer-specific information.

The rate at which the above steps are repeated depends on the variance of the channel conditions and the applications' requirements.

In general, the different parameters can be divided into four main types:

1. Directly tunable (DT) parameters that can be directly set as a result of the optimization process
2. Indirectly tunable (IT) parameters that may be modified as a result of the setting of the DT parameters
3. Descriptive (D) parameters that can only be read but not tuned by the optimizer
4. Abstracted (A) parameters that are abstractions of the previous types of parameters

More details about this framework can be found in Ref. [12]. The mechanism described in this section works in three layers, namely the Physical, Data-Link (MAC), and Application layers.

The cross-layer cooperation mechanism for IEEE 802.16e networks is split into two parts, namely the BS part and the mobile station (MS) part, residing at the BS and the MSs of an IEEE 802.16e network, respectively (Figure 5.2).

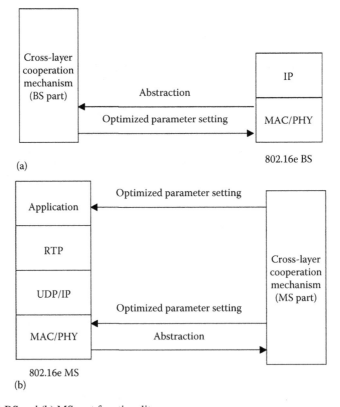

FIGURE 5.2   (a) BS and (b) MS part functionality.

The BS part accepts an abstraction of layer-specific information, namely, the channel conditions, QoS parameters, and transmission power capabilities of all active connections, provided by the BS PHY and MAC layers (Step 1: Layer abstraction). According to this information, a decision algorithm determines the desired modulation schemes, encoding modes, and transmission power levels of each MS, separately for each direction (uplink or downlink) (Step 2: Optimization). Finally, the BS part configures the corresponding layers with the determined parameters (Step 3: Layer reconfiguration) (Figure 5.2a). If the decision part of the BS requires media encoding changes at the Application layer or transmission power changes at the Physical layer of the MS, it communicates with its counterpart entity at the MS through the MS MAC layer. The MS part may either accept the BS part's suggestions regarding encoding mode adjustments or refine them, based on its better knowledge of the status of the MS's active connections and instruct the MS Application layer accordingly (Figure 5.2b). More details on the specific operation of the BS and MS parts are provided in the next section.

5.4.2 THE PROPOSED MECHANISM

The proposed cross-layer cooperation mechanism for IEEE 802.16e networks is split into two parts, namely the BS part and the MS part, residing at the BS and the MSs, respectively [11]. An outline of the proposed mechanism operation for both uplink and downlink directions is shown in Figure 5.3.

The BS part collects information regarding each connection's performance status (i.e., packet timeout rate, mean delay), channel-state conditions, and transmission power on both directions (uplink and downlink). Based on this information, a decision algorithm residing inside the BS part determines the encoding mode, modulation order, and transmission power level of each MS. Channel conditions on the uplink are known from the BS PHY layer, while on the downlink they can be obtained through IEEE 802.16(e) signaling [32,33] (arrow (1) in Figure 5.3a). Packet timeout rate and mean delay for all active connections in both directions can be provided by the BS MAC layer (arrow (2) in Figure 5.3a and b). Transmission power levels on the downlink can be obtained by the BS PHY layer, while on the uplink they can be derived by signaling messages defined in the IEEE 802.16(e) standards [32,33] (arrow (1) in Figure 5.3b) and described below.

The cross-layer cooperation mechanism decisions regarding encoding mode adjustments are transferred to the MS part through the BS MAC layer (arrow (3) in Figure 5.3a and b) using a specially defined MAC management message referred to as Rate Adjustment Request (RATE-ADJ-REQ), described below (arrow (4) in Figure 5.3a and b). The MS MAC layer transfers the received rate modification request to the MS part of the cross-layer cooperation mechanism that is responsible for the communication with the Application layer (arrow (5) in Figure 5.3a and b). The MS part decides on the connections that should be affected and sends proper cross-layer messages to the Application layer (arrow (6) in Figure 5.3a and arrow (6a) in Figure 5.3b). On the uplink direction, the Application layer can perform the proper adjustments (e.g., tune the data encoder to produce the required overall media encoding rate). In the downlink direction, the MS has to notify the distant traffic sources for the necessary encoding adjustments, using an Application layer Real-Time Control Protocol (RTCP) feedback message, to suggest a new encoding mode to the senders (arrow (7) in Figure 5.3a).

The BS MAC layer is also responsible for transferring the power control messages containing the BS part instructions regarding the transmission power levels on the uplink. These instructions are transferred to the MS MAC layer (arrow (4) in Figure 5.3b) through an IEEE 802.16e power control message (e.g., Power Control Mode Change Response – PMC_RSP) [33]. The MS MAC layer will in turn inform the MS part accordingly (arrow (5) in Figure 5.3b). Again, the MS part is responsible for distributing the available transmission power to its connections (arrow (6b) in Figure 5.3b).

All the management messages exchanged between the network entities for the operation of the cross-layer cooperation mechanism are either available in the IEEE 802.16e and RTCP standards or can be easily defined, as described below:

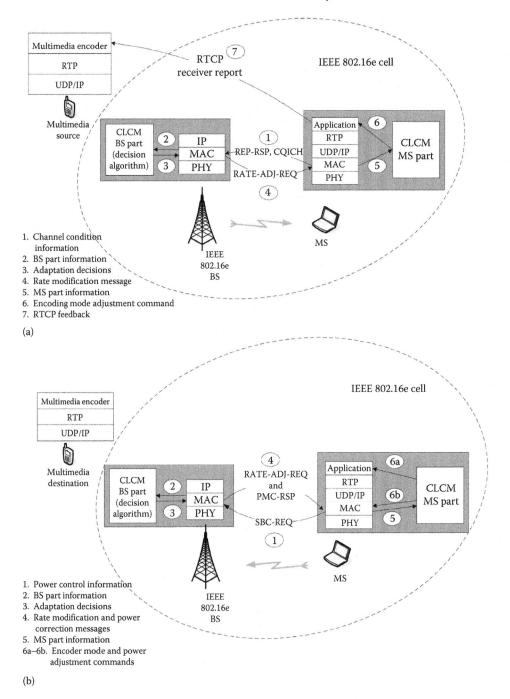

FIGURE 5.3 Cross-layer cooperation mechanism for (a) downlink and (b) uplink directions.

Channel quality information: To increase the robustness of downlink transmissions (from BS to MSs), the BS needs information regarding the quality of the signal received by the MSs. For doing so, each MS sends channel quality measurements to the BS using standard IEEE 802.16(e) signaling either periodically, through the Channel Quality Information Channel (CQICH), or on demand, through the Channel Measurement Report Request and Response (REP-REQ, REP-RSP) messages [32,33] (arrow (1) in Figure 5.3a).

Rate modification message: The BS part decisions regarding recommended media encoding rate adjustments are transferred to the MS part using the Rate Adjustment Request message (RATE-ADJ-REQ) defined in Ref. [11] (arrow (4) in Figure 5.3a and b). This message contains the BS part recommendation in the form of a target overall mean media encoding rate either for the downlink or the uplink. Its syntax is shown in the following table.

RATE-ADJ-REQ message

Syntax	Size
RATE-ADJ-REQ_Message_Format(){	
Management message type = 67	8 bits
Direction	1 bit
Total rate recommended	32 bits
}	

Message type 67 is determined as "reserved" in both Refs. [32,33], meaning that it is left unused for future purposes. The "Direction" parameter declares uplink or downlink direction. The "Total Rate Recommended" parameter contains the recommendation of the BS part in kb/s. Considering that this refers to the total transmission rate for an MS, the size of 32 bits allows rates up to approximately 4.3 Gb/s per direction.

Encoding mode adjustments: For the downlink direction, the MS application has to inform the traffic source for media encoding rate adjustments. This can be performed through a standard Application layer RTCP Receiver Report (RR) message. Application layer messages are part of the extensions of Real-time Transport Protocol (RTP) for RTCP-based feedback, as defined in Ref. [34], and can be used to transport application-defined data directly from the receiver to the sender application. Such a message can be sent either immediately or at the next regular time instance, depending on the RTCP mode and the bandwidth available for RTCP messages [34].

Power control information exchange: Each MS has to provide the BS with the necessary information regarding its maximum available and its current transmission power. This communication can be achieved through messages such as the Subscriber Station Basic Capability Request message (SBC-REQ) and the Channel Measurement Report Response (REP-RSP) message (arrow (1) in Figure 5.3b). Based on the results of the decision algorithm, the BS part has to notify the MSs of the required transmission power modifications (arrow (4) in Figure 5.3b). These notifications are performed using any of the power correction messages described in Refs. [32,33]. Power correction is performed using messages such as the Subscriber Station Basic Capability Request (SBC-REQ), the Power Control Mode Change Response (PMC_RSP), the Ranging Response (RNG-RSP), and the Fast Power Control (FPC) messages, the Power Control IE format and the UL-MAP_Fast_Tracking_IE field of the Uplink Access Definition (UL-MAP) field of the MAC layer time frame.

5.4.3 DECISION ALGORITHM

At the beginning, the required notation has to be defined.

Transmissions between one BS and K connections using N OFDMA subcarriers are considered. An OFDMA frame consists of L time slots and contains packets from K^* connections, $K^* \leq K$. Assuming a connection $k, k \in \{1, 2, ..., K\}$ the most important QoS parameters used are the following:

1. $R_{err}(k)$ is the PER of connection k, that is, the percentage of packets that are lost due to channel errors.
2. $R_{timeout}(k)$ is the packet timeout rate of connection k, that is, the percentage of packets that are lost due to expiration.
3. $R_{loss}(k) = R_{err}(k) + R_{timeout}(k)$ is the total packet loss rate experienced by connection k.
4. d_k is the mean delay experienced by connection k.

5. $d_{\max_k}$ is the maximum acceptable delay of connection k.

6. $\delta_k = \frac{R_{\text{err}}(k)}{R_{\text{loss}}(k)}$ is the percentage of packet errors with respect to the total loss rate of connection k. It is the main QoS parameter used in the decision algorithm as it represents the packet errors contribution to the overall connection packet loss rate and allows the determination of the nature of the packet losses experienced by a connection.

7. $\delta_{\text{low}}, \delta_{\text{med}}$, and δ_{high} are the thresholds based on which the algorithm decides on the appropriate adaptation actions. If $\delta_k > \delta_{\text{high}}$, the connection is considered to be suffering almost exclusively by packet errors that are the result of significantly unfavorable channel conditions. Otherwise, if $\delta_{\text{med}} < \delta_k < \delta_{\text{high}}$ the connection is still considered to be suffering mainly by hostile transmission conditions, but a considerable part of its total packet losses is the result of packet timeouts. If $\delta_{\text{low}} < \delta_k < \delta_{\text{med}}$, most of the connection packet losses are the result of packet timeouts but a significant percentage of them is the result of packet errors caused by poor channel quality. Finally, if $\delta_k < \delta_{\text{low}}$ the connection is considered to be suffering almost exclusively by unacceptable delays that lead to packet timeouts.

8. β_k is a threshold indicating whether the delay of connection k is close to its maximum acceptable bound $d_{\max_k}$ (i.e., if $\frac{d_k}{d_{\max_k}} > \beta_k$).

An outline of the decision algorithm executed at the BS part for a connection k is as follows (Figure 5.4).

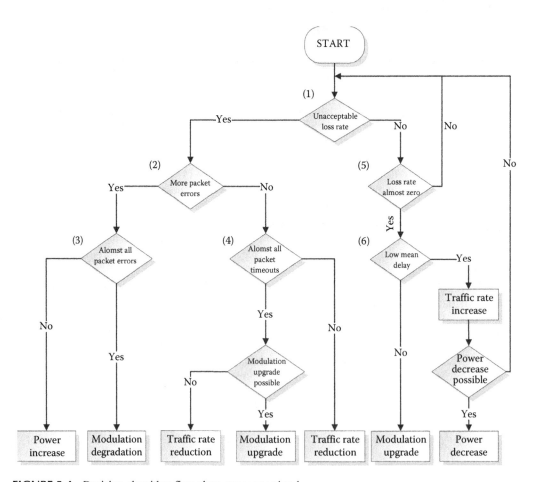

FIGURE 5.4 Decision algorithm flow chart per connection k.

The algorithm is initiated every time a connection faces unacceptable packet loss rates that lead to a degradation of the provided QoS (1). The actions to be taken depend on the nature of these losses:

1. In case $\delta_k > \delta_{med}$ (2), most of the losses are due to packet errors caused by bad channel conditions. The decision depends on the packet errors contribution to the overall packet losses experienced by the specific connection:

   a. If $\delta_k > \delta_{high}$ (3), that is, the connection loss rate is almost exclusively caused by packet errors, the algorithm concludes that the transmission conditions are significantly unfavorable, resulting in high BERs. In this case, the most appropriate action is a decrease in the modulation order that will allow the connection to achieve higher channel error resilience and increase robustness against interference. Thus, the BS part selects the highest modulation order that will restore the loss rate to acceptable values, according to a BER versus Carrier to Interference-plus-Noise Ratio (CINR) curve, such as the one included in Ref. [35], and instructs the MAC layer accordingly. The MAC layer sends the required primitives to the PHY for the modulation order change and informs the MS through the Downlink Access Definition (DL-MAP), for downlink changes and UL-MAP, for uplink changes, fields of the next MAC time frame.

   b. Otherwise ($\delta_{med} < \delta_k < \delta_{high}$), a significant percentage of packet losses are caused by packet timeouts. The algorithm assumes that the connection is under congestion conditions. In this case, a decrease in the modulation order that results in significant reduction of the transmission rate would most probably cause an immediate increase in the connection packet timeouts. Therefore, the BS part will instruct a transmission power increase, which will improve the connection CINR without reducing the transmission rate.

2. In case $\delta_k < \delta_{med}$ (2), most of the losses are the result of unacceptable delays that cause packet timeouts. The action to be performed depends on the contribution of these timeouts to the overall packet losses:

   a. If $\delta_k < \delta_{low}$ (4), the overall loss rate is caused almost exclusively by packet timeouts. The algorithm can safely conclude that the transmission rate is very low and unable to satisfy the connection transmission speed requirements. In this case, the BS part instructs an increase in the modulation order, which will increase the transmission rate and reduce the losses caused by timeouts. Again, the error rate induced by the candidate modulation order can be predicted using a BER versus CINR curve.

   b. In the case where $\delta_{low} < \delta_k < \delta_{med}$, a significant percentage of packet losses are caused by errors due to the poor channel conditions, that do not allow a modulation order increase. The BS part instructs the MS part for a media encoding rate reduction to moderate timeouts.

To achieve an efficient performance under all possible conditions, the algorithm has to take adaptation decisions also when the conditions for a specific connection are improved. Thus, when the loss rate decreases significantly (5), the algorithm may decide to switch to a higher modulation order that will increase the transmission rate, reduce the connection power to achieve power efficiency, or increase the media encoding rate and improve the QoS. The specific action depends on the mean delay:

1. If $\frac{\delta_k}{\delta_{max_k}} > \beta_k$ (6), the mean delay is close to the connection delay bound. The algorithm instructs for a modulation order upgrade, which will immediately increase the transmission rate and reduce the mean delay.

2. Otherwise, if the mean delay is relatively low compared to the delay bound $\frac{\delta_k}{\delta_{max_k}} < \beta_k$ (6), the algorithm instructs for a media encoding rate increase to improve the QoS provided to the user. However, this increase should be performed carefully, as it may lead to increased traffic and possible delay-bound violations. In any case, the percentage of this increase should be relevant to the difference between the current value of the mean delay and the delay bound. Additionally, if the channel conditions that the connection experiences are such that they result in a significantly low BER, the algorithm may also instruct for a reduction of the transmitting or receiving power of the specific connection to reduce power consumption, achieve power efficiency, and reduce possible interference to neighboring nodes.

It should be noted that the performance of the proposed cross-layer mechanism is strongly affected by the specific values of the thresholds $\delta_{low}, \delta_{med}, \delta_{high}$, and β_k, which activate modulation order, transmission power, and media encoding rate adaptations and determine the mechanism's "sensitivity" regarding adaptation decisions.

The use of the cross-layer mechanism induces some overhead in the RTCP Receiver Report messages exchanged between the multimedia sources and the MSs to perform encoding rate control. However the amount of this overhead is limited and, as can be seen in the following section, it results in significant improvement of the system performance.

5.4.4 SIMULATION MODEL AND RESULTS

The proposed mechanism performance was evaluated using a simulation model constructed in C++. The aim was to compare its performance against the cross-layer mechanism introduced in Ref. [11] and a typical IEEE 802.16e system that performs adaptive modulation at the PHY layer and media encoding adjustments at the Application layer separately and independent of each other. The components comprising the simulator are the following (Figure 5.5):

Traffic generator: It generates one video variable-length traffic frame every 40 ms for each connection, starting at a random instance within the first 40 ms of a simulation run. Each connection maximum encoding rate depends on the feedback received from the cross-layer mechanism.

Traffic scheduler: It performs the resource allocation to the connections. It receives input traffic from the Traffic Generator, information regarding the connection modulation order changes by the Cross-Layer Mechanism, and channel conditions information from the Channel Model. In our simulation

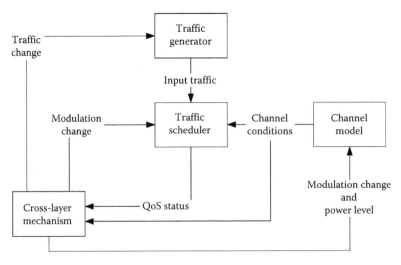

FIGURE 5.5 Simulation model outline.

we use the "Frame Registry Tree Scheduler" (FRTS) described in Ref. [36] and an OFDMA allocator Ref. [37] which aims at allocating to each MS an amount of resources that satisfy its QoS requirements while guaranteeing favorable channel conditions.

Channel model: It simulates the system channel conditions by providing short-term fading. Bit errors are randomly produced for each connection, according to its modulation scheme information provided by the Traffic Scheduler, with a mean rate according to the BER versus CINR curve included in Ref. [35]. Path loss is assumed to be $P_r(dBm) = P_0 + L + 10n \log(d)$ where P_r is the received power, P_0 is the transmitted power, L is the link budget [38], d is the distance between BS and MS, and n is the path loss exponent that, assuming a dense urban environment, equals 4 [39]. Short-term channel fading envelopes are considered to be Rayleigh distributed assuming the well-known Jakes spectra model with a maximum Doppler frequency shift of 128.2 Hz. Carrier frequency is set to 2.3 GHz and the number of subcarriers N is set to 64. The MSs are randomly distributed in the area of an IEEE 802.16e cell of 1 km radius and move within the cell range with a velocity of 60 km/hour.

Cross-Layer mechanism: It executes the proposed cooperation mechanism's decision algorithm described in the previous section and decides on each connection's modulation order, transmission power level, and media encoding rate based on the QoS information received by the Traffic Scheduler and channel condition information received by the Channel Model. The Cross-Layer Mechanism decisions are transferred to the Traffic Scheduler, the Channel Model, and the Traffic Generator, which perform the appropriate adjustments.

For the system employing the cross-layer mechanism introduced in Ref. [11], the Cross-Layer Mechanism does not perform power control and decides only on the modulation order and media encoding rate of each connection.

For the legacy system, the operations performed by the Cross-Layer Mechanism are split into two parts that operate independently. The *Physical Layer Adapter* is the entity that instructs the Traffic Scheduler for modulation order adjustments based solely on the channel conditions information provided by the Channel Model. On the other hand, the *Application Layer Adapter* instructs the Traffic Generator for traffic changes based on the connections' total packet loss rates. The BS receives channel conditions information by each MS periodically. The Traffic Generator receives RTCP feedback by each MS every 5 s according to Ref. [40].

The simulation scenario considers an increasing number of MSs, each one with one downlink video connection. The systems performance is measured in terms of throughput, packet loss rate, mean delay, and total transmitted power. To focus on the improvement process, the BS is considered as the traffic source for all connections and minimum delay is assumed for the transmission of signaling messages at the radio interface. For comparison purposes (see Ref. [11]), the modulation schemes used are QPSK, 16-QAM and 64-QAM, and the code rate equals 1/2. The system's maximum transmission power was set to 20W. The time frame length is set to 1 ms, and the maximum transmission rate is 120 Mb/s (for a modulation order equal to 64-QAM). To achieve lower processing complexity in the simulation model, a percentage equal to 7% of this data rate is reserved for the above connections, while the rest is assumed dedicated to other kinds of traffic. The values of δ_{low}, δ_{med}, δ_{high}, and β_k were set to 0.05, 0.5, 0.99, and 0.95, as they resulted in improved performance in multiple executions of various simulation scenarios.

In all simulations, the initial states and channel conditions are the same for all systems.

Figure 5.6 illustrates the total packet loss rates of the legacy system, the system employing the heuristic cross-layer mechanism described in Ref. [11], and the system employing the proposed cross-layer mechanism. Packet losses are the sum of packet errors and packet timeouts. All systems managed to efficiently adapt to the wireless channel quality fluctuations by instructing proper modulation order adjustments and maintained the PER in values around 10^{-3}. Thus, the dominant part of the packet loss rate is the packet timeouts. As can be seen, the legacy system is characterized by inability to combat unacceptable packet losses in cases of congestion, while the proposed cross-layer mechanism achieves considerably improved packet loss rates as the number of MSs

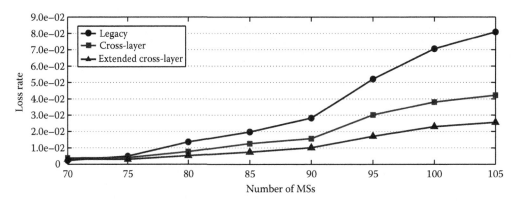

FIGURE 5.6 Packet loss rate with and without the use of the proposed mechanism.

increases. This is a result of the fact that in the legacy system adaptive modulation is performed based solely on the channel state information received by the MSs and is totally unaware of the loss rates experienced by the MSs. Additionally, RTCP feedback messages are exchanged between the MSs and the real-time sources every 5 s, and thus are unable to quickly detect unacceptable packet losses and enable the MSs to efficiently react by reducing the media encoding rate on time. On the other hand, the coordinated adaptation of modulation, media encoding rate, and transmission power of each connection employed by the cross-layer mechanism provides the system with the ability to accommodate to frequent channel quality fluctuations and heavy traffic situations. Figure 5.6 also depicts the performance improvement of the described cross-layer mechanism against the heuristic cross-layer mechanism described in Ref. [11]. This performance improvement is a result of the power control functionality introduced by the cross-layer mechanism, which avoids frequent modulation order decreases in heavy congestion situations and prefers to combat packet loss rates by increasing the transmission power.

Figure 5.7 depicts the created and served MAC layer rates per MS for the three systems under consideration. All the systems have similar created rates that follow a declining course due to the fact that the operation of the RTCP becomes more intensive as the number of MSs increases. However,

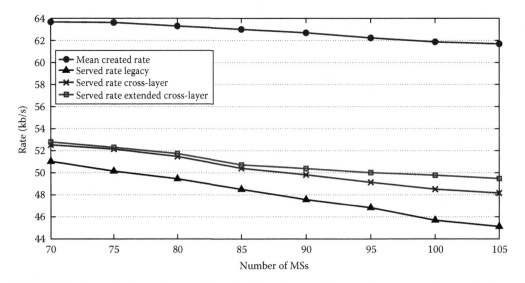

FIGURE 5.7 Created and served rates with and without the use of the cross-layer cooperation mechanism.

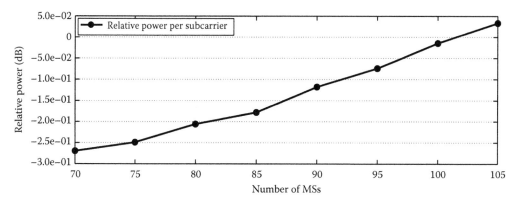

FIGURE 5.8 Relative power per subcarrier.

the legacy system does not have the ability to efficiently handle these encoding rates while the system employing the cross-layer mechanism manages to maintain higher served rates. Note once again that the cross-layer mechanism has slightly improved served rates compared to the heuristic cross-layer mechanism of Ref. [11]. Mean delay is only slightly reduced with the use of the proposed mechanism, because this is mainly affected by the deadline-based scheduler (FRTS) [36] operation, which schedules traffic as close to its deadline as possible in any case.

Figure 5.8 depicts the relative transmission power per subcarrier in the system employing the proposed cross-layer mechanism. Although the use of power adaptations is more intensive, as the number of the MSs increases and congestion becomes more intense, the mean power consumption per subcarrier is lower than in the case of the legacy system, with the only exception of the heavy congestion situation when the number of MSs is 105. This is because the power decrease performed when the channel quality is favorable is more than the power increase during heavy congestion situations.

5.5 CONCLUSIONS

In this chapter, we discussed a variety of recently proposed cross-layer cooperative designs and organized them into seven different categories based on their main objectives. The discussion of each category showed that cross-layer designs can be an efficient approach to tackle issues that emerge in cooperative wireless networks. An increasing number of proposals demonstrate the feasibility of these techniques for a plethora of problems.

In the second part, we focused on a specific cross-layer solution that falls into the QoS and throughput improvement category and provides improved performance in WiMAX networks. This solution combines the operation of adaptive modulation and transmission power control at the Physical layer and multi-encoding rate scheme at the Application layer. Extensive performance evaluation results showed that the described mechanism results in considerably improved performance compared with a legacy system that performs adaptive modulation at the Physical layer and RTCP-based encoding rate adaptation at the Application layer separately and independently of each other.

REFERENCES

[1] Srivastava, V. and Motani, M., Cross-layer design: A survey and the road ahead, *IEEE Communications Magazine*, 43(12), 112–119, Dec. 2005.

[2] Jurdak, R., *Wireless Ad Hoc and Sensor Networks: A Cross-Layer Design Perspective*, Springer, New York, 2007.

[3] Lin, X., Shroff, N.B., and Srikant, R., A tutorial on cross-layer optimization in wireless networks, *IEEE Journal on Selected Areas in Communications*, 24(8), 1452–1463, Aug. 2006.

[4] Kawadia, V. and Kumar, P.R., A cautionary perspective on cross-layer design, *IEEE Wireless Communications*, 12(1), 3–11, Feb. 2005 [see also *IEEE Personal Communications*].

[5] Raisinghani, V.T. and Iyer, S., Cross-layer feedback architecture for mobile device protocol stacks, *IEEE Communications Magazine*, 44(1), 85–92, Jan. 2006.

[6] Wang, X. and Kar, K., Cross-layer rate optimization for proportional fairness in multihop wireless networks with random access, *IEEE Journal on Selected Areas in Communications*, 24(8), 1548–1559, Aug. 2006.

[7] Ge, W., Zhang, J., and Shen, S., A cross-layer design approach to multicast in wireless networks, *IEEE Transactions on Wireless Communications*, 6(3), 1063–1071, Mar. 2007.

[8] van der Schaar, M. and Turaga, D.S., Cross-layer packetization and retransmission strategies for delay-sensitive wireless multimedia transmission, *IEEE Transactions on Multimedia*, 9(1), 185–197, Jan. 2007.

[9] Villalon, J., Cuenca, P., Orozco-Barbosa, L., Yongho Seok, and Turletti, T., Cross-layer architecture for adaptive video multicast streaming over multirate wireless LANs, *IEEE Journal on Selected Areas in Communications*, 25(4), 699–711, May 2007.

[10] Haratcherev, L., Taal, J., Langendoen, K., Lagendijk, R., and Sips, H., Optimized video streaming over 802.11 by cross-layer signaling, *IEEE Communications Magazine*, 44(1), 115–121, Jan. 2006.

[11] Triantafyllopoulou, D., Passas, N., Salkintzis, A., and Kaloxylos, A., A heuristic cross-layer mechanism for real-time traffic over IEEE 802.16 networks, *Wiley's International Journal of Network Management*, 17(5), 347–361, 2007 (special issue on Management solutions for QoS support over the entire audio-visual service distribution chain).

[12] Khan, S., Peng, Y., Steinbach, E., Sgroi, M., and Kellerer, W., Application-driven cross-layer optimization for video streaming over wireless networks, *IEEE Communications Magazine*, 44(1), 122–130, Jan. 2006.

[13] Jiang, H. and Zhuang, W., Cross-layer resource allocation for integrated Voice/Data traffic in wireless cellular networks, *IEEE Transactions on Wireless Communications*, 5(2), 457–468, Feb. 2006.

[14] Fei Yu., Krishnamurthy, V., and Leung, V.C.M., Cross-layer optimal connection admission control for variable bit rate multimedia traffic in packet wireless CDMA networks, *IEEE Transactions on Signal Processing*, 54(2), 542–555, Feb. 2006 [see also *IEEE Transactions on Acoustics, Speech, and Signal Processing*].

[15] Ying Jun Zhang and Letaief, K.B., Cross-layer adaptive resource management for wireless packet networks with OFDM signaling, *IEEE Transactions on Wireless Communications*. 5(11), 3244–3254, Nov. 2006.

[16] Long, C., Guan, X., and Li, B., Cross-layer congestion control, scheduling and power control design in multihop networks with random access, *IEEE International Conference on Multimedia and Expo*. pp. 1125–1128, Toronto, Ontario, Canada, July 2006.

[17] Mohanty, S. and Akyildiz, I.F., A cross-layer (Layer 2 + 3) handoff management protocol for next-generation wireless systems, *IEEE Transactions on Mobile Computing*, 5(10), 1347–1360, Oct. 2006.

[18] Chen, Y.W. and Hsieh, F.Y., A cross layer design for handoff in 802.16e network with IPv6 mobility, *IEEE Wireless Communications and Networking Conference, WCNC 2007*. pp. 3844–3849, Hong Kong, March 11–15, 2007.

[19] Rutagemwa, H., Pack, S., Shen, X., and Mark, J.W., Cross-layer design and analysis of wireless profiled TCP for vertical handover, *IEEE International Conference on Communications, ICC '07*. pp. 4488–4493, Glasgow, Scotland, June 24–28, 2007.

[20] Muraleedharan, R. and Osadciw, L.A., Security: Cross layer protocol in wireless sensor network, *Proceedings of 25th IEEE International Conference on Computer Communications. INFOCOM 2006*. pp. 1–2, Barcelona, Catalunya, Spain, April 2006.

[21] Xiao, M., Wang, X., and Yang, G., Cross-layer design for the security of wireless sensor networks, *The 6th World Congress on Intelligent Control and Automation, WCICA 2006*. vol. 1, pp. 104–108, 2006.

[22] Mao, Y. and Wu, M., Tracing malicious relays in cooperative wireless communications, *IEEE Transactions on Information Forensics and Security*, 2(2), 198–212, June 2007.

[23] Liu, P., Tao, Z., Lin, Z., Erkip, E., and Panwar, S., Cooperative wireless communications: A cross-layer approach, *IEEE Wireless Communications*, 13(4), 84–92, Aug. 2006 [see also *IEEE Personal Communications*].

[24] Sadek, A.K., Liu, K.J.R., and Ephremides, A., Cooperative multiple access for wireless networks: Protocols design and stability analysis, *40th Annual Conference on Information Sciences and Systems, 2006*. pp. 1224–1229, Princeton, NJ, March 22–24, 2006.

[25] Ng, T.C.Y. and Yu, W., Joint optimization of relay strategies and resource allocations in cooperative cellular networks, *IEEE Journal on Selected Areas in Communications*, 25(2), 328–339, Feb. 2007.

[26] Kwon, H., Kim, T.H., Choi, S., and Lee, B.G., A cross-layer strategy for energy-efficient reliable delivery in wireless sensor networks, *IEEE Transactions on Wireless Communications*, 5(12), 3689–3699, Dec. 2006.

[27] Akyildiz, I.F., Vuran, M.C., and Akan, O.B., A cross-layer protocol for wireless sensor networks, *40th Annual Conference on Information Sciences and Systems*, pp. 1102–1107, Piscataway, NJ, Mar. 22–24, 2006.

[28] Karvonen, H. and Pomalaza-Raez, C., A cross layer design of coding and awake/sleep periods in WSNS, *IEEE 17th International Symposium on Personal, Indoor and Mobile Radio Communications*, pp. 1–5, Helsinki, Finland, Sept. 2006.

[29] Chen, L., Low, S.H., Chiang, M., and Doyle, J.C., Cross-layer congestion control, routing and scheduling design in ad hoc wireless networks, *Proceedings 25th IEEE International Conference on Computer Communications, INFOCOM 2006*. pp. 1–13, Barcelona, Catalunya, Spain, Apr. 2006.

[30] Iannone, L. and Fdida, S., Evaluating a cross-layer approach for routing in wireless mesh networks, Telecommunication Systems Journal (Springer) Special issue: Next Generation Networks—Architectures, Protocols, Performance, 31, 173—193, March 2006.

[31] Chen, L. and Heinzelman, W.B., QoS-aware routing based on bandwidth estimation for mobile ad hoc networks, *IEEE Journal on Selected Areas in Communications*, 23(3), 561–572, Mar. 2005.

[32] IEEE Std 802.16, IEEE standard for local and metropolitan area networks. Part 16: Air interface for fixed broadband wireless access systems, Oct. 2004.

[33] IEEE Std 802.16e, IEEE standard for local and metropolitan area networks Part 16: Air Interface for fixed and mobile broadband wireless access systems. Amendment 2: Physical and medium access control layers for combined fixed and mobile operation in licensed bands and corrigendum 1, Feb. 2006.

[34] RFC 4585, Extended RTP profile for Real-time Transport Control Protocol (RTCP)-based feedback (RTP/AVPF), July 2006.

[35] Poulin, D., Piggin, P., Yaniv, R., and Andelman, D., Correction to Rx SNR, Rx sensitivity, and Tx relative constellation error for OFDM and OFDMA systems, IEEE 802.16 working group contribution, document number IEEE C802.16maint-05/112r8, Sept. 2005.

[36] Xergias, S.A., Passas, N., and Merakos, L., Flexible resource allocation in IEEE 802.16 wireless metropolitan area networks, *The 14th IEEE Workshop on Local and Metropolitan Area Networks, LANMAN 2005*. p. 6, Chania, Crete, Greece, Sept. 18–21, 2005.

[37] Xenakis, D., Tsolkas, D., Xergias, S., Passas, N., and Merakos, L., A dynamic subchannel allocation algorithm for IEEE 802.16e networks, 3rd *International Symposium on Wireless Pervasive Computing*, pp. 165–169, Santorini, Greece, May 2008.

[38] Andrews, J.G., Ghosh, A., and Muhamed, R., *Fundamentals of WiMAX, Prentice Hall Communication Engineering and Emerging Technologies Series*, Upper Saddle, NJ, 2007.

[39] Rappaport, T., *Wireless Communications, Principles and Practice, Prentice Hall Communication Engineering and Emerging Technologies Series*, Englewood Cliffs, NJ, 1996.

[40] RFC 3550, RTP: A transport protocol for real-time applications, July 2003.

6 Power Allocation in Cooperative Wireless Networks

Kamel Tourki

CONTENTS

We consider a multiple access (MAC) fading channel with two users communicating with a common destination, where each user mutually acts as a relay for the other one and transmits his own information as opposed to having dedicated relays. We wish to evaluate the usefulness of relaying from the point of view of the system's throughput (sum rate) rather than from the sole point of view of the user benefiting from the cooperation, as is typically done. We do this by allowing a trade-off between relaying and fresh data transmission through a resource allocation framework. Specifically, We propose a cooperative transmission scheme allowing each user to allocate a certain amount of power for his own transmitted data while the rest is devoted to relaying. The underlying protocol is based on a modification of the so-called nonorthogonal amplify-and-forward (NAF) protocol [1]. We develop capacity expressions for our scheme and derive the rate-optimum power allocation in closed form for centralized and distributed frameworks. In the distributed scenario, partially statistical and partially instantaneous channel information is exploited.

The centralized power allocation algorithm indicates that even in a mutual cooperation setting like ours, on any given realization of the channel, cooperation is never truly mutual, that is, one of the users will always allocate zero power to relaying the data of the other one, and thus act selfishly. But in a distributed framework, our results indicate that the sum rate is maximized when both mobiles act selfishly.

6.1 INTRODUCTION

In many wireless applications, wireless users may not be able to support multiple antennas due to size, complexity, power, or other constraints. The wireless medium brings along its unique challenges such as fading and multiuser interference, which can be combated with the concept of cooperative diversity [2–5]. In traditional cooperative diversity setups, a user is unilaterally designated to act as a relay for the benefit of another one, at least for a given period of time. In certain scenarios, the relay is an actual component of the infrastructure with no own data to be delivered to the network [6–9]. In cellular-type multiuser networks, however, there will be a compromise to strike by all users between transmitting their own information and helping others by relaying their data to the destination [10–18]. A simplified instance of this scenario is given by a MAC with two or more users trying to reach a common destination (for example, base station). Since each user wishes to send its own information, it must allocate resource (the total of which is constrained at each user) wisely between its own data transmission and the data it will relay for the benefit of some other user (Figure 6.1).

In this chapter we consider resource control in the form of power allocated by a user across its own data and its relay data. The underlying protocol considered here is similar to the one considered by Azarian et al. [1], which itself evolved from the early work by Laneman et al. [2]. There, the authors imposed the half-duplex constraint on the cooperating nodes and proposed several cooperative transmission protocols. All the proposed schemes in Ref. [2] used a time-division multiple-access (TDMA) strategy, where the two partners relied on the use of orthogonal signaling to repeat each other's signals. In our scheme, the relay and own transmission operations take place in orthogonal resource slots but share a common average power resource where the average is computed over two frames. Note that other (than power) types of resource division could also be considered, such as bandwidth [17]. Recently nonorthogonal signaling strategies have been proposed, for example [1], in which a relay transmits delayed information by a user while this user simultaneously transmit fresh data. In this nonorthogonal amplify-and-forward (NAF) scheme, the diversity-multiplexing trade-off is studied, showing the superiority of the NAF scheme over the orthogonal counterpart. However in Ref. [1] and much previous work, the relay network model is unbalanced in the sense that the transmission of own data by the relay is not considered, and the source node is not invited to act as a

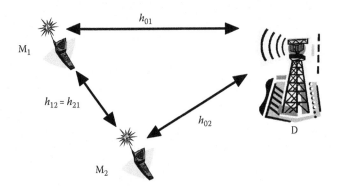

FIGURE 6.1 Cellular model.

relay either. In multiuser networks, it is desirable from a global capacity point of view that each user allocates a fraction of its resource toward cooperation. Just how big this fraction should be is one of the questions addressed by this chapter.

We consider the problem of maximizing the sum rate for this cooperative MAC channel, as a function of the power allocation toward own and relay data, given certain knowledge of the channel for both users. We derive the optimum power allocation policy in closed form for certain scenarios of interest. We consider both centralized and distributed cases. In the centralized case first explored in Ref. [19], the base initially collects instantaneous Channel State Information (CSI) from both users and computes the optimum power allocation vector on behalf of the two users.

In the distributed case, it is assumed that the complete fast-varying CSI cannot be exchanged between the users and the base. In that case, the users individually come up with a power allocation strategy based on a mix of local CSI and nonlocal statistical CSI. We show that in fact, when the optimum policy is used, one of the users always acts completely selfishly. Interestingly, this type of selfish behavior by some users in multiuser cooperative MAC was noted by Kaya and Ulukus [20], but in a different context with decode-and-forward signaling.

Also, the problem of distributed power allocation was addressed in Ref. [21]; however, in this paper the algorithm is used to optimize the bit error ratio (BER) performance. Furthermore, the power allocation is done across users rather than across relay and data transmission operations.

Then we investigate the system gain (sum rate) of mutual cooperation in two different network geometries. We show the system gain depends on the level of symmetry in the user positions.

This chapter is organized as follows. In Section 6.2 we present the cooperative protocols in wireless networks and in Section 6.3, we describe the system model. In Section 6.4 the sum-rate expression is derived and the optimal power allocation algorithm is presented for the centralized framework. In Section 6.5, the distributed framework is investigated and the simulation results are presented. We conclude the chapter in Section 6.6.

Notations: All boldface letters indicate vectors (lower case) or matrices (upper case). The operator det( ) is the determinant of matrix, with ( )^H denoting its conjugate-transpose and ( )* denoting its conjugate. $\mathbb{E}[.]$ is the expectation operator.

6.2   COOPERATIVE PROTOCOLS IN WIRELESS NETWORKS

The basic ideas behind cooperative communication can be traced back to the groundbreaking work of Cover and El Gamal on the information-theoretic properties of the relay channel [6]. This work analyzed the capacity of the three-node network consisting of a source, a destination, and a relay. It was assumed that all nodes operate in the same band, so the system can be decomposed into a broadcast channel from the viewpoint of the source and a MAC from the viewpoint of the destination. Many ideas that appeared later in the cooperation literature were first presented in Ref. [6].

However, in many respects the cooperative communication we consider is different from the relay channel. First, recent developments are motivated by the concept of diversity in a fading channel, whereas Cover and El Gamal mostly analyze capacity in an additive white Gaussian noise (AWGN) channel. Second, in the relay channel, as mentioned in the introduction of this chapter, the relay's sole purpose is to help the main channel, whereas in cooperation the total system resources are fixed, and users act both as information sources as well as relays. We now review two of the main cooperative signaling methods.

6.2.1   DECODE-AND-FORWARD

This method is perhaps closest to the idea of a traditional relay. In this method a user attempts to detect the partners' bits and then retransmits the detected bits. The partners may be assigned mutually by the base station or via some other technique. For the purposes of this chapter we consider two users partnering with each other, but in reality the only important factor is that each user has a partner that provides a second (diversity) data path. The easiest way to visualize this is via pairs, but it is also possible to achieve the same effect via other partnership topologies that remove the strict constraint of pairing. Partner assignment is a rich topic whose details are beyond the scope of this chapter.

It is possible that detection by the partner is unsuccessful, in which case cooperation can be detrimental to the eventual detection of the bits at the base station. Also, the base station needs to know the error characteristics of the interuser channel for optimal decoding. To avoid the problem of error propagation, Laneman et al. [22] proposed a hybrid decode-and- forward method where, at times when the fading channel has high instantaneous signal-to-noise ratio (SNR), users detect-and-forward their partners' data, but when the channel has low SNR, users revert to a noncooperative mode. In Ref. [23], the authors considered a regenerative relay in which the decision to cooperate is based on an SNR threshold and considered the effect of the possible erroneously detected and transmitted data at the relay.

6.2.2 AMPLIFY-AND-FORWARD

Another simple cooperative signaling is the amplify-and-forward (or nonregenerative) method. Each user in this method receives a noisy version of the signal transmitted by its partner. As the name implies, the user then amplifies and retransmits this noisy version. The base station combines the information sent by the user and partner and makes a final decision on the transmitted bit. Although noise is amplified by cooperation, the base station receives two independently faded versions of the signal and can make better decisions on the detection of information.

This method was proposed and analyzed by Laneman et al. [2,22,24]. It has been shown that for the two-user case, this method achieves diversity order of two, which is the best possible outcome at high SNR. In amplify-and-forward it is assumed that the base station knows the interuser channel coefficients to do optimal decoding, so some mechanism of exchanging or estimating this information must be incorporated into any implementation. Another potential challenge is that sampling, amplifying, and retransmitting analog values are technologically nontrivial. Nevertheless, amplify-and-forward is a simple method that lends itself to analysis and thus has been very useful in furthering our understanding of cooperative communication systems.

6.3 SYSTEM MODEL

We consider a two-user fading Gaussian MAC, where both the receiver and the transmitters receive noisy versions of the transmitted messages. Each receiver maintains channel-state information and employs coherent detection. The channels between users (interuser channels) and from each user to the destination (uplink channels) are mutually independent. Time is divided into two consecutive frames. Each frame is further divided into two half-frames T_1 and T_2 as shown in Table 6.1. We use a combination of TDMA and nonorthogonal signaling: In the first half of frame 1, user 1 sends its first half packet (containing $\frac{N}{2}$ bits) while user 2 listens. In the second half, user 2 relays the overheard data with power level β, while user 1 simultaneously sends fresh information (its second-half packet) with power level $1 - \alpha$ where α is chosen in [0, 1]. In frame 2 we proceed just as in frame 1, but with the roles of user 1 and 2 α, β reversed. Thus, we maintain a constant average power across the two frames, for each user, regardless of the choice of α, β.

6.3.1 WIRELESS NETWORKS

6.3.1.1 User-Based Cooperation

An example of user cooperation can be found in the work of Sendonaris et al. [3,4]. This work presents analysis and a simple code-division multiple access (CDMA) implementation of decode-and-forward cooperative signaling. In this scheme, two users are paired to cooperate with each other.

TABLE 6.1

Power Allocation Coefficients over Two Frames for Two Users Transmitting to a Base Using TDMA Scheme

	T_1	T_2	T_1	T_2
User 1	1	$1 - \alpha$	0	α
User 2	0	β	1	$1 - \beta$

Note: Power levels are used to either send own or relay data. T_1 (resp. T_2) is first (resp. second) half of the frame.

Each user has its own spreading code, denoted $c_1(t)$ and $c_2(t)$. The two user's data bits are denoted $b_i^{(n)}$ where $i = 1, 2$ are the user indices and n denotes the time index of information bits. Factors a_{ij} denote signal amplitudes and hence represent power allocation to various parts of the signaling. Each signaling period consists of three bit intervals.

Denoting the signal of user 1 $X_1(t)$ and the signal of user 2 $X_2(t)$, $X_1(t) = \left[a_{11}b_1^{(1)}c_1(t), a_{12}b_1^{(2)}c_1(t), a_{13}b_1^{(2)}c_1(t) + a_{14}b_2^{(2)}c_2(t) \right]$ $X_2(t) = \left[a_{21}b_2^{(1)}c_2(t), a_{22}b_2^{(2)}c_2(t), a_{23}b_1^{(2)}c_1(t) + a_{24}b_2^{(2)}c_2(t) \right]$.

In other words, in the first and second intervals, each user transmits its own bits. Each user then detects the other user's second bit (each user's estimate of the other's bit is denoted $\hat{b}_i$). In the third interval, both users transmit a linear combination of their own second bit and the partner's second bit, each multiplied by the appropriate spreading code. The transmit powers for the first, second, and third intervals are variable, and by optimizing the relative transmit powers according to the conditions of the uplink and interuser channels, this method provides adaptability to channel conditions.

The powers are allocated through the factors $a_{i,j}$ such that an average power constraint is maintained. Roughly speaking, whenever the interuser channel is favorable, more power will be allocated to cooperation, whereas whenever the interuser channel is not favorable, cooperation is reduced.

6.3.1.2 Coded Cooperation

Coded cooperation [14,25] is a method that integrates cooperation into channel coding. Coded cooperation works by sending different portions of each user's code word via two independent fading paths. The basic idea is that each user tries to transmit incremental redundancy to its partner. Whenever that is not possible, the users automatically revert to a noncooperative mode. The key to the efficiency of coded cooperation is that all this is managed automatically through code design, with no feedback between the users.

The users divide their source data into blocks that are augmented with cyclic redundancy check (CRC) code. In coded cooperation, each of the users' data are encoded into a codeword that is partitioned into two segments, containing N_1 bits and N_2 bits, respectively. It is easier to envision the process by a specific example: consider that the original codeword has $N_1 + N_2$ bits; puncturing this codeword down to N_1 bits, we obtain the first partition, which itself is a valid (weaker) codeword. The remaining N_2 bits in this example are the puncture bits. Of course, partitioning is also possible via other means, but this example serves to give an idea of the intuition behind coded cooperation.

Likewise, the data transmission period for each user is divided into two time segments of N_1 and N_2 bit intervals, respectively. We call these time intervals frames. For the first frame, each user transmits a code word consisting of the N_1-bit code partition. Each user also attempts to decode the transmission of its partner. If this attempt is successful (determined by checking the CRC code), in the second frame the user calculates and transmits the second code partition of its partner, containing N_2 code bits. Otherwise, the user transmits its own second partition, again containing N_2 bits. Thus, each user always transmits a total of $N = N_1 + N_2$ bits per source block over the two frames. We define the level of cooperation as N_2/N, the percentage of the total bits for each source block the user transmits for its partner.

In general, various channel coding methods can be used within this coded cooperation framework. For example, the overall code may be a block or convolutional code or a combination of both. The code bits for the two frames may be selected through puncturing, product codes, or other forms of concatenation.

6.3.2 NETWORK GEOMETRY

We anticipate that cooperation will be performed differently as a function of the positions of the users with respect to destination. We study two particular different network geometries, denoted by symmetric and asymmetric or linear, depicted respectively by Figures 6.2 and 6.3. In the asymmetric case, we model the path loss, that is, the mean channel powers σ_{ij}^2, as a function of the relative relay position d by

$$\sigma_{01}^2 = 1, \quad \sigma_{12}^2 = d^{-\nu}, \quad \sigma_{02}^2 = (1-d)^{-\nu} \tag{6.1}$$

where ν is the path loss exponent and $0 < d = d_{12} < 1$. The distances are normalized by the distance d_{01}. In these coordinates, the user 1 can be located at $(0,0)$, and the destination can be located at $(1,0)$, without loss of generality. User 2 is located at $(d,0)$ [24]. In the symmetric case, all channels are drawn with same the unit variance.

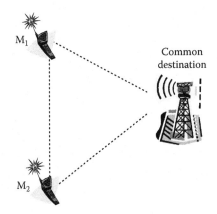

FIGURE 6.2 Symmetric network.

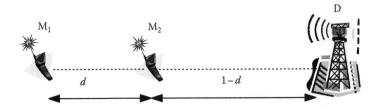

FIGURE 6.3 Asymmetric (or linear) network.

6.3.3 CHANNEL MODEL

We model all channels as Rayleigh block flat fading with AWGN. Channel coefficients h_{ij} are modeled as zero-mean, circularly symmetric complex Gaussian random variables with different variances. Noises are modeled as zero-mean mutually independent, circularly symmetric, complex Gaussian random sequences with variance N_0.

6.3.4 SIGNAL MODEL

The signal received by the common destination during the first frame (first and second half) is given by

$$\begin{cases} y_1(n) = h_{01}x_1(n) + z_0(n) \\ y_1\left(n + \frac{N}{2}\right) = \sqrt{1-\alpha}h_{01}x_1\left(n + \frac{N}{2}\right) + \sqrt{\beta}h_{02}A_1\left[h_{21}x_1(n) + w_2(n)\right] + z_0\left(n + \frac{N}{2}\right) \end{cases} \tag{6.2}$$

During the second frame, the received signal is

$$\begin{cases} y_2(n) = h_{02}x_2(n) + z_0(n) \\ y_2\left(n + \frac{N}{2}\right) = \sqrt{1-\beta}h_{02}x_2\left(n + \frac{N}{2}\right) + \sqrt{\alpha}h_{01}A_2\left[h_{12}x_2(n) + w_1(n)\right] + z_0\left(n + \frac{N}{2}\right) \end{cases} \tag{6.3}$$

where $n = 1, \ldots, \frac{N}{2}$ and h_{ij} captures the effects of fading between transmitter j and receiver i.

Thus, in Equations 6.2 and 6.3, α and β can be seen as cooperation levels for user 1 and user 2 respectively. $x_{j \in \{1,2\}}(n) \in C$ is the nth coded symbol, $w_{i \in \{1,2\}}(n)$ and $z_0(n)$ are, respectively, the noise sample (of variance $N_{i \in \{1,2\}}$) observed by the transmitter $j \in \{1, 2\}$ and the noise sample (of variance N_0) observed by the destination. h_{21} and h_{12} represent the interuser channel gains, and h_{01} and h_{02} denote the user-destination channel gains, which are maintained constant during $T_1 + T_2$. $A_1 \leq \sqrt{\frac{P_2}{|h_{21}|^2 P_1 + N_2}}$ and $A_2 \leq \sqrt{\frac{P_1}{|h_{12}|^2 P_2 + N_1}}$ are the relay repetition gains, where $P_{j \in \{1,2\}}$ is the sample energy. Clearly, the choice $\alpha = 0, \beta = 1$ would result in a classical 3 node relay scenario with user 1 acting as the unique source of information and user 2 being a dedicated relay node. We remark that Equations 6.2 and 6.3 are reduced to equations of an orthogonal direct transmission (noncooperative protocol) if $\alpha = \beta = 0$ and to an amplify-and-forward protocol if $\alpha = \beta = 1$ [2].

We hereby focus our attention on how the relay operation can improve the sum rate of the cooperative MAC system, by proper choices of α and β. We argue that sum rate (rather than the traditional diversity-oriented measures) is a valuable performance metric when degrees of freedom for diversity can be acquired at other layers of the protocol stack, such as via scheduling.

6.4 CENTRALIZED POWER ALLOCATION

6.4.1 ANALYSIS OF SUM RATE

In the proposition below, we assume a central control unit (located, for example, at the base) with knowledge of all CSI. We develop the expression for the sum rate for the above protocol and power allocation system in a way similar to developments by Laneman et al. and others.

PROPOSITION 1

For the Gaussian memoryless MAC with user cooperation, if the rate pair (R1, R2) is achievable, then the sum rate $R1 + R2 \leq I_{\alpha,\beta}$ where

$$I_{\alpha,\beta} \triangleq \log_2\left[1 + \gamma_{01} + (1-\alpha)\frac{K_1}{l_1(\beta)} + f(\beta\gamma_{02}, \gamma_{21})\right]$$
$$+ \log_2\left[1 + \gamma_{02} + (1-\beta)\frac{K_2}{l_2(\alpha)} + f(\alpha\gamma_{01}, \gamma_{12})\right] \quad (6.4)$$

where

$$K_1 = \left[\gamma_{01}^2 + \gamma_{01}\right]\left[\gamma_{21} + 1\right]$$
$$K_2 = \left[\gamma_{02}^2 + \gamma_{02}\right]\left[\gamma_{12} + 1\right]$$
$$l_1(\beta) = 1 + \gamma_{21} + \beta\gamma_{02} \quad (6.5)$$
$$l_2(\alpha) = 1 + \gamma_{12} + \alpha\gamma_{01}$$
$$f(x,y) = \frac{xy}{x+y+1}$$

and γ_{ij} is defined as $|h_{ij}|^2\frac{P_j}{N_i}$ where P_j is the power of the transmitted signal from user j, N_i is the noise power at the receiver i, and $i,j \in \{1,2\}$.

Proof: For simplicity, we formulate Equations 6.2 and 6.3, respectively, as

$$\underbrace{\begin{bmatrix} y_1(n) \\ y_1(n+\frac{N}{2}) \end{bmatrix}}_{\mathbf{y_1}} = \underbrace{\begin{bmatrix} h_{01} & 0 \\ \sqrt{\beta}A_1h_{02}h_{21} & \sqrt{1-\alpha}h_{01} \end{bmatrix}}_{\mathbf{M_1}} \underbrace{\begin{bmatrix} x_1(n) \\ x_1(n+\frac{N}{2}) \end{bmatrix}}_{\mathbf{x_1}}$$
$$+ \underbrace{\begin{bmatrix} 0 & 1 & 0 \\ \sqrt{\beta}A_1h_{02} & 0 & 1 \end{bmatrix}}_{\mathbf{B_1}} \underbrace{\begin{bmatrix} w_2(n) \\ z_0(n) \\ z_0(n+\frac{N}{2}) \end{bmatrix}}_{\mathbf{z_1}} \quad (6.6)$$

and

$$\underbrace{\begin{bmatrix} y_2(n) \\ y_2(n+\frac{N}{2}) \end{bmatrix}}_{\mathbf{y_2}} = \underbrace{\begin{bmatrix} h_{02} & 0 \\ \sqrt{\alpha}A_2h_{01}h_{12} & \sqrt{1-\beta}h_{02} \end{bmatrix}}_{\mathbf{M_2}} \underbrace{\begin{bmatrix} x_2(n) \\ x_2(n+\frac{N}{2}) \end{bmatrix}}_{\mathbf{x_2}}$$
$$+ \underbrace{\begin{bmatrix} 0 & 1 & 0 \\ \sqrt{\alpha}A_2h_{01} & 0 & 1 \end{bmatrix}}_{\mathbf{B_2}} \underbrace{\begin{bmatrix} w_1(n) \\ z_0(n) \\ z_0(n+\frac{N}{2}) \end{bmatrix}}_{\mathbf{z_2}} \quad (6.7)$$

and, without loss of generality, we compute the maximum average mutual information during $T_1 + T_2$.

$$
\begin{aligned}
I(\mathbf{x}_1, \tilde{\mathbf{y}}_1) &= I(\mathbf{x}_1; \mathbf{M}_1) + I(\mathbf{x}_1; \mathbf{y}_1/\mathbf{M}_1) \\
&= I(\mathbf{x}_1; \mathbf{y}_1/\mathbf{M}_1) \\
&\leq \log_2 \det\left(\mathbf{I}_2 + \mathbf{M}_1 \Lambda_{\mathbf{x}_1} \mathbf{M}_1^{\mathrm{H}} \sum_{\mathbf{n}_1}^{-1}\right)
\end{aligned}
\tag{6.8}
$$

where $\mathbf{n}_1 = \mathbf{B}_1 \mathbf{z}_1$ and $\Lambda_{\mathbf{x}_1} = \mathbf{E}(\mathbf{x}_1 \mathbf{x}_1^{\mathrm{H}}) = P_1 \mathbf{I}_2$.
Therefore, $\sum_{\mathbf{n}_1} = \mathbf{B}_1 \mathbf{E}(\mathbf{z}_1 \mathbf{z}_1^{\mathrm{H}}) \mathbf{B}_1^{\mathrm{H}}$ and equal to

$$
\sum_{\mathbf{n}_1} = \begin{bmatrix} N_0 & 0 \\ 0 & N_0 + \beta(A_1)^2 |h_{02}|^2 N_2 \end{bmatrix}
\tag{6.9}
$$

$$
\mathbf{M}_1 \mathbf{M}_1^{\mathrm{H}} = \begin{bmatrix} A & B \\ C & D \end{bmatrix}
\tag{6.10}
$$

where

$$
\begin{aligned}
A &= |h_{01}|^2 \\
B &= \sqrt{\beta} A_1 h_{01} (h_{02})^* (h_{21})^* \\
C &= \sqrt{\beta} A_1 (h_{01})^* h_{02} h_{21} \\
D &= (1-\alpha)|h_{01}|^2 + \beta(A_1)^2 |h_{02} h_{21}|^2
\end{aligned}
\tag{6.11}
$$

Therefore,

$$
\begin{aligned}
\log_2 \det\left(\mathbf{I}_2 + P_1 \mathbf{M}_1 \mathbf{M}_1^{\mathrm{H}} \sum_{\mathbf{n}_1}^{-1}\right) = \log_2 \Bigg[&1 + |h_{01}|^2 \frac{P_1}{N_0} + \frac{(1-\alpha)|h_{01}|^2 P_1 + \beta(A_1)^2 |h_{02} h_{21}|^2 P_1}{N_0 + N_2 \beta(A_1)^2 |h_{02}|^2} \\
&+ \frac{(1-\alpha)|h_{01}|^4 P_1^2}{N_0(N_0 + N_2 \beta(A_1)^2 |h_{02}|^2)} \Bigg]
\end{aligned}
\tag{6.12}
$$

and after substitutions and algebraic manipulations, we obtain

$$
\log_2 \det\left(\mathbf{I}_2 + P_1 \mathbf{M}_1 \mathbf{M}_1^{\mathrm{H}} \sum_{\mathbf{n}_1}^{-1}\right) = \log_2\left[1 + \gamma_{01} + (1-\alpha)\frac{K_1}{l_1(\beta)} + f(\beta \gamma_{02}, \gamma_{21})\right]
\tag{6.13}
$$

Therefore, Equation 6.4 is straightforward.

Note that this expression requires channel information at the receiver but not the transmitter. However the optimization with respect to power control coefficients α and β will require full channel knowledge. We can consider in the sequel that $P_1 = P_2 = P$ and $\gamma_{21} = \gamma_{12} = \gamma$ since the same frequency is used in both directions of interuser communication.

6.4.2   OPTIMIZATION OF RELAY POWER ALLOCATION

We now address the problem of optimizing the power allocated by each user toward either transmission of its own data or relay data. The objective function taken here is the multiuser sum rate defined in Equation 6.4. We start by characterizing the sum rate in some border points of the power region.

The lemma below comes handy in the more general characterization of the optimal power allocation policy.

LEMMA 1

We characterize the sum rate over the feasible power allocation region by

$$
\begin{cases}
I_{\alpha,0} > I_{\alpha,1} & \forall \, \alpha \in [0,1] \\
I_{0,\beta} > I_{1,\beta} & \forall \, \beta \in [0,1]
\end{cases}
\tag{6.14}
$$

Proof: We show without loss of generalities that $I_{\alpha,0} > I_{\alpha,1}$.

$$
I_{\alpha,0} = \log_2\left[1 + \gamma_{01} + (1-\alpha)\frac{K_1}{l_1(0)}\right] + \log_2\left[1 + \gamma_{02} + \frac{K_2}{l_2(\alpha)} + f(\alpha\gamma_{01},\gamma)\right]
\tag{6.15}
$$

$$
I_{\alpha,1} = \log_2\left[1 + \gamma_{01} + (1-\alpha)\frac{K_1}{l_1(1)} + f(\gamma_{02},\gamma)\right] + \log_2\left[1 + \gamma_{02} + f(\alpha\gamma_{01},\gamma)\right]
\tag{6.16}
$$

In order to demonstrate that $I_{\alpha,0} > I_{\alpha,1}$ it is enough to show that

$$
\underbrace{\left[1 + \gamma_{01} + (1-\alpha)\frac{K_1}{l_1(0)}\right]\left[1 + \gamma_{02} + \frac{K_2}{l_2(\alpha)} + f(\alpha\gamma_{01},\gamma)\right]}_{\pi_1}
$$

$$
> \underbrace{\left[1 + \gamma_{01} + (1-\alpha)\frac{K_1}{l_1(1)} + f(\gamma_{02},\gamma)\right]\left[1 + \gamma_{02} + f(\alpha\gamma_{01},\gamma)\right]}_{\pi_2}
\tag{6.17}
$$

where

$$
\begin{cases}
\pi_1 = (1+\gamma_{01})(1+\gamma_{02}) + (1+\gamma_{01})\frac{K_2}{l_2(\alpha)} + (1+\gamma_{01})f(\alpha\gamma_{01},\gamma) \\
\quad + (1-\alpha)\frac{K_1}{l_1(0)}(1+\gamma_{02}) + (1-\alpha)\frac{K_1}{l_1(0)}\frac{K_2}{l_2(\alpha)} + (1-\alpha)\frac{K_1}{l_1(0)}f(\alpha\gamma_{01},\gamma) \\
\pi_2 = (1+\gamma_{01})(1+\gamma_{02}) + (1+\gamma_{01})f(\alpha\gamma_{01},\gamma) + (1-\alpha)\frac{K_1}{l_1(1)}(1+\gamma_{02}) \\
\quad + (1-\alpha)\frac{K_1}{l_1(1)}f(\alpha\gamma_{01},\gamma) + f(\gamma_{02},\gamma)(1+\gamma_{02}) + f(\gamma_{02},\gamma)f(\alpha\gamma_{01},\gamma)
\end{cases}
\tag{6.18}
$$

First we have $l_1(0) < l_1(1)$, so $\frac{1}{l_1(0)} > \frac{1}{l_1(1)}$. Therefore,

$$
\begin{cases}
(1-\alpha)\frac{K_1}{l_1(0)}(1+\gamma_{02}) > (1-\alpha)\frac{K_1}{l_1(1)}(1+\gamma_{02}) \\
(1-\alpha)\frac{K_1}{l_1(0)}f(\alpha\gamma_{01},\gamma) > (1-\alpha)\frac{K_1}{l_1(1)}f(\alpha\gamma_{01},\gamma)
\end{cases}
\tag{6.19}
$$

and after some manipulations, we obtain

$$
(1-\alpha)\frac{K_1}{l_1(0)}\frac{K_2}{l_2(\alpha)} + (1+\gamma_{01})\frac{K_2}{l_2(\alpha)} = (1+\gamma_{01})\frac{K_2}{l_2(\alpha)}\left[1 + (1-\alpha)\gamma_{01}\right]
\tag{6.20}
$$

and

$$f(\gamma_{02}, \gamma)(1 + \gamma_{02}) + f(\gamma_{02}, \gamma)f(\alpha\gamma_{01}, \gamma) = \frac{\gamma K_2}{(1 + \gamma + \gamma_{02})l_2(\alpha)} + \frac{\alpha\gamma\gamma_{01}\gamma_{02}}{l_2(\alpha)} \tag{6.21}$$

We remember that $K_2 - \gamma_{02}^2\gamma = \gamma_{02}(1 + \gamma + \gamma_{02})$; therefore,

$$f(\gamma_{02}, \gamma)(1 + \gamma_{02}) + f(\gamma_{02}, \gamma)f(\alpha\gamma_{01}, \gamma) = \frac{K_2}{l_2(\alpha)} \frac{\gamma(1 + \alpha\gamma_{01})}{(1 + \gamma + \gamma_{02})} - \frac{\alpha\gamma_{01}\gamma^2\gamma_{02}^2}{(1 + \gamma + \gamma_{02})l_2(\alpha)} \tag{6.22}$$

We can say that $0 < \frac{\gamma}{1+\gamma+\gamma_{02}} < 1$ and $0 < 1 + \alpha\gamma_{01} \leq (1 + \gamma_{01})$. So, $\frac{K_2}{l_2(\alpha)} \frac{\gamma(1+\alpha\gamma_{01})}{(1+\gamma+\gamma_{02})} \leq (1 + \gamma_{01})\frac{K_2}{l_2(\alpha)}$ then $\frac{K_2}{l_2(\alpha)} \frac{\gamma(1+\alpha\gamma_{01})}{(1+\gamma+\gamma_{02})} \leq (1 + \gamma_{01})\frac{K_2}{l_2(\alpha)}[1 + (1 - \alpha)\gamma_{01}]$. Therefore, we can finally say that

$$\frac{K_2}{l_2(\alpha)} \frac{\gamma(1 + \alpha\gamma_{01})}{(1 + \gamma + \gamma_{02})} - \alpha\frac{\gamma_{01}\gamma^2\gamma_{02}^2}{(1 + \gamma + \gamma_{02})l_2(\alpha)} \leq (1 + \gamma_{01})\frac{K_2}{l_2(\alpha)}[1 + (1 - \alpha)\gamma_{01}] \tag{6.23}$$

and from Equations 6.19, 6.20, and 6.23 we conclude that $I_{\alpha,1} < I_{\alpha,0}$.
The relation $I_{1,\beta} < I_{0,\beta}$ is straightforward.

This shows that, from the point of view of global throughput performance, the system would rather have all users act selfishly.

6.4.3   POWER ALLOCATION ALGORITHM

We now proceed to give a complete characterization of the optimal power allocation policy for an arbitrary realization of the multiuser channels.

PROPOSITION 2

The optimal power allocation that maximizes the sum rate Equation 6.4 is given by

1. $\alpha = \alpha_* \neq 0$ and $\beta = 0$ if $\begin{cases} \gamma > \gamma_{02}^2 + \gamma_{02} \\ \gamma_{01} > \frac{(1+\gamma_{02})^2(1+\gamma)}{\gamma - (\gamma_{02}^2 + \gamma_{02})} - 1 \end{cases}$

2. $\alpha = 0$ and $\beta = \beta_* \neq 0$ if $\begin{cases} \gamma > \gamma_{01}^2 + \gamma_{01} \\ \gamma_{02} > \frac{(1+\gamma_{01})^2(1+\gamma)}{\gamma - (\gamma_{01}^2 + \gamma_{01})} - 1 \end{cases}$

3. $\alpha = 0$ and $\beta = 0$ if neither condition above is met.

where optimal values α_*, β_* are detailed in the appendix and shown below.

Proof: In order to seek (α_*, β_*) for which $I_{\alpha,\beta}$ is maximized,

$$(\alpha_*, \beta_*) = \arg\max_{\alpha,\beta\in[0,1]} I_{\alpha,\beta} \tag{6.24}$$

we must solve this system of equations:

$$\begin{cases} \dfrac{\partial I_{\alpha,\beta}}{\partial\alpha} = 0 \\[2mm] \dfrac{\partial I_{\alpha,\beta}}{\partial\beta} = 0 \end{cases} \tag{6.25}$$

The partial derivatives of $I_{\alpha,\beta}$, $\frac{\partial I_{\alpha,\beta}}{\partial \alpha}$ and $\frac{\partial I_{\alpha,\beta}}{\partial \beta}$, respectively, of α and β give

$$\frac{\partial I_{\alpha,\beta}}{\partial \alpha} = \frac{1}{\ln(2)}\left[\frac{\dfrac{-K_1}{l_1(\beta)}}{1 + \gamma_{01} + (1-\alpha)\dfrac{K_1}{l_1(\beta)} + f(\beta\gamma_{02},\gamma)} + \frac{(1-\beta)K_2\dfrac{-\partial l_2(\alpha)}{\partial \alpha}}{1 + \gamma_{02} + (1-\beta)\dfrac{K_2}{l_2(\alpha)} + f(\alpha\gamma_{01},\gamma)}\right]$$

(6.26)

and

$$\frac{\partial I_{\alpha,\beta}}{\partial \beta} = \frac{1}{\ln(2)}\left[\frac{\dfrac{-K_2}{l_2(\alpha)}}{1 + \gamma_{02} + (1-\beta)\dfrac{K_2}{l_2(\alpha)} + f(\alpha\gamma_{01},\gamma)} + \frac{(1-\alpha)K_1\dfrac{-\partial l_1(\beta)}{\partial \beta} + \dfrac{\gamma\gamma_{02}(1+\gamma)}{[l_1(\beta)]^2}}{1 + \gamma_{01} + (1-\alpha)\dfrac{K_1}{l_1(\beta)} + f(\beta\gamma_{02},\gamma)}\right]$$

(6.27)

after some simplifications, α_* and β_* are determined as solutions of

$$\begin{cases} A_1\alpha^2 + 2\alpha A_2 - C - B_2\beta - B_1\beta^2 = 0 \\ A_1'\beta^2 + 2\beta A_2' - C' - B_2'\alpha - B_1'\alpha^2 = 0 \end{cases}$$

(6.28)

where

$$\begin{aligned}
A_1 &= K_1\gamma_{01}^2(1 + \gamma + \gamma_{02}) \\
A_2 &= K_1\gamma_{01}(1+\gamma)(1 + \gamma + \gamma_{02}) \\
B_1 &= K_2\gamma_{01}\gamma_{02}(1 + \gamma + \gamma_{01}) \\
B_2 &= K_1K_2(2 + \gamma + \gamma_{01}) + \gamma_{01}\gamma_{02}(1 + \gamma + \gamma_{01})(\gamma(1+\gamma) - K_2) \\
C &= K_1\left[\gamma\frac{K_1}{\gamma_{01}} - \frac{K_2}{\gamma_{02}}(1+\gamma) - K_2(2 + \gamma + \gamma_{01})\right]
\end{aligned}$$

(6.29)

and

$$\begin{aligned}
A_1' &= K_2\gamma_{02}^2(1 + \gamma + \gamma_{01}) \\
A_2' &= K_2\gamma_{02}(1+\gamma)(1 + \gamma + \gamma_{01}) \\
B_1' &= K_1\gamma_{01}\gamma_{02}(1 + \gamma + \gamma_{02}) \\
B_2' &= K_1K_2(2 + \gamma + \gamma_{02}) + \gamma_{01}\gamma_{02}(1 + \gamma + \gamma_{02})(\gamma(1+\gamma) - K_1) \\
C' &= K_2\left[\gamma\frac{K_2}{\gamma_{02}} - \frac{K_1}{\gamma_{01}}(1+\gamma) - K_1(2 + \gamma + \gamma_{02})\right]
\end{aligned}$$

(6.30)

therefore, the system (Equation 6.25) becomes

$$\begin{cases} \dfrac{\widetilde{\alpha}^2}{B_1} - \dfrac{\widetilde{\beta}^2}{A_1} = \kappa_1 \\[3mm] \dfrac{\widetilde{\beta}^2}{B_1'} - \dfrac{\widetilde{\alpha}^2}{A_1'} = \kappa_2 \end{cases}$$

(6.31)

where

$$
\begin{cases}
\tilde{\alpha} = \alpha + \dfrac{A_2}{A_1} \\[2ex]
\tilde{\beta} = \beta + \dfrac{B_2}{2B_1}
\end{cases}
\tag{6.32}
$$

and

$$
\begin{cases}
\kappa_1 = \dfrac{C}{A_1 B_1} + \dfrac{1}{B_1}\left(\dfrac{A_2}{A_1}\right)^2 - \dfrac{1}{A_1}\left(\dfrac{B_2}{2B_1}\right)^2 \\[2ex]
\kappa_2 = \dfrac{C'}{A_1' B_1'} + \dfrac{1}{B_1'}\left(\dfrac{A_2'}{A_1'}\right)^2 - \dfrac{1}{A_1'}\left(\dfrac{B_2'}{2B_1'}\right)^2
\end{cases}
\tag{6.33}
$$

In Equation 6.31, we have two equations of hyperboles. When we replace $\tilde{\alpha}$ in the second equation by its expression derived from the first one to solve this system, we obtain

$$
\tilde{\beta}^2 \left(\dfrac{1}{B_1'} - \dfrac{A_1}{A_1' B_1}\right) = \underbrace{\kappa_2 + \dfrac{B_1}{A_1'}\kappa_1}_{\neq 0}
\tag{6.34}
$$

and because we have

$$
\dfrac{B_1}{A_1} = \dfrac{A_1'}{B_1'}
\tag{6.35}
$$

it is straightforward that there are no solutions, graphically traduced by the no intersection between these hyperboles where Equation 6.35 shows the equality of the slopes of the asymptotes, unless on the plans $\mathcal{P}_{\alpha,0} = \{\beta = 0, \forall \alpha\}$, $\mathcal{P}_{\alpha,1} = \{\beta = 1, \forall \alpha\}$, $\mathcal{P}_{0,\beta} = \{\alpha = 0, \forall \beta\}$ and $\mathcal{P}_{1,\beta} = \{\alpha = 1, \forall \beta\}$.

Using Lemma 1, we are interested only in $I_{\alpha,0}$ and $I_{0,\beta}$. Therefore, at most one user cooperates, so

$$
\begin{cases}
\alpha_* = \arg\max_{\alpha \in [0,1]} I_{\alpha,0} \\
\beta_* = \arg\max_{\beta \in [0,1]} I_{0,\beta}
\end{cases}
\tag{6.36}
$$

The derivatives of $I_{\alpha,0}$ and $I_{0,\beta}$, $\frac{dI_{\alpha,0}}{d\alpha}$ and $\frac{dI_{0,\beta}}{d\beta}$ give

$$
\begin{cases}
\tilde{\alpha}^2 = \dfrac{C}{A_1} + \left(\dfrac{A_2}{A_1}\right)^2 \\[2ex]
\tilde{\beta}^2 = \dfrac{C'}{A_1'} + \left(\dfrac{A_2'}{A_1'}\right)^2
\end{cases}
\tag{6.37}
$$

Therefore, α_* exists when

$$
\begin{cases}
\dfrac{C}{A_1} + \left(\dfrac{A_2}{A_1}\right)^2 > 0 \\[2ex]
-\left(\dfrac{A_2}{A_1}\right) + \sqrt{\dfrac{C}{A_1} + \left(\dfrac{A_2}{A_1}\right)^2} \in [0,1]
\end{cases}
\tag{6.38}
$$

and it becomes easy to lead to

$$
\begin{cases}
\gamma > \gamma_{01}^2 + \gamma_{01} \\[2ex]
\gamma_{02} > \dfrac{(1 + \gamma_{01})^2 (1 + \gamma)}{\gamma - (\gamma_{01}^2 + \gamma_{01})} - 1
\end{cases}
\tag{6.39}
$$

and the same method is intended for β_*.

Interpretations: We remark that zero or at most one user out of the two cooperates with the other one. Hence, the two users will never both take the role of relay on a given channel realization. In fact the user with "worse" channel conditions always acts selfishly and concentrates all its power for its own data, whereas the other user will graciously help the selfish user or possibly be selfish also. Of course, the roles of selfish users and cooperative users will be alternating randomly as the channel changes, so that in the long run both users are going to participate in the cooperation at some points and benefit from it at some other points. Interestingly, this result deprives the otherwise appealing concept of mutual cooperation from much of its sense. However, a truly mutual cooperation remains possible on the basis of averaging across many realizations of the fading channel.

We now examine the interesting particular case of an instantaneously symmetric channel:

LEMMA 2

In the particular case, when $\gamma = \gamma_{01} = \gamma_{02}$, the two users act selfishly.

Algorithm 1 Power allocation with instantaneous CSI

The implementation of the algorithm below requires a centralized power allocation procedure done by for example, the base.

The following intermediate quantities are computed:

$A_1 = K_1 \gamma_{01}^2 (1 + \gamma + \gamma_{02})$

$A_2 = K_1 \gamma_{01} (1 + \gamma)(1 + \gamma + \gamma_{02})$

$C = K_1 \left[\gamma \frac{K_1}{\gamma_{01}} - \frac{K_2}{\gamma_{02}} (1 + \gamma) - K_2 (2 + \gamma + \gamma_{01}) \right]$

$A_1' = K_2 \gamma_{02}^2 (1 + \gamma + \gamma_{01})$

$A_2' = K_2 \gamma_{02} (1 + \gamma)(1 + \gamma + \gamma_{01})$

$C' = K_2 \left[\gamma \frac{K_2}{\gamma_{02}} - \frac{K_1}{\gamma_{01}} (1 + \gamma) - K_1 (2 + \gamma + \gamma_{02}) \right]$

$\text{cond1} = \frac{K_2}{1+\gamma}$, $\text{condp1} = \frac{K_1}{1+\gamma}$, $\text{cond2} = \frac{(1+\gamma_{02})^2(1+\gamma)}{\gamma - (\gamma_{02}^2 + \gamma_{02})} - 1$ and $\text{condp2} = \frac{(1+\gamma_{01})^2(1+\gamma)}{\gamma - (\gamma_{01}^2 + \gamma_{01})} - 1$.

if $\gamma > \text{cond1}$ & $\gamma_{01} > \text{cond2}$, **then**

$$\alpha_* = -\frac{A_2}{A_1} + \sqrt{\frac{C}{A_1} + \left(\frac{A_2}{A_1}\right)^2}$$

user 1 cooperates with a level given by α_*, resulting in a sum rate of $I_{\alpha_*,0}$.

else

if $\gamma > \text{condp1}$ & $\gamma_{02} > \text{condp2}$, **then**

$$\beta_* = -\frac{A_2'}{A_1'} + \sqrt{\frac{C'}{A_1'} + \left(\frac{A_2'}{A_1'}\right)^2}$$

user 2 cooperates with a level given by β_*, resulting in a sum rate of I_{0,β_*}.

else

Decision: No cooperation, $[\alpha_*, \beta_*] = [0, 0]$, sum-rate = $I_{0,0}$.

6.4.4   SIMULATIONS RESULTS

We report results for path loss exponent $\nu = 4$ and we model all channels as Rayleigh block flat fading with AWGN. Figures 6.4 through 6.7 show the outage capacity behavior for the new cooperation scheme with centrally optimized power allocation, compared with the regular MAC channel with no cooperation. We look at the sum rate performance but also plot for information the corresponding individual user rate performance. We take SNR $= \frac{P}{N_0} = 10\,\text{dB}$. In the symmetric case , Figure 6.4 shows a marginal improvement in sum rate due to cooperation in the MAC channel, although

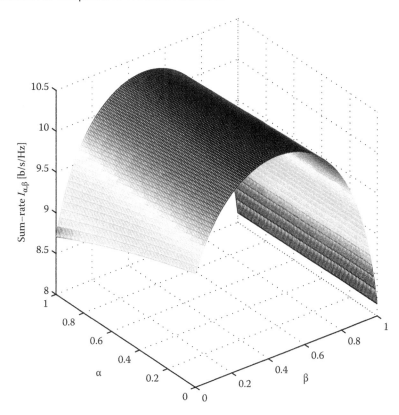

FIGURE 6.4 Outage capacity in centralized case and symmetric network in which we consider equal channel gains ($\sigma^2 = 1$). In this situation, individual users' worst-case rates are slightly improved, while the sum rate is almost unchanged.

the worst case behavior of the individual user rate is improved (for low outage probabilities, less than 0.04).

This is due to the fact that in many cases for the symmetric network both users will tend to act selfishly (as per our result in Proposition 2), and when they do not, the gains made by the user benefiting from cooperation tend to be offset by the losses incurred by the relaying user (the roles of benefitor and relayer alternating randomly with new channel realizations).

Figures 6.5 through 6.7 show the simulation results for an asymmetric (linear) network when user 2 is located between user 1 and the destination at $(0.1, 0)$, $(0.5, 0)$, and $(0.9, 0)$, respectively. The gains due to centrally optimized power allocation in the cooperation are clearly more significant for the user further away from the base.

However, this gain also translates into a sum-rate (system) gain. For instance when $d = 0.1$, the sum rate benefits from cooperation by 0.33 bit/s/Hz and the user 1 benefits by up to 1 bit/s/Hz. User 2, which is closer to the destination than user 1, still benefits on average, but to a lesser extent.

When user 2 is located close to the destination ($d = 0.9$) both the sum rate and user 1's rate benefit from the cooperation, whereas user 2 almost never uses user 1 as a relay. Still, this user has negligible loss of rate in relaying because the amount of power it allocates to relaying user 1's data corresponds to a very small rate loss, while a significant gain to user 1 who undergoes severe channel conditions.

6.5 DISTRIBUTED POWER ALLOCATION

In the distributed framework, each node has a hybrid channel-state information. Therefore, instead of considering the global knowledge of the instantaneous channel realizations, each mobile has only

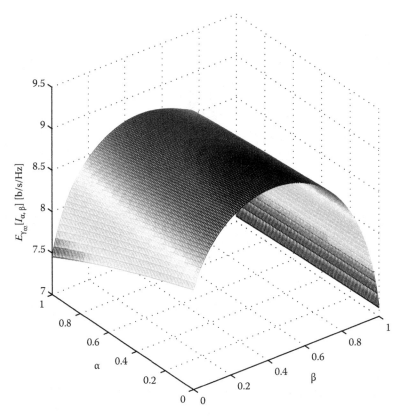

FIGURE 6.5 Outage capacity in centralized case and asymmetric network, with user 2 located at (0.1,0), that is, close to the user 1; Both users as well as the sum rate benefit from cooperation, especially mobile 1.

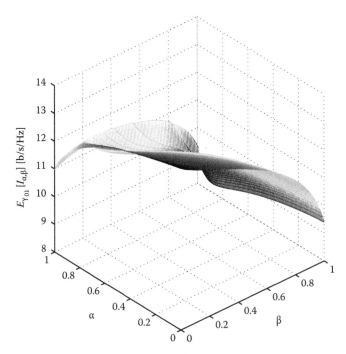

FIGURE 6.6 Outage capacity in centralized case and asymmetric network, with user 2 located at (0.5,0), that is, halfway between user 1 and destination; Only user 1 benefits from cooperation; however, the sum rate is also improved.

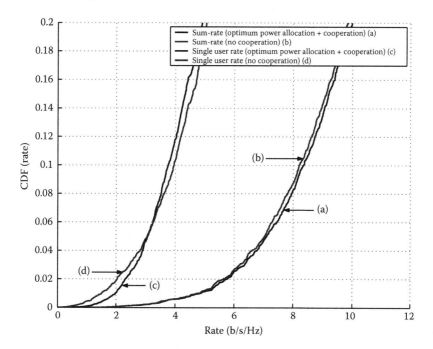

FIGURE 6.7 Outage capacity in centralized case and asymmetric network with user 2 located at (0.9,0), that is, close to the destination; both the sum-rate and user 1's rate benefit from the cooperation, while user 2 almost never uses user 1 as a relay.

a local CSI knowledge, that is, each mobile has the perfect knowledge of its links with the base station and the other mobile, but only a statistical knowledge of the link between the other mobile and the base station.

In this framework, each user is optimizing on an individual basis the amount of power allocated to relaying the other user's data. We draw the reader's attention to the fact that, although the users optimize their power allocation in a distributed manner, they will do so in a cooperative fashion since each user has the average sum rate as an objective function to maximize rather than its own individual rate. To perform the optimization, user 1 (resp. user 2) must estimate the complete vector $[\alpha, \beta]$, following which it implements a power control based on α (resp. β) while the other variable plays an auxiliary role only.

Before proceeding to give the optimal distributed power allocation solution, we provide a characterization of the objective function that each user sets out to maximize. Let $\bar{I}_i$ the expected sum rate seen by mobile i. By construction, the expected sum rate is function of the local CSI, known deterministically by mobile (i) and averaged over all realizations of the channel gains, which are nonlocally observable by this mobile. In this chapter, the example of distributed scenario considered assumes that mobile i has instantaneous knowledge of γ_{0i} and γ, whereas only the statistics of $\gamma_{0j}, j \neq i$ are known to this mobile. Thus, we can define

$$\bar{I}_1(\alpha, \beta) = \mathbb{E}_{\gamma_{02}}\left[I_{\alpha,\beta}\right] \tag{6.40a}$$

$$\bar{I}_2(\alpha, \beta) = \mathbb{E}_{\gamma_{01}}\left[I_{\alpha,\beta}\right] \tag{6.40b}$$

where $\mathbb{E}$ is the expectation operator.

In order to seek the optimal distributed power allocation, we start by developing the expressions in Equation 6.40.

LEMMA 3

For the mobile 1, $\bar{I}_1(\alpha, \beta)$ is defined $\forall \alpha$ as

$$
\begin{cases}
\log_2(a) - \Phi(0) - \log_2(l_2(\alpha)), \text{if } \beta = 0 \\[2mm]
-\left[\dfrac{\exp\left(\dfrac{a}{b(1)}\right)Ei\left(-\dfrac{a}{b(1)}\right) - \ln(a)}{\ln(2)}\right] + \left[\dfrac{\exp\left(\dfrac{1}{\xi(1)}\right)Ei\left(-\dfrac{1}{\xi(1)}\right)}{\ln(2)}\right] \\[4mm]
\quad -\left[\dfrac{\exp\left(\dfrac{c}{d(1)}\right)Ei\left(-\dfrac{c}{d(1)}\right) - \ln(c)}{\ln(2)}\right] - \log_2(l_2(\alpha)), \quad \text{if } \beta = 1 \\[4mm]
-\left[\dfrac{\exp\left(\dfrac{a}{b(\beta)}\right)Ei\left(-\dfrac{a}{b(\beta)}\right) - \ln(a)}{\ln(2)}\right] + \left[\dfrac{\exp\left(\dfrac{1}{\xi(\beta)}\right)Ei\left(-\dfrac{1}{\xi(\beta)}\right)}{\ln(2)}\right] \\[4mm]
\quad - \log_2(l_2(\alpha)) - \Phi(\beta), \quad \text{if } 0 < \beta < 1
\end{cases}
\tag{6.41}
$$

where $l_2(\alpha)$ has been previously defined and

$$
\begin{aligned}
a &= (1 + \gamma_{01})(1 + (1-\alpha)\gamma_{01}) \\[1mm]
\xi(\beta) &= \frac{\beta\bar{\gamma}_{02}}{1+\gamma} \\[1mm]
b(\beta) &= \xi(\beta)(1 + \gamma + \gamma_{01}) \\[1mm]
c &= (1+\gamma)(1 + \alpha\gamma_{01}) \\[1mm]
d(\beta) &= \left[(2-\beta)(1+\gamma) + \alpha\gamma_{01}\right]\bar{\gamma}_{02} \\[1mm]
f(\beta) &= (1-\beta)(1+\gamma)\left(\bar{\gamma}_{02}\right)^2 \\[1mm]
\Delta(\beta) &= [d(\beta)]^2 - 4cf(\beta) \\[1mm]
\Lambda_1(\beta) &= \frac{d(\beta) - \sqrt{\Delta(\beta)}}{2f(\beta)} \\[1mm]
\Lambda_2(\beta) &= \frac{d(\beta) + \sqrt{\Delta(\beta)}}{2f(\beta)} \\[1mm]
\Phi(\beta) &= \left[\frac{\exp\left(\Lambda_1(\beta)\right)Ei\left(-\Lambda_1(\beta)\right)}{\ln(2)}\right] + \left[\frac{\exp\left(\Lambda_2(\beta)\right)Ei\left(-\Lambda_2(\beta)\right)}{\ln(2)}\right] - \log_2(c)
\end{aligned}
\tag{6.42}
$$

where $Ei(.)$ is the exponential integral defined as $Ei(x) = \int_{-\infty}^{x}\frac{\exp(t)}{t}dt$. The notation of the dependence on α in Equation 6.42 is omitted.

We deduce the relation for mobile 2 by changing γ_{01} in Equation 6.41 by γ_{02}, $\bar{\gamma}_{02}$ in Equation 6.41 by $\bar{\gamma}_{01}$, α by β, and vice versa.

Proof: We start by some mathematical analysis and according to Ref. [26]

$$\int_0^\infty \log_2 (A + B\lambda) \exp(-\lambda) \, d\lambda = - \left[\frac{\exp\left(\dfrac{A}{B}\right) Ei\left(-\dfrac{A}{B}\right)}{\ln(2)} \right] + \log_2(A) \qquad (6.43)$$

where $Ei(.)$ is the exponential integral defined as $Ei(x) = \int_{-\infty}^x \frac{\exp(t)}{t} dt$, and

$$\int_0^\infty \log_2 \left(A + B\lambda + C\lambda^2\right) \exp(-\lambda) \, d\lambda = - \left[\frac{\Upsilon - \ln(A)}{\ln(2)} \right] \qquad (6.44)$$

where

$$\Upsilon = \exp(R_1) Ei(-R_1) + \exp(R_2) Ei(-R_2)$$

$$R_1 = \frac{B - \sqrt{B^2 - 4AC}}{2C} \qquad (6.45)$$

$$R_2 = \frac{B + \sqrt{B^2 - 4AC}}{2C}$$

$\bar{I}_{\gamma_{02}}(\alpha, \beta)$ will have different forms according to the value of β, and the notation of the dependence on α in Equation 6.46 is omitted. Therefore, if

$$\beta = 0, I_{\alpha,0} = \log_2(a) + \log_2\left(c + d(0)\lambda + f(0)\lambda^2\right) - \log_2(l_2(\alpha))$$
$$\beta = 1, I_{\alpha,1} = \log_2(a + b(1)\lambda) - \log_2(1 + \xi(1)\lambda) + \log_2(c + d(1)\lambda) - \log_2(l_2(\alpha))$$

else

$$I_{\alpha,\beta} = \log_2(a + b(\beta)\lambda) - \log_2(1 + \xi(\beta)\lambda) + \log_2\left(c + d(\beta)\lambda + f(\beta)\lambda^2\right) - \log_2(l_2(\alpha))$$

where

$$a = (1 + \gamma_{01})(1 + (1 - \alpha)\gamma_{01})$$

$$\xi(\beta) = \frac{\beta \bar{\gamma}_{02}}{1 + \gamma}$$

$$b(\beta) = \xi(\beta)(1 + \gamma + \gamma_{01})$$

$$c = (1 + \gamma)(1 + \alpha\gamma_{01}) \qquad (6.46)$$

$$d(\beta) = \left[(2 - \beta)(1 + \gamma) + \alpha\gamma_{01}\right] \bar{\gamma}_{02}$$

$$f(\beta) = (1 - \beta)(1 + \gamma) \left(\bar{\gamma}_{02}\right)^2$$

$$\lambda = \frac{\gamma_{02}}{\bar{\gamma}_{02}}$$

Therefore, using the relations (Equations 6.43 and 6.44) we have the result in Lemma 1.

6.5.1 Optimal Distributed Power Allocation

Taking Equation 6.40a and 6.40b respectively, for mobile 1 and 2 as the objective functions, the optimal distributed power allocation problem can be stated as

$$\begin{cases} (\alpha_*^1, \beta_*^1) = \arg\max_{\alpha,\beta} \bar{I}_1(\alpha, \beta) \\ (\alpha_*^2, \beta_*^2) = \arg\max_{\alpha,\beta} \bar{I}_2(\alpha, \beta) \end{cases} \qquad (6.47)$$

where mobile 1 (resp. mobile 2) is mainly concerned with α_*^1 (resp. β_*^2). Therefore, the distributed power allocation vector resulting from our scheme will be (α_*^1, β_*^2). Due to the complex expressions in Equation 6.41, it appears difficult at first glance to give a closed-form solution for the optimal power allocation vector. However, the following result helps reach a surprisingly simple solution to our problem.

PROPOSITION 3

$\forall \alpha_0 \neq 0, \forall \beta_0 \neq 0, (\alpha_0, \beta_0)$ *cannot be an optimal solution for power allocation for mobile 1 nor mobile 2.*

Proof: A subtle technical point is under which conditions the first-order derivative of $\mathbb{E}_{\gamma_{02}}[I_{\alpha,\beta}]$ can be taken inside the expectation operator:

$$\begin{cases} \dfrac{\partial}{\partial \alpha}\mathbb{E}_{\gamma_{02}}[I_{\alpha,\beta}] = \mathbb{E}_{\gamma_{02}}\left[\dfrac{\partial I_{\alpha,\beta}}{\partial \alpha}\right] \\ \dfrac{\partial}{\partial \beta}\mathbb{E}_{\gamma_{02}}[I_{\alpha,\beta}] = \mathbb{E}_{\gamma_{02}}\left[\dfrac{\partial I_{\alpha,\beta}}{\partial \beta}\right] \end{cases} \tag{6.48}$$

For simplicity, and without loss of generality, we consider mobile 1.

THEOREM 1

If the following two conditions hold at a point α (resp. β), then $\bar{I}_1(\alpha, \cdot)$ (resp $\bar{I}_1(\cdot, \beta)$) is differentiable at α (resp β) and Equations 6.48 hold [27]:

(i) The function $I_{.\beta}$ (resp $I_{\alpha,.}$) is differentiable at α (resp β) w.p.l.
(ii) There exists a positive valued random variable $K(\gamma_{02})$ such that $\mathbb{E}_{\gamma_{02}}(K(\gamma_{02}))$ is finite and the inequality

$$|I_{\alpha_1,\beta} - I_{\alpha_2,\beta}| \leq K(\gamma_{02})|\alpha_1 - \alpha_2| \tag{6.49}$$

resp.

$$|I_{\alpha,\beta_1} - I_{\alpha,\beta_2}| \leq K(\gamma_{02})|\beta_1 - \beta_2| \tag{6.50}$$

holds w.p.l for all α_1, α_2 (resp. β_1, β_2) in a neighborhood of α (resp. β).
Note $I_{.\beta}$ satisfies the conditions (i) and (ii).
Now, suppose that $\alpha_0 \neq 0$ and $\beta_0 \neq 0$, therefore (α_0, β_0) is an optimal power allocation for mobile 1 if

$$\begin{cases} \dfrac{\partial}{\partial \alpha}\mathbb{E}_{\gamma_{02}}[I_{\alpha,\beta}]\,|_{\alpha=\alpha_0} = 0 \\ \dfrac{\partial}{\partial \beta}\mathbb{E}_{\gamma_{02}}[I_{\alpha,\beta}]\,|_{\beta=\beta_0} = 0 \end{cases} \tag{6.51}$$

Therefore Equations 6.40 become

$$\begin{cases} \int \dfrac{\partial I_{\alpha,\beta}}{\partial \alpha}|_{\alpha=\alpha_0}p_{\gamma_{02}}\mathrm{d}\gamma_{02} = 0 \\ \int \dfrac{\partial I_{\alpha,\beta}}{\partial \beta}|_{\beta=\beta_0}p_{\gamma_{02}}\mathrm{d}\gamma_{02} = 0 \end{cases} \tag{6.52}$$

However, (α_0, β_0) cannot maximize $I_{\alpha,\beta}$ (see Proposition 2); therefore, $\frac{\partial I_{\alpha,\beta}}{\partial \alpha}|_{\alpha=\alpha_0} \neq 0$ or $\frac{\partial I_{\alpha,\beta}}{\partial \beta}|_{\beta=\beta_0} \neq 0$, and γ_{02} is exponentially distributed ($p_{\gamma_{02}}(\gamma_{02}) > 0 \forall \gamma_{02}$), so this leads to a contradiction with Equation 6.52 because at least one of the two equations cannot be held.

Consequently, in a distributed scenario with hybrid CSI at the users, we find that from the point of view of each mobile, at least one of the two users should not cooperate. In order to determine which one shall cooperate, mobile 1 considers the two following cases: either it shall not cooperate, or the other mobile shall not cooperate. In each case, a power allocation coefficient is determined. Thus, for mobile 1, the following problem is solved:

$$\begin{cases} \alpha_*^1 = \arg\max_\alpha \bar{I}_1(\alpha, 0) \\ \beta_*^1 = \arg\max_\beta \bar{I}_1(0, \beta) \end{cases} \tag{6.53}$$

and for mobile 2, the following problem is solved:

$$\begin{cases} \alpha_*^2 = \arg\max_\alpha \bar{I}_2(\alpha, 0) \\ \beta_*^2 = \arg\max_\beta \bar{I}_2(0, \beta) \end{cases} \tag{6.54}$$

The following lemma helps further simplify the problem:

LEMMA 4

In Equation 6.53 $\alpha_*^1 = 0$ and in Equation 6.53 $\beta_*^2 = 0$.

Proof: Suppose that $\alpha_0 \neq 0$; therefore, α_0 will be allocated to mobile 1 if

$$\frac{\partial \bar{I}_{\gamma_{02}}(\alpha, 0)}{\partial \alpha}\bigg|_{\alpha=\alpha_0} = 0 \tag{6.55}$$

Therefore, we derive the following expression of $\frac{\partial \bar{I}_{\gamma_{02}}(\alpha,0)}{\partial \alpha}$

$$-\frac{1}{\ln(2)}\left[\frac{\gamma_{01}}{1+(1-\alpha)\gamma_{01}} + \frac{\gamma_{01}}{1+\gamma+\alpha\gamma_{01}}\right] - \frac{\partial \Phi(0)}{\partial \alpha} \tag{6.56}$$

Using the relation below

$$\frac{\partial(\exp(\varepsilon)Ei(-\varepsilon))}{\partial \alpha} = \left(\frac{\partial \varepsilon}{\partial \alpha}\right)\left[\exp(\varepsilon)Ei(-\varepsilon) + \frac{\partial \varepsilon}{\partial \alpha}\right] \tag{6.57}$$

Replacing ε by $\Lambda_1(0)$ and $\Lambda_2(0)$ in Equation 6.55, it becomes easy to show that

$$\frac{\partial \bar{I}_{\gamma_{02}}(\alpha, 0)}{\partial \alpha} < 0, \quad \forall \alpha \neq 0 \tag{6.58}$$

and because $\bar{I}_{\gamma_{02}}(\alpha, 0)$ is differentiable in 0, decreases in $[0, 1]$, therefore $\alpha_*^1 = 0$.

PROPOSITION 4

In the distributed power allocation problem represented in Equation 6.47, the optimal strategy is for each user to act selfishly.

Proof: Lemma 4 indicates that from the point of view of mobile 1 (resp. mobile 2), the optimal power allocation strategy takes the form $(0, \beta_*^1)$ (resp. the form $(\alpha_*^2, 0)$, hence neither mobile is going to cooperate. Note that this result is valid for any individual value of the average SNR on all links, thus for any position of the nodes. Our results mean that, to gain from relaying in terms of sum rate, access to instantaneous knowledge of all channel links is crucial.

6.5.2 SIMULATION RESULTS

We provide examples of this behavior through 3D plots of the sum rate for particular instances of the channels. As depicted in Figures 6.8 and 6.9, mobile 1 has $(0, 0.447)$ as the optimal power allocation, while $(0, 0)$ is the optimal power allocation for mobile 2. Therefore, the distributed algorithm allocates zero power for each mobile to relay the data of the other one and thus act selfishly.

Figures 6.8 through 6.10 serve to illustrate the shape of the sum rate and the function of the power allocation coefficients in both centralized and distributed frameworks. For SNR = 10 dB and when mobile 2 is placed at $(0.1, 0)$, Figure 6.10 gives an example where mobile 2 must allocate a fraction $\beta_* = 0.446$ to relay information for mobile 1, while mobile 1 allocates zero power to relay as this maximizes the sum rate in the centralized framework.

In the distributed framework, we plot the expected sum rates "seen" respectively by user 1 and 2 in Figures 6.8 and 6.9. In each case, the expected sum rate is maximized locally by assigning

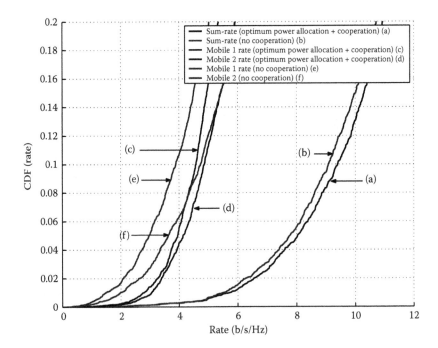

FIGURE 6.8 Expected sum-rate seen by mobile 1 (distributed processing) when the mobile 2 is located at $(0.1, 0)$ in linear network and SNR equal to 10 dB. The relations in Equation 6.53 give $(0, 0.447)$ as optimal power allocation for mobile 1, therefore it does not cooperate.

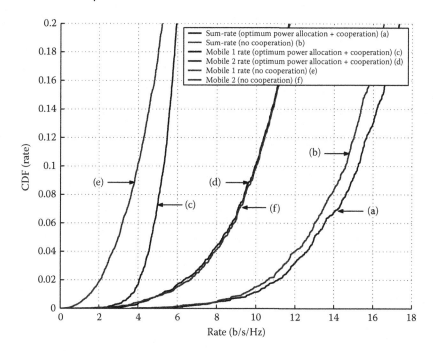

FIGURE 6.9 Expected sum-rate seen by mobile 2 (distributed processing) when it is located at (0.1, 0) in linear network and SNR equal to 10 dB. The relations in Equation 6.54 give (0, 0) as optimal power allocation for mobile 2, therefore it should not cooperate.

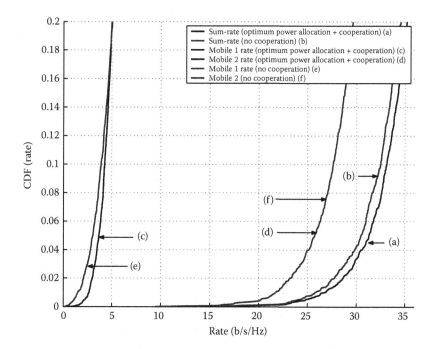

FIGURE 6.10 The sum-rate, in centralized case and linear network, when the mobile 2 is located at (0.1, 0) and SNR equal to 10 dB. The centralized algorithm gives (0, 0.446) as optimal power allocation indicating that only mobile 2 will cooperate.

zero power to relaying, resulting in a completely selfish behavior by both users, as predicted by our Proposition 4.

6.6   CONCLUSIONS

We have addressed the problem of mutual cooperation between data-carrying users in a wireless MAC channel. Via a power allocation framework, we proposed a scheme allowing each user to balance transmission of own data with a relaying operation to maximize the sum rate received at the base station. We have characterized analytically the optimum cooperation levels, and we showed that cooperation is never mutual on an instantaneous basis as at most one user acts as a relay for the other but never both at the same time.

In the case where gathering complete Channel State Information at the Transmitter (CSIT) is problematic, we have addressed the problem of distributed power allocation in the cooperative MAC channel, where the users adjust their cooperation power levels as a function of a mix of local channel-state information and statistical nonlocal channel information. We showed that, unlike in the centralized case, both users should act selfishly in the distributed framework.

REFERENCES

[1]   K. Azarian, H. E. Gamal, and P. Schniter, On the achievable diversity-multiplexing tradeoff in half-duplex cooperative channels, *IEEE Transactions on Informations Theory*, 51(12), 4152–4172, Dec. 2005.

[2]   J. N. Laneman, D. N. Tse, and G. W. Wornell, Cooperative diversity in wireless networks: Efficient protocols and outage behavior, *IEEE Transaction on Information Theory*, 50(12), 3062–3080, Dec. 2004.

[3]   A. Sendonaris, E. Erkip, and B. Aazhang, User cooperation diversity, part I: System description, *IEEE Transactions on Communication*, 51(11), 1927–1938, Nov. 2003.

[4]   A. Sendonaris, E. Erkip, and B. Aazhang, User cooperation diversity, part II: Implementation aspects and performance analysis, *IEEE Transactions on Communication*, 51(11), 1939–1948, Nov. 2003.

[5]   T. M. Cover and C. S. K. Leung, An achievable rate region for the multiple-access channel with feedback, *IEEE Transactions on Informations Theory*, IT-27(3), 292–298, May 1981.

[6]   T. M. Cover and A. A. E. Gamal, Capacity theorems for relay channel, *IEEE Transactions on Informations Theory*, IT-25(5), 572–584, Sept. 1979.

[7]   G. Kramer, M. Gastpar, and P. Gupta, Cooperative strategies and capacity theorems for relay networks, *IEEE Transactions on Informations Theory*, 51(9), 3037–3063, Sept. 2005.

[8]   C. T. K. Ng and A. J. Goldsmith, Capacity and power allocation for transmitter and receiver cooperation in fading channels, in *Proceedings of IEEE International Conference on Communications*, Istanbul, Turkey, June 2006.

[9]   C. T. K. Ng, N. Jindal, A. J. Goldsmith, and U. Mitra, Capacity gain from two-transmitter and two-receiver cooperation, *IEEE Transactions on Information Theory*, 53(10), 3822–3827, Oct. 2007.

[10]   M. A. Khojastepour, A. Sabharwal, and B. Aazhang, Improved achievable rates for user cooperation and relay channels, in *Proceedings of IEEE International Symposium on Information Theory*, Chicago, IL, June 2004.

[11]   K. Azarian, Y.-H. Nam, and H. E. Gamal, Multi-user diversity without transmitter csi, in *Proceedings of IEEE International Symposium on Information Theory*, Adelaide, Australia, Sept. 2005.

[12]   Z. Yang and A. Høst-Madsen, Rateless coded cooperation for multiple-access channels in the low power regime, in *Proceedings of IEEE International Symposium on Information Theory*, Seattle, WA, June 2006.

[13]   A. Høst-Madsen, Capacity bounds for cooperative diversity, *IEEE Transactions on Information Theory*, 52(4), 1522–1544, Apr. 2006.

[14]   T. E. Hunter, S. Sanayei, and A. Nosratinia, Outage analysis of coded cooperation, *IEEE Transactions on Information Theory*, 52(2), 375–391, Feb. 2006.

[15]   A. Nosratinia and T. E. Hunter, Grouping and partner selection in cooperative wireless networks, *IEEE JSAC Special Issue on Cooperative Communications and Networking*, 25(2), 369–378, Feb. 2007.

[16]   A. Murugan, K. Azarian, and H. E. Gamal, Cooperative lattice coding and decoding in half-duplex channels, *IEEE JSAC Special Issue on Cooperative Communications and Networking*, 25(2), 268–279, Feb. 2007.

[17] T. C.-Y. Ng and W. Yu, Joint optimization of relay strategies and resource allocations in cooperative cellular networks, *IEEE Journal on Selected Areas in Communications*, 25(2), 328–339, Feb. 2007.

[18] D. Chen, K. Azarian, and J. N. Laneman, A case for amplify-forward relaying in the block-fading multiaccess channel, *IEEE Transactions on Information Theory*, 54(8), 3728–3733, Aug. 2008.

[19] K. Tourki, D. Gesbert, and L. Deneire, Cooperative diversity using per-user power control in the multiuser mac channel, in *Proceedings of IEEE International Symposium on Information Theory*, Nice, France, June 2007.

[20] O. Kaya and S. Ulukus, Power control for fading multiple access channels with user cooperation, in *Proceedings of IEEE International Conference on Wireless Networks, Communications, and Mobile Computing*, Maui, Hawaii, June 2005.

[21] J. Adeane, M. R. D. Rodrigues, and I. J. Wassell, Centralised and distributed power allocation algorithms in cooperative networks, in *Proceedings of the 6th IEEE International Workshop on Signal Processing Advances for Wireless Communications*, New York, June 2005.

[22] J. N. Laneman, G. W. Wornell, and D. N. C. Tse, An Efficient Protocol for Realizing Cooperative Diversity in Wireless Networks, in *Proceedings of IEEE International Symposium on Information Theory*, Washington, DC, June 2001.

[23] K. Tourki, M.-S. Alouini, and L. Deneire, Blind Cooperative Diversity using Distributed Space-Time Coding in Block Fading channels, in *Proceedings of IEEE International Conference on Communications*, Shanghai, Chine, Mai, 2008.

[24] J. N. Laneman and G. W. Wornell, Energy-efficient antenna sharing and relaying for wireless networks, in *Proceedings of IEEE Wireless communication and Networking Conference (WCNC)*, Chicago, IL, Sept. 2000.

[25] A. Nosratinia, T. E. Hunter, and A. Hedayat, Cooperative Communication in Wireless Networks, *IEEE Communications Magazine*, 42(10), 74–80, Oct. 2004.

[26] I. S. Gradshteyn and I. M. Ryshik, *Table of Integrals, Series, and Products*. 6th edn., Academia Press, San Diego, 2000.

[27] V. I. Norkin, G. C. Pflug, and A. Ruszczynski, A branch and bound method for stochastic global optimization. *Mathematical Programming*: Series A and B, 83(3), 425–450, Nov. 1998.

7 Joint Power Allocation and Partner Selection in CD Systems

Veluppillai Mahinthan, Lin Cai, Jon W. Mark, and Xuemin (Sherman) Shen

CONTENTS

Maximizing the cooperative diversity (CD) energy gain in wireless cellular networks requires joint efforts from several layers of the layered communication model. Based on the physical layer CD scheme used, the optimal solution depends on the optimal power allocation strategy and the optimal matching of active users in a radio cell, aimed to enhance the performance of not only a particular user but also the whole network. In this chapter, we study how to appropriately match users for two-user CD systems deploying an optimal power allocation strategy. Optimal power allocation solutions, which can minimize the total energy consumptions for the cooperating pair, are derived. We further study the location of the optimal partner for a user and develop a novel matching algorithm with performance very close to the optimal maximum-weighted (MW) matching algorithm, but with much lower computational complexity.

7.1 INTRODUCTION

Future wireless networks are expected to provide much higher data rates, energy efficiency, and reliability in a cost-effective manner. With the state-of-the-art channel coding, energy and bandwidth efficiency in point-to-point communications can be made to approach the Shannon limit. On the other hand, battery life becomes the bottleneck for wireless devices. To meet the ever-increasing demand for higher data rates and longer battery life, a promising approach to further improve the energy and bandwidth efficiency is diversity reception. Cooperation among a group of users to transmit and relay the same signal can emulate a multiple transmit antennae environment to achieve spatial diversity gains. With the broadcast nature of the wireless channel, when a source transmits signals to a destination, neighboring users can also receive the signals. These neighboring users can relay the signals to the destination. The performance of CD schemes depends heavily on the interuser channel condition. For example, (i) the interuser channels incur transmission errors in practice, (ii) the implementation complexity increases with the number of users participating in the cooperation, and (iii) the spectral efficiency of the wireless channel decreases with the number of participating users. Thus, in this chapter, we consider the cooperation between two users, that is, two sources relaying for each other.

Wireless mobile devices are normally battery powered. It is important to minimize the energy consumption to maximize the time the wireless device can be functional without recharging or replacing the battery. Although CD energy gain for a single group of users has become an active research topic, how much energy gain can be achieved for a network that employs a CD scheme and how the diversity gain can be maximized for the whole network are important issues. In a wireless mobile network, user mobility further complicates the grouping problem. The mobile users' velocities and moving directions can change over time, which affects the CD gain of the network.

Each individual user has its own preference in choosing its partner. The objective of an individual user (maximizing its own energy gain by cooperation) may conflict with the objective of the network (maximizing the energy gain of the whole network). In a mobile environment, the best grouping at the current time instant may not be the best at a future time instant. Thus, our objective is to group active users in a radio cell to maximize the energy gain of the whole network by taking user mobility into consideration. This optimization problem involves the particular physical layer CD scheme used, the transmission power level of each user to satisfy their quality of service (QoS) requirements, and the partner selection or matching strategy for the whole network.

The physical layer CD scheme used and the power level of each user together should decide the degree of CD energy gain of the cooperating pair of users or the energy gain of the individual user. The power needed for cooperative transmission or the power gain of cooperative transmission over noncooperative transmission can be used as weights for the matching (partner selection) algorithm at the network layer.* In addition, considering user mobility, the power allocation and matching algorithm should have low overhead, low computational complexity, and easy implementation.

Most of the power allocation strategies for CD schemes reported in the literature [1–6] are minimizing error performance of the CD schemes for fixed total power, by appropriately distributing the power among the source and relays. These schemes mainly focus on error probability reduction or coverage extension. Different from the previous work, we propose optimum power allocation strategies by minimizing the power consumption of the pair of users to ensure their error probability requirements (which might be different for different applications).

For the partner selection problem, in Refs. [7,8] a cooperating partner is chosen based on whether the cooperation can result in improved frame error rate or throughput. In a multi-hop wireless sensor networking environment, a partner selection strategy is given in Ref. [9] based on the number of packets transmitted. The partner selection schemes in Refs. [7–9] aim at optimizing the performance of a single user or a pair of users. On the other hand, in Refs. [10–13], centralized and distributed

---

* In this chapter, the terms partner selection and partner matching are used interchangeably.

matching algorithms, such as minimal-weighted, Greedy, and random matching, are presented for CD networks, and average outage probability given by matching is studied. In contrast, we propose a partner selection algorithm based on energy saving, which enhances not only individual user performance but also the network performance. The proposed partner selection algorithm can be used in a centralized or distributed network environment. Furthermore, it performs better than the Greedy matching algorithm in both static and mobile user networks [14].

In this chapter, we first formulate the generalized power allocation framework for the two-user CD schemes proposed in the literature. Depending on the physical layer limitations and the operating environment of the CD system, additional power allocation constraints are taken into account. We then study how to match the users to maximize the energy gain of the whole network. Finally, open research issues are discussed.

7.2   SYSTEM MODEL

Figure 7.1 shows a wireless cellular network where the base station (BS) of a radio cell supports N mobile users. We consider the scenario in which a user is capable of cooperating with another user, that is, cooperation between two active users. The BS and the mobile devices each has a single antenna. The uplink signals transmitted by the sender and relayed by the partner are combined at the BS. The CD scheme thus emulates a "two inputs one output" (2I1O) situation.

Generally, the error probability (P_e), which can be bit error probability (BEP) (P_b), symbol error probability (P_s), or outage probability (P_o), of a CD system depends on the received signal-to-noise ratio (SNR) at the relay and the destination. Without loss of generality, consider a CD system in which user 1 (u_1) cooperates with user 2 (u_2). The generalized error probability of u_1, P_{e1}, can be written for high SNRs when u_1 cooperates with u_2

$$P_{e1} = \frac{A}{\bar{\gamma}_1 \bar{\gamma}_{1,2}} + \frac{B}{\bar{\gamma}_1 \bar{\gamma}_2} + \frac{C}{\bar{\gamma}_1 \bar{\gamma}_2 \bar{\gamma}_{1,2}} \tag{7.1}$$

where the average received SNR at the destination for the transmitted signal from the source (u_1) is $\bar{\gamma}_1 = \sigma_1^2 \frac{E_{b1}^S}{N_0}$; the average received SNR at the destination for the transmitted signal from the relay (u_2)

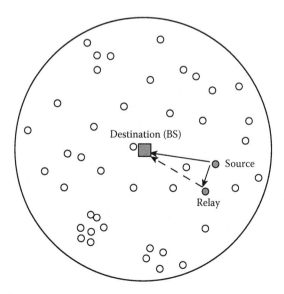

FIGURE 7.1   Cellular network with user cooperation.

is $\bar{\gamma}_2 = \sigma_2^2 \frac{E_{b2}^R}{N_0}$; and the average received SNR at u_2 of the transmitted signal from u_1 is $\bar{\gamma}_{1,2} = \sigma_{1,2}^2 \frac{E_{b1}^S}{N_0}$. The σ^2's are the variances of the respective Rayleigh fading channel coefficients, which are defined as the average channel state information (CSI). E_{b1}^S and E_{b2}^R are the energies spent by the source (u_1) and the relay (u_2) in transmitting one bit for u_1. The parameters A, B, C, and P_e depend on the CD systems [3,6,15–21]. Exceptionally, a few other CD systems proposed in the literature have a more complicated error performance equation than Equation 7.1. In that case, the technique used in Ref. [14] for fixed CD scheme of Ref. [22] can be used with the same power allocation framework presented in the next section [23–25].

Define $k_1 = \frac{\sigma_{1,2}^2}{\sigma_1^2}$ and $k_2 = \frac{\sigma_{2,1}^2}{\sigma_2^2}$. Due to the broadcast nature of the channel, $\bar{\gamma}_{1,2} = k_1 \bar{\gamma}_1$ and $\bar{\gamma}_{2,1} = k_2 \bar{\gamma}_2$. Furthermore, $\sigma_2^2 = \frac{k_1}{k_2}\sigma_1^2$ because of the reciprocity of the interuser channel ($\sigma_{1,2}^2 = \sigma_{2,1}^2$). Therefore, Equation 7.1 can be written as

$$P_{e1}(E_{b1}^S, E_{b2}^R) = \frac{Ak_1 N_0^2 \sigma_1^2 E_{b2}^R + Bk_1 k_2 \sigma_1^2 N_0^2 E_{b1}^S + Ck_2 N_0^3}{k_1^2 \sigma_1^6 (E_{b1}^S)^2 E_{b2}^R}. \tag{7.2}$$

Similarly, for u_2,

$$P_{e2}(E_{b2}^S, E_{b1}^R) = \frac{Ak_2 N_0^2 \sigma_1^2 E_{b1}^R + Bk_1 k_2 \sigma_1^2 N_0^2 E_{b2}^S + Ck_2 N_0^3}{k_1^2 \sigma_1^6 (E_{b2}^S)^2 E_{b1}^R} \tag{7.3}$$

where E_{b1}^R and E_{b2}^S are the energies spent by the relay (u_1) and the source (u_2) in transmitting one bit for u_2, respectively.

7.3 POWER ALLOCATION

The error probability of a CD system primarily depends on the quality of the interuser channel and user-to-destination channel and the transmitting power of the cooperating users. To minimize the requested power for a guaranteed error probability, the transmitting power should be allocated optimally between the source and the relay according to the channel qualities and the CD system employed. In this context, we first provide a generalized optimization framework and then substantiate the analysis by calculating the optimal powers for the "amplify-and-forward" (AF) CD scheme in Ref. [3] and the regenerate-and-forward CD scheme in Ref. [20] under different operational constraints.

The objective of power allocation is to minimize the combined energy consumption at the sender and the relay, subject to BEP and other appropriate constraints. Accordingly, we seek expressions for the energy consumption at the sender and the relay as a function of BEP. Since power and energy are directly related, in what follows, we use the term power allocation even though the optimization problem is formulated as a minimization of the energy.

7.3.1 GENERALIZED OPTIMIZATION FRAMEWORK

To minimize the power consumption for the pair, the generalized optimization framework can be formulated as follows:

$$\min \left(E_{b1}^S + E_{b2}^S + E_{b1}^R + E_{b2}^R \right) \tag{7.4}$$

$$\text{s.t. } P_{e1}(E_{b1}^S, E_{b2}^R) \le e_1 \quad \text{and} \quad P_{e2}(E_{b2}^S, E_{b1}^R) \le e_2, \tag{7.5}$$

$$E_{b1}^S = \lambda_1 E_{b1}^R \quad \text{and} \quad E_{b2}^S = \lambda_2 E_{b2}^R, \tag{7.6}$$

$$E_{b1}^S = \mu_2 E_{b2}^R \quad \text{and} \quad E_{b2}^S = \mu_2 E_{b1}^R, \tag{7.7}$$

where e_1 and e_2 are the maximum tolerable (guaranteed) error probability for u_1 and u_2, respectively, and the operational constraints λ_1, λ_2, μ_1, and μ_2 given in Equations 7.6 and 7.7 depend on the CD system and the operating environment.

Depending on the operational constraints, we can simplify the formulation as in the following four cases.

Case I

For Case I, if constraints 7.6 and 7.7 can be relaxed, the bit energies can be allocated independently and the λ's and μ's can take any arbitrary values. Since P_{e1} is a function of E_{b1}^S and E_{b2}^R only, and P_{e2} is a function of E_{b2}^S and E_{b1}^R only, the power allocation problem can be decomposed into two independent optimization problems:

$$\min \left(E_{b1}^S + E_{b2}^R \right)$$
$$\text{s.t. } P_{e1}(E_{b1}^S, E_{b2}^R) \leq e_1, \tag{7.8}$$

and

$$\min \left(E_{b2}^S + E_{b1}^R \right)$$
$$\text{s.t. } P_{e2}(E_{b2}^S, E_{b1}^R) \leq e_2. \tag{7.9}$$

Case I is applicable when each user can set different power levels for the source and relay bits [3,6,15,16,19].

Case II

In this case, constraint 7.6 can be relaxed, that is, the λ's can take any arbitrary values. Therefore, E_{b1}^S only depends on E_{b2}^R, and E_{b2}^S only depends on E_{b1}^R. The power allocation problem can be reformulated into two independent optimization problems:

$$\min \left(E_{b1}^S + E_{b2}^R \right)$$
$$\text{s.t. } P_{e1}(E_{b1}^S, E_{b2}^R) \leq e_1 \quad \text{and} \quad E_{b1}^S = \mu_1 E_{b2}^R \tag{7.10}$$

and

$$\min \left(E_{b2}^S + E_{b1}^R \right)$$
$$\text{s.t. } P_{e2}(E_{b2}^S, E_{b1}^R) \leq e_2 \quad \text{and} \quad E_{b2}^S = \mu_2 E_{b1}^R. \tag{7.11}$$

In Case II, constraint 7.7 is due to operating environment or application of the CD scheme. For example, for the schemes in [3,6,15,16,19], we can set $\mu_1 = \mu_2 = 1$ to ensure that the energy spent by both users is equal or set $\mu_1 = 1/\mu_2 = k_2/k_1$ to ensure equal SNR reception at the destination.

Case III

In this case, constraint 7.7 can be relaxed (μ's can take any value), and the power allocation problem can be given by

$$\min \left(E_{b1}^S + E_{b2}^S + E_{b1}^R + E_{b2}^R \right)$$
$$\text{s.t. } P_{e1}(E_{b1}^S, E_{b2}^R) \leq e_1 \quad \text{and} \quad P_{e2}(E_{b2}^S, E_{b1}^R) \leq e_2, \tag{7.12}$$
$$E_{b1}^S = \lambda_1 E_{b1}^R \quad \text{and} \quad E_{b2}^S = \lambda_2 E_{b2}^R$$

This case is applicable when the CD system has a constraint that energy allocated for the source's bit should be equal to the relaying bit in each user [20,22].

Case IV

Case IV is equivalent to our original optimization problem in Equations 7.4 through 7.7. From Equations 7.6 and 7.7, λ's and μ's should satisfy the relationship given by $\lambda_1 \lambda_2 = \mu_1 \mu_2$ [20].

In the following section, we discuss the optimal operating point for different communication scenarios based on these four cases.

7.3.2 POWER ALLOCATION SCENARIOS AND OPTIMAL OPERATING POINTS

Optimally allocating transmitting power to the sender and the relay can enhance system performance at the expense of increased system complexity. It is thus important to consider power allocation strategies that offer a good performance-complexity trade-off by imposing certain suitable constraints. Implementation complexity of a power allocation strategy is a function of the operational environment of the CD system. For example, in a sensor network, the objective is to maximize the lifetime of the network. It is therefore necessary to minimize the energy consumption of those nodes that consume energy at the maximum rate. Since all nodes start with the same initial energy, equal power transmission (EPT) can prolong the lifetime of a pair of nodes, subject to the BEP constraint. On the other hand, for a cellular network, the system performance is limited by the intercell interference, which can be reduced by power control. In this case, a power allocation strategy that constrains the BS receiver to have equal SNR reception would have the effect of reducing the intercell interference.

With the general framework set forth in the above four cases, in this section, we provide five scenarios that are specific to a particular CD scheme and operational constraints. The first two scenarios belong to an AF CD system [3] in which optimal power allocation and EPT constraint are considered. The last three scenarios belong to a regenerate and forward CD system based on quadrature amplitude modulation (QAM) [20]. The bit energies of the sources and relaying bits of the QAM CD system should be equal to reduce the I-Q imbalance in the radio frequency (RF) circuitry. Since most of the CD systems in the literature do not have this constraint, previous works of power allocation for CD systems do not apply to the CD system of [20]. Thus, we are motivated to study power allocation strategies for this scheme in Scenarios 3, 4, and 5 such that distinct operational constraints are considered.

Scenario 1

Consider the AF system reported in Ref. [3] with optimal power allocation. This scenario fits into Case I such that the λ's and the μ's can take any arbitrary values. For this CD system, $A = \frac{3}{16}$, $B = \frac{3}{16}$, $C = 0$, and $P_e = P_b$.

Notice that given the error probability requirement of a user, the transmission power of the user is always a nonincreasing continuous convex function of the relay power of its partner, and vice versa, as shown in Figure 7.2. To minimize the energy consumption $(E_{b1}^S + E_{b2}^R)$ with constraint $P_{e1} \le e_1$, the optimal operating point is the tangent point of the dotted curve touched by the line with slope -1, that is, point $S_{1,1}$ in Figure 7.2. This is because all the points on the line with slope -1 correspond to the same value of $(E_{b1}^S + E_{b2}^R)$, and all other points on the dotted curve have a higher total energy level than that of point $S_{1,1}$. Similarly, the solution for problem (Equation 7.9) is the tangent point of the solid curve touched by the line with slope -1, that is, point $S_{1,2}$ in Figure 7.2.

We verify the optimal operating point by substituting $A = 3/16$, $B = 3/16$, and $C = 0$ in Equation 7.2. The minimal relay power can be obtained as a function of the transmission power and the guaranteed BEP, e_1:

$$E_{b2}^R = \frac{3k_2 N_0^2 E_{b1}^S}{\left(16 k_1 e_1 \sigma_1^4 (E_{b1}^S)^2 - 3 N_0^2\right)}. \tag{7.13}$$

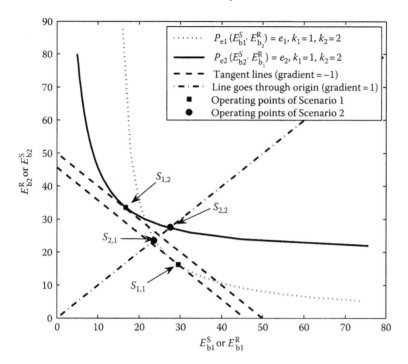

FIGURE 7.2 Optimal operating points of Scenario 1 and 2 when $k_1 = 1$ and $k_2 = 2$.

By solving $\frac{dE_{b2}^R}{dE_{b1}^S} = -1$, the optimal solutions of E_{b1}^S and E_{b2}^R can be written as

$$E_{b1}^S = \frac{N_0}{\sigma_1^2}\left(\frac{6 + 3k_2 + \sqrt{9k_2^2 + 72k_2}}{32k_1e_1}\right)^{1/2} \quad \text{and} \tag{7.14}$$

$$E_{b2}^R = \left(\frac{2k_2}{k_2 + \sqrt{k_2^2 + 8k_2}}\right)\frac{N_0}{\sigma_1^2}\left(\frac{6 + 3k_2 + \sqrt{9k_2^2 + 72k_2}}{32k_1e_1}\right)^{1/2}. \tag{7.15}$$

Similarly, for user 2,

$$E_{b1}^R = \frac{3k_1k_2N_0^2E_{b2}^S}{\left(16k_1^2e_2\sigma_1^4(E_{b2}^S)^2 - 3k_2N_0^2\right)}. \tag{7.16}$$

By solving $\frac{dE_{b1}^R}{dE_{b2}^S} = -1$, the optimal solutions of E_{b2}^S and E_{b1}^R can be written as

$$E_{b2}^S = \frac{N_0}{\sigma_1^2}\left(\frac{6k_2 + 3k_1k_2 + \sqrt{9k_2^2k_1^2 + 72k_2^2k_1}}{32k_1^2e_2}\right)^{1/2} \quad \text{and} \tag{7.17}$$

$$E_{b1}^R = \left(\frac{2k_1}{k_1 + \sqrt{k_1^2 + 8k_1}}\right)\frac{N_0}{\sigma_1^2}\left(\frac{6k_2 + 3k_1k_2 + \sqrt{9k_2^2k_1^2 + 72k_2^2k_1}}{32k_1^2e_2}\right)^{1/2}. \tag{7.18}$$

Scenario 2

Consider the AF system reported in Ref. [3] with EPT constraint. In this scenario, the λ's can take any arbitrary values and $\mu_1 = \mu_2 = 1$. In this situation, it is preferable to let the relaying power at the partner be equal to the source's transmission power (Case II). For this CD system, $A = \frac{3}{16}$, $B = \frac{3}{16}$, $C = 0$, and $P_e = P_b$.

Obviously, for Equation 7.10, the optimal operating point is the intersection of the equal power line (the line with slope 1 and passing through the origin) and the dotted curve, that is, point $S_{2,1}$ in Figure 7.2. The optimal solution is given by

$$E_{b1}^S = E_{b2}^R = \sqrt{\frac{3N_0^2(1 + k_2)}{16k_1 e_1 \sigma_1^4}} \tag{7.19}$$

For Equation 7.11, the intersection point $S_{2,2}$ in Figure 7.2 corresponds to the optimal solution given by

$$E_{b2}^S = E_{b1}^R = \sqrt{\frac{3N_0^2 k_2(1 + k_1)}{16k_1^2 e_2 \sigma_1^4}} \tag{7.20}$$

Scenario 3

In this scenario, we consider the regenerate and forward CD scheme reported in Ref. [20] for 4-QAM modulation. Similar to Scenario 2, the λ's can take any arbitrary values and $\mu_1 = \mu_2 = 1$ (Case II). By substituting $A = K_N/16$, $B = 3/16$, $C = -3K_N/64$, and $P_{e1} = e_1$ in Equation 7.2

$$E_{b2}^R = \frac{\left(12k_1 k_2 \sigma_1^2 N_0^2 E_{b1}^S - 3K_N k_2 N_0^3\right)}{\left(64k_1^2 e_1 \sigma_1^6 (E_{b1}^S)^2 - 8K_N k_1 N_0^2 \sigma_1^2\right)}. \tag{7.21}$$

Similarly, for user 2 in Equation 7.3,

$$E_{b1}^R = \frac{\left(12k_1 k_2 \sigma_1^2 N_0^2 E_{b2}^S - 3K_N k_2 N_0^3\right)}{\left(64k_1^2 \sigma_1^6 e_2 (E_{b2}^S)^2 - 8K_N k_2 N_0^2 \sigma_1^2\right)}. \tag{7.22}$$

Since $E_{b1}^S = E_{b2}^R$, and $E_{b2}^S = E_{b1}^R$, the optimal operating points are given by $S_{3,1}$ and $S_{3,2}$ in Figure 7.3 for user 1 and user 2, respectively.

The coordinates of the intersection point of $E_{b1}^S - E_{b2}^R = 0$ and Equation 7.21, point $S_{3,1}$, can be obtained from the solutions of a third-order polynomial which can be derived by substituting $E_{b2}^R = E_{b1}^S$ in Equation 7.21. Similarly, the coordinates of point $S_{3,2}$ can be given by the solutions of a third-order polynomial which can be derived by substituting $E_{b1}^R = E_{b2}^S$ in Equation 7.22. The coordinates of points $S_{3,1}$ and $S_{3,2}$ can be solved numerically.

Scenario 4

Consider the CD systems as in scenario 3 with $\lambda_1 = \lambda_2 = 1$ and μ's can any take any arbitrary value (Case III). Thus, $E_{b1}^S = E_{b1}^R$ and $E_{b2}^S = E_{b2}^R$.

The optimization problem can be solved as follows. As mentioned earlier, the transmission power of the user is always a nonincreasing continuous convex function of the relay power of its partner. For a given BEP, if E_{b1}^S is smaller, E_{b2}^R must be larger, and vice versa. As shown in Figure 7.3, to guarantee $P_{e1} \leq e_1$ and $P_{e2} \leq e_2$, the energy levels of the two users should be in the intersection of the region above* the dotted and the solid curves, which is shown as the shaded area. To select the optimum operating point that minimizes the energy consumption ($E_{b1}^S + E_{b2}^R$) with the constraints $P_{e1} \leq e_1$ and $P_{e2} \leq e_2$, we should consider the following three points.

---

* When a point is "above" another point, we mean that the total energy consumption of the two users corresponding to the former one is higher.

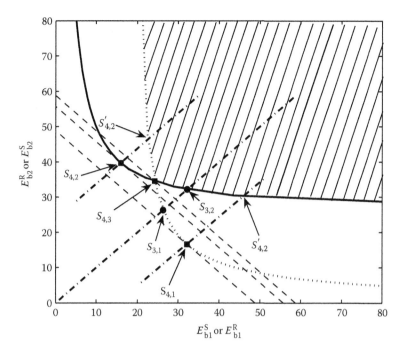

FIGURE 7.3 Optimal operating points of Scenario 3, 4, and 5 when $k_1 = 1$ and $k_2 = 2$.

Point $S_{4,1}$: The optimal operating point that satisfies $P_{e1} = e_1$ only is the tangent point of the dotted curve touched by the line with slope -1, that is, point $S_{4,1}$ in Figure 7.3. This is because all the points on the line with slope -1 correspond to the same value of $(E_{b1}^S + E_{b2}^R)$, and all other points on the dotted curve have higher total energy levels than that of point $S_{4,1}$. If point $S_{4,1}$ also satisfies $P_{e2} \leq e_2$, point $S_{4,1}$ is the solution for the optimal power allocation (OPA) problem.

Point $S_{4,2}$: If P_{e2} is larger than e_2 at point $S_{4,1}$, we can find the operating point to minimize the total energy with the constraint $P_{e2} = e_2$ only, which is the tangent point of the solid curve touched by the line with slope -1, that is, point $S_{4,2}$ in Figure 7.3. If point $S_{4,2}$ also satisfies $P_{e1} \leq e_1$, point $S_{4,2}$ is the optimal solution for the OPA problem.

Point $S_{4,3}$: If points $S_{4,1}$ and $S_{4,2}$ cannot ensure both $P_{e1} \leq e_1$ and $P_{e2} \leq e_2$, according to the following theorem, the two curves must intersect at a point, $S_{4,3}$, which is the optimal operating point of the OPA problem.

THEOREM 1

Let point $S_{4,1}$ be the tangent point of $P_{e2}(E_{b2}^S, E_{b1}^R) = e_2$ (the solid curve in Figure 7.3) touched by the line with slope -1 and point $S_{4,2}$ be the tangent point of $P_{e1}(E_{b1}^S, E_{b2}^R) = e_1$ (the dotted curve in Figure 7.3) touched by the line with slope -1. If $P_{e1}(E_{b1}^S, E_{b2}^R)_{S_{4,1}} > e_1$ and $P_{e2}(E_{b2}^S, E_{b1}^R)_{S_{4,1}} > e_2$, the two functions $P_{e1}(E_{b1}^S, E_{b2}^R) = e_1$ and $P_{e2}(E_{b2}^S, E_{b1}^R) = e_2$ must intersect at a point labeled $S_{4,3}$. $S_{4,3}$ is located between the two parallel lines with slope 1, one passing through point $S_{4,1}$ and the other passing through point $S_{4,2}$, and $S_{4,3}$ is the optimal solution of the OPA problem in Equation 7.4.

Proof: The line passing through point $S_{4,1}$ with slope 1 intersects the solid curve at point $S'_{4,1}$ which must be above point $S_{4,1}$; otherwise, point $S_{4,1}$ will satisfy the condition $P_{e1} \leq e_1$, which contradicts the condition of the theorem. The line passing through point $S_{4,2}$ with slope 1 intersects the solid curve at point $S'_{4,2}$, which must be above point $S_{4,2}$; otherwise, point $S_{4,2}$ will satisfy the condition $P_{e2} \leq e_2$, which contradicts the condition of the theorem. Given $P_{e1}(E^S_{b1}, E^R_{b2}) = e_1$ and $P_{e2}(E^S_{b1}, E^R_{b2}) = e_2$ are continuous convex functions, $S_{4,1}S'_{4,2}$ and $S_{4,2}S'_{4,1}$ must intersect at one and only one point, that is, point $S_{4,3}$. On the solid curve, from point $S_{4,1}$ to point $S'_{4,2}$, $E^S_{b1} + E^R_{b2}$ monotonically increases, and any point between $S_{4,1}$ and $S_{4,3}$ cannot satisfy $P_{e1}(E^S_{b1}, E^R_{b2}) \leq e_1$. Thus, $S_{4,3}$ is the point with the minimum $E^S_{b1} + E^R_{b2}$ and satisfies both constraints, and it is the optimum solution.

The coordinates of point $S_{4,2}$ can be obtained by solving $\frac{dE^R_{b2}}{dE^S_{b1}} = -1$ of $P_{e1}(E^S_{b1}, E^R_{b2})$. The coordinates of point $S_{4,1}$ can be obtained by solving $\frac{dE^R_{b1}}{dE^S_{b2}} = -1$ of $P_{e2}(E^R_{b1}, E^S_{b2})$. The coordinates of the intersection point of $P_{e1} = e_1$ and $P_{e2} = e_2$, point $S_{4,3}$, can be obtained from the solutions of a fifth-order polynomial, obtained by substituting $E^R_{b2} = E^S_{b2}$ in Equation 7.21 and $E^R_{b1} = E^S_{b1}$ in Equation 7.22. The coordinates of points $S_{4,1}$, $S_{4,2}$, and $S_{4,3}$ can be solved numerically.

Scenario 5

In Scenario 5, the regenerate and forward CD scheme reported in Ref. [20] is considered with equal energy constellation and EPT constraint. This belongs to Case IV with $\lambda_1 = \lambda_2 = 1$ and $\mu_1 = \mu_2 = 1$, that is, $E^S_{b1} = E^R_{b1} = E^S_{b2} = E^R_{b2}$. Consider intersecting points of the line that has slope 1 and goes through origin with $P_{e1} \leq e_1$ and $P_{e2} \leq e_2$ ($S_{3,1}$ and $S_{3,2}$ in Figure 7.3). In this scenario, the point in the boundary of the shaded area is selected as the optimal operating point ($S_{3,2}$ in Figure 7.3). This is a special case of Scenario 3.

The optimal operating point of each scenario was derived for the given channel conditions (k_1 and k_2). The location of the cooperating partner who requires minimal energy to achieve a given error probability in a cell is unknown. This can be optimized by using an intelligent partner selection strategy. In the next section, we formulate the partner selection as a matching problem and describe the available matching algorithms.

7.4 MATCHING ALGORITHMS

The problem of maximizing the CD energy gain of any pair of users in a radio cell by optimally grouping all the active users can be formulated as a nonbipartite MW-matching problem, which can be solved by the state-of-the-art MW-matching algorithm in polynomial time $O(n^3)$. Another 1/2-approximation matching algorithm, called Greedy matching algorithm, can be solved in polynomial time $O(n^2 \log n)$. Due to user mobility and intermittent traffic, the matching algorithm should be periodically executed in real time. Thus, it is important to reduce the computational and implementation complexity of the matching algorithm without compromising too much energy gain. We have proposed a worst-link-first (WLF) algorithm in Ref. [14], which gives the user with the worse channel condition (and thus the higher energy consumption rate) a higher priority to choose its partner. The computational complexity of the proposed WLF algorithm is $O(n^2)$.

With user mobility, both the user locations and the optimal matching vary with time, and frequently updating the matching will introduce significant overhead. We propose how to incorporate the mobility information in the matching algorithm to reduce the overhead. It is shown that, by intelligently incorporating user mobility, the MW-matching problem and the WLF matching algorithms can maintain high cell energy gain with low overhead.

7.4.1 PROBLEM FORMULATION

Choosing pairs of cooperating users is known as matching on graphs [10,26]. Let $\mathcal{G} = \{\mathcal{V}, \mathcal{E}\}$ be a graph, where $\mathcal{V}$ is a set of vertexes and $\mathcal{E} \subseteq \mathcal{V} \times \mathcal{V}$ is a set of edges between vertexes. Each mobile

TABLE 7.1
List of Symbols

E_{bi}^{no}	Bit energy spent by user i in NCD
E_{bi}^{R}	Bit energy spent for relaying bit at user i when cooperating
E_{bi}^{S}	Bit energy spent for source bit at user i when cooperating
e_i	Maximum tolerable BEP of user i
$e_{i,j}$	Edge between user i and user j
$\mathcal{G} = \{\mathcal{V}, \mathcal{E}\}$	A graph, where $\mathcal{V}$ is a set of vertices and $\mathcal{E} \subseteq \mathcal{V} \times \mathcal{V}$ is a set of edges between vertices
G_{CD}	Cooperative diversity energy gain
n and N	Number of active users in a cell
N_0	Noise power spectral density
K	Total number of users paired in a cell
P_b	Bit error probability
P_e	Error probability
P_{ei}	Error probability of user i
P_o	Outage probability
P_s	Symbol error probability
R	Radius of the cell centered at the BS
S	A *matching* subset of $\mathcal{E}$ if no two edges in S share the same vertices
T	The period of matching algorithm being executed
v	Velocity of the user
V_{max}	The max velocity of mobile users
V_{norm}	The normalized max velocity of mobile users
$w(e_{i,j})$	The weight of the edge linking users i and user j.
$\bar{\gamma}_i$	Average SNR at BS for user i's transmission
$\bar{\gamma}_{i,j}$	Average SNR at user j for user i's transmission
σ_i^2	Variance of the fading channel coefficient for the channel from user i to BS
$\sigma_{i,j}^2$	Variance of the fading channel coefficient for the channel from user i to user j
$\lambda_1, \lambda_2, \mu_1$ and μ_2	Operational constraints

user in a cell is represented as a vertex. The (i,j)th edge, $e_{i,j} \in \mathcal{E}$, has a weight $w(e_{i,j})$, which equals the energy gained by cooperation between users i and j over no cooperation. If there is no cooperative energy gain, the two users will just use the noncooperative scheme, and the weight of the edge linking them is zero. Thus, the weight is always nonnegative, and a positive weight represents the energy gain of cooperation over no cooperation.

A subset S of $\mathcal{E}$ is called a matching subset if there are no two edges in S sharing the same vertex. The overall energy gain in the network is the sum of the positive weights of all edges in S. For easy reference, the notations used throughout the paper are listed in Table 7.1.

7.4.2 MAXIMUM-WEIGHTED-MATCHING PROBLEM

Maximizing the energy gain by cooperation is equivalent to maximizing $w(S) = \sum_{e_{i,j} \in S} w(e_{i,j})$ among all possible matchings, which is a non-bipartite weighted-matching problem. The number of matchings with $|S| = \lfloor \frac{n}{2} \rfloor!$ equals $n!/(2^{\lfloor \frac{n}{2} \rfloor} \lfloor \frac{n}{2} \rfloor!)$, which exponentially increases with n, where $n = |\mathcal{V}|$ is the cardinality of the set $\mathcal{V}$ or the number of users in the network. Comparing all possible matching by brute force search is time consuming when the number of active users is large.

The MW-matching problem algorithm developed in Ref. [27] can yield the optimal solution for the nonbipartite weighted-matching problem in polynomial time $O(n^3)$. This state-of-the-art algorithm can be used to obtain the upper bound of energy gain in a wireless network.

7.4.3 GREEDY MATCHING

Another heuristic matching algorithm is the Greedy matching algorithm [28].

Greedy Matching Algorithm:

1. The BS selects a user pair i and j such that energy gain $w(e_{i,j})$ is the largest among $w(e)$ for $e \in \mathcal{E}$. $e_{i,j}$ is added to the matching set.
2. Remove all edges incident to $e_{i,j}$ from $\mathcal{E}$.
3. Repeat steps 1 and 2 until the number of unmatched users is less than two.

The Greedy matching algorithm requires sorting the weights of all edges in $\mathcal{E}$, so its computational complexity is $O(n^2 \log n)$.

7.4.4 WORST-LINK-FIRST MATCHING

With user mobility and intermittent traffic, the matching algorithm should be periodically executed in real time. Thus, it is important to further reduce the computational complexity of the matching algorithm without compromising too much energy gain.

Since the user with the worse channel quality (far from the BS) consumes more energy than the one with a better channel quality (near the BS) in a conventional transmission system, cooperation generally gives more energy gain to the far user than the near user. Therefore, when considering the radio cell, those users with worse channel quality and higher energy consumption rate should be given a higher priority. This motivates us to develop the following WLF matching algorithm in Ref. [14].

WLF Matching Algorithm:

1. The BS selects an unmatched user i with the worst channel quality among all unmatched users.
2. The BS selects an unmatched user j such that energy gain $w(e_{i,j})$ provided by the cooperation of user i and user j over no cooperation is the maximum one among all $w(e_{i,k})$, where k is an unmatched user other than i. $e_{i,j}$ is added to the matching set.
3. Repeat steps 1 and 2 until the number of unmatched users is less than two.

The computational complexity of the WLF algorithm is $O(n^2)$.

7.4.5 RANDOM MATCHING

The random matching algorithm is the simplest one and is used as a benchmark. The algorithm randomly selects an unmatched user i and matches it with another unmatched user j, until there are fewer than two unmatched users remaining. Although the computational complexity of the random matching algorithm is $O(n)$, due to the randomness in matching, a significant number of pairs cannot obtain positive energy gain by cooperation. Therefore, random matching provides limited energy gain.

In the following sections, numerical results are presented for the four matching algorithms in both static and mobile user networks.

7.5 PERFORMANCE EVALUATION IN STATIC USER NETWORK

Consider a wireless network where the coordinates of the BS are $(0, 0)$. N users are randomly placed on a unit disk centered at the BS as given in Figure 7.1, with their coordinates x and y uniformly distributed in $[-1, 1]$. The average CSIs (σ^2's) are inversely proportional to d^α, where d is the distance between the sender and the receiver, and the path loss exponent α takes the value 3 in the simulation.

The guaranteed BER is $e_1 = e_2 = 10^{-3}$. Different user deployments are generated by using different random seeds of the random number generator. We change the number of active users in the network from 10 to 100 to consider both the low and high user-density situations.

It is assumed that the BS can track the user locations and thus determine their pair-wise distances and CSIs. From the CSIs, the average energy required for no cooperation and cooperation schemes are calculated. Matching is performed by the BS according to the N^2 weights, and the users are grouped according to the matching results. The weights used in the matching algorithms are the CD energy gain of each pair of users, which is given by $w(e_{i,j}) = E_{bi}^{no} + E_{bj}^{no} - E_{bi}^{S} - E_{bi}^{R} + E_{bj}^{S} - E_{bj}^{R}$ where E_{bi}^{no} and E_{bi}^{no} are the bit energy required for user i and user j, respectively, in the noncooperative transmission. Since cooperation is not always beneficial, a pair can cooperate if there is positive cooperative energy gain, that is, $w(e_{i,j}) > 0$. Otherwise, they communicate with the BS using the noncooperative scheme.

The average energy gain of the cell, which is the energy gain of a cell with user cooperation over a cell without user cooperation, is defined as

$$G_{CD} = 10\log_{10}\left(\frac{\sum_{i=1}^{N} E_{bi}^{no}}{\sum_{i=1}^{K}(E_{bi}^{S} + E_{bi}^{R}) + \sum_{i=K+1}^{N} E_{bi}^{no}}\right), \tag{7.23}$$

where the first K users are paired to have cooperation and the remaining $(N - K)$ users have no partners. Since $\sum_{i=1}^{N} E_{bi}^{no}$ is a constant independent of matching, G_{CD} is maximized when $\sum_{i=1}^{K}(E_{bi}^{S} + E_{bi}^{R})$ is minimized.

The number of users without a partner and the average cell energy gains with the four matching algorithms of Scenario 3 are shown in Figures 7.4 and 7.5, respectively. All the results are obtained by averaging the performance parameters over 25 different user deployments randomly generated. As shown in Figure 7.4, with the MW, Greedy, and WLF matching algorithms, the number of users without a partner is almost constant, independent of the number of active users in the network. Therefore, the chance for a user without a partner is low for a high-density network. On the other hand,

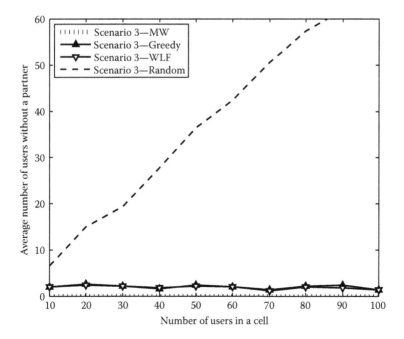

FIGURE 7.4 Average number of users without a partner versus number of users in the cell for Scenario 3.

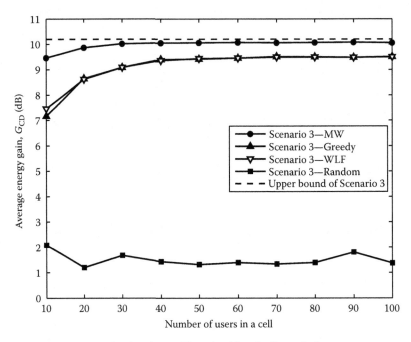

FIGURE 7.5   Average energy gain of each matching algorithm for Scenario 3.

with the random matching algorithm, the number of users without partner increases proportionally with the number of users in the network. This is because each user has a cooperative region and only users in the cooperative region grouped together can obtain positive cooperative energy gain. The probability of two randomly chosen users is within each other's cooperative region is constant, independent of the network density [14].

Figure 7.5 shows the average cell energy gains for the four matching algorithms. The gains of the MW, Greedy, and WLF matching algorithms increase with the number of users. This is because, as shown in Figure 7.4, in a lower-density network, the chance for a user without a suitable partner is higher, so the average cell energy gain is lower. To approach the analytical upper bound, the network should have a sufficiently large number of users, so that every user can be grouped with an ideal partner [14]. However, the higher energy gain regions become smaller; therefore, the energy gain of the cell increases slower when the number of active users in the cell is larger. In contrast, the random matching algorithm provides almost constant gain, independent of the number of users in the network.

From the numerical results, if a BS does not have the knowledge of the CSIs and just randomly matches users for cooperation, only about 1.5 dB average energy gain over no cooperation can be achieved for Scenario 3. If the CSIs were available or could be estimated, the WM, Greedy, and WLF matching algorithms would achieve 7–10 dB average energy gain for Scenario 3. Similar results are observed for the other scenarios.

Although the WLF algorithm does not guarantee the worst-case performance, extensive simulations demonstrate that the performance of the WLF algorithm is close to the Greedy algorithm, and the average energy gains in a cell with the WLF and Greedy algorithms are about 1 dB less than that with the MW algorithm. The WLF algorithm is less complex and easier to implement than the MW and Greedy algorithms: the latter two algorithms require the matching gains of any pair of active users ($n(n - 1)/2$ pairs), which are difficult to obtain. With the WLF, the BS can choose an unmatched active user with the farthest distance to the BS (or the worst channel condition to the BS) first. Then, the BS selects an unmatched user in the high-dB-gain region to be its partner. In addition,

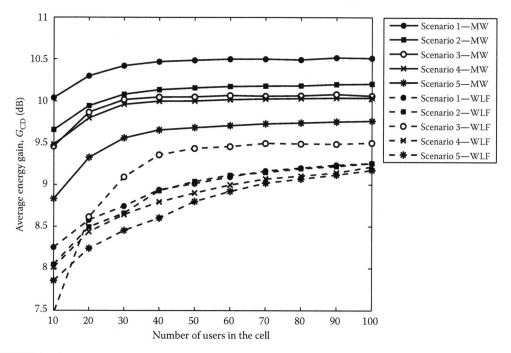

FIGURE 7.6 Average energy gain provided by the WLF and MW algorithms for each Scenario.

the WLF algorithm can potentially be implemented in a distributed manner: each user chooses its desired partner; if there is any conflict, the user farther away from the BS (or has worse channel condition to the BS) has a higher priority.

In Figure 7.6, we compare the performance of the MW and WLF algorithms for all the scenarios. Trends of the curves are similar despite the power allocation and CD schemes used. When we consider the MW algorithm with an AF CD scheme (Scenarios 1 and 2), Scenario 1 performs better than Scenario 2, because all the operational constraints are relaxed for Scenario 1. For the case of regenerate-and-forward CD scheme with MW, the average energy gain of Scenario 5 is worse than that of Scenarios 3 and 4. This is due to the fact that Scenario 5 has more operational constraints than the other two scenarios. Even though Scenarios 3 and 4 have one operational constraint each, Scenario 3 provides more energy gain than Scenario 4 because a single operating point of Scenario 4 makes the cooperating user to spend more energy than required.

With the WLF matching algorithm, the difference in performance between the power allocation scenarios is negligible except in Scenario 3, especially when the number of users in the cell is large. This is because the WLF algorithm tends to choose a partner close to the user for cooperation to maximize their cooperative energy gain. When the two users are close, cooperative energy gain of Scenarios 1, 2, 4, and 5 are close to each other. However, operating points of Scenario 3 make the WLF algorithm to choose partners quite different from others. The behavior of Scenario 3 motivates us to improve the WLF algorithm considering other weighting mechanisms, which are given in Section 7.7.

7.6 PERFORMANCE EVALUATION IN MOBILE USER NETWORK

In mobile networks, user mobility complicates the optimal matching problem. Because users may move in different directions with different velocities, and the velocities and directions change over time, their absolute and relative locations keep on changing. The currently best matching strategy may be less attractive or even no longer applicable after a while. Therefore, the matching algorithm should be periodically executed according to the current user locations and channel conditions.

7.6.1 MATCHING ALGORITHM CONSIDERING MOBILITY

Although frequently updating the matching can accurately track the locations and channel conditions of random and high-mobility users, it introduces significant overhead to not only the BS, but also the mobile users. Furthermore, mobile users need to synchronize with their new cooperative partners frequently. To reduce the overhead without significantly sacrificing the performance, it is proposed to predict the future CD energy gain of mobile users based on their current location and mobility information and match users accordingly. How a BS detects the location and speed of active mobile users has been extensively studied in the literature, and the technologies have been widely used for wireless 911 service and other location-dependent services.

The procedure to implement the WLF matching algorithm taking mobility into consideration, which is periodically executed every T seconds, is as follows.

WLF Matching Procedure:
```
1  at time t, sort V according to CSIs
2  for each i ∈ V
3      MaxW = 0; partner(i) = 0
4      for (j = i + 1; j < N; j++)
5         if v_j ∈ V
6             w(e_{i,j}) = f(w(e_{i,j}(t)), T)
7             if w(e_{i,j}) > MaxW
8                 partner(i) = j;  MaxW = w(e_{i,j})
9      remove i and partner(i) from V; add e_{i,partner(i)} to S
```

At time t, the set of N active users, $\mathcal{V}$, is sorted according to their channel conditions (CSIs), such that the user with the worst channel condition is considered first, as shown in Line 1. All grouped users are removed from $\mathcal{V}$ (Line 9). For each unmatched user i, the BS calculates the CD gain of i and another unmatched user j, $w(e_{i,j})$, (Lines 4, 5, 6). Note that $w(e_{i,j})$ is a function of $w(e_{i,j}(t))$ and T. $w(e_{i,j}(t))$ is the energy gain according to users i and j's current channel conditions or user locations (at time t). Assuming that the velocities and directions of i and j remain the same in the next T seconds, the BS can predict their future locations and channel conditions. $w(e_{i,j}(t+\delta))$ is the predicted energy gain according to the predicted user locations and channel conditions at time $t + \delta$. Function f in Line 6 calculates the average energy gain during t to $t + T$:

$$f(w(e_{i,j}(t)), T) = 1/T \int_{t}^{t+T} w(e_{i,j}(x)) \mathrm{d}x. \tag{7.24}$$

To simplify the calculation, when T is small, $f(w(e_{i,j}(t)), T)$ can be approximated as $[w(e_{i,j}(t)) + w(e_{i,j}(t+T))]/2$. Similarly, the MW and Greedy algorithms can be modified by using the average CD gain during $[t, t+T]$ as the weight.

There are certain implications that the system designers may consider. First, to reduce the overhead by lengthening T, the prediction of future user channel conditions and locations become less accurate, which will degrade the overall cell energy gain. Second, even if all active users keep their current velocities and directions for a long time, less frequently updating the matching will also reduce the overall cell energy gain. This can be illustrated as follows. Observe l consecutive time slots, $t_1, t_2, ..., t_l$, where each slot has a very short duration ϵ. Assume that the user locations and channel conditions remain the same in each slot. If the MW-matching problem algorithm is executed at each slot, the energy gain in that slot will always be the highest among any matchings. Therefore, by executing the MW algorithm at each slot, the energy gain is always better than or equal to matching once for a period of l time slots. Third, if the matching algorithm is executed only once per T s, and user mobility estimation is accurate, using $w(e_{i,j}) = f(w(e_{i,j}(t)), T) = 1/T \int_t^{t+T} w(e_{i,j}(x))dx$ in the MW algorithm also leads to optimal matching for overall energy gain of the cell during t to $t+T$. This is because the overall energy gain of the cell during t to $t+T$ is $\int_t^{t+T} \sum_{e_{i,j} \in S} w(e_{i,j}(x))dx = \sum_{e_{i,j} \in S} \int_t^{t+T} w(e_{i,j}(x))dx$, which is maximized using the MW algorithm.

7.6.2 NUMERICAL RESULTS

We consider a wireless mobile network in which the users are uniformly distributed over a $(4R)^2$ square area, as shown in Figure 7.7. The BS is located at the center of the square, covering all active users in the disk centered at the BS with radius R. The mobile users move at constant velocities and the directions of motion are independent and identically distributed (i.i.d.) with uniform distribution in the range $[0, 2\pi)$. If a mobile user reaches the edge of the square, it will be bounced back and move with the same velocity. The velocity is a uniformly distributed random variable in the range $[0, V_{max}]$. In the simulation, a user chooses a direction and a velocity and moves in that direction (unless being bounced back) at a constant velocity for a time duration t_d, which is also a uniformly distributed random variable in the range $(0, t_{max})$ slots. After t_d, the process is repeated. The matching algorithm will be executed every T seconds. The grouped pairs will cooperate with each other till new matching results separate them or when any of them moves out of the cell or when there is no longer any gain between them. We use the following parameters in the simulations. The number of active users in the square area is 200. The normalized velocity V_{norm}, which is defined by $V_{norm} = \frac{V_{max}T}{R}$, is set to 0, 0.25, 0.5, 0.75, and 1, which cover the static, low, medium, and high mobility cases, respectively.

The energy gain achieved by the WLF matching algorithm for the CD scheme with and without mobility is shown in Figure 7.8 for Scenario 3. It can be seen that for $V_{norm} = 1.0$, from the time after the matching ($t = 0$) to the time just before the next matching ($t = T$), the cell energy gain with the WLF algorithm without mobility information quickly drops from 9 to 1 dB. On the other hand, with the same user deployment, the WLF algorithm with mobility information maintains a high cell energy gain (above 7 dB). The simulation results confirm that if we intelligently apply

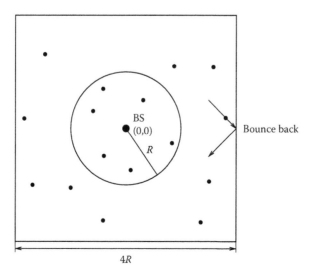

FIGURE 7.7   Mobility model.

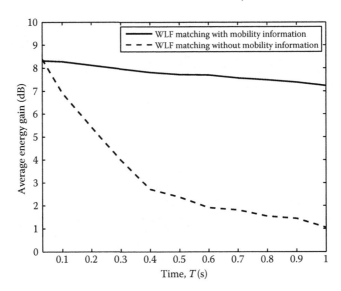

FIGURE 7.8 Average energy gain of the WLF matching with and without mobility information for Scenario 3.

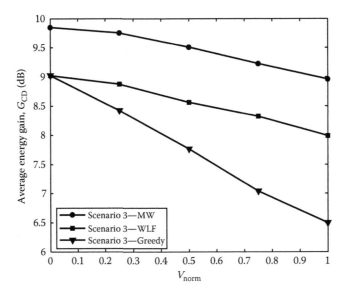

FIGURE 7.9 Average energy gain versus normalized velocity for Scenario 3.

the mobility information in the matching algorithm, a significant cell energy gain can be achieved for mobile networks. Similar results are obtained for other scenarios, with the MW, WLF, and Greedy algorithms. In the following, we focus on matching algorithms with mobility information and compare their performance metrics for Scenario 3.

The percentage of in-cell users participating in the cooperation (specifically in the matching process) is approximately $(1 - 0.3V_{norm})$. In the low-mobility situation ($V_{norm} < 0.25$), more than 90% of the in-cell users participate in the cooperation. It is reduced to 70% for the high-mobility situation, that is, $V_{norm} = 1$. On the other hand, the average energy gain decreases as V_{norm} increases, as shown in Figure 7.9. With other scenarios or matching algorithms, the same trend can be observed for the average energy gain versus the V_{norm} curve. This is due to two factors. First, the percentage of participating users remaining in the cell for a given duration T decreases as V_{norm} increases. Second, with high mobility, even if the matching is ideal at the beginning of a slot, it becomes less favorable or even impractical at the end of the slot.

Figure 7.9 demonstrates the trade-off between the performance and overhead. If the BS updates the matching more frequently, that is, if T is shorter, V_{norm} can be reduced and higher cell energy gain can be achieved; otherwise, the BS updates the matching less frequently with less overhead and less energy gain. The simulation results show that the WLF algorithm outperforms the Greedy algorithm in mobile networks. The performance of the WLF and MW algorithms degrades gracefully when V_{norm} is higher, and the performance of Greedy matching algorithm degrades quickly with higher mobility. Similar behavior has been noticed for other scenarios also.

7.7   IMPROVED WLF ALGORITHM

In the previous sections, the energy saved by cooperation between users i and j over no cooperation, $w(e_{i,j}) = E_{\text{bi}}^{\text{no}} + E_{\text{bj}}^{\text{no}} - E_{\text{bi}}^{\text{S}} - E_{\text{bi}}^{\text{R}} - E_{\text{bj}}^{\text{S}} - E_{\text{bj}}^{\text{R}}$, is used as the weight of a pair of users. As mentioned in Section 7.4.1, the weight is always nonnegative. If there is no energy saved, the two users will just use the noncooperative scheme, and the weight of the edge linking them is zero.

Since WLF tends to maximize the energy saving of a single pair of users, it does not necessarily maximize the energy gain of the network. Thus, we propose to use the maximum energy spent by the pair of users as the weight for matching, and we refer to this scheme as WLF-MinMaxEnergy matching. The WLF-MinMaxEnergy algorithm gives the highest priority to the worst-link user to choose the partner, subject to minimizing their maximum energy usage. This approach tends to minimize the energy usage of all users and improves the average energy gain of the network. The matching algorithm of WLF-MinMaxEnergy is similar to that of WLF, except that the weights used for the pair of users being matched are different.

7.7.1   WORST-LINK-FIRST BY MINIMIZING MAXIMUM ENERGY

To minimize the energy consumption of the worst-link users, we use the maximum energy spent by the cooperating users as the weight, $w(e_{i,j}) = \max(E_{\text{bi}}^{\text{S}} + E_{\text{bi}}^{\text{R}}, E_{\text{bj}}^{\text{S}} + E_{\text{bj}}^{\text{R}})$. The steps of the proposed WLF-MinMaxEnergy matching algorithm are as follows.

WLF-MinMaxEnergy Matching Algorithm:

1. The BS selects an unmatched user i with the worst channel quality among all unmatched users
2. The BS selects an unmatched user j such that $\max(E_{\text{bi}}^{\text{S}} + E_{\text{bi}}^{\text{R}}, E_{\text{bj}}^{\text{S}} + E_{\text{bj}}^{\text{R}})$ is minimized among all $\max(E_{\text{bi}}^{\text{S}} + E_{\text{bi}}^{\text{R}}, E_{\text{bk}}^{\text{S}} + E_{\text{bk}}^{\text{R}})$, where k is an unmatched user other than i
3. Repeat steps 1 and 2 until the number of unmatched users is less than two

Both the WLF and WLF-MinMaxEnergy algorithms have computational complexity of $O(n^2)$ and both can potentially be implemented in a distributed manner. The numerical results demonstrate that the performance of the WLF-MinMaxEnergy algorithm is close to that of the optimal MW matching algorithm, and it outperforms the WLF algorithm for all the scenarios.

7.7.2   NUMERICAL RESULTS

We simulate a wireless network as mentioned in Section 7.5 and present the numerical results of the three matching algorithms for Scenarios 1, 2, 4, and 5 in a static user network. The weights used in the MW and WLF matching algorithms are the CD energy gains of each pair of users, and the weights used in the WLF-MinMaxEnergy matching algorithm are the maximum energy levels of each

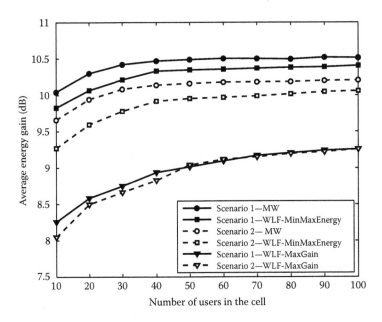

FIGURE 7.10 Average energy gain provided by the MW, WLF, and WLF-MinMaxEnergy algorithms for Scenarios 1 and 2.

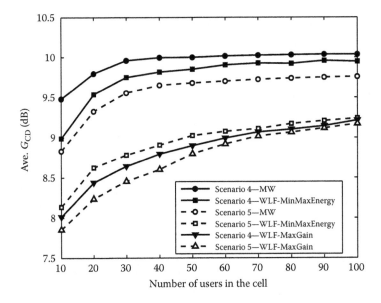

FIGURE 7.11 Average energy gain provided by the MW, WLF, and WLF-MinMaxEnergy algorithms for Scenarios 4 and 5.

cooperating pairs. The average energy gains versus the number of users are shown in Figures 7.10 and 7.11 for both AF (Scenarios 1 and 2) and regenerate-and-forward (Scenarios 4 and 5) CD schemes, respectively.

In Figure 7.10, the solid curves are for Scenario 1 and the dotted curves for Scenario 2. In Figure 7.11, the solid curves are for Scenario 4 and the dotted curves for Scenario 5. With the WLF matching algorithm, the performance difference between Scenario 1 and Scenario 2 is negligible in Figure 7.10, especially when the number of users in the cell is large. Similar behavior can be observed in Figure 7.11 for Scenarios 4 and 5. This is because, with the WLF algorithm, the BS tends to choose a partner close to the user for cooperation to maximize their cooperative energy gain.

On the other hand, the figures show that the performance of the WLF-MinMaxEnergy is close to that of the optimal MW matching algorithm, for Scenarios 1, 2, and 4. This is because the BS tends to choose a partner close to the mid-point between the user and the BS [29]. The numerical results also demonstrate that the WLF-MinMaxEnergy outperforms the WLF. This can be explained by using the following example. Let users u_1 and u_2 be collocated at the point one unit distance away from the BS and users u_3 and u_4 be collocated in the mid-point between u_1 (or u_2) and the BS. Using the WLF algorithm, u_1 and u_2 will be grouped because the energy saved by them, $w(e_{1,2})$, is the maximum among all pairs ($w(e_{1,3})$, $w(e_{1,4})$, $w(e_{2,3})$, $w(e_{2,4})$, and $w(e_{3,4})$), and u_3 and u_4 will be grouped thereafter. Using WLF-MinMaxEnergy, u_1 will be grouped with u_3 (or u_4), and the remaining two users will be grouped together. Even though $w(e_{1,2}) > w(e_{1,3}) = w(e_{2,4})$, the total energy saving, $w(e_{1,3}) + w(e_{2,4})$, is larger than $w(e_{1,2}) + w(e_{3,4})$, because $w(e_{3,4}) \ll w(e_{1,3}) = w(e_{2,4})$. The performance of the WLF-MinMaxEnergy for Scenario 5 is worse than MW and slightly better than WLF. This is due to the fact that $w(e_{i,j}) = \max(E_{bi}^S + E_{bi}^R, E_{bj}^S + E_{bj}^R) = E_{bi}^S + E_{bi}^R = E_{bj}^S + E_{bj}^R$ for Scenario 5. Therefore, the improvement over the weighting scheme of WLF ($w(e_{i,j}) = E_{bi}^{no} + E_{bj}^{no} - E_{bi}^S - E_{bi}^R - E_{bj}^S - E_{bj}^R$) is minimal. Similar behavior can be observed for Scenario 3.

Generally, the proposed WLF-MinMaxEnergy matching performs closer to the optimum MW matching and better than the WLF (and Greedy) matching, as shown in Figures 7.10 and 7.11. Furthermore, the average G_{CD} increases with the number of users in the low user-density region and saturates in the higher user-density region.

In summary, the numerical results demonstrate the importance of the combination of power allocation and partner selection. To achieve high energy gain for the cell, the power allocation with the WLF-MinMaxEnergy matching algorithm is more desirable.

7.8   CONCLUSIONS AND OPEN ISSUES

In this chapter, we have introduced a generalized power allocation framework for CD systems. Using the optimization framework, we have derived optimal operating points for different power allocation and cooperative communication scenarios. Then, we formulated the partner selection problem as a nonbipartite weighted matching problem. We have also proposed a novel and simple WLF matching algorithm, which can achieve close to optimal cooperative energy gain in a wireless cellular network. Then, we combined the power allocation and partner selection to optimize the energy saving of the cooperative communication network. The energy gain provided by the optimal MW algorithm, the Greedy algorithm, the proposed WLF algorithm, and the random matching algorithm, with computational complexity of $O(n^3)$, $O(n^2 \log n)$, $O(n^2)$, and $O(n)$, respectively, have been presented. We have further studied how to optimally match mobile users considering user mobility. Simulation results demonstrate that, by intelligently applying user mobility information in the matching algorithm, a high energy gain with moderate overhead is achievable in mobile networks. It is conjectured that our study provides much-needed insights into the trade-off between matching overhead and energy gain in a wireless network, which is an important step toward practically deploying CD schemes in wireless networks. Finally, we have demonstrated the effectiveness and the efficiency of the proposed WLF-MinMaxEnergy matching algorithm using the optimal power allocation strategy. It is shown that a 9–10 dB CD gain can be achieved, which is equivalent to prolonging the cell phone battery recharge time by about ten times.

To fully explore user CD gain in different wireless systems, there are many open issues that need to be further explored. In this chapter, the partner selection algorithms assumed that the global information (inter-user CSIs, user-to-BS CSIs, requested power, etc.) of users is available at the BS. In reality, estimating all inter-user channels may not be feasible. It would be of practical importance to investigate partner selection with partial information. The proposed WLF algorithm is a potential candidate for this scenario and it could be integrated with channel estimation techniques. On the other hand, proposing a fully distributed partner selection algorithm for infrastructureless wireless networks is a challenging issue requiring further investigation. In addition, partner selection for multiuser (more than two) cooperation is a difficult and rich area for future research.

REFERENCES

[1] J. Luo, R. S. Blum, L. J. Cimini, L. J. Greenstein, and A. M. Haimovich, Decode-and-forward cooperative diversity with power allocation in wireless networks, *IEEE Global Telecommunications Conference (GLOBECOM 2005)*, vol. 5, 28 Nov.–Dec. 2005.

[2] Y. Zhao, R. Adve, and T. J. Lim, Improving amplify-and-forward relay networks: Optimal power allocation versus selection, *IEEE Inter. Symposium on Inform. Theory (ISIT)*, Seattle, WA, July 2006.

[3] J. Adeane, M. R. D. Rodrigues, and I. J. Wassell, Optimum power allocation in cooperative networks, in *Proc. 12th Inter. Conf. on Telecommunications (ICT 2005)*, Cape Town, South Africa, May 2005.

[4] A. Host-Madsen and J. Zhang, Capacity bounds and power allocation for wireless relay channels, *IEEE Trans. Inform. Theory*, 51(6), 2020–2040, June 2005.

[5] E. G. Larsson and Y. Cao, Collaborative transmit diversity with adaptive radio resource and power allocation, *IEEE Commun. Lett.*, 9(6), 511–513, June 2005.

[6] W. Su, A. K. Sadek, and K. J. R. Liu, SER performance analysis and optimum power allocation for decode-and-forward cooperation protocol in wireless networks, in *Proc. IEEE WCNC 2005)*, vol. 2, pp. 984–989, New Orleans, LA, March 2005.

[7] Z. Lin, E. Erkip, and A. Stefanov, Cooperative regions and partner choice in coded cooperative systems, *IEEE Trans. Commun.*, 54(7), 1323–1334, July 2006.

[8] Z. Lian, E. Erkip, and A. Stefanov, Cooperative regions for coded cooperative systems, in *Proc. IEEE GLOBECOM '04*, Nov. 2004.

[9] N. Shastry, J. Bhatia, and R. S. Adve, Theoretical analysis of cooperative diversity in wireless sensor networks, in *Proc. IEEE Globecom 2005*, St. Louis, MO, Nov.–Dec. 2005.

[10] J. N. Laneman, Cooperative diversity in wireless networks: Algorithms and architectures, PhD thesis, MIT, 2002.

[11] A. Scaglione, D. L. Goeckel, and J. N. Laneman, Cooperative communications in mobile ad hoc networks, *IEEE Signal Proc. Mag.*, 54(7), 18–29, Sep. 2006.

[12] J. Cai, X. Shen, J. W. Mark, and A. S. Alfa, Semi-distributed user relaying algorithm for amplify-and-forward wireless relay networks, *IEEE Trans. Wireless Commun.*, 7(4), 1348–1357, April 2008.

[13] A. Nosratinia and T. E. Hunter , Grouping and partner selection in cooperative wireless networks, *IEEE JSAC*, 25(2), 369–378, Feb. 2007.

[14] V. Mahinthan, L. Cai, J. W. Mark, and X. Shen, Maximizing cooperative diversity energy gain for wireless networks, *IEEE Trans. Wireless Commun.*, 6(7), 2530–2539, July 2007.

[15] J. N. Laneman, D. N. C. Tse, and G. W. Wornell, Cooperative diversity in wireless networks: Efficient protocols and outage behavior, *IEEE Trans. Inform. Theory*, 50(12), 3062–3080, Dec. 2004.

[16] J. Luo, R. S. Blum, L. J. Cimini, L. J. Greenstein, and A. M. Haimovich, Link-failure probabilities for practical cooperative relay networks, in *Proc. 61th IEEE Vehicular Tech. Confe.*, vol. 3, pp. 1489–1493 May–June 2005.

[17] E. Zimmermann, P. Herhold, and G. Fettweis, On the performance of cooperative diversity protocols in practical wireless systems, in *Proc. 58th IEEE Vehicular Tech. Confe.*, Oct. 2003.

[18] E. Zimmermann, P. Herhold, and G. Fettweis, A novel protocol for cooperative diversity in wireless networks, in *Proc. The Fithe European Wireless Conference—Mobile and Wireless System Beyond 3G*, Barcelona, Spain, Feb. 2004.

[19] P. Herhold, E. Zimmermann, and G. Fettweis, Cooperative multi-hop transmission in wireless networks, *Elsevier J. Comput. Networks*, 49, 299–324, June 2005.

[20] V. Mahinthan, J. W. Mark, and X. Shen, Performance analysis and power allocation of cooperative diversity systems, *IEEE Trans. Wireless Commun.*, in press.

[21] S. Nagaraj and M. Bell, A coded modulation technique for cooperative diversity in wireless networks, in *Proc. ICASSP '05*, Vol. 3, pp. 525–528, March 2005.

[22] V. Mahinthan, J. W. Mark, and X. S. Shen, A cooperative diversity scheme based on quadrature signaling, *IEEE Trans. Wireless Commun.*, 6(1), 41–45, Jan. 2007.

[23] T. Himsoon, W. Su, and K. J. R. Liu, Differential transmission for amplify-and-forward cooperative communications, *IEEE Signal Processing Lett.*, 12(9), 597–600, Sep. 2005.

[24] Z. Dawy, Power allocation in wireless multihop networks with application to virtual antenna array, in *Proc. IEEE PIMRC 2004*, Vol. 3, pp. 1682–1688, 5–8 Sept. 2004.

[25] Z. Han, T. Himsoon, W. P. Siriwongpairat, and K. J. R. Liu, Energy-efficient cooperative transmission over multiuser OFDM networks: Who helps whom and how to cooperate, in *Proc. IEEE WCNC'05*, New Orleans, LA, March 2005.

[26] R. K. Ahuja, T. L. Magnanti, and J. B. Orlin, *Network Flows: Theory, Algorithms, and Applications*, Prentice-Hall, Englewood Cliffs, NJ, 1993.

[27] H. N. Gabow, An efficient implementation of Edmonds' algorithm for maximum matching on graphs, *J. ACM.*, 23(2), 221–234, April 1976.

[28] D. Avis, A survey of heuristics for the weighted matching problem, *Networks*, 13, 475–493, 1983.

[29] V. Mahinthan, L. Cai, J. W. Mark, and X. Shen, Partner selection based on optimal power allocation in cooperative diversity systems, *IEEE Trans. Veh. Technol.*, 57(1), 511–520, Jan. 2008.

8 Topology Control in Cooperative Wireless Ad Hoc Networks

Vasileios Karyotis and Symeon Papavassiliou

CONTENTS

In this chapter, we present a framework for topology control (TC) as a cooperation mechanism in ad hoc networks and then describe several TC approaches within this framework. We present a classification of the current TC mechanisms, discuss the functionality and the key principles of the individual existing protocols, and provide a qualitative comparison of their characteristics according to several distinct features and performance parameters. Throughout our discussion special emphasis is placed on the main objectives of TC, namely reducing interference and energy consumption, while maintaining connectivity and adapting quickly to the dynamic changes of the ad hoc networking environment. The overall evaluation of the relative performances makes it apparent that no single mechanism can strike an absolute balance among all requirements, and application-specific considerations are needed for choosing an appropriate solution.

8.1 INTRODUCTION

The topology of a wireless network describes mainly connectivity information between its node and edge sets. In general, links are determined by some protocol operation (routing, application) subject to a transmission scheme, usually in a fixed manner [1]. However, dynamic approaches, such as TC, can be employed for adapting online network topologies, thus making them more appropriate for the wireless ad hoc environment.

TC is a mechanism to determine the transmission ranges of wireless nodes, to generate a network with desired connectivity properties and better resource utilization. The main objectives of TC are to increase the traffic-carrying capacity of the network (by increasing spatial reuse) and to reduce energy consumption, while maintaining connectivity and adapting quickly to variations of the topology. TC is a cross-layer approach that resides at the intersection of the Medium Access Control (MAC) and Network layers. Decisions of a TC module require feedback from the MAC layer and affect the operation of the routing protocol. In addition, some TC approaches exploit various MAC features, whereas others operate fully at the Network layer.

Furthermore, TC is inherently a cooperative mechanism. Nodes employing such adaptation algorithms rely heavily on their neighbors to perform some of their own networking tasks, most notably routing. Specifically, because a TC mechanism reduces a node's transmission range to a value smaller than the maximum possible, the node loses connectivity with some of its formerly distant neighbors. Thus, continuous cooperation is required with the remaining neighbors, to forward packets on behalf of the node employing TC to previously existent one-hop neighbors. For this procedure to be repeated for all network nodes, cooperation mechanisms are necessary to ensure communication and convergence of operation. TC itself may not be realized in a network lucking cooperation, as some approaches work in a pure or semicentralized fashion. In addition, cooperation among nodes in the immediate (hop-wise) neighborhood is essential for the convergence of TC protocols and the construction of the final network topology. Apart from packet forwarding, cooperative ad hoc networks may also depend on TC for improving their traffic-carrying capacity. This is achieved by implicitly reducing contention in a geographic area through transmission power adjustment. Furthermore, cooperation can also be achieved by several TC protocols mainly through load balancing and role rotation mechanisms [2].

Because the number of wireless cooperative ad hoc networks may increase significantly and nowadays many of them exhibit a delay-tolerant nature, TC can be used to define a new operational paradigm. TC-enabled networks exploit the advantages of node cooperation in better disseminating local information and using it to operate more efficiently in terms of energy resources and interference [1]. However, as nodes reduce their transmission radii, the average path length increases, meaning that the network essentially trades off delay for energy utilization and increased bandwidth. On the other hand, scaling issues and fault-tolerance considerations can significantly impact the applicability of TC and the specific design principles, affecting finally the performance of a cooperative network. Thus, finding the appropriate balance between TC and network requirements depends on the degree of desired and feasible cooperation and the availability of network and computing resources.

Several taxonomies have been proposed in the literature for TC [3,4], classifying various approaches either in distributed/centralized or homogeneous (equal transmission radii)/nonhomogeneous (unequal radii). In this chapter, the presentation of various mechanisms for TC focuses mainly on the design principles of the corresponding approaches, to reveal the main philosophy of each class of mechanisms, identify commonalities and differences among them, and to categorize them accordingly. Among our objectives are to understand and compare the main design characteristics of the different mechanisms, to identify their corresponding strengths and drawbacks, and present a comparative qualitative study. The rest of the chapter is organized as follows. In Section 8.2, MAC-based TC approaches are presented, while in Section 8.3, Network layer mechanisms are described. In this section, different design principles are identified and based on them the respective Network layer mechanisms are classified. Section 8.4 contains a comparative evaluation of the presented protocols according to different selected metrics, and finally Section 8.5 provides conclusions and some insight into possible future research topics and open issues.

8.2   MAC-LAYER TOPOLOGY CONTROL METHODS

TC at the MAC-layer mainly aims at reducing the idle energy consumption of a wireless node by turning off its radio for a time period (functionality available by the IEEE 802.11 standard [5]). MAC-layer approaches are able to support mobility, and their operation is fully asynchronous. The goal of this section is to provide an overview of the recent progress in the field of MAC-based TC protocol design and development in cooperative mobile ad hoc networking environments.

8.2.1   GEOGRAPHIC ADAPTIVE FIDELITY PROTOCOL

The geographic adaptive fidelity (GAF) protocol [2], assumes perfect node location information, symmetry of transmission ranges, and deterministic wireless propagation. GAF identifies routing-equivalent nodes and turns off the radios of all but one for each equivalence class for a specific time period. This reduces the idle/promiscuous consumption of a node, which in most cases represents a great portion of the overall energy consumption [6].

Nodes running GAF use location information to embed a virtual grid over the deployment area. The virtual grid is defined so that for two adjacent grid cells all nodes in a cell can communicate with all nodes in a neighboring one and vice versa. All nodes in a specific cell are equivalent in terms of routing properties.

In GAF, a node can be in one of three states, namely discovery, active, or sleeping as shown in Figure 8.1. Initially, a node starts in the discovery state, where it exchanges discovery messages for a period of T_d seconds. Upon expiration of this period, it enters the active state. Timer T_a defines the active period before returning to the discovery state. While active, the node keeps on broadcasting a discovery message every T_d seconds. A node can switch to the sleeping state from both the discovery and active states, when another node takes over the routing, according to a ranking procedure.

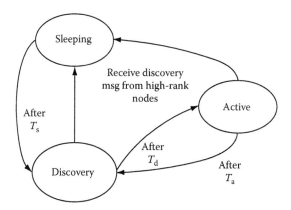

FIGURE 8.1   GAF protocol state diagram.

A node in the sleeping state powers off its radio and cancels all timers but one, T_s, after which it returns to the discovery state.

A ranking procedure is intended to maximize the network lifetime, as a mechanism for rotating roles around the nodes of a grid. To achieve this, GAF assigns higher rank to nodes with longer expected lifetime. A node in the active state has a higher rank than a node in the discovery state. This intends to drive the grid quickly to a state with only one node active. Node labels are used to break ties. Role rotation can also be used for load balancing, making it possible for nodes with higher residual energy (i.e., higher rank) in each grid to become the new active nodes.

GAF is realized in two versions (basic and mobility assisted, referred to as GAF-b and GAF-ma, respectively), both of which perform better than conventional routing protocols like ad hoc on demand distance vector [7] and Dynamic Source Routing (DSR) [8]. GAF-b sustains more nodes alive for a longer time than GAF-ma, while for a higher node density, GAF-b seems to be even more suitable, especially for higher movement rates. In terms of network capacity, all GAF variations seem to perform the same.

8.2.2 SPAN PROTOCOL

As with GAF, the principle of Span [9] is that in a relatively dense network, only a fraction of the nodes needs to be active to carry traffic. However, Span does not need geographical information for its operation. The active nodes are denoted as coordinators and they form a forwarding backbone. Coordinator rotation takes place to balance resource expenditure. To reduce the number of redundant coordinators, a node delays the announcement of its willingness in becoming a coordinator (to prevent multiple nodes from becoming simultaneously coordinators) by taking into account the residual power and the number of neighbors it can connect.

The basic operations of Span are to select the coordinators, advertise them in the network, and maintain the backbone they formed through withdrawal and announcement processes. Enough coordinators need to be elected so that each node is in the range of at least one coordinator. The number of coordinators should be minimized by using local only information.

Coordinator announcement is achieved by periodic, local decisions of a node and application of the coordinator eligibility rule. If two neighbors of a noncoordinator node cannot reach each other directly or through a coordinator, the node should become a coordinator. The contention to become a coordinator can be resolved through a randomized back-off delay mechanism. The value of the delay is important, because it determines the redundancy or lack of coordinators in the network. A coordinator should withdraw if every pair of nodes of its neighbors can reach each other directly or through some other coordinator. This can be extended to achieve some degree of fairness. A node withdraws, if every pair of neighbor nodes can reach each other through some other neighbor, even if those neighbors are not currently coordinators.

Simulations in Ref. [9] have shown that in static networks, Span does not significantly degrade network capacity and can forward more packets than 802.11-psm (power saving mode) [5] under high load. It increases latency slightly, and the degree of mobility does not have substantial impact on its performance. Additionally, the energy conservation is significant compared to the two 802.11 modes (pure/power saving); however, as the node density increases the energy savings do not increase proportionally.

8.3 NETWORK-LAYER TOPOLOGY CONTROL METHODS

While the design freedom for TC at the MAC layer is relatively restricted, at the Network layer, design options are much more diverse. In this section, we first identify several generic design principles and then, based on them, we classify the presented TC approaches. Then we discuss the main characteristics of each protocol and highlight their respective strengths and drawbacks.

TABLE 8.1

Classification of Network Layer TC Approaches

Category	Techniques				
Cone angles	PCAM 1	PCAM 2	PCAM 3	PCAM 4	CBTC
Relay regions	MECN	SMECN			
Tree based	LMST	MST-red.	MIP-red.	STH	iSTH
Node degree	KNeigh	XTC	MPR		
Clustering	COMPOW	CLUSTERPOW			
Special graphs	DG	RNG	GG	Yao	Ord/S/YaoGG
Heterogeneous TC	DRNG	DLSS			
Constrained optimization	CONNECT	BICONN-AUG	LINT	LILT	

Assuming a typical network graph describing the topology of an ad hoc or sensor network, the cone angles of a node are the angles formed between its neighboring transmission links. The first design principle at the Network layer is based on enforcing (if possible) several constraints on the values of a node's cone angles, in a totally distributed and asynchronous fashion. Another approach that works distributively and effectively is based on the relay regions, defined as the geographical areas that can be reached more efficiently (energy-wise) through relay nodes. Building energy efficient tree structures can also be used for controlling the topology of a network. In such cases, different link metrics can be used for tree construction and yield connected, energy-efficient, or low-interference topologies. Special form network graphs may be considered as an extension of the idea to use graph theoretic structures for TC and are shown to work rather effectively. Furthermore, the node degree itself can be used as a TC design parameter. A similar approach to the MAC-layer based techniques is clustering, which essentially exploits spatial node redundancy to achieve power conserving TC. Finally, standard optimization problems may be formed according to various objectives and constraints, the solutions of which yield topologies with highly preferable TC properties.

Various heuristics may accompany the above design principles, further improving the performance of the proposed protocols and yielding desirable properties that otherwise would be absent. Table 8.1 contains a taxonomy of the presented TC mechanisms based on their design principles. The first column presents the various classes of TC techniques (according to design principles), while each row of the table enumerates the protocols and algorithms that belong in each category. The first row contains techniques using the cone angles, while the second row contains mechanisms based on relay regions. The third class (row) involves mechanisms based on graph theoretic tree structures and similarly the fourth one consists of protocols based on the degree of a node. Clustering and graphs with special properties are then presented followed by techniques for heterogeneous TC and constrained optimization TC.

It should be noted here that this chapter does not intend to provide a full survey and quantitative performance comparison of all the various existing TC schemes. It rather presents a structured and comprehensive effort to identify, report, and summarize the work in progress in the area of TC in cooperative ad hoc networks. However, a qualitative comparison of the various presented protocols is provided. Its main objective is the understanding and comparison of the key design principles of the different approaches and their evaluation based on a set of qualitative criteria and requirements, such as their complexity, performance, scalability, their distributed nature, and their capability of supporting mobility.

8.3.1 Cone Angle Techniques

8.3.1.1 Pure Cone Angles Methods

The simplest operation for a cone angle technique is to constrain the value of the inherent node angles. In this section we first describe the operation of four different variations of pure cone angles methods (PCAMs) [10] and then present their properties.

In PCAM 1, every node chooses its transmission radius to have the minimum possible value provided that all cone angles have values less than π. Such requirement demands that there exist at least three neighbors in the coverage area of a node. However, not any arrangement of three nodes can yield cone angles with values less than π. Nevertheless, the greater the coverage radius, the more neighbors will become available. The values of cone angles are in inversely direct proportion to the transmission radius.

PCAM 2 is simply an extension of the previous one. Initially, PCAM 1 is executed. Then, if the coverage radius of node u is chosen to be r and the maximum length of an incoming edge is t, then u makes the coverage radius equal to $\max\{r, t\}$. It essentially creates a bidirectional edge, wherever PCAM 1 would create a unidirectional or bidirectional edge.

In PCAM 3, a node chooses its coverage radius as the minimum possible value making all its cone angles smaller than or equal to θ, where θ is a fixed number in $(0, \frac{2\pi}{3})$. Compared to PCAM 1, a node will now have to transmit at a higher power level to ensure that more nodes are within its neighborhood, thus decreasing its cone angles further.

Finally, in PCAM 4 every node chooses its transmission power r to the minimum possible value, with the constraint that its cone angles are all smaller than or equal to θ $\left(0 < \theta < \frac{2\pi}{3}\right)$, like in PCAM 3. Transmission radius r is compared to a given threshold value δ, and the final range is set to the $\max\{r, \delta\}$.

For the remainder of this section, let us assume a Poisson point process of rate $\lambda = 1$ node per unit area. Then for PCAM 1 the probability mass function (p.m.f.) of the node degree Z is proven to be $\Pr(Z = n) = \frac{n-2}{2^{n-1}}$, for $Z > 2$ and zero elsewhere [10]. A direct computation yields $E(Z) = 5$ and $\mathrm{Var}(Z) = 4$. The probability density function of the transmission radius is $f_R(x) = 2\pi x e^{-\pi x^2} + \pi^2 x^3 e^{-\pi x^2/2} - 2\pi x e^{-\pi x^2/2}$. The expectation and variance of the radius are computed to be $E(R) = \frac{\sqrt{2}+1}{2}$ and $\mathrm{Var}(R) = \frac{5}{\pi} - \frac{3+2\sqrt{2}}{4}$, respectively. The percentage of nodes with large coverage radii diminishes exponentially fast to zero for large radii; however, the probability density of the radius is unbounded. This is an immediate consequence of the fact that a node should not be isolated even in the case of an infinitely large network. A PCAM 1 topology contains a unique infinitely large and strongly connected subgraph that every node can reach through directed paths [10].

The graph created by PCAM 1 is actually a subset of the corresponding graph created by PCAM 2. The distribution of the coverage radius for the latter is shown to be extremely difficult to compute [10]. However, a theoretical upper bound on the expectation and variance of the radius has been derived, which can be used for the comparison of the performance. Specifically, the expected value and variance of the coverage radius are $E(Z) = 5$ and $\mathrm{Var}(Z) = 4$.

A network constructed by PCAM 3 is strongly connected and it enables geographical routing with no dead ends for any source–destination node pair. For any two nodes u, v, the length distortion for the ordered pair is upper bounded by $\left[1 - \sqrt{2 - 2\cos(\theta/2)}\right]^{-1}$. To the other extreme, hop distortion can be arbitrarily large. This happens in case of high node density, where every path connecting two nodes might have many edges of very small lengths. The probability density functions for both the node degree and the coverage radius decrease exponentially fast [10]. This means the expectations and variances of the node degree and the coverage radius are finite.

The induced graph of PCAM 4 is a supergraph of that of PCAM 3, and therefore it will be strongly connected as well. Moreover, the upper bound on the length distortion remains the same as in the case of PCAM 3, because nodes do not change locations. On the contrary, the length of a path between two connected nodes can only decrease, which in turn decreases the length distortion. Thus, the previous bound also holds for PCAM 4, but the upper bound can be made tighter: $\frac{2}{\delta}\left[1 + \frac{1}{1-\sqrt{2-2\cos(\theta/2)}}\right]$. The probability density functions for node degree and coverage radius decrease exponentially fast.

8.3.1.2 Cone-Based Topology Control

In cone-based TC (CBTC) approach a node needs only directional information of its neighbors to operate in a distributed fashion [11–13]. CBTC with parameter α, CBTC(α), first builds a connected graph with the wireless nodes as its vertex set and then it eliminates nonefficient edges. Each node u beacons its neighborhood with growing power $P_t(u)$. If u discovers a new neighbor v, it will be added into the neighborhood set $N(u)$ of u. Node u continues to increase its transmission power until $N(u)$ is large enough such that for any cone with angle α there is at least one neighbor $v \in N(u)$ or until the power level reaches the maximum transmission power. The decision on whether or not the requirement is satisfied can be easily made. For every node v in its $N(u)$, u creates a cone area of angle α, with v in the middle of the area. If the union of these cover-cone areas spans the whole 2π angle, the criterion is satisfied.

CBTC(α) mainly aims at preserving connectivity properties of the induced graph, while reducing communication power. Consequently, several optimizations have been proposed to improve its operation. In Ref. [11], if node u has two neighbor nodes $v, w \in N(u)$, such that the power needed to send from u to w directly is not less than the total power to send through v, w can be eliminated from $N(u)$. In Refs. [12,13], shrink-back for boundary nodes (transmission power of boundary nodes is chosen to be lower than the maximum), asymmetric edge removal, and pairwise edge removal are used to improve CBTC(α) by removing redundant edges that do not harm connectivity. The protocol can be extended so that if two nodes $v, w \in N(u)$ satisfy $P_t(u, v) \leq P_t(u, w)$ and $P_t(u, v) + P_t(v, w) \leq q \cdot P_t(u, w)$, for $q \geq 1$, then w is removed from $N(u)$ and by symmetry u from $N(w)$. If multiple nodes satisfy the above criteria, then the node with minimum $P_t(u, \cdot)$ is chosen. In that case, parameter q is equally important to α for the operation of the algorithm.

Authors in Refs. [12,13] have proven that for $\alpha \leq \frac{5\pi}{6}$, CBTC($\alpha$) preserves connectivity. Furthermore, the same choice of α preserves connectivity after shrink-back and pairwise edge-removal optimizations, while $\alpha \leq \frac{2\pi}{3}$ suffices for preserving connectivity in the asymmetric edge-removal case. In Ref. [11], it was shown that if q is larger or equal to 2, then the node degree of any node is at the most 6.

8.3.2 RELAY REGIONS TECHNIQUES

Relay regions approaches aim at producing energy efficient communication networks, by exploiting the inherent topological properties of a network and by taking into account the wireless propagation environment. Essentially, the relay region denotes the area for which it is more efficient to forward messages through a relay node.

Assuming three nodes i (source), j (destination), and r (relay) on a two-dimensional plane, the relay region $R_{i \to r}$ of the transmitting–relaying node pair (i, r) is defined as the set of points (x, y) of the plane

$$R_{i \to r} \equiv \left\{ (x, y) \mid P_{i \to r \to (x,y)} < P_{i \to (x,y)} \right\} \tag{8.1}$$

where

$P_{i \to r \to (x,y)}$ denotes the power required to transmit information from i through r to a node at point (x, y)

$P_{i \to (x,y)}$ denotes the power required to transmit from i to a node at point (x, y) directly

The shape of the relay regions in planar topologies depends heavily on the wireless propagation model. For instance, the relay region of the simple arrangement of i, j, r in an environment with path loss $\alpha = 4$ is shown in Figure 8.2. For an $1/d^\alpha$ propagation law ($\alpha > 2$), if $B(i, r)$ is the boundary

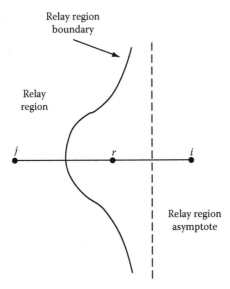

FIGURE 8.2 Relay region for path loss $\alpha = 4$.

of $R_{i \rightarrow r}$, then in the limit, $B(i, r)$ has the form of a vertical asymptotic line going over the midpoint of the segment between the source and relay node as shown in Figure 8.2. For transmit node i, we denote the membership relation of node $(x_j, y_j) \in R_{i \rightarrow k}$ by $j \in R(r)$. Then if $j \in R(r)$, $d_{jr} < d_{ij}$ and $d_{ir} < d_{ij}$. Furthermore, a node cannot belong to its own relay region. Also, if a node belongs to the relay region of another node, the second cannot belong to the relay region of the first.

The enclosure area of a source node i is defined as the nonempty solution ϵ_i of the set of equations:

$$\epsilon_i = \bigcap_{k \in N(i)} R^c_{i \rightarrow k} \bigcap D_\aleph \qquad (8.2)$$

$$N(i) = \{n \in \aleph \mid (x_n, y_n) \in \epsilon_i, n \neq i\} \qquad (8.3)$$

Set $\aleph$ represents the network nodes and $D_\aleph$ their deployment region. $N(i)$ represents the neighbor set of i. For bounded deployment regions, the enclosure areas will also be bounded. From Equation 8.2, it is deduced that the enclosure region is the union of the complements of the relay regions of the nodes in the neighborhood set of source node i that also lie in the deployment region. The enclosure region for a node i with three neighbors k, l, and m is depicted in Figure 8.3.

The enclosure of a source node specifies an area beyond which it is not power efficient to search for more neighboring nodes. Thus, a source node needs only to maintain connectivity with nodes in its enclosure region. The enclosure graph of a set of nodes $\aleph$ is the graph whose vertex set is $\aleph$ and the edge set $\bigcup_{j \in \aleph} \bigcup_{k \in R(j)} l_{j \rightarrow k}$, where edge $l_{i \rightarrow k}$ is the directed communication link from i to k. It is essentially the graph with vertices the nodes of the network and edges that connect each node with all its neighbors as defined by Equation 8.3.

If A denotes the set of all nodes that node i has identified up to a point as nodes found in the enclosure region and its neighboring set, then if $j, k \in A$, whenever $k \in R(j)$, a directed edge from j to k is added and denoted by $e_{j \rightarrow k}$. The relay graph $G(i)$ of node i is the digraph with vertex set A and edges defined by $\bigcup_{j \in A} \bigcup_{k \in R(j)} e_{j \rightarrow k}$. $G(i)$ has vertices the neighbors of a node and edges the directed arcs from the node's neighbors to their own neighbors, provided a node pointed by an edge lies in the relay region of the first one. The relay graph $G(i)$ of node i, has no cycles, as shown in Ref. [14].

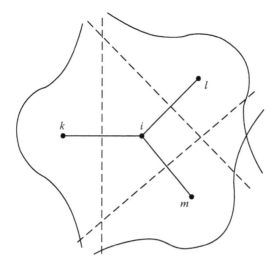

FIGURE 8.3   Enclosure area of node i.

8.3.2.1   Minimum-Energy Communication Network Approach

Minimum-energy communication network (MECN) mechanism [14] aims at creating a graph that contains a minimum energy path for every pair of nodes. MECN operates in two phases. Specifically, a node first eliminates any nodes lying in its relay regions by local search and selects only links to the remaining nodes in its immediate neighborhood to find global minimum-power paths to a sink node. The local search is actually the procedure to construct the enclosure and neighbor set of each node individually. The enclosure computation requires knowledge of the positions of nearby nodes, which is provided by the relay graph. A node finds out the positions of nearby nodes by sending out beacon messages and receiving responses. It then calculates the relay regions for these nodes. However, only nodes not residing in the relay regions of previously found nodes are maintained. Thus, the relay graph is updated every time a new node is discovered. If a node falls in the relay region of a previously found node, it is marked "dead" and the previously found node is denoted as "blocking." Otherwise, it is marked as "alive." The final set of alive nodes is the set of neighbors and they enclose the source node. In the case of serial blocking, namely when a node k is blocked by a node j and then a new node p blocks j but not k, then node k can reenter the enclosure, as it will not be blocked anymore. In a chain of blocking, when a node that blocks the first blocking node of the chain is discovered, all the remaining nodes in the chain are added again in the enclosure.

As soon as the neighborhood set is determined, MECN proceeds with the second phase, which consists of finding the optimal links on the enclosure graph. Each node broadcasts its cost to its neighbors. The cost is determined as the minimum power necessary to establish a path to the sink node. Given the neighbor costs, a node determines the minimum cost to the sink node by computing $C_{i,n} = \mathrm{Cost}(n) + P_t(i,n) + P_r(n)$, where $P_t(i,n)$ is the power required to transmit from i to n and $P_r(n)$ is the additional receiver power that i's connection to n would induce at n. Then node i computes $\mathrm{Cost}(i) = \min_{n \in N(i)} C_{i,n}$ and picks the link corresponding to the minimum cost neighbor. The computation is repeated periodically, and the algorithm converges to the set of links after a finite number of iterations.

The induced topology has several desirable properties. More specifically, given a deployment region of wireless nodes, the enclosure graph of the set of nodes $\aleph$ is strongly connected, as shown in Ref. [14]. Furthermore, the search procedure for the enclosure graph terminates with the solution $(\epsilon_i, N(i))$, namely with the enclosure and the neighborhood regions of node i, which is also unique. The latter allows the algorithm to run asynchronously in every node.

The MECN protocol works well with stationary nodes; however, it can be easily modified to be applicable in mobile networks also. Each node can run the algorithm, sleep for an interval, and then wake up to look for changes. This interval is called a cycle period. Simulation results [14] for mobile networks have shown that the greater the displacement of a node in an iteration, the greater the consumption per node for a search. On a final note, the power consumption per node scales better than linearly with a maximum displacement in the range that can track the correct costs.

The SMECN (Small MECN) protocol presented in Ref. [15] is similar to MECN but produces a more efficient topology. Using a process similar to MECN for neighbor discovery and construction of a node's local neighborhood, SMECN computes a subgraph of the enclosure graph computed by MECN, provided that the search regions during the local search are circular (no obstacles between nodes). During neighbor discovery, node locations are also distributed among the network nodes. SMECN increases transmission power gradually until the enclosing region of a node is a subset of the broadcasting area (that can be reached by a single message broadcast of known power). The main difference between the two algorithms is the computation of the enclosure region n of a node. In SMECN $n = \bigcap_{v \in A} (F(u, p_{max}) - R_{u \to v})$ at the end of each iteration, while in MECN $n = \bigcap_{v \in A - NonNbrs} (F(u, p_{max}) - R_{u \to v})$, where $F(u, v)$ is the region around node u that can be reached with power P and *NonNbrs* is the set of original neighbors not included in the final topology. In SMECN, contrary to MECN, once a node enters *NonNbrs* it is never removed.

The graph induced by SMECN has the minimum energy property and presents lower average node degree than that in MECN. Network nodes die out slower and disconnect later in SMECN, as the node's consumption is lower. In addition, MECN produces greater average end-to-end delay and a lower packet delivery ratio.

8.3.3 Tree-Based Techniques

TC methods of this category exploit properties of graph-theoretic tree structures, most notably the minimum spanning tree (MST) or shortest path tree. The following approaches incorporate such structures in the TC framework.

8.3.3.1 Local Minimum Spanning Tree

Local minimum spanning tree (LMST) is a distributed algorithm based on the construction of MSTs [16]. It consists of three phases, namely, information exchange, topology construction, and determination of transmit power, and it could be enhanced with an optional optimization phase denoted by construction of bidirectional links.

Information exchange involves broadcasting node ID and location information in its vicinity through a beacon packet. In the sequence, LMST is essentially an extension of Prim's algorithm. The latter does not produce unique solutions in the event of multiple edges with equal cost. A link-weight function depending on the Euclidean distance of two nodes and the arithmetic order of their IDs can be defined to distinguish unique edge weights. After Prim's algorithm operation, all the nodes in the computed MST that are one hop away in a node's neighbor list are preserved and the rest are discarded. In the final phase, a node measuring the receiving power of HELLO messages determines the specific power levels required to reach its neighbors. Topology G_0 derived by LMST is a digraph. Two more topologies can be induced at the (optional) fourth phase. G_0^+ is the topology where all unidirectional links in G_0 become bidirectional and G_0^- is the topology where all unidirectional links are deleted. Essentially, LMST runs at every node and starts by computing the local MST from each node's perspective. Then it adds all the nodes in the computed MST that are one hop away in its neighbor list and discards the rest. Bidirectional topologies can be derived in the sequence.

In Ref. [16], it is proven that the node degree in G_0 is bounded by 6 and connectivity is preserved. The degree of any node in G_0^+ or G_0^- is also bounded by 6 and both preserve connectivity of G. Furthermore, LMST has been compared against MECN and the maximum power transmit scheme. Corresponding results show that both LMST and MECN reduce significantly the average node degree, as well as maximum and minimum node degrees. LMST appears superior in terms of maximum and average node degree, even though the average value of the difference is only marginal. The average node degree of MECN increases slightly when the node density increases, while the average node degree of LMST is practically constant. In addition, the average node degree under LMST does not differ much for the two approaches (pure and bidirectional). Although LMST expends less power to transmit to the farthest neighbor, simulation results have shown that even though its power consumption is lower than the maximum-power scheme, it is slightly higher than that of MECN, despite the smaller average link length.

8.3.3.2 Minimum Spanning Tree and Minimum Incremental Power

The MST-reduced algorithm, presented in Ref. [17] is based on the structure of the MST, which approximates the optimal solution of a static network by a factor of 2. Assuming a $P_r(i) = P_t(i)r^{-\alpha}$ path loss model, where $P_r(i)$ and $P_t(i)$ are the receipt and transmit power of node i, respectively, if all nodes have the same receiving thresholds, the transmission link cost can simply be r^α. The receiving/processing component of the energy consumption can be ignored, as it will be constant for all nodes. The power assignment at node i to reach node j must satisfy the relation $P_t(i) \geq |(i,j)|^\alpha$. For a set of nodes V, a vertex r is chosen and used as the root of an MST. For each one of the remaining nodes v, there exists a path from v to r and each edge on the path from v to r is assigned a direction toward the root. The direction of each edge is inverted and the resulting tree is denoted by T. If for each vertex v, $m(v)$ denotes the longest out-edge at v in T so that $|m(v)| = \max_{e \in \text{out}_{T(v)}} |e|$, all $m(v)$'s form disjoint paths from the internal vertices to leaves. These are exactly the critical paths.

MST-reduced starts by constructing an MST (i.e., using Kruskal's method) for vertex v. Then it finds the critical paths and creates cycles on them rather than using backtracking edges $e(v)$, to reduce the cost of the MST. The maximum energy expenditure of MST-reduced is the minimum among all spanning trees. For every critical path if the cost of the backtracking is greater than that of the direct link from the starting node of the critical path to the leaf, then the direct link is added with direction from the leaf to the initial node of the critical path to complete a cycle and reduce cost. The comparison with the case of an ordinary MST shows that it outperforms the latter and usually approaches the performance of the optimal solution. Simulations show that its performance lies indeed within a factor of 2 of the optimal [17].

On the other hand, minimum incremental power (MIP)-reduced protocol [17] is based on the MIP tree [18]. Initially, the total assigned power P of all nodes is assumed zero and the set of nodes in the final topology S contains only one node. The next node u to enter S is such that when connected to a node v in S, it forces P to increase by the minimum possible value. The step value is $P_t(v) + \delta_{P_t(u)}$, where $P_t(v)$ is the power expenditure at v and $\delta_{P_t(u)}$ is the increased power needed for u to reach v. This procedure is repeated until $S = V$.

8.3.3.3 Shortest-Path Tree Heuristic

The shortest-path tree heuristic (STH) [19] constitutes an approximation to the NP-hard problem of minimum power configuration (MPC) [19,20]. In brief, MPC defines a framework for minimizing the total power consumption of nodes, while maintaining connectivity.

Let us assume a network graph $G(V, E)$ and a set of traffic demands denoted by $I = \{(s_i, t_j, r_{ij}) | s_i \in S, t_j \in T\}$, where S is the set of source nodes, T the set of sink nodes, and r_{ij} the traffic rate at which source i sends information to sink j. The MPC consists of finding a subgraph $G'(V', E')$ and a path $f(s_i, t_j)$ in G' for each traffic demand in I, such that the total power cost $P(G')$ is minimal. Namely, G' corresponds to the subset that satisfies the traffic demands with minimum power consumption. The total power cost is equal to the sum of the costs along the shortest path of each traffic demand and the total nodal costs (fixed consumption components).

STH tries to balance the flow-dependent cost and the fixed nodal cost, using a combined cost metric. The computation of the shortest path can be done by Dijkstra's algorithm or any other shortest path algorithm. Because no edge of the initial graph is removed, connectivity is preserved. The topology is energy efficient, as the algorithm yields shortest paths, where the costs are functions of transmission powers. The approximation ratio of STH is proven in Ref. [19] to be upper bounded by $|S|$.

8.3.3.4 Incremental Shortest-Path Tree Heuristic

In the previous approach, the weight function was different for each source node. This means that the shortest path from a source to a destination is not affected by already established shortest paths for other sources. Sharing such a path can make the technique more efficient. The weight function can be modified, so that once a path is established and another path is about to use the links of the initial, the edge weights on the existing path links should not include the nodal costs, as they have already been used. The new algorithm should find the minimal incremental cost of the new paths. The incremental shortest-path tree heuristic (iSTH) algorithm implements this idea [19].

The method results in a smaller number of nodes used to route packets from a source to a destination. Thus, more nodes can turn their radio off and conserve energy. iSTH performs better when idle power dominates the total consumption. This happens when the network load is low. Once a shortest path is found the weights on its links become zero. To find another one, it is sufficient to find the shortest path to a node of the existing path (minimum-weight Steiner tree).

8.3.4 NODE DEGREE-BASED TECHNIQUES

Approaches in this category rely on the capability of cooperation between neighboring nodes for collecting accurate local information. In most cases, this is the minimum required information for successful TC and distributed operation.

8.3.4.1 KNeigh Protocol

KNeigh protocol is based on the assumption that a node is able to determine the distance to a neighbor (with a certain error margin) [21,22]. Initially, every node broadcasts its ID at maximum power to perform neighbor discovery. A node saves both neighboring IDs and their estimated distances. Then, every node computes its k-closest neighbors and rebroadcasts this information enabling the creation of symmetric neighbor lists. The broadcast transmit power of each node is set to the minimum value needed to reach the farthest node in the final configuration. An additional optimization step might follow, to remove remaining energy-inefficient links. According to this, if there is a third node w between nodes u, v, so that it is more energy efficient to transmit using the two-hop path instead of the direct, then edge (u, v) is removed, like in CBTC.

By the end of the operation of KNeigh, the degree of each node is upper bounded by 6. Simulation results in Ref. [21] have demonstrated that KNeigh is 20% more energy efficient than CBTC and presents better performance with respect to the average node degree (after optimizations). Without optimizations, the gap increases in favor of KNeigh for energy efficiency, while for the average node degree the KNeigh and CBTC perform similarly. Distance estimation errors have marginal impact on the operation of the protocol. Furthermore, it has been demonstrated in Ref. [22] that LMST achieves not only a slightly lower average degree than KNeigh but also lower consumption when the optimization is applied. However, LMST yields longer path lengths, followed by KNeigh and then CBTC. As a final note, KNeigh is lightweight (only $2n$ messages are exchanged among n nodes) and has low implementation complexity. On its downside, it guarantees connectivity only with high probability, which means that in the worst case the final topology could be disconnected.

8.3.4.2 XTC Protocol

The XTC protocol [23] can be considered as a generalization of KNeigh, with the difference that when neighbors are ordered KNeigh considers only the first k, while in XTC the entire neighbor set is considered. Furthermore, neighbor ordering is based more on link quality for XTC, rather than distance as in the KNeigh protocol.

Initially, every node establishes an ordering $\prec$ of the neighbor set according to a quality metric (in Ref. [23] the considered metric is the received signal strength). Subsequently, $w \prec_u v$ denotes that link (u, w) has a relatively higher quality than (u, v). The ordered list is rebroadcast so that nodes compute their final neighbor lists locally. A node u considering a neighbor v checks whether there exists a node w with $w \prec_u v$ such that $w \prec_v u$, in which case edge (u, v) is discarded. When all neighbors are examined, XTC terminates with the induced topology.

XTC is a lightweight protocol requiring only $2n$ messages to be exchanged. It is proven to preserve connectivity and yield symmetric topologies. In the event that the link-weight metric is distance-based (as in KNeigh), XTC yields a planar topology with node degrees upper bounded by 6, which is a subgraph of the relative neighbor graph (RNG) [1,4].

8.3.4.3 Minimum Power Routing Approach

The minimum power routing (MPR) technique [24] is similar to XTC. Nodes rank their neighbors based on the average received power levels. Only the first N are maintained in the final neighbor list of direct neighbors. The transmit power is properly adjusted to reach only those N direct neighbors. To ensure global topological view, final connectivity information is broadcast.

The N selected nodes form a group (cluster). Either power adjustment within the cluster or assignment of a common power level for all nodes could be applied. For every chosen management strategy, a node needs to establish the minimum power level to reach its farthest node or to reach at most the jth ranked node. Transmission power is used as a link cost and the objective is to solve a minimization problem on the aforementioned cost function within the specified bounds of the transmission power vector. Based on the global topology view formed, a node creates all possible routes from source to destination and then employs a shortest path algorithm for the minimum cost route from source to destination.

According to performance evaluation results reported in Ref. [24], for the nonadaptation power management scheme the average node throughput decreases as the connectivity range N increases. Even though the increased node degree allows packets to traverse fewer number of hops (and hence reduces retransmission attempts), this is canceled out by increased MAC contention. The average power consumption increases monotonically as N increases. In the case of power adjustment within a cluster, the interference levels are significantly reduced. The adjustment scheme appears to be better for low connectivity ranges, whereas the nonadjusting appears superior for higher ones. Similar behavior is observed for the power consumption. For a degree higher than ten, the power consumption seems to become constant, due to the fact that the routes will already have many intermediate hops with small link lengths, and additional nodes will not offer significant improvement.

8.3.5 CLUSTERING TECHNIQUES

The key principles of these techniques aim at grouping nodes into clusters so that spatial redundancy can be exploited. The corresponding techniques differ from the MAC-based approaches in that the clustered nodes do not elect a representative and then turn their radio off. They all continue to work normally but in a more localized fashion. Such methods are mainly more appealing to nonuniform node deployments.

Representative examples of these techniques are Common Power (COMPOW) [25] and clustered COMPOW (CLUSTERPOW) [26]. The COMPOW protocol operates all nodes at a common power level, reducing the power control problem to the choice of one optimal value. Common power levels ensure bidirectional links. In the case of randomly deployed nodes, homogeneous node density, and common broadcast power level, it is proven in Ref. [27] that one should minimize the transmission range to achieve a higher per node throughput. Thus, the common transmission power level should be the lowest possible, provided that it preserves the connectivity of the network.

The implementation of the protocol demands a discrete set of transmission levels. Any proactive routing protocol can be used in conjunction with COMPOW. The chosen protocol maintains multiple routing tables of the network at the application level, one for each transmission level. The table, which corresponds to the lower transmission power while preserving the connectivity, is selected as the master routing table and copied in the kernel of the operating system.

Overall, COMPOW is able to preserve connectivity, yield power-aware routes, reduce MAC-layer contention, and conserve energy on a per node basis. The main drawback is that it assumes static nodes while its operation can be severely affected by mobility. Significant overhead is produced from the need to maintain multiple instances of the routing tables and the necessary message exchanges of the peer entities.

Several generalizations of COMPOW have been proposed in the literature. Again, the goal is to adjust the transmit level of nodes, so that most of the intra-cluster communication takes place at lower transmit power levels, and higher power transmits are reserved only for the inter-cluster connections. The power levels should be sufficient to preserve the connectivity while increasing the network lifetime as much as possible. The clustering problem is one of classifying nodes hierarchically into equivalence classes and could be thought of as a degenerate case of TC.

CLUSTERPOW is able to support power control, clustering, and routing [26]. It generally produces routes consisting of hops of different transmit power. The algorithm is executed locally at every node, in a distributed way. With a discrete set of available power levels, the protocol uses the lowest possible levels, such that the destination is reachable (in multiple hops), while no power level is larger than p.

The software architecture of CLUSTERPOW is almost identical to that of COMPOW. One routing daemon for each power level is run at the application level, building its own routing table. The table is built by communicating with peer daemons through HELLO packets exchanged at the same power level. The kernel routing table is constructed by consulting the lowest power routing table in which the destination is reachable.

The clustering is implicit, because the choice of a power level and next hop for each destination creates the equivalence classes. It is also adaptive, because it is based on multiple instances of the routing protocol, which by definition ensures adaptation to topology changes. As the algorithm runs at every node separately, clustering is also distributed. Moreover, the routes discovered correspond to a nonincreasing sequence of transmit power levels and are loop free. Starting with a connected network graph, the protocol provides a connected graph as well. A node reduces the transmit power to a level such that the destination is still reachable.

8.3.6 SPECIAL-GRAPH-BASED TECHNIQUES

Special form graphs have been used for TC, as their distributed operation makes them suitable for adaptive and local maintenance of the network topology. In the following, we present the most popular approaches of this type and describe how they could be combined to produce more complex and sophisticated techniques.

In the unit disk graph (UDG) [4], every two nodes are connected with an edge if and only if their distance is at most equal to their common transmission radius. UDG is typically considered the underlying graph model for these approaches. Assuming the wireless network is represented as a UDG, a power spanner is a subgraph, such that there is a positive real constant ρ (power stretch factor), so that the power consumption of the shortest path between any two nodes in the subgraph is at most ρ times the power consumption of the path connecting the two nodes in the original graph.

The Delaunay graph (DG) on UDG [4], the key principle of which is depicted in Figure 8.4a, is the unique triangulation of the UDG vertices, which is such that the circumcircle of every triangle contains no vertices of the UDG in its interior. Similarly, the RNG on UDG [4], depicted in Figure 8.4b, contains an edge (u, v) such that the intersection of the circles centered at u, v with radius at the edge (u, v) does not contain any vertex of UDG. The *Gabriel graph* (GG) on UDG (shown in Figure 8.4c)

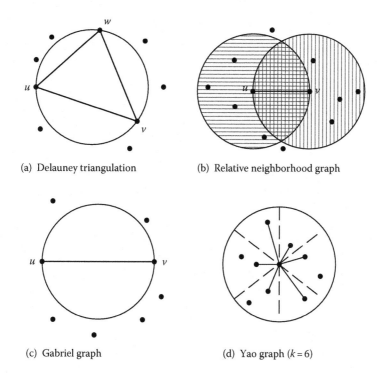

(a) Delauney triangulation   (b) Relative neighborhood graph

(c) Gabriel graph   (d) Yao graph ($k = 6$)

FIGURE 8.4   Special form graphs.

is the graph containing an edge (u, v), if and only if no other node of UDG lies in the circle with (u, v) as diameter. Finally, the Yao graph with parameter $k \geq 6$ (YGk) (shown in Figure 8.4d for $k = 6$) is the graph that for each node preserves (if available) only the closest neighbor in each of k equally spaced cones around it. The GG and RNG graphs are connected and planar. Furthermore, the GG has a power stretch factor of $\rho = 1$.

An extension of the Yao and Gabriel graphs is the degree-$(k + 5)$ OrdYaoGG graph, which is based on the construction of a Yao graph on top of an underlying GG [28]. Initially, the GG topology is constructed and after a ranking procedure, nodes construct a Yao graph ($k > 6$) on top of the GG topology based on their ranking. Nodes decrease their power to the level at which they can reach their farthest neighbor in OrdYaoGG. The final topology of OrdYaoGG is a planar power spanner of UDG whose node degree is upper bounded by $k + 5$, where $k > 6$ is an adjustable parameter [28].

A modification of the OrdYaoGG is the SYaoGG graph. Again a Yao graph is constructed on top of a GG, where now no ranking procedure is employed. The construction of the Yao graph is simply determined by the node IDs. The structure of SYaoGG is a k-degree bounded power spanner, which has a lower node degree bound than that in OrdYaoGG, but it yields a higher power stretch factor.

In simulations presented in Ref. [28] all the stretching factors are small, slightly higher than the unit value. This means that the sparser structures, even though they reduce the interference significantly, still yield near optimal power routes compared with the initial network. In terms of power consumption, the average value of node transmission energy of each topology decreases as the node density increases. The three schemes, OrdYaoGG, SYaoGG, and YaoGG, have an identical profile of node power expenditure. The maximum and average node degree of both SYaoGG and OrdYaoGG is lower than that in YaoGG and other planar power spanners, while between the two, SYaoGG gives a lower maximum and average node degree. As far as the communication cost overhead is concerned, simulation data showed that the constructions of the two graphs did not cost more messages as the density of nodes increased, while SYaoGG costs fewer messages than OrdYaoGG.

8.3.7 TOPOLOGY CONTROL FOR HETEROGENEOUS NETWORKS

An implicit assumption of all the previous schemes is that the maximum transmission power is the same for all network nodes (homogeneous networks). However, this is not the case in practice, where devices with different capabilities are usually employed (heterogeneous networks) [29]. Consequently, most of the previous schemes cannot be directly applied to such cases. In this section, two localized protocols specifically designed for heterogeneous networks are presented and analyzed, namely the directed relative neighborhood graph (DRNG) and directed local spanning subgraph (DLSS). Both these schemes consist of three phases: information collection, topology construction, and construction of topology with only bidirectional links (the latter is optional) [29]. In the information collection step, each node collects information on its visible neighborhood, namely the set of nodes it can reach by transmitting at maximum power. In the final stage, bidirectional topologies are constructed by addition or removal of edges.

The two protocols differ in the topology construction step (second phase). DRNG is similar to the RNG graph construction. More specifically, in DRNG a directed edge from $u \rightarrow v$ exists if and only if no other node w in the visible neighborhood of u exists such that the link weight of edges (u, w), (w, v) are smaller than the weight of link (u, v) and the Euclidean distance $d(u, v) \leq P_t(w)$. The link weight function is defined as in the LMST protocol for uniqueness purposes. Note that the first two requirements define an undirected RNG graph. Basically, XTC and DRNG share the same idea (even though they were independently developed). As for the DLSS protocol, a node v is a neighbor of u, if and only if v lies on u's directed local spanning tree (computed using Prim's algorithm). DLSS is a natural extension of LMST protocol.

It has been shown in Ref. [29] that both DRNG and DLSS preserve strong connectivity of the initial graph. Furthermore, after the optional phase, both protocols preserve bidirectionality in addition to strong connectivity. The in-degree of any node in the induced topologies of DRNG and DLSS is upper bounded by 6 and the out-degree is bounded by a constant that depends on the minimum and maximum transmission radii (logical node degrees). Simulation results have shown that on average, fewer edges of the initial graph remain in DRNG and DLSS than in MECN. DLSS outperforms DRNG, MECN, and no TC schemes, in terms of the out-degree and the largest maximum logical degree, average radius, and average link length.

8.3.8 CONSTRAINED OPTIMIZATION FORMULATION

The problem of TC can be formulated as a constrained optimization problem with the requirement to preserve the connectivity or biconnectivity [30]. In such a case, the objective could be the minimization of transmission power.

A node is associated with a vector denoting different tunable parameters, such as the power level and antenna direction. The transmit power of a node u is denoted by $P_t(u)$. The propagation function $\gamma(l_i, l_j)$ gives the loss at location l_j when the packet originates at l_i. Assuming a receiver sensitivity of S, it is necessary to transmit at $P_t(l_i) - \gamma(l_i, l_j) \geq S$ for successful reception. The least power function $\lambda(d) = \gamma\left(d(l_i, l_j)\right) + S$, giving the minimum power needed to communicate over a distance d, can be defined. The k-vertex/edge connectivity of a graph is defined as existence of k vertex/edge-disjoint paths among every pair of vertices. Taking into account connectivity ($k = 1$) or biconnectivity ($k = 2$), two constrained optimization problems can be defined. Given a network and a least-power function λ, find a per-node minimal assignment of transmit powers, such that the induced graph is connected and $\max_{u \in G} P_t(u)$ is minimized (connected MinMax power, CMP problem). If the network G, the least-power function, and the initial power assignment are given such that G is connected, the problem of finding the per-node minimal set of power increases $\delta(u)$, such that the induced graph is biconnected and $\max_{u \in G}\{P_t(u) + \delta(u)\}$ is minimum. This is called the biconnectivity augmentation with MinMax power (BAMP) problem.

Algorithm CONNECT is a greedy approach for CMP. Initially, each node is considered an isolated component. Node pairs are selected in nondecreasing order of distance. If they belong to different components the transmit power of each node is increased, so they can both reach each other. The two components become connected and the procedure is repeated until the entire network is connected. Cases where the power increase adds more than one edge to the induced graph (side-effect edges possibly forming a loop) are properly avoided. The node's power is reduced to the maximum point possible without losing the connectivity of the induced graph. The procedure is done for a discrete set of power levels to make it faster and more efficient.

The equivalent of CONNECT for the BAMP problem is BICONN-AUGMENT. Initially CON-NECT runs to acquire a connected network. Next, BICONN-AUGMENT searches for biconnected components in the induced topology. Node pairs are selected in nondecreasing order of their distance and in case they are in different biconnected components, they are joined by the missing link to make it bidirectional. Both CONNECT and BICONN-AUGMENT are optimal for the CMP and BAMP problems in the case of a static topology, as proven in Ref. [30]. Their worst-case running time is $O(n^2 \log n)$; however, they both constitute centralized algorithms.

For the case of mobile nodes, two heuristics are proposed to solve the CMP and BAMP problems. The first one, local information no topology (LINT) [30], is based on the node degree. LINT maintains three parameters, the threshold d_l, the threshold d_h, and the desired d_d node degrees. It periodically checks the current node degree and if it is greater than d_h, it reduces its operational power. If the current value is lower than d_l, the node increases its power; otherwise, no action takes place. The LINT protocol does not ensure network connectivity, as two nodes might move away from each other and lose connection due to the upper node degree bound.

The local information link-state topology (LILT) protocol [30] is designed to provide biconnectivity. It consists of two parts, namely neighbor reduction protocol and neighbor addition protocol. The first is very similar to LINT, while the second phase is triggered by link-state events. It overrules the upper bound degree to increase a node's power if the topology change indicated leads to undesirable connectivity. Each time a node receives a routing update, it determines the state of the network. It can be one of disconnected, connected but not biconnected, or biconnected. If the network is biconnected, no action takes place. If it is connected but not biconnected, the node determines its distance from the closest point whose removal partitions the network (articulation point). Then, it sets a randomized timer according to an exponential function of the computed distance, and if the network is still connected but not biconnected when the timer fires, the articulation point node increases its power to the maximum possible. Nodes that are closer to the articulation point (the edge whose removal disconnects the network) are more likely to remove it by adding the inverse direction path.

For the static topology case, BICONN-AUGMENT gives the best results among the four in terms of throughput and adapts well to changes of the node density. For low node densities, CONNECT faces some problems, while for high densities, its induced topology resembles that of BICONN-AUGMENT and thus approaches its performance. Nevertheless, these are both superior to the case where no TC is applied. As far as the power consumption is concerned, BICONN-AUGMENT expends significantly more power than CONNECT, because it has to extend some links (by increasing node consumptions) to biconnect isolated nodes. This happens especially for low node densities. Among the two, BICONN-AUGMENT seems to perform better than CONNECT for higher node densities, whereas at lower node densities they both present tradeoffs among throughput and energy consumption.

The protocols supporting mobility do not seem to have significant improvements on the performance. In terms of delay, they only have marginal gains compared to the no TC scheme and only for average degrees of node mobility. Between the two, LILT appears to be slightly superior in terms of the delay, while with respect to network throughput, LINT performs better than LILT. The main reason for this is that the link-state database of LILT may be outdated (even for a minor interval), which might cause undesirable actions. The situation deteriorates as the node density increases and the topology changes increase as well.

8.4   EVALUATION AND IMPLEMENTATION ISSUES

8.4.1   QUALITATIVE COMPARISON

In the following section we present a set of parameters that we use for the protocols' qualitative comparison. For each parameter we also provide the set of possible values. Most of the studies considered do not explicitly address each of these considerations. The following is a comprehensive list of parameters that attempts to provide for a fair and meaningful comparison basis between the approaches presented. This comparison constitutes a first effort to characterize and identify the qualitative behavior of the various TC protocols and is by no means exhaustive or quantitative.

1. Distributed Operation: {yes, no}
2. Complexity: {high, medium, low}
3. Performance: {high, medium, low}
4. Scalability: {high, medium, low}
5. Mobility: {yes, no}

Regarding the distributed operation, a technique is evaluated on the basis of whether the corresponding protocol runs on each network node or requires a centralized entity to perform the operation. The complexity metric characterizes not only the computational complexity of the involved protocols, but also implementation complexity and respective message overheads. Thus, a protocol with very low computational complexity and very high overhead, overall, would qualify as "medium" with respect to this metric. Similarly, the performance of the presented approaches is a combined metric of their effectiveness in reducing the energy consumption (also reduction of the transmission radii), maintenance of connectivity, and also the increase in delay or biconnectivity maintenance. Thus, performance is an overall characterization of the success of a TC approach with respect to the objectives that were defined at the beginning of this chapter. Scalability deals with the ability of the methods to maintain their performance as the number of network nodes increases, independently of whether the density of the network increases or not. This metric characterizes the performance of a protocol regarding the population of the network alone and regardless of the effect of the deployment area (covered by the density of the network). As far as mobility is concerned, almost all protocols are able to support it in the classic sense of successive time snapshots of the network. However, due to the complexity of some protocols such an approach might be difficult, especially due to frequently varying node movements. Mobility here discriminates these cases, and a positive indication means that the corresponding approach supports seamless mobility, although a negative one can be interpreted as mobility in the snapshot sense only or no mobility support at all. In the literature, not all of the protocols are assessed according to these metrics. Therefore, if the corresponding information is absent or cannot be accurately obtained, we denote it by "N/Av." Finally, it should be noted that most of the values used in this qualitative evaluation have relative meaning (i.e., this is a comparative evaluation), and no absolute values are used. In Table 8.2, the qualitative comparison of the presented approaches is shown.

8.4.2   QUANTITATIVE COMPARISON AND IMPLEMENTATION ISSUES

It is noted that based on the previous discussions, the cone-angle-based and the relay-region-based TC approaches appear to be among the dominating ones in the literature. Therefore, in this section we present a quantitative evaluation of the corresponding performances of these approaches and discuss some related implementation issues.

Operation and properties of the PCAM family have been proven for infinite networks, i.e., in the limit where the number of nodes grows to infinity, exploiting percolation theory. However, for practical purposes the techniques need to be studied for finite topologies and bounded deployment

TABLE 8.2

Comparison of TC Methods

TC Class	Protocol	Distributed	Simplicity	Performance	Scalability	Mobility
MAC based	GAF	Yes	Medium	Medium	Medium	Yes
	Span	Yes	Medium	Medium	Medium	Yes
Cone angle based	PCAM 1	Yes	Low	Medium	Medium	N/Av.
	PCAM 2	Yes	Low	Medium	Medium	N/Av.
	PCAM 3	Yes	Low	Medium	High	N/Av.
	PCAM 4	Yes	Low	High	High	N/Av.
	CBTC	Yes	Medium	High	High	No
Relay regions	MECN	Yes	High	High	High	No
	SMECN	Yes	High	High	High	No
Tree based	LMST	Yes	Medium	High	High	No
	MST-red.	No	Medium	High	High	No
	MIP-red.	No	Medium	High	High	No
	STH	No	Medium	Medium	Medium	No
	iSTH	No	Medium	Medium	Medium	No
Node degree based clustering	KNeigh	Yes	Low	High	Medium	No
	XTC	Yes	Low	High	High	N/Av.
	MPR	Yes	High	N/Av.	Medium	Yes
	COMPOW	Yes	High	N/Av.	N/Av.	No
	CLUSTERPOW	Yes	High	N/Av.	N/Av.	No
Special form graphs	DG	Yes	Low	Medium	N/Av.	N/Av.
	RNG	Yes	Low	Medium	N/Av.	N/Av.
	GG	Yes	Low	Medium	Medium	N/Av.
	YGk	Yes	Low	Medium	Medium	N/Av.
Heterogeneous TC	DRNG	Yes	Low	High	High	No
	DLSS	Yes	Medium	High	High	No
Constrained optimization	CONNECT	No	Medium	High	High	No
	BICONN-AUG	No	High	High	Low	No
	LINT	Yes	Low	Medium	High	Yes
	LILT	Yes	Low	Medium	High	Yes

regions. In this section, we compare the finite implementations of PCAM family to the CBTC and MECN algorithms, as the last two have been studied and used as reference extensively.

Among the four PCAM approaches presented in Section 8.4.1, the fourth (i.e., PCAM 4) can be realized in many variations, depending on the value of parameter δ. For the purposes of this study, δ = 150 m and δ = 250 m were chosen. The performance differences of the two variants with respect to the average number of neighbors and the average transmission radius of the induced topology reveal a slight advantage of PCAM with δ = 150 m (denoted by PCAM 4-150). Performance differences, however, are below 1 percent. Similarly, two variations of CBTC (CBTC(90), CBTC(120)) were used, which exhibited almost identical performance with respect to the previously used metrics. However, CBTC(120) is considered superior, as it imposes less strict requirements on the cone angles. This in turn produces less computational overhead and fewer edge nodes at the end of the protocol operation. Consequently, in the following, we compare PCAM 4-150 and CBTC(120) with the rest of the PCAM family and MECN approaches.

For the simulation studies, the network deployment region was assumed square. To study the impact of increasing network densities, the area was fixed with size length $L = 1000$ m and node

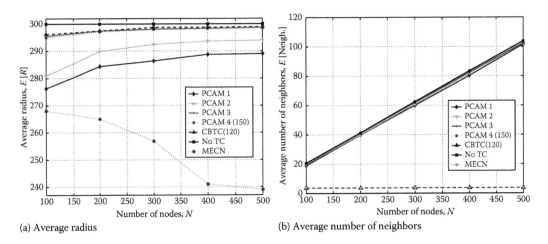

FIGURE 8.5 Comparison of cone angles with relay regions (increasing density).

population was gradually increased. For the case of constant density, a density value of 10^{-4} nodes/m^2 was chosen and maintained, for square regions with side lengths 1000, 1400, 1700, and 1900 m and network sizes of 100, 196, 289, 361, and 441 nodes, respectively. For the second phase of CBTC, $q = 2$ was considered.

Figure 8.5 presents the variation of the average transmission radius (Figure 8.5a) and the average number of neighbors (Figure 8.5b) for PCAM 1-4, CBTC(120), MECN, and the case where no TC is applied. It is deduced from Figure 8.5a that the most efficient approach in terms of reducing the transmission range (and consequently reducing the energy consumption) is the MECN protocol, which however, is the least efficient in terms of implementation, computational and communication overhead. The performance of PCAM 3, 4 and CBTC is only marginally better than the no TC case; however, CBTC provides guaranteed connectivity as opposed to the other two, where some nodes might only have one-way connectivity to a fully connected component of the induced graph. As far as the average number of neighbors is concerned, all protocols have approximately the same linear behavior for larger network densities. Among them, CBTC and MECN reduce the number of neighbors significantly, while retaining connectivity, which in turn indicates lower channel contention in the local neighborhood of a node.

Figure 8.6 presents the fluctuation of the same quantities for constant network density but increasing network sizes. Similar observations with the case of increasing density apply here as well. MECN appears superior when compared to cone angles with respect to energy conservation, but it lacks the implementation simplicity of the cone-angle-based approaches. As far as the average number of nodes is concerned, MECN and CBTC demonstrate the lowest values, having significant differences from the other approaches. This essentially reflects an inherent and intuitive tradeoff, where very simple and operationally efficient approaches (PCAM family) have inferior performance compared with more complex ones (CBTC, MECN). Depending on the nature of the application and the available resources, PCAM might be chosen for a network with limited resources and large size (e.g., sensor network), while MECN appears more suitable for a network with smaller size and more resources (e.g., ad hoc networks).

8.5   CONCLUSIONS AND OPEN ISSUES

In this chapter TC in the framework of cooperative wireless networks was studied and various approaches at the MAC and Network layer were presented and compared. An inherent tradeoff

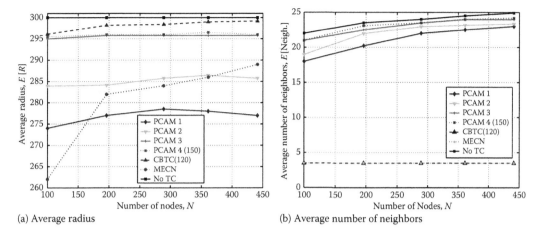

(a) Average radius   (b) Average number of neighbors

FIGURE 8.6 Comparison of cone angles with relay regions (constant density).

between simple operation and degree of efficiency emerged as the determining factor for the selection of a technique for different application environments and requirements.

Several research, design, and implementation problems remain open for investigation in the framework of TC and cooperative networks. Among the most challenging ones is the traffic-carrying capacity of a cooperative ad hoc network. This problem has not yet been studied analytically and lack of closed form solutions motivates further investigation in this area. In addition, even though TC is a form of cooperation itself, it has not been studied how additional cooperation between the set of nodes that have employed TC and those that transmit at full power can be established for reducing interference and increasing spatial reuse.

Furthermore, more realistic wireless propagation models need to be taken into account by the TC protocols, some of which have not considered at all the impact of the channel variation. Cooperation for channel state information collection and distribution among network nodes so that real-time TC is achieved is of high research and practical interest.

Similarly, investigating the impact of mobility and corresponding models on the operation of the presented protocols requires extensive analytical, simulation, and experimental work. Especially in the case of cooperative networks, mobility may not only involve a single node movement but also group mobility. Taking into account that in ad hoc networks communications are often among teams that tend to coordinate their movements, issues associated with group mobility and the corresponding cooperation requirements that arise from such behaviors will gain significant attention in the near future.

REFERENCES

[1] J. Hou, N. Li, and I. Stojmenovic, Topology constriction and maintenance in wireless sensor networks, in *Handbook of Sensor Networks: Algorithms and Protocols*, I. Stojmenovic (editor), Chapter 10, John Wiley & Sons, West Sussex, U.K., September 2005.

[2] Y. Xu, J. Heidemann, and D. Estrin, Geographically-informed energy conservation for ad hoc routing, *Proceedings of the 7th ACM Conference on Mobile Computing and Networking (MobiCom)*, Rome, Italy, pp. 70–84, July 2001.

[3] G. Srivastava, P. Boustead, and J. F. Chicharo, A comparison of topology control algorithms for ad-hoc networks, *Proceedings of Australian Telecommunications, Networks and Applications Conference, (ATNAC)*, Melbourne, Australia, December 2003.

[4] P. Santi, Topology control in wireless ad hoc and sensor networks, *ACM Computing Surveys (CSUR)*, 37(2), pp. 164–194, June 2005.

[5] Wireless LAN Media Access Control (MAC) and Physical Layer (PHY) Specification, IEEE 802.11 Draft Version 4.0, May 1996.

[6] L. M. Feeney and M. Nilsson, Investigating the energy consumption of a wireless network interface in an ad hoc networking environment, *Proceedings of the 20th Annual Joint Conference of the IEEE Computer and Communications Societies (INFOCOM)*, Anchorage, AL, vol. 3, pp. 1548–1557, April 2001.

[7] C. E. Perkins and E. M. Royer. Ad hoc on-demand distance vector routing, *Proceedings of the 2nd IEEE Workshop on Mobile Computing Systems and Applications* (WMCSA), New Orleans, LA, pp. 90–100, February 1999.

[8] D. B. Johnson and D. A. Maltz. Dynamic source routing in ad hoc wireless networks, in *Mobile Computing*, T. Imielinski and H. Korth (editors), Chapter 5, pp. 153–181, Kluwer Academic Publishers, Boston, MA, 1996.

[9] B. Chen, K. Jamieson, H. Balakrishnan, and R. Morris, Span: An energy -efficient coordination algorithm for topology maintenance in ad hoc wireless networks, *Proceedings of the 7th ACM Conference on Mobile Computing and Networking (MobiCom)*, Rome, Italy, pp. 85–96, July 2001.

[10] A. Jiang and J. Bruck, Monotone percolation and the topology control of wireless networks, *Proceedings of the 24th Annual Joint Conference of the IEEE Computer and Communications Societies (INFOCOM)*, Miami, FL, vol. 1, pp. 327–338, March 2005.

[11] R. Wattenhoffer, L. Li, P. Bahl, and Y.-M. Wang, Distributed topology control for power efficient operation in multihop wireless ad hoc networks, *Proceedings of 20th Annual Joint Conference of the IEEE Computer and Communications Societies (INFOCOM)*, Anchorage, AL, vol. 3, pp. 1388–1397, April 2001.

[12] L. Li, J. Y. Halpern, P. Bahl, Y.-M. Wang, and R. Wattenhoffer, Analysis of a cone-based distributed topology control algorithm for wireless multi-hop networks, *Proceedings of the 20th ACM Symposium on Principles of Distributed Computing (PODC)*, Newport, RI, pp. 264–273, August 2001.

[13] L. Li, J. Y. Halpern, P. Bahl, Y.-M. Wang, and R. Wattenhoffer, A cone-based distributed topology-control algorithm for wireless multi-hop networks, *IEEE/ACM Transactions on Networks*, 13(1), 147–159, February 2005.

[14] V. Rodoplu and T. H. Meng, Minimum energy mobile wireless networks, *IEEE Journal on Selected in Areas Communication*, 17(8), 1333–1344, August 1999.

[15] L. Li and J. Y. Halpern, Minimum-energy mobile wireless networks revisited, *Proceedings of the IEEE International Conference on Communications (ICC)*, Atlanta, GA, vol. 3, pp. 1633–1639, June 1998.

[16] N. Li, J. C. Hou, and L. Sha, Design and analysis of an MST-based topology control algorithm, *Proceedings of the 22nd Annual Joint Conference of the IEEE Computer and Communications Societies (INFOCOM)*, San Francisco, CA, vol. 3, pp. 1702–1712, March–April 2003.

[17] M. X. Cheng, M. Cardei, J. Sun, X. Cheng, L. Wang, Y. Xu, and D. Du, Topology control of ad hoc networks for energy efficiency, *IEEE Transactions on Computers*, 53(12), 1629–1635, December 2004.

[18] X. Cheng, B. Narahari, R. Simha, M. X. Cheng, and D. Liu, Strong minimum energy topology in wireless sensor networks: NP-completeness and heuristics, *IEEE Transactions on Mobile Computers*, 2(3), 248–256, July–September 2003.

[19] G. Xing, C. Lu, Y. Zhang, Q. Huang, and R. Pless, Minimum power configuration in wireless sensor networks, *Proceedings 6th ACM International Symposium on Mobile Ad Hoc Networking and Computing (MobiHoc)*, Urbana-Champaign, IL, pp. 390–401, May 2005.

[20] W. T. Chen and N. F. Huang, The strongly connecting problem on multihop packet radio networks, *IEEE Transactions on Communications*, 37(8), 293–295, March 1989.

[21] D. Blough, M. Leoncini, G. Resta, and P. Santi, The KNeigh protocol for symmetric topology control in ad hoc networks, *Proceedings 4th ACM International Symposium on Mobile Ad Hoc Networking and Computing (MobiHoc)*, Annapolis, MD, pp. 141–152, June 2003.

[22] D. Blough, M. Leoncini, G. Resta, and P. Santi, The k-neighbors approach to interference bounded and symmetric topology control in ad hoc networks, *IEEE Transactions on Mobile Computers*, 5(9), 1267–1282, September 2006.

[23] R. Wattenhoffer and A. Zollinger, XTC: A practical topology control algorithm for ad hoc networks, *Proceedings of 18th International Parallel and Distributed Processing Symposium (IPDPS)*, Santa Fe, NM, pp. 216–223, April 2004.

[24] T. A. Elbatt, S. V. Krishnamurthy, D. Connors, and S. Dao, Power management for throughput enhancement in wireless ad hoc networks, *Proceedings of the IEEE International Conference on Communications (ICC)*, New Orleans, LA, vol. 3, pp. 1506–1513, June 2000.

[25] S. Narayanaswamy, V. Kawadia, R. S. Sreenivas, and P. R. Kumar, Power control in ad hoc networks: Theory, architecture, algorithm and implementation of the COMPOW protocol, *Proceedings of the 8th European Wireless Conference*, Florence, CA, February 2002.

[26] V. Kawadia and P. R. Kumar, Power control and clustering in ad hoc networks, *Proceedings of the 22nd Annual Joint Conference of the IEEE Computer and Communications Societies (INFOCOM)*, vol. 1, pp. 459–469, March–April 2003.

[27] P. Gupta and P. R. Kumar, The capacity of wireless networks, *IEEE Transactions on Information Theory*, 46(3), pp. 388–404, March 2000.

[28] W. Song, Y. Wang, X. Li, and O. Frieder, Localized algorithms for energy efficient topology in wireless ad hoc networks, *Proceedings of the 5th ACM International Symposium on Mobile Ad Hoc Networking and Computing (MobiHoc)*, Tokyo, Japan, pp. 98–108, May 2004.

[29] N. Li and J. C. Hou, Localized topology control algorithms for heterogeneous wireless networks, *IEEE/ACM Transactions on Networking*, 13(6), 1313–1324, December 2005.

[30] R. Ramanathan, and R. Rosales-Hain, Topology control of multihop wireless networks using transmit power adjustment, *Proceedings of the 19th Annual Joint Conference of the IEEE Computer and Communications Societies (INFOCOM)*, Tel-Aviv, Israel, vol. 2, pp. 404–413, March 2000.

9 Game Theory and Cooperation Analysis

Almudena Alcaide, Juan M. E. Tapiador,
Julio C. Hernández-Castro, and Arturo Ribagorda

CONTENTS

Network component cooperation is a crucial requirement for operational multi-hop wireless ad hoc networks, as selfish and noncooperative behavior, can critically damage network performance. In this chapter we use some game theoretical concepts for the modeling and formal analysis of operations in this type of network. Our analysis will develop from fully cooperative conditions to more complex scenarios where network reliability, connectivity, and mobility make it impossible to consider entirely cooperative frameworks.

9.1   INTRODUCTION

Game Theory is a branch of applied mathematics consisting of a set of analytical tools that serve to model the behavior of complex interactions among rational (i.e., self-interested) entities. In recent years, this field has been applied to analyze and solve different problems in a broad number of areas such as cryptographic protocols [1,2,14], peer-to-peer (P2P) systems [12,16], as well as wireless communications and mobile networking [3–6].

However, this field, widely known by economists and theoretical biologists, has not been made accessible to computing scientists in general. The main objective of this chapter is to make basic and advanced Game Theory concepts available to computing network researchers, avoiding mathematical notation whenever possible and focusing on explaining how cooperative wireless interactions can be modeled by making use of this formalism. Any network component such as nodes, operators, agents, or even a wireless connection can be modeled as a player in a game. The fact that entities are considered to be rational does not prevent them from displaying cooperative and noncooperative behaviors toward a global common purpose. The presented analysis will serve as an example to model any kind of wireless networking operation at any level of the protocol stack. Additionally, we also address some fundamental issues related to the overall performance. In particular, the fundamental questions are

- What motivates a network component to behave according to a given network protocol?
- Could nodes be given the capability to assess risk at each decision stage of a particular interactive event?
- How can network nodes evaluate their environment and dynamically decide their future actions?
- Can participant agents form conjectures about current network conditions?
- How can nodes take network reliability into account when evaluating a possibly disadvantageous situation?

This chapter treats each one of these questions from a theoretical point of view, any previous familiarity with Game Theory not being necessary. We model a basic forwarding operation between two different network components and analyze cooperation when executed in increasingly complex network environments.

The problem to analyze is whether cooperation can exist based on network agent rational behavior or whether external incentive mechanisms are needed to enforce cooperation. Some researchers have claimed to find particular instances in which cooperation emerges naturally without the need for incentive mechanisms [7,15,17]. Unfortunately, strong restrictions are imposed on network topologies and randomized connection times for cooperation behavior to emerge without any external incentives.

The chapter is organized as follows. In Section 9.2, the main basic Game Theory concepts (game strategy, game equilibrium, utility functions, and utility values) are introduced. Section 9.3 is devoted to modeling one of the most resource-demanding operations in wireless networks, namely packet forwarding. A basic packet forwarding operation can be represented as a game (the Relay Game) in which any network component (nodes, operators, agents, and the wireless connection itself) can be modeled as a player. We start by studying cooperation in relation to such a simple game played among blind and fully cooperative components following a single fixed strategy. Usually, entirely cooperative environments represent networks controlled by a single authority (e.g., military networks, transport operation systems, etc.) in which nodes cooperate for critical common purposes. In most other cases, network components are managed by different authorities and cooperation toward a common purpose cannot be taken for granted.

In Sections 9.4 and 9.5, players reflect realistic scenarios, gradually gaining the capability to exhibit complex behaviors and choose from different possible strategies. The aforementioned

Relay Game is analyzed again but using more advanced Game Theoretical concepts such as Games of Imperfect Information and Games of Incomplete Information or Bayesian games. The formal analytical model used in these sections has already been applied to the analysis of similar games in different contexts such as rational exchange cryptographic protocols and P2P file sharing systems (see e.g., [1,2,12]).

Finally, in Section 9.6 the main conclusions and the related open issues are outlined.

9.2 INTRODUCTION TO GAME THEORY

The basic theory came into being in 1944, with the classic *Theory of Games and Economic Behavior* by John von Neumann and Oskar Morgenstern [8]. In general, Game Theory provides a formal modeling approach to situations in which decision makers interact with other players. It analyzes and represents such situations as games, where players choose different actions in an attempt to maximize their returns. Although some Game Theoretic analysis appears similar to Decision Theory, Game Theory studies the decisions made in environments where players interact. In other words, Game Theory studies choice of optimal behavior when costs and benefits of each option depend upon the choices of other individuals.

Basic Game Theoretical concepts become relevant to the analysis of wireless networking from many different angles:

- First, it provides a formal framework in which to analyze network component behavior in a variety of wireless operations in different environments and structures.
- Furthermore, Game Theory also assists in evaluating actions taken by network participants in indefinitely repeated scenarios in which nodes are able to remember previous experiences. This is particularly interesting, as in general, network components usually perform the same operation an indeterminate number of times. One can assume that it is only at high levels (session layer and above) of the operational stack where network participants are aware of operations being executed for the last time.
- Finally, Game Theory enables us to formally represent noncooperative components and analyze the impact they might have on overall network performance.

We now introduce a few basic related concepts and some mathematical notation needed for the concluding analysis. Since we cannot cover at length any of the notions presented here, the reader is referred to Refs. [9,13] for excellent Game Theory tutorial books.

9.2.1 GAME STRATEGIES AND UTILITY FUNCTIONS

Informally, a game is considered to be a collection of players who play different moves (legal moves are those that comply with the game rules), aiming at maximizing their individual payoff, obtained when the game is ended. Usually, players are able to conform to different game strategies and following one or another strategy dictates which move to make at each given turn in the game.

The aforementioned concepts can be formalized bringing in the following notation:

DEFINITION 1

Game: A game is defined by the tuple: $G = \langle E, S, \overrightarrow{v} \rangle$ *where*

- $E = \{E_i\}$, $i \in \{1, \ldots, n\}$ *is a set of n players*
- $S = S_1 \times \cdots \times S_n$ *is a set of strategy profiles and*
- $\overrightarrow{v} = [v_1(), \ldots, v_n()]$ *is a vector of payoff functions*

In a static game, players make their moves simultaneously. By contrast, in a dynamic game, players take turns to move, so their actions may depend on what actions other players have taken in previous turns.

DEFINITION 2

Pure strategy: In a game G, a pure strategy for player E_i, denoted by s_i, is a complete contingency plan for player E_i. It describes the series of actions that this player would take at each possible decision point in the game G. $s_i \in S_i$, where S_i represents the set of all possible strategies for player E_i in G.

DEFINITION 3

Strategy profile: A strategy profile is a vector of strategies $(s_1, \ldots, s_n)$, one for each player E_i of a game. The set of strategy profiles, denoted by S, is the Cartesian product of the strategy spaces over all players: $S = S_1 \times \cdots \times S_n$.

A convention is to describe as s_{-i} the strategies chosen by all other players except for a given player E_i. Any given strategy profile in a game may then be represented by the tuple (s_i, s_{-i}).

Note that specifying a strategy profile univocally determines the outcome of the game.

DEFINITION 4

Payoff function: In a game G, a payoff function v_i (also called utility function*) is defined for each player E_i. The domain of v_i is the set of strategy profiles S, and the range of the function is the set of real numbers, so that, for each strategy profile $(s_i, s_{-i}) \in S$, $v_i(s_i, s_{-i})$ represents the player E_i's payoff when E_i plays strategy s_i and the other players follow strategies s_{-i}.*

9.2.2 Game Equilibrium

A Nash equilibrium, named after Nobel Prize awarded John Nash, is a strategy profile $s^* = (s_i^*, s_{-i}^*)$ in which no player has an incentive to unilaterally modify his or her strategy. In other words, given the other players' strategies s_{-i}^*, player E_i cannot increase his or her payoff by choosing a strategy different from s_i^*. The current set of strategies s^* and the corresponding payoff values constitute a Nash equilibrium [10,11]. Next is the formal definition for such a concept:

DEFINITION 5

Nash equilibrium: Given a game $G = \langle E, S, \vec{v} \rangle$, a strategy profile $s^* \in S$ represents a Nash equilibrium if and only if for every player E_i, $i \in \{1, \ldots, n\}$:

$$v_i(s_i^*, s_{-i}^*) \geq v_i(s_i, s_{-i}^*) \quad \forall s_i \in S_i.$$

A Nash equilibrium is said to represent a solution for a given game. Moreover, once such a state is identified, rational (self-interested) players are forced to follow the set of strategies to reach such an outcome. By definition, changing their strategy unilaterally will not result in better payoffs.

By similarity, in a game used to exemplify a networking operation, a Nash equilibrium strategy profile allows us to predict the final outcome of such an operation. Cooperative actions naturally emerge toward those points of equilibrium. Moreover, this will set the basis for further conclusive network participant analysis, in operations that are repeated an undetermined number of times.

Finally, we define a subgame of a dynamic game and we also define a refinement of the Nash equilibrium used in dynamic games.

DEFINITION 6

Subgame of a dynamic game: A subgame of a dynamic game G is a smaller game that takes any given point in the larger game as the start and carries on until the end.

DEFINITION 7

Subgame perfect equilibrium: A strategy profile is a subgame perfect equilibrium if it represents a Nash equilibrium in every sub-game of the original game.

Informally this means that if players, following a subgame perfect equilibrium strategy, play any smaller game that consists of only one part of the larger game, then their behavior would still represent a Nash equilibrium in that sub-game.

9.3 WIRELESS INTERACTIONS ON FULLY COOPERATIVE ENVIRONMENTS

The previous definitions provide a formal framework in which to analyze wireless network operations and in particular, a simple packet forwarding transaction. In the first part of the analysis we represent such an operation as a two-player game in which players are considered to be fully cooperative toward common purposes. We refer to this game as the Relay Game and the entirely cooperative environment as a blind context.

9.3.1 THE BLIND RELAY GAME

The Relay Game, (also known as The Forwarding Game [7] or The Forwarder's Dilemma [5]) is intended to represent a basic wireless forwarding operation between two different network components. These two components, represented by players P_1 and P_2, are supposed to operate a direct link that enables them to communicate without intermediaries. Moreover, both players have symmetrical roles to play and these are

- Packets sent from P_1 onto P_2 must be forwarded by P_2 to a particular recipient r_1.
- In a similar way, packets sent from P_2 onto P_1 must be forwarded by P_1 to a particular recipient r_2.
  It is assumed that there is no direct connection between senders and the appropriate recipients r_1 and r_2.

In most cases, since fixed network structures are absent or rare, these symmetrical roles can be assured along consecutive networking operations. The Relay Game is illustrated in Figure 9.1.

FIGURE 9.1   Relay Game.

From this point onward and throughout the chapter, we refer to network components and players indistinctly. Moreover, the terms Relay Game and forwarding operation are also used without distinguishing between them.

The first instance of the general Relay Game is denoted as the Blind Relay Game when this takes place between two blind players that are programmed to follow a fixed given script. In such a context, a script gives a certain player P_1 (node/device/operator/wireless connection) a single fixed strategy consisting of forwarding packets to the corresponding recipient r_2 immediately after receiving them from another player P_2. In due course, player P_2 is asked to follow a similar fixed strategy, relaying all packets sent by player P_1 onto the appropriate recipient r_1. Note that, in the case of either P_1 or P_2 representing a wireless connection, this particular instance of the game considers the connection to be fully reliable and resilient.

9.3.2   Game Strategies and Equilibrium

As already mentioned, in the instance of a Relay Game being played blindly, both players have a fixed strategy consisting of forwarding incoming packets. Trivially, a single system equilibrium is reached when both participants follow their only possible strategy. Such a point of equilibrium (both participants cooperate) corresponds to a solution of the Blind Relay Game and therefore an outcome of the forwarding operation represented in such a game.

9.4   WIRELESS INTERACTIONS ON SEMI-COOPERATIVE ENVIRONMENTS

A less trivial scenario occurs when different network components belong to different authorities, each one of them holding a dissimilar and possibly contrary set of goals. In this type of environment, rational network components (nodes, connections, etc.) aim at maximizing their own payoff values even if this means disobeying global network protocols. When considering an instance of the Relay Game being played in this kind of environment, nodes are no longer programmed to follow a fixed script but they are given a choice of actions they can play at each turn in the game, depending on several factors. For the sake of simplicity, we consider an instance of the Relay Game in which participants can only choose between two possible actions: Forward and Not Forward an incoming packet. We refer to this context as a semi-cooperative environment. Finally, we also assume that player P_1 is the first one in choosing his or her action, followed by player P_2.

9.4.1   The Relay Game in Semi-Cooperative Environments

The following formalisms enclose the aforementioned semi-cooperative scenario in relation to the Relay Game.

- *Players' set of actions.* The set of actions for players P_1 and P_2 is defined as

$$A = \{F, NF\}$$

  where F represents the action Forward and NF represents the action Not Forward.
- *Pure strategies for player P_1.* The complete set of pure strategies for player P_1, denoted as S_1, is defined as

$$S_1 = \{F, NF\}$$

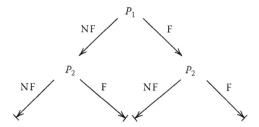

FIGURE 9.2   Relay Game in semi-cooperative environments, represented in an extensive form.

- *Pure strategies for player P_2*. The complete set of *pure* strategies for player P_2, denoted as S_2, is defined as the set of tuples:

$$S_2 = \{(NF, NF), (F, F), (NF, F), (F, NF)\}$$

where the first component of each tuple represents the action taken by player P_2 if player P_1 plays F and the second component the action taken by player P_2 if P_1 plays NF.

For instance, pure strategy (F, NF) represents a strategy in which player P_2 copies whatever P_1 has played. By contrast, pure strategy (NF, NF) represents a unifying strategy in which P_2 plays *NF* regardless of P_1's move.

- *Strategy profiles*. We define the set of strategy profiles for the Relay Game played in semi-cooperative environments as the set of tuples:

$$S = \{(s_1, s_2) : s_1 \in S_1, s_2 \in S_2\}$$

Note how each strategy profile univocally determines the outcome of the game. The Relay Game in semi-cooperative environments is illustrated in its extensive form in Figure 9.2.

- *Payoff values*. We define a set of payoff values to be assigned to each player P_i at each possible outcome of the Relay Game.

  1. We define values b_1 and b_2 as the benefits obtained by players P_1 and P_2, respectively, if both players have forwarded each other's packets.
  2. For each player P_i we define a value d_i to represent the benefit obtained when P_i's packets are forwarded and by contrast, P_i does not relay any of his or her opponent's.
  3. For each player P_i, we assign the value of zero to represent the benefit obtained when neither of the players forward any of the incoming packets.
  4. Finally, for each player P_i, let $c_i > 0$ be the cost of a forwarding operation.

Any game analysis depends on the relationship that exists between values b_i, c_i, and d_i. For each one of the entities, these values determine favorable strategies and points of equilibrium in the game. Also, they allow us to further distinguish between incentivated and nonincentivated semi-cooperative environments. We study the outcome of the Relay Game in each possible scenario.

9.4.1.1   The Relay Game in Semi-Cooperative Incentivated Environments

In this type of environment there exists a mechanism by which participants are given incentives to perform actions guided toward a common global goal. For instance, nodes executing a packet forwarding operation could be incentivated by a reputation scheme that gradually blocks off selfish nodes from future network operations. Other similar schemes are designed to remunerate honest nodes, using for example a micropayment scheme.

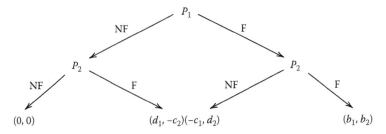

FIGURE 9.3 Relay Game in semi-cooperative and incentivated environments ($b_i > d_i$). The vectors assigned to each terminal node represent the payoff values for P_1 (first component) and P_2 (second component) when P_1 and P_2 follow the path of actions to finish the protocol at that end.

Semi-cooperative incentivated environments can be formally defined by establishing the following relationship between payoff values b_i and d_i. In incentivated environments, the following inequality is satisfied for each player P_i:

$$b_i > d_i$$

Next is a table expressing the payoff values obtained by players P_1 and P_2 when the Relay Game is played in a semi-cooperative and incentivated environment. In addition, see Figure 9.3 for a graphical representation of the game in its extensive form.

$P_1 \backslash P_2$	(NF, NF)	(F, F)	(NF, F)	(F, NF)
F	$(-c_1, d_2)$	(b_1, b_2)	$(-c_1, d_2)$	(b_1, b_2)
NF	$(0, 0)$	$(d_1, -c_2)$	$(d_1, -c_2)$	$(0, 0)$

Consider an instance of the Relay Game in which player P_1 receives a packet from P_2 and has to choose a strategy from his or her set of possible options S_1. In due course, player P_2 is be asked to forward an incoming packet received from P_1 and at that time, having observed P_1's previous move, player P_2 decides which action to take from his or her set of possible options S_2.

It is easy to conclude from the table above that strategy profile $s^* = (s_1^*, s_2^*) = \left(F, (F, NF)\right)$ verifies Definitions 5 and 7, that is, it constitutes a Nash equilibrium of the dynamic Relay Game in semi-cooperative and incentivated environments. Note that the incentive scheme guides P_1 to cooperate and player P_2 to copy P_1's previous action.

To this point, we have formally shown how cooperation emerges in packet forwarding operations, in incentivated conditions. This type of network performs in a manner similar to privately owned networks. Although nodes are in theory able to display noncooperative behavior, they are rationally forced to cooperate in favor of better payoff values.

9.4.1.2 The Relay Game in Semi-Cooperative Nonincentivated Environments

Similar analysis can be carried out on nonincentivated environments although with a completely different result. Semi-cooperative incentivated environments can be formally defined by establishing the following relationship between player payoff values b_i and d_i. In nonincentivated environments the following equality is satisfied for each player P_i:

$$b_i = d_i$$

In a nonincentivated scenario, a player P_2 playing the Relay Game has no motivation to forward incoming packets from P_1. The table below illustrates such a scenario, and Figure 9.4 represents this instance of the game in its extensive form.

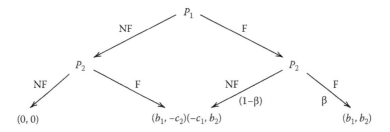

FIGURE 9.4 Relay Game in semi-cooperative and nonincentivated environments. Parameter β can be considered as the level of cooperation player P_1 expects from player P_2. Vectors assigned to each terminal node represent the payoff values for P_1 (first component) and P_2 (second component) when P_1 and P_2 follow the path of actions to finish the protocol at that end.

$P_1 \backslash P_2$	(NF,NF)	(F,F)	(NF,F)	(F,NF)
F	$(-c_1, b_2)$	(b_1, b_2)	$(-c_1, b_2)$	(b_1, b_2)
NF	$(0, 0)$	$(b_1, -c_2)$	$(b_1, -c_2)$	$(0, 0)$

In this type of scenario, player P_1 can create a conjecture over player P_2's cooperation level. In other words, player P_1 can estimate, based maybe on previous transactions with the same node, whether player P_2 will forward or not P_1's incoming packets. In the case of player P_2 representing the actual wireless connection, P_1 could base his or her conjecture on estimating network reliability.

Given the payoffs table above, we study cooperation from P_1's point of view. Let parameter $\beta \in [0, 1]$ be P_1's estimation on the conditional probability of P_2 forwarding a packet once P_2 has observed P_1's previous action. In other words, parameter β can be considered as the leve of cooperation player P_1 expects from player P_2. Let $(1 - \beta)$ be the complementary probability.

The following two equations represent the Expected Gain player P_1 hopes to obtain when playing one or another of his or her strategies and taking into account the conjecture over P_2's future action. That is, function $EG(P_1, \beta, \sigma)$ will represent the payoff value P_1 expects to obtain from playing strategy $\sigma \in S_1$, taking into account that β represents an estimation of the conditional probability of P_2 choosing F and $(1 - \beta)$ represents an estimation of the probability of P_2 choosing NF.

$$EG(P_1, \beta, NF) = (1 - \beta) * 0 + \beta * b_1$$
$$EG(P_1, \beta, F) = (1 - \beta) * (-c_1) + \beta * b_1$$

Note that given β as the conjecture player P_1 has on player P_2's future action, we can conclude that

$$EG(P_1, \beta, F) \geq EG(P_1, \beta, NF) \Leftrightarrow$$
$$(1 - \beta) * (-c_1) \geq 0 \Leftrightarrow \beta \geq 1$$

Note, however, that $\beta \in [0, 1]$. Therefore, for player P_1 to choose strategy F as an equally profitable strategy as NF, parameter β must be set to one. In other words, P_1 plays action F, if and only if P_1 is absolutely sure that P_2 is going to respond forwarding the packet as well. In any other case, player P_1 is better off playing NF. Parameter $\beta = 1$ is equivalent to reducing P_2's strategy space to one single possible fixed strategy and that is F.

As a conclusion, strategy profiles $s_1^* = \left(\text{NF}, (\text{F, NF})\right)$ with $\beta < 1$ and $s_2^* = \left(\text{F}, (\text{F, NF})\right)$ with $\beta = 1$ verify Definitions 5 and 7, that is, they constitute a Nash equilibrium of the dynamic Relay Game in semi-cooperative and nonincentivated environments, where parameter β represents player P_2's level of cooperation.

We have just seen how cooperation in nonincentivated environments emerges in the Relay Game when the second player P_2 has a fixed cooperative strategy to play. Otherwise, player P_1 will rationally drop any cooperative behavior.

9.4.2 GAMES OF IMPERFECT INFORMATION

Cooperation in forwarding wireless network operations can be further analyzed when representing the following variant of the Relay Game played in previous sections. Before, we had only considered the Relay Game as a dynamic game in which P_2 observes P_1's move and then makes a decision on which strategy to choose. Two more new scenarios emerge when P_2 is asked to choose an action without being aware of P_1's previous move. Conditions under which cooperation emerges in these types of scenarios can be studied using Games of Imperfect Information.

In a Game of Imperfect Information, players do not always know other players' previous actions. In particular, since player P_2 is unsure about whether player P_1 previously forwarded incoming packets, P_2 has to assess risk, evaluating cost and benefits of the corresponding forwarding operation. Figure 9.5 represents the Relay Game as a Game of Imperfect Information. The dashed line across the nodes indicates P_2's uncertainty about which node of the game he or she is at.

9.4.3 THE RELAY GAME AS A GAME OF IMPERFECT INFORMATION ON SEMI-COOPERATIVE ENVIRONMENTS

The following formalisms describe the aforementioned semi-cooperative scenario in relation to the Relay Game played as a Game of Imperfect Information

- *Players' set of actions.* The set of actions for players P_1 and P_2 is defined as before and with the same semantics:

$$A = \{F, NF\}$$

  where
    F represents the action Forward
    NF represents the action Not Forward.

- *Mixed strategies for player P_i.* A mixed strategy for player P_i, denoted as m_i, is defined as a tuple:

$$m_i = (F, p)$$

  where
    p represents the probability of player P_i to follow strategy F
    $(1 - p)$ represents the probability of P_i to follow strategy NF

  Note that P_2 cannot elaborate any combined strategies as P_2 does not know what P_1 has previously done.

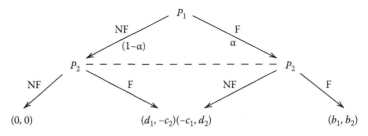

FIGURE 9.5 Relay Game represented as a Game of Imperfect Information, in semi-cooperative and nonincentivated environments.

- *Probabilistic strategy profile:* A probabilistic strategy profile for the Relay Game in semi-cooperative environments, denoted as m, is a tuple:

$$m = (m_1, m_2)$$

where
m_1 is a mixed strategy for player P_1
m_2 is a mixed strategy for player P_2

We now carry out similar analysis as before exploring the outcome of the new game in incentivated and nonincentivated contexts.

9.4.3.1 The Relay Game as a Game of Imperfect Information on Semi-Cooperative Incentivated Environments

The Imperfect Information Relay Game in semi-cooperative and incentivated environments is illustrated in its extensive form in Figure 9.6. Similar to previous analysis, the following property defines an incentivated environment. For each player P_i

$$b_i > d_i$$

Next is the corresponding payoff table. Payoff values are taken from the general Relay Game definition described in Section 9.4.1:

$P_1 \backslash P_2$	F	NF
F	(b_1, b_2)	$(-c_1, d_2)$
NF	$(d_1, -c_2)$	$(0, 0)$

None of the players know how the game is going to proceed. The expected gain from playing one strategy or another depends on what the other player is doing. In this case, players P_1 and P_2 can only form conjectures about each other's actions and levels of cooperation.

We define parameter α as the conjecture, from player P_2's point of view on P_1's level of cooperation. In other words, let parameter α be the estimated probability of P_1 choosing strategy F and $(1 - \alpha)$ the estimated probability of P_1 choosing strategy NF. In a similar way, we define parameter β as the conjecture, from player P_1's point of view on P_2's level of cooperation. In simple terms, let parameter β be the estimated probability of P_2 choosing strategy F and $(1 - \beta)$ the probability of P_2 choosing strategy NF.

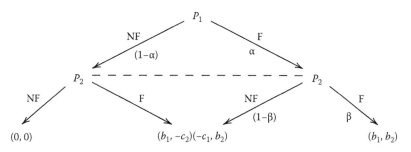

FIGURE 9.6 Relay Game represented as a Game of Imperfect Information, in semi-cooperative and incentivated environments ($b_i > d_i$).

With both entities having defined their conjectures, it is easy to compute their expected gain values taking the payoffs from the table above. The following two equations represent the expected gain that player P_1 hopes to obtain when playing one or another of his strategies $\{F, NF\}$.

$$EG(P_1, \beta, NF) = \beta * d_1$$
$$EG(P_1, \beta, F) = (1 - \beta) * (-c_1) + \beta * b_1$$

From these equations we can conclude that

$$EG(P_1, \beta, F) > EG(P_1, \beta, NF) \Leftrightarrow$$
$$(1 - \beta) * (-c_1) + \beta * b_1 > \beta * d_1 \Leftrightarrow \beta > c_1/(c_1 + b_1 - d_1)$$

The best response player P_1 can give in this scenario is summarized in the following best-response function. Best-response function denoted $\alpha^*(\beta)$ takes β as a parameter and computes the best probability distribution over the set of P_1's strategies to respond to P_2:

$$\alpha^*(\beta) = \begin{cases} 1 & \text{if } \beta > c_1/(c_1 + b_1 - d_1) \\ 0 & \text{if } \beta < c_1/(c_1 + b_1 - d_1) \end{cases}$$

Note how, for player P_1, function $\alpha^*(\beta)$ defines the best probability distribution over his or her set of possible actions. The values $\alpha^*(\beta) = 1$ and $\alpha^*(\beta) = 0$ represent actions F and NF, respectively.

Player P_2 carries out a similar analysis. The following two equations represent the expected gain player P_2 hopes to obtain when playing one or another of his or her strategies $\{F, NF\}$.

$$EG(P_2, \alpha, NF) = \alpha * d_2$$
$$EG(P_2, \alpha, F) = (1 - \alpha) * (-c_2) + \alpha * b_2$$

From these equations we can conclude that

$$EG(P_2, \alpha, F) > EG(P_2, \alpha, NF) \Leftrightarrow$$
$$(1 - \alpha) * (-c_2) + \alpha * b_2 > \alpha * d_2 \Leftrightarrow \alpha > c_2/(c_2 + b_2 - d_2)$$

The best response that player P_2 can give in this type of scenario is summarized in the following best-response function. Best-response function $\beta^*(\alpha)$ takes α as a parameter and computes the best probability distribution over the set of P_2's strategies to respond to P_1:

$$\alpha^*(\beta) = \begin{cases} 1 & \text{if } \alpha > c_2/(c_2 + b_2 - d_2) \\ 0 & \text{if } \alpha < c_2/(c_2 + b_2 - d_2) \end{cases}$$

Again, note how for player P_2, function $\beta^*(\alpha)$ defines the best probability distribution over his or her set of possible actions. The values $\beta^*(\alpha) = 1$ and $\beta^*(\alpha) = 0$ represent actions F and NF, respectively.

By applying Definitions 5 and 7, the points where both best-response functions intersect represent all Nash equilibria in the game. For different levels of player cooperation α and β, the intersection points are those from which players will not want to deviate as they will not obtain better payoff values by unilaterally doing so. See Figure 9.7 for a graphical representation. The following tuples represent the probabilistic strategy profiles for all Nash equilibria points in the Relay Game with Imperfect Information on semi-cooperative and incentivated scenarios:

$$(m_1, m_2) = \big((F, 1), (F, 1)\big),$$
$$\big((F, 0), (F, 0)\big),$$
$$\big((F, c_1/(c_1 + b_1 - d_1)), \quad (F, c_2/(c_2 + b_2 - d_2))\big)$$

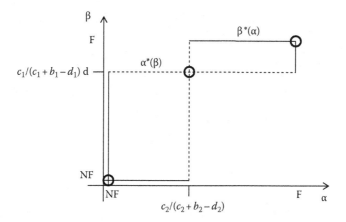

FIGURE 9.7 Nash Equilibrium points in the Relay Game of Imperfect Information in semi-cooperative and incentivated environments.

The first two equilibria take place when both components behave in the same way, that is, when entities' cooperation levels are equal. For the third equilibrium point, participant components have to evaluate the cost of a particular forwarding operation, the expected benefit when both players collaborate, and the possible gain when only one of them displays a nonselfish behavior.

9.4.3.2 The Relay Game as a Game of Imperfect Information on Semi-Cooperative and Nonincentivated Environments

Similar analysis can be carried out for nonincentivated scenarios but again, with totally different results. Similar to the previous analysis we formulate a property to distinguish between incentivated and nonincentivated environments. In the case of a nonincentivated context the following applies for each player P_i in the Relay Game:

$$b_i = d_i$$

Next is the corresponding payoff table with values taken from the general scenario described in Section 9.4.1. Figure 9.6 also represents the game in its extensive form.

$P_1 \backslash P_2$	F	NF
F	(b_1, b_2)	$(-c_1, b_2)$
NF	$(b_1, -c_2)$	$(0, 0)$

The analysis of this scenario is simpler than the previous one. Player P_2 has a dominant strategy to play, which will deliver him or her better payoff value than any other. This strategy is NF. Player P_1 is also aware of this situation and opts for the best possible response, which is also strategy NF. The probabilistic strategy profile $m = (m_1, m_2) = \big((F, 0), (F, 0)\big)$ is the only Nash equilibrium found in the game and therefore the only possible rational outcome.

At this point there is still a possible network configuration in which to analyze cooperativeness in the Relay Game. This new scenario arises when entities ignore other participant's payoff values, disabling them to analyze situations like the previous one, where P_1 can figure out player P_2's next move by knowing the utility values he or she would obtain for each different action. In the next section we explore these types of scenarios.

9.5 WIRELESS INTERACTIONS ON MULTIPLEX ENVIRONMENTS

In Section 9.4.1 we defined semi-cooperative environments as those in which entities had a choice of strategies (Forwarding and Not forwarding incoming packets) at each given point in the game. Moreover, in previous semi-cooperative conditions, entities' set of payoff values were known by all participants. A new type of environment is now defined in which entities still hold a choice of strategies but, by contrast, entities do not have information about other participants' payoff functions. We refer to this new type of environment as Multiplex environments. Games of Incomplete Information or Bayesian Games are used to model such new scenarios.

9.5.1 GAMES OF INCOMPLETE INFORMATION OR BAYESIAN GAMES

In a Bayesian Game, each player is allowed to have some private information that affects the overall game but which is not known by others. This information is typified for each particular entity in such a way that individuals are able to recognize their own type but can only conjecture about the type of their opponents. However, players might have initial beliefs about the type of their opponents and can update their beliefs on the basis of the actions they have played.

First, we informally describe the two most relevant concepts in Bayesian Games: player type and player beliefs.

- *Player type.* In games of Incomplete Information or Bayesian Games, each player is given a type from a particular type space. In general, types are assigned to players in relation to a particular characteristic such as computing capability, connection timeout, transmission rate, payoff function, or simply player identity. The type of a player determines univocally that player's payoff function so that different types will be associated with different payoff functions. Type distribution is modeled by introducing *Nature* as a player in a game. Nature randomly chooses a type for each player according to the probability distribution across each player's type space.
- *Player beliefs.* In Bayesian Games, players hold a set of conjectures over their opponent types. An initial set of conjectures could be created from previous experience operating in similar networks, reputation schemes, or other factors. Player beliefs can be modified during the game after observing their opponents' actions.

9.5.2 THE MULTIPLEX RELAY GAME

The last instance of the general Relay Game takes place between two rational players P_1 and P_2 who can be of different types: cooperative and noncooperative. Player P_1 does not know what type P_2 is and vice versa. Informally, a cooperative player is a network component that is incentivated (via a higher payoff) to collaborate toward a network common purpose. In this regard cooperative network components are identical to incentivated players defined in Section 9.4.1. By contrast, a noncooperative node displays disrupting behavior toward global network welfare, obtaining a higher payoff for such a misconduct. In other words, a noncooperative component is be incentivated (via a higher payoff) to misbehave, affecting overall network performance. This approach allows us to estimate the impact such components have on the network when deceiving honest participants. We refer to this instance of the Relay Game as the Multiplex Relay Game. The following definitions formalize the aforementioned concepts for the Multiplex Relay Game.

- *Player's type-space.* We assume that each player P_i of the Multiplex Relay Game has a type $t_i \in \mathcal{T}_i$, where $\mathcal{T}_i$ is the type-space for P_i. We define $\mathcal{T}_i = \{C, NC\}$, where C denotes a cooperative player, while NC denotes a noncooperative one.

    We denote a cooperative player P_1 as P_1^C and a noncooperative player P_1 as P_1^{NC}. In a similar way, P_2^C and P_2^{NC} are cooperative and noncooperative players, respectively.

In practice, the assignment of types to players is carried out by introducing a fictitious player: Nature. In the course of the game, Nature randomly chooses a type for each player according to a probability distribution over each player's type-space.

Since each type is associated to a different payoff function, uncertainty about the opponent's type has severe implications on the decision-making process, particularly in computing the best-response strategy during the game execution.

- P_1's *belief system.* Assuming player P_1 is the first in choosing an action, at that point P_1 has no information about the type of opponent player P_2. P_1 can only make an estimation of how likely it is that player P_2 is a cooperative component as opposed to a noncooperative one. The belief that player P_1 has about player P_2's type is represented by a probability distribution over the type-space $\mathcal{T}_2$. We define the following conjecture or belief system:

$$\theta_c = \text{prob}(P_2^C)$$
$$\theta_{nc} = \text{prob}(P_2^{NC})$$
$$\text{s.t. } \theta_c + \theta_{nc} = 1$$

From this point on, player P_1 has to include into his or her analysis the probability of player P_2 being of one type or another (P_2^C or P_2^{NC}).

- *Players' set of actions.* The set of actions for players P_1 and P_2 is defined as before and with the same semantics:

$$A = \{F, NF\}$$

where
  F represents the action Forward
  NF represents the action Not Forward

- *Player's pure strategies.* A pure strategy for player p_i is a tuple:

$$s_i = (s_i^c, s_i^{nc})$$

where the first component s_i^c represents a strategy for player P_i of type C whereas the second component s_i^{nc} represents a strategy for player P_i of type NC. Both, s_i^c and s_i^{nc} are in set A.

- *Strategy profile.* A strategy profile in the Multiplex Relay Game is a vector $s = (s_1, s_2)$ of pure strategies, one for each player. Note how specifying a strategy profile univocally determines the outcome of the game.

- *Payoff values.* Utility values for the general form of the Multiplex Relay Game are shown below. With semantics identical to previous sections, for each player P_i, value b_i represents the benefit obtained when both players cooperate, value c_i represents the cost of a forwarding operation, and d_i is the value obtained when one of the players misbehaves. Figure 9.8 also represents the game in its extensive form.

$P_1 \backslash P_2$	F	NF
F	(b_1, b_2)	$(-c_1, d_2)$
NF	$(d_1, -c_2)$	$(0, 0)$

- *Payoff functions.* As introduced before, player types C and NC have associated different payoff functions. We establish the following criteria.

  For each player P_i type cooperative we have

$$b_i > d_i$$

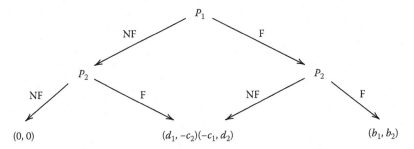

FIGURE 9.8 General form of a Multiplex Relay Game.

For each player P_i type noncooperative we have

$$b_i < d_i$$

9.5.3 POSSIBLE INTERACTIONS

In multiplex environments, network forwarding operations can be implemented in several ways. In particular, we have four possible instances of the game that can take place in such contexts. The following table illustrates all possibilities and displays the distinctive payoff values for each participant.

	P_2^C	P_2^{NC}
P_1^C	$b_1 > d_1$	$b_1 > d_1$
	$b_2 > d_2$	$b_2 < d_2$
P_1^{NC}	$b_1 < d_1$	$b_1 < d_1$
	$b_2 > d_2$	$b_2 < d_2$

Each one of the instances described in the table above has a different outcome. A player is able to recognize his or her type but has no information on the type of her opponent and therefore on the outcome of the game. Because player P_1 is the first one to choose an action, we explore what the expected gain would be from all possible strategies depending on what type P_1 is and the conjecture on what type of opponent P_2 is. The following equations assist in the analysis:

$$EG(P_1^C, \theta_c, F) = \theta_c * b_1 + (1 - \theta_c) * (-c_1)$$
$$EG(P_1^C, \theta_c, NF) = 0$$

From these equations we can conclude that

$$EG(P_1^C, \theta_c, F) \geq EG(P_1^C, \theta_c, NF) \Leftrightarrow$$
$$\theta_c \geq c_1/(b_1 + c_1)$$

In the same way, when P_1 is a noncooperative component we can carry out the following calculations:

$$EG(P_1^{NC}, \theta_c, F) = \theta_c * b_1 + (1 - \theta_c) * (-c_1)$$
$$EG(P_1^{NC}, \theta_c, NF) = 0$$

From these equations we can conclude that

$$EG(P_1^{NC}, \theta_c, F) \geq EG(P_1^{NC}, \theta_c, NF) \Leftrightarrow$$
$$\theta_c \geq c_1/(b_1 + c_1)$$

We have just formally established the fact that P_1, regardless of his or her type, applies the same criteria to decide which strategy to choose. This decision depends on evaluating the cost of the transaction, benefit, and the estimation of how likely it is to interact with a noncollaborative network component. Since P_2 knows her type and can also observe P_1's move, P_2 can respond to P_1's strategy as described in the analysis of previous sections.

9.6 CONCLUSION AND OPEN ISSUES

This chapter illustrates the complexity and scale in the issue of cooperation in wireless ad hoc networks. The absence of fixed infrastructures on which to support security mechanisms to ensure certain properties related to cooperation makes it difficult to analyze this type of environment.

Some of the security aspects with regard to wireless ad hoc network operations such as data confidentiality, integrity, and message and entity authentication can be preserved with the use of cryptography. However, the absence of fixed infrastructures makes it very difficult to ensure certain properties related to cooperation such as fairness or rationality.

Although some of the results prove that cooperation based solely on rational behavior exists, in practice this remains unclear.

Moreover, many wireless mobile networks base their operations on multi-hop and multi-agent transactions for which cooperation between several participants becomes critical. In this kind of environment, collaboration between network components could also be ensured studying component rational behavior. However, the formalism of such an approach is still awaiting implementation.

REFERENCES

[1] A. Alcaide, J.M. Estévez-Tapiador, J.C. Hernández Castro, and A. Ribagorda. An extended model of rational exchange based on dynamic games of imperfect information, *ETRICS'06*, LNCS Vol. 3995/2006, pp. 396–408. June 2006. Springer-Verlag, Berlin/Heidelberg.

[2] A. Alcaide, J.M. Estévez-Tapiador, J.C. Hernández Castro, and A. Ribagorda. Bayesian rational exchange, *International Journal of Information Security*. 7(2), 190–195, Sep 2008.

[3] A.B. MacKenzie, L. Dasilva, and W. Tranter. *Game Theory for Wireless Engineers*, Morgan and Claypool Publishers, San Rafael, CA, 2006.

[4] E. Altman, T. Boulogne, R. El Azouzi, T. Jimenez, and L. Wynter. A survey on net-working games in telecommunications, *Computers and Operations Research*, 33(2), 286–311, Feb. 2006.

[5] L. Buttyán and J.-P. Hubaux. *Security and Cooperation in Wireless Networks*, Cambridge University Press, Cambridge, U.K. (ISBN-13: 9780521873710), 2007.

[6] M. Félegyházi and J.-P. Hubaux. Game theory in wireless networks: A tutorial, Technical report: LCA-REPORT-2006-002.

[7] M. Félegyházi, J.-P. Hubaux, and L. Buttyán. Nash equilibria of packet forwarding strategies in wireless ad hoc networks, *IEEE Transactions on Mobile Computing*, 5(5), May 2006.

[8] J. von Neumann and O. Morgenstern. *Theory of Games and Economic Behavior*, Princeton University Press, Princeton, NJ, ISBN 13:978-0-691-13061-3. 1944.

[9] R. Gibbons. *Game Theory for Applied Economists*, Princeton University Press, Princeton, NJ, 1992.

[10] J.F. Nash. Non-cooperative games, PhD dissertation, Princeton, NJ, 1950.

[11] J.F. Nash. Equilibrium points in n-person games, *Proceedings of the National Academy of the USA*, 36(1):48–49, 1950.

[12] E. Palomar, A. Alcaide, J.M. Estevez-Tapiador, and J.C. Hernandez-Castro. Bayesian analysis of secure P2P sharing protocols, in *Proceedings of the Information Security Workshop (OTM 2007)*, LNCS, vol. 4805/2007, pp. 1701–1717.

[13] D. Fudenberg and J. Tirole. Game Theory. MIT Press, Cambridge, MA, 1991, ISBN 0-262-06141-4.

[14] L. Buttyán. Building blocks for secure services: Authenticated key transport and rational exchange protocols. PhD thesis. Laboratory of Computer Communications and Applications, Swiss Federal Institute of Technology. Lausanne.

[15] V. Srinivasan, P. Nuggehalli, C. Chiasserini, and R.R. Rao. Cooperation in wireless ad hoc networks. In *Proceedings of IEEE Infocom*, San Francisco, March 30–April 3, 2003.

[16] C. Buragohain, D. Agrawal, and S. Suri. *A Game Theoretic Framework for Incentives in P2P Systems*. P2P '03: *Proceedings of the 3rd International Conference on Peer-to-Peer Computing*, 2003. ISBN 0-7695-2023-5. IEEE Computer Society.

[17] A. Urpi, M. Bonuccelli, and S. Giordano. Modelling cooperation in mobile ad hoc networks: A formal description of selfishness. *Proceedings of WiOpt 2003*, March 3–5, 2003, Sophia-Antipolis, France.

10 Cooperative Cognitive Radio

O. Simeone, Y. Bar-Ness, and U. Spagnolini

CONTENTS

As a novel paradigm for the interaction among radio terminal, cooperation is informing most of the current proposals for next-generation wireless systems. Among such proposals, much attention is currently being devoted to idea of cognitive radio (CR): this terms refers, broadly speaking, to intelligent radio terminals that are able to leverage side information acquired from the environment to best adapt the transmission strategy to the user's needs. This chapter aims at investigating the value of cooperation for CR applications by addressing both a general view of the problem and specific case studies. The presentation focuses on typical models for CR scenarios that are characterized by the coexistence in the same spectral resource of licensed and unlicensed users. In particular, the following two models are considered: (1) commons model: according to this scenario, the "intelligence" (i.e., cognition capability) resides in the unlicensed terminals, that, based on some side information acquired from the environment, attempt to fill the spectrum voids left by the activity of the licensed users (these are, in turn, oblivious to the presence of unlicensed terminals); and (2) property-rights model: unlike the commons model, in this scenario, the "cognitive" activity is performed by the licensed terminals that have the possibility to lease the spectrum to passive unlicensed users. For both scenarios, the advantages of cooperation are discussed by means of analysis and numerical results, along with open issues.

10.1   INTRODUCTION

The debate around the concept of cognitive radio (CR) has by now broadened its scope to include substantially different technologies and solutions. The identifying features, common to different schools of thought on the subject, are (a) the coexistence on the same spectral resource of both licensed (or primary) and unlicensed (or secondary) terminals and services; and (b) the deployment of "intelligent" radio terminals whose software control processes leverage side information on the radio environment for the achievement of specific goals related to the needs of the user or network. The term CR was coined in 1999 by Joseph Mitola III, who defines a CR as [1]: "A radio that employs model based reasoning to achieve a specified level of competence in radio-related domains." Recently, the term CR has been mostly used in a narrower sense. In particular, the Federal Communications Commission (FCC) suggests the following definition [2]: A CR is a radio that can change its transmitter parameters based on the interaction with the environment in which it operates. The motivation behind the interest of both academia and industry on CR is in fact the evidence that (1) current spectrum allocation granting exclusive use to licensed services is highly inefficient [2], and (2) new wireless communication technologies allow effective spectrum sharing. Among the different debated positions, two main approaches to CR have emerged [3–6]:

1. Commons model: According to this framework, primary terminals are oblivious to the presence of secondary users, thus behaving as if no secondary activity was present. Secondary users, instead, sense the radio environment in search of spectrum holes (portions of the bandwidth where primary users are not active) and then exploit the detected transmission opportunities;
2. Property-rights model (or spectrum leasing): Here primary users own the spectral resource and possibly decide to lease part of it to secondary users in exchange for appropriate remuneration [7,8].

Another interesting model that is not explicitly considered here is that of interruptible leasing (see, e.g., Ref. [5]): this scenario applies to bandwidths used by the government or local enforcement agencies for emergency communications. When not needed, the bandwidth can be leased to secondary users and quickly reclaimed only in times of emergency. We remark that this solution poses different technical challenges with respect to the first two models that essentially lie in the design of sensing algorithms for nonstationary sources with short detection times in the spirit of change detection theory [9].

Recent standardization efforts within the working groups IEEE 802.22, 802.21, and 802.11h provide evidence of the great interest surrounding the CR paradigm. Specifically, the 802.22 working group is pursuing the development of a wireless regional area network standard for secondary use of the spectrum that is currently allocated to the television service [10,11]. The 802.21 standard, instead, is based on the concept of vertical handover that is strongly related with the idea of dynamic spectrum access usually associated with CR [12]. Finally, the 802.11h standard contains a dynamic

frequency selection procedure that has been developed according to a CR framework [13]. Beside
the standardization bodies, several other agencies and companies have significantly invested in
the concept of CR, including defense advanced research projects agency and the FCC (see also
Ref. [14]).

In parallel, much activity is being devoted within the community to cooperative technologies that
enable radio terminals to interact (collaborate) so as to enhance their performance. Many different
forms of cooperation have been investigated so far that are suitable for different network topologies
and operating conditions. A recent overview of the state of the art can be found in Ref. [15].

The scope of this chapter is the application of cooperative technologies to CR for both commons
and property-rights models. For the commons model, we discuss in Section 10.2 how cooperation
has been recently proved to be an effective technique for improving both the tasks of sensing the
primary activity, through distributed detection techniques, and of exploiting efficiently the transmission opportunities left open by the primary activity, through cooperative transmission. Moreover,
cooperative relaying of primary packets through secondary nodes is shown to be a promising technique to enhance the secondary throughput by increasing the idle periods of the primary nodes. The
property-rights model is discussed in Section 10.3, where it is shown that cooperation can be a viable
means for secondary nodes to remunerate incumbent primary nodes willing to grant the use of the
owned spectral resource. The rationale is that the primary nodes may accept to lease their bandwidth
for a fraction of time if, in exchange for the concession, they benefit from enhanced quality of service
(QoS) (e.g., in terms of rate or probability of outage) thanks to cooperation with the secondary nodes.

10.2   COOPERATIVE CR IN THE COMMONS MODEL

In this section, we first review the commons model for CR in Section 10.2.1, then we present
an overview of the opportunities for exploiting cooperative techniques in such a scenario in
Section 10.2.2, and finally we focus on the performance of a specific cooperative scheme that
prescribes relaying of primary traffic through secondary users (cognitive relaying) in Section 10.2.3.

10.2.1   COMMONS MODEL

According to the commons model, secondary users actively attempt to fill the spectrum holes left
open by the primary activity through sensing and opportunistic transmission under the following two
requirements:

- Activity of the secondary nodes should be transparent to the primary transmission, or, more
  realistically, comply with some specified QoS constraint on the primary activity
- Primary users should be completely oblivious to the secondary activity

The main challenges in the implementation of the commons model appear to be

- Primary activity detection at the secondary nodes (spectrum sensing): Secondary users
  need to monitor the available spectrum (or a portion of it, if there are energy consumption
  constraints) to be able to detect spectral holes. A typical way to address the problem is to
  look for primary transmissions by using a signal detector. The trade-off between probability
  of a false alarm and missed detection then amounts in this scenario to trading missed
  transmission opportunities for an increased interference level on the primary [6,16]. The
  main issue is to enable quick and effective detection at all the secondary nodes that can
  potentially interfere with the primary transmission. Note that given the receiver-centric
  definition of interference, a major problem is also that of detecting or locating primary
  receivers [18]. The impact of practical limitations in spectrum sensing on the performance
  of CR from a system perspective have been investigated in Ref. [19];

- Transmission opportunity exploitation: Once a spectral hole has been identified (within the error probability set by the receiving operating curve of the detector), secondary users need to exploit the transmission opportunity to satisfy two conflicting objectives: (1) making their activity transparent to the primary users (according to a defined criterion for transparency in terms of QoS constraint); (2) maximizing their own performance in terms of the desired QoS indicator (e.g., rate, delay, etc.). Secondary users might be competing for the resource or cooperating to improve efficiency and fairness of resource sharing. In both cases, analysis can be usefully carried out using concepts from various fields such as information theory [24] and game theory [25].

To address satisfactorily such challenges, transmission protocols that are able to provide the necessary flexibility for spectrum access while still guaranteeing QoS constraints for the primary transmission need to be devised. To properly account for the complexity of the problem, the design should account for a cross-layer view of the problem, encompassing different communications layers [16]. As an example, consider a secondary ad hoc network, where secondary nodes communicating in an ad hoc fashion need to design routes that account for the primary QoS constraints. Routes should be adaptively modified as soon as a new primary user becomes active to accommodate the new QoS needs by, for example, reducing the interference in a given area of the network. There exists a trade-off between the effectiveness of such an approach at the network layer and the complexity of sophisticated techniques at the Medium Access Control (MAC)/Physical layer in terms of sensing, opportunistic channel access, and, for example, power control: "better" routes require less stringent constraints on the performance of the lower layers.

It should be reminded that the view discussed here that separates the design problem for commons-model cognitive networks into sensing and transmission opportunity exploitation is arguably the most common and practical, but not the only one. In fact, in the information-theoretic literature, analysis of commons-model scenarios generally assumes that secondary nodes are allowed to transmit at the same time as the primary nodes. The typical underlying assumption is that secondary nodes have information about the primary message before it is actually transmitted to the intended destination (e.g., through auxiliary channels). This information can then be leveraged through a superposition coding scheme whereby the primary interference on the own transmission is precanceled via dirty paper coding and cooperation with the primary is exploited to compensate for the additional noise created by the secondary transmission [17] (see also Refs. [5,20,21] and references therein). In the presence of fading channels, the performance of this approach is highly dependent on the presence of perfect channel state information (CSI) at the transmitter, which might be difficult to attain [22]. Another approach is that of imposing a strict interference (i.e., receive power) constraint at specific primary locations (see Ref. [32]).

The next section is devoted to explaining how cooperative techniques can be deployed to meet the challenges discussed here for a commons model with sensing/transmission structure.

10.2.2 COOPERATION AND CR IN THE COMMONS MODEL

The performance of both primary activity detection and transmission opportunity exploitation (see Section 10.2.1) can be shown to improve drastically if cooperation is deployed. In the following, we consider the two problems separately and provide a brief description of different available solutions.

10.2.2.1 Cooperative Sensing in CR

To enhance the performance of primary (transmitter) detectors, a natural solution that is able to cope with different shadowing/fading conditions is cooperation. The basic idea is to employ distributed detection at the secondary nodes: each node measures (e.g., through an energy detector) the local received signal, then local signals are exchanged (possibly to a central decision point), and finally a global decision on the primary activity is achieved (Figure 10.1a). This approach is clearly robust

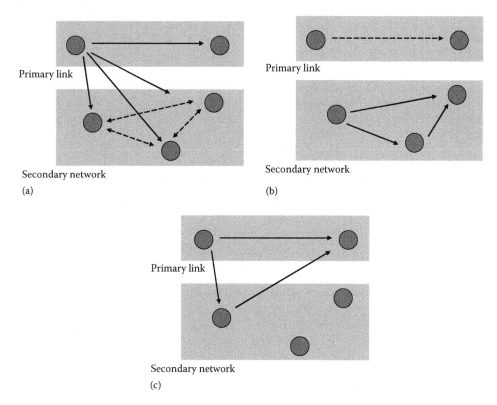

FIGURE 10.1 CR wireless network with one primary link and a secondary network of three nodes. Three examples of applications of cooperative technology to CR are shown: (a) cooperative sensing; (b) cooperative transmission between secondary terminals; and (c) cognitive relay.

to possible imbalance of the channel qualities of different secondary users and has show been to achieve a drastic improvement of the receiving operating curve [27].

An alternative approach to cooperative sensing is interestingly based on cooperative transmission among the secondary users and is discussed in the next section (see Figure 10.1b).

10.2.2.2 Cooperative Transmission in CR

Cooperative transmission in its basic forms refers to the information theoretic model of the relay channel, where one node (the relay) forwards the transmission from another node (the source) toward the intended destination. Performance advantages achievable from collaboration arise from (1) power gains that can be harnessed if the relay happens to be in a convenient location, typically halfway between source and destination; (2) diversity gains that leverage the double path followed by the signal (direct source–destination and relay transmissions) [15,26]. In the context of CR, cooperative transmission can give rise to two different basic scenarios, as explained in the following.

10.2.2.3 Cooperative Transmission between Secondary Users

In this scenario, a secondary user acts as relay from the transmission of another (source) secondary terminal (see Figure 10.1b). General considerations valid for cooperative transmission can be applied in this case (see, e.g., Refs. [15,26]) with the important caveat that secondary nodes need to continuously monitor the channel for possible transmissions by the primary. Interestingly, Ref. [28] has proposed to use cooperative transmission as a means to enhance the sensing process as well. The main idea is to let the secondary relay node amplify and forward the received signal because the latter contains not only the transmission from the secondary source, but also, if present, the signal from the primary. This forwarding then allows the secondary destination to improve the local detection of the primary user in a scenario where the relay is placed approximately halfway between primary and secondary destination. As a final remark, we point out that such an approach to sensing is particularly well suited for an interruptible leasing model of CR (see Section 10.1), where it is especially important to continuously monitor the spectrum and vacate the bandwidth as soon as possible (see also Ref. [29]).

10.2.2.4 Cognitive Relay

Apart from cooperation between secondary users, a different form of cooperative transmission can be envisioned where a secondary user has the possibility to relay the traffic of a primary transmitter toward the intended destination (see Figure 10.1c). The rationale of this choice is that helping the primary to increase its throughput entails (for a fixed demand of rate by the primary) a diminished transmission time of the primary, which in turns leads to more transmission opportunities for the secondary [30]. Therefore, while cooperation between secondary users aims at increasing the secondary throughput for a given spectral hole, cognitive relaying pursues an enhanced throughput by increasing the probability of transmission opportunities (see also Refs. [5,17] where cognitive relaying is shown to increase the capacity region). In Section 10.2.3, we discuss a simple scenario with one primary and one secondary link where the transmitter may act as relay (also referred to as cognitive interference channel; see Figure 10.2) to show the advantages of cognitive relaying.

10.2.3 COGNITIVE RELAY: A CASE STUDY

As introduced in the previous section, cognitive relaying appears to a be a promising approach for improving the throughput of secondary nodes by increasing the transmission opportunities. In this section, we further investigate this concept, by considering the case-study scenario in Figure 10.2.

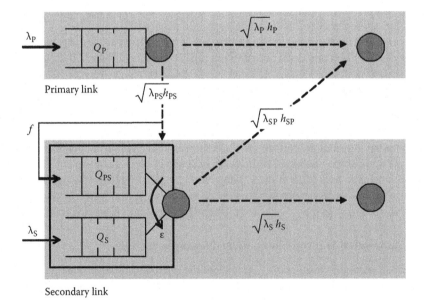

FIGURE 10.2 CR scenario with one primary and one secondary link where the secondary transmitter has the capability of relaying primary traffic (cognitive relay).

10.2.3.1   System Model

Referring to Figure 10.2, both primary and secondary transmitting nodes are equipped with an infinite queue in which incoming packets are stored. All packets have the same number of bits, and their transmission time coincides with a time slot, which we consider as our reference time unit. The arrivals of packets at each transmitting station are independent and stationary processes, with λ_P (packets/slot) being the mean arrival rate at the primary queue and λ_S (packets/slot) the mean arrival rate at the secondary queue. The primary transmitter accesses the channel whenever it has a packet in its queue $Q_P(t)$ at the beginning of the slot t, being oblivious to the presence of a secondary link. On the contrary, the secondary transmitter sends a packet to its destination in a given slot only if it senses an idle channel according to the spectrum sensing scheme described above (and if it has a packet to transmit in its queues and $Q_{PS}(t)$; see below).

Whenever a primary packet is not correctly received by the intended destination but is instead decoded at the secondary transmitter, the latter has the choice to store the packet in a separate queue $Q_{PS}(t)$ for later forwarding to the secondary transmitter (cognitive relaying). Here we assume that, under the conditions mentioned above, the secondary transmitter accepts the primary packet with a probability f according to a packet acceptance control (PAC) mechanism (which can be seen as a form of active queue management [31]). Moreover, whenever an idle slot is detected, the secondary transmitter transmits a packet from the queue $Q_S(t)$ containing its own packets with scheduling probability ε and from the queue $Q_{PS}(t)$ with complementary probability $1 - \varepsilon$.

As for the signaling protocols, we consider that each receiving node sends the respective transmitting node an ACK message in case of a correct reception or a NACK message in case of an erroneous reception. A packet reception error requires retransmission. Notice that, in this analysis, the overhead introduced in the system by the transmission of ACK–NACK messages is considered negligible and therefore not modeled. Notice that, for a primary packet accepted by the secondary, the primary might receive two acknowledgments for the same packet. In this case, it will simply consider the packet as correctly received if at least one acknowledgment is positive.

Independent Rayleigh block-fading channels are assumed between every pair of nodes. In particular, the complex channel gains on the ith link (where subscript i identifies transmitter–receiver pairs as illustrated in Figure 10.2) at the tth slot read $\sqrt{\gamma_i}h_i(t)$ where $h_i(t)$ is a zero-mean unit-variance stationary process and γ_i is the average (time-invariant) channel power gain. Moreover, the primary transmitter employs, without loss of generality, unit power, while the secondary transmits with power $P_S \leq 1$. Packet transmission is considered successful if the instantaneous signal-to-noise ratio is above given thresholds β_P for the primary link and β_S for the secondary. Notice that in case of missed detection of primary activity, the secondary interferes with the primary transmission, and in this case primary transmission is successful if the signal-to-noise-plus-interference ratio is above the threshold β_P.

10.2.3.2   Design of Cognitive Relaying: Degrees of Freedom and Constraints

According to the principle of the commons model for CR (recall Section 10.2.1), the primary transmitter in Figure 10.2 selects its arrival rate λ_P within its own stability region, being oblivious to the presence of the secondary. On the contrary, the secondary transmitter adapts its transmission mode to best accomplish two conflicting goals: (1) making its activity transparent to the primary link and (2) maximizing its own stable throughput μ_S (defined as the number of packets per slot such that the secondary queue is stable for any $\lambda_S \leq \mu_S$). As previously discussed, transparency of the activity of the secondary must be measured in terms of given QoS parameters on the primary transmission. In the considered cross-layer view of the problem that accounts for both MAC and physical layers, interesting QoS measures exist, for example, stability of the queues or average delays. Here we focus for simplicity on the former. In other words, we say that the activity of the secondary node is transparent to the primary if the queue of the latter remains stable irrespective of the secondary transmissions. The transmission mode of the secondary transmitter is here defined by the transmit power P_S and the threshold α of the spectrum sensing detector (see below) at the physical layer and the scheduling probability ε and the packet acceptance ratio f at the MAC layer.

To decide the transmission mode, the secondary transmitter is assumed to know the average (or long-term) channel parameters (γ_P, γ_S, γ_{PS}, γ_{SP}), under the premise that, before starting transmission, the secondary node has estimated these parameters during an observation period of several time slots. Notice that γ_S, γ_{PS}, and γ_{SP} can be estimated from energy measurements on the respective channels (considering feedback from the receivers, e.g., for ACK/ NACK), whereas the knowledge of γ_P can be inferred by observing the fraction of ACK/ NACK messages on the primary link (see Ref. [32] for a similar idea).

Remark: Cognitive relaying aims at enhancing the secondary throughput with the increase in transmission opportunities for the secondary. It should be noted that this is achieved by increasing the overall energy consumed by the secondary (i.e., power times the number of transmitted packets), because the latter has to deliver not only its traffic but also some packets from the primary. Here, we consider only power constraints and thus do not tackle this issue. However, in an energy-limited scenario, the performance of cognitive relaying requires further investigation.

10.2.3.3   Packet Acceptance Control

Cognitive relaying prescribes the possibility for the secondary transmitter to forward primary packets that have not been successfully received at the intended destination but have instead been decoded at the primary transmitter. This event is clearly likely to happen if the direct average channel power γ_P is sufficiently smaller that the power gain γ_{PS} between primary and secondary transmitters. However, to be able to actually deliver the primary packets to the destination, thus avoiding overflow of the queue $Q_{PS}(t)$, the channel gain γ_{SP} needs to be sufficiently large [30]. The solution studied here prescribes a PAC mechanism where the secondary, based on the knowledge of the average channels (see Section 10.2.3.2), accepts only a fraction of packets f out of the primary packets that have not been successfully received at the intended destination but have instead been decoded at the secondary transmitter. It is expected that an optimal f will be small whenever channel gain γ_{SP} is not sufficiently large to avoid congestion of queue $Q_{PS}(t)$. Further discussion on this point is provided below.

10.2.3.4   Spectrum Sensing

According to the basic principle of CR, at the beginning of each slot, the secondary transmitter senses the channel to find whether it is employed by the primary or not. Here, we explicitly model the detection process by using an energy detector [33]. More precisely, at the beginning of each slot, the secondary node measures m samples (at symbol rate) of the received signal to detect the activity of the primary. The secondary transmitter compares the output of the energy detector with a particular threshold α to decide whether the primary node is transmitting or not. Notice that in the discussion above and in the numerical results below, we assume that the secondary transmitter is able to recover the primary symbol timing from the received signal. However, when symbol synchronization is not feasible or too expensive, the energy detector could be implemented after appropriate oversampling with no performance loss. The secondary transmitter is interested in choosing the optimum detection threshold α as the result of a trade-off between limiting the interference on the primary (probability of missed detection) and increasing secondary channel access (probability of false alarm) [6].

10.2.3.5   Numerical Results

In this section, we present some numerical results to get insight into the performance of cognitive relaying and, in particular, into the conditions under which this solution provides performance advantages. Notice that optimizing the transmission mode (P_S, ε, f, α) of the secondary amounts to

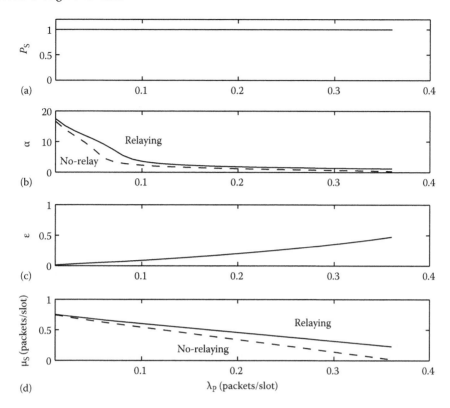

FIGURE 10.3 Optimal choice for power P_S, detection threshold α and scheduling probability ε, and corresponding maximum stable throughput μ_S versus the arrival rate λ_P selected by the primary node ($\beta_P = \beta_S = 4$ dB, $\gamma_P = 4$ dB, $\gamma_S = \gamma_{PS} = \gamma_{SP} = 10$ dB, $m =5$).

solving the design problem discussed here. We refer the reader to Ref. [34] for technical details. We set the number of training symbols for the energy detector to $m = 5$, the outage thresholds to $\beta_P = \beta_S = 4$ dB, and $\gamma_S = 10$ dB.

Figure 10.3 compares the performance of the simple cognitive scenario in Figure 10.2 for two cases where the secondary can (solid lines) and cannot (dashed lines) act as a relay for the primary traffic. In particular, we compare the two scenarios by evaluating the optimum choices for secondary transmit power P_S, detection threshold α, and scheduling probability ε (only applicable to relaying) and the maximum stable throughput of the secondary node μ_S. Here we set $f = 1$ (no packet admission control) for simplicity, and other parameters are as follows: $\gamma_P = 4$ dB, $\gamma_{PS} = \gamma_{SP} = 10$ dB. Clearly, the secondary maximum stable throughput μ_S decreases with the primary arrival rate λ_P according to the hierarchical model of CR. In particular, the arrival rate $\lambda_P \simeq 0.36$ packets/slot coincides with the maximum primary stable throughput μ_P, which would lead to $\mu_S = 0$ if no relaying was deployed (see Ref. [30] for details). Moreover, in this example, the primary-to-secondary and secondary-to-primary channel power gains (γ_{PS} and γ_{PS}, respectively) are sufficiently larger than the direct primary channel γ_P, so that cognitive relaying leads to relevant secondary throughput gains, especially for large primary rate λ_P (notice, for instance, that a nonzero throughput μ_S is achieved even for $\lambda_P > 0.36$ packets/slot). Finally, it can be seen that with cognitive relaying the optimal detection threshold α is larger than that in the case of no relaying: this implies that relaying allows the secondary to choose a less conservative point on the receiving operating curve of its detector and thus to attempt access to the channel more often (notice also that the optimal power P_S equals the maximum value 1 in both cases).

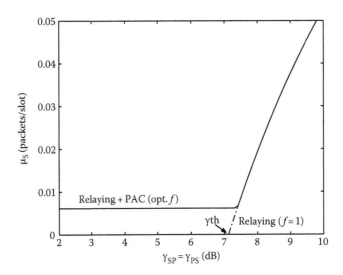

FIGURE 10.4 Maximum throughput of the secondary user μ_S for a fixed $\lambda_P = \mu_P^{max}$ versus the value of the channel parameter γ_{SP} for $f = 1$ (relaying) and optimized f (PAC) ($\beta_P = 4$ dB, $\beta_S = 4$ dB, $\gamma_P = 7$ dB, $\gamma_S = 10$ dB, $m = 5$, and $\lambda_P = 0.36$).

We are now interested in evaluating the limits of cognitive relaying. Consider a scenario where the direct primary channel is $\gamma_P = 7$ dB and the primary-to-secondary and secondary-to-primary channels γ_{PS} and γ_{SP} take different values. Figure 10.4 shows the maximum throughput of the secondary μ_S for cognitive relaying and $\lambda_P = 0.36$ (see discussion earlier) considering both the case $f = 1$ (no packet access control) and optimized packet acceptance ratio f. It is seen that for channel gains γ_{SP} and γ_{PS} large enough ($> \gamma th \simeq \gamma_P = 7$ dB), the secondary is able to efficiently relay traffic from the primary, thus creating transmission opportunities for its own traffic and increasing its own throughput. However, if γ_{SP} and γ_{SP} are not larger than this threshold ($< \gamma th \simeq \gamma_P$) and there is no PAC ($f = 1$), the secondary is not able to deliver all the extra traffic coming from the primary, and the throughput of the secondary is zero. However, optimizing the packet acceptance ratio f allows to obtain a nonzero secondary throughput even for small channel gains. The conclusions here are further corroborated by Figure 10.5, where we consider again $\gamma_{PS} = \gamma_{SP}$ and evaluate the minimum γ_{SP} and γ_{PS} (i.e., γth) that allows a nonzero throughput for relaying without PAC ($f = 1$), for different values of γ_P. In particular, $\gamma th/\gamma_P$ is shown to increase with γ_P, suggesting that advantages of cognitive relaying are more difficult to harness for larger γ_P. Moreover, the lower figure shows that the corresponding secondary throughput of PAC for this threshold condition is decreasing with γ_P.

10.2.4 CONCLUDING REMARKS

In this section, we have discussed some aspects of the interplay of cooperation and CR technologies in the commons model. In the first part, an overview of the advantages of cooperation in the context of CR has been presented, while in the second we have focused on an interesting instance of cooperative transmission, where a secondary node acts as a relay for the primary traffic. It has been shown that, under certain conditions on the network topology, cognitive relaying is effective in enhancing secondary throughput by increasing the number of transmission opportunities.

10.3 COOPERATIVE CR IN THE PROPERTY-RIGHTS MODEL

In this section, we discuss the role cooperative technology can play in a CR scenario that complies with the property-rights model. We recall that the property-rights model can be considered as dual

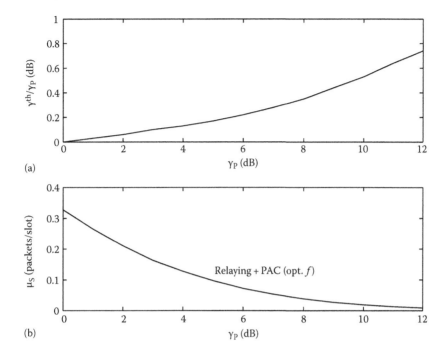

FIGURE 10.5 (a) Minimum value of $\gamma_{SP} = \gamma_{PS}$ that allows a nonzero throughput for cognitive relaying with $f = 1$ (i.e., γth), versus γ_P. (b) Secondary throughput μ_S with PAC for $\gamma_{SP} = \gamma_{PS} = \gamma$th versus γ_P ($\beta_P = 4$ dB, $\beta_S = 4$ dB, $\gamma_S = 10$ dB, $\lambda_P = 0.36$, $m = 5$).

with respect to the commons model discussed in the previous section in that it shifts the "intelligence" and the computational burden from the secondary to the primary users. In particular, primary users here are assumed to be aware of the presence of secondary users (as opposed to oblivious as assumed in the commons model) so that, based on side information on the radio environment and on the available unlicensed users, they actively decide when and to whom to lease part of the spectrum they own [7,8].

10.3.1 COOPERATION AND CR IN THE PROPERTY-RIGHTS MODEL

As discussed here, the property-rights model prescribes spectrum leasing by the licensed users to the unlicensed users. Much of the discussion on this scenario has then focused on defining appropriate protocols and pricing strategies to be applied to enable such transactions between licensed and unlicensed users [7,8]. In this section, we discuss how cooperative technology can provide a key tool to solve this problem by exploiting the specific features of wireless communications. In particular, we advocate a scenario where cooperation (relaying) is used by the secondary terminals as a means of remuneration to the incumbent primary in exchange for spectrum leasing. The rationale is that the primary nodes will be willing to lease the owned bandwidth for a fraction of time if, in exchange for the concession, they benefit from enhanced QoS thanks to cooperation with the secondary nodes. In turn, clearly, secondary nodes have the choice on whether to cooperate or not with the primary on the basis of the offer of the licensed users in terms of amount of cooperation required and the corresponding fraction of time leased.

To illustrate this idea, we consider a single primary link sharing the spectrum with an ad hoc network of secondary nodes as depicted in Figure 10.6. In the secondary network, each transmitter has information to deliver to a given secondary receiver (interference channel). The primary link may

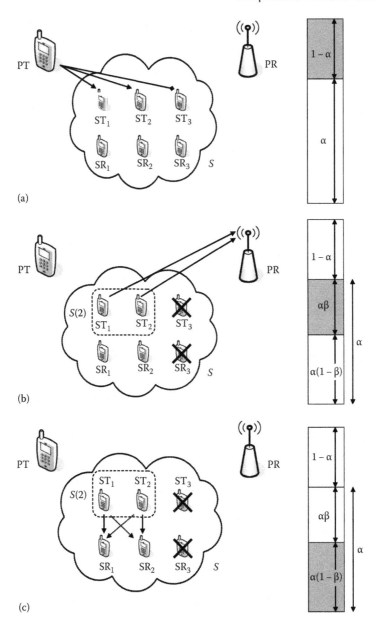

FIGURE 10.6 Secondary spectrum access through cooperation-based spectrum leasing, for $K = 3$ secondary transmitters and receivers: (a) primary transmission; (b) space-time coded cooperation; and (c) secondary transmission.

lease the owned bandwidth for a fraction of time to a subset of secondary transmitters in exchange for cooperation (relaying) in the form of transmission via distributed space-time coding (DSTC) [35].

In the fraction of time available for the secondary activity, the subset of selected secondary transmitters compete for transmission to their respective receivers by performing decentralized power control. Accordingly, each secondary node seeks to maximize a utility function that accounts for the cost/benefits trade-off between the expense in terms of transmitted power and the achievable QoS. For instance, an excessively small fraction of time leased for secondary transmission may not compensate for the cost of transmission and the secondary may then decide not to transmit and thus not cooperate with the primary. The outcome of this competitive and decentralized behavior can be described by the basic solution concept in game theory, namely the "Nash equilibrium" (NE) [36]. Game-theoretic analysis of the interference channel with distributed power control has been reported in Refs. [37,38].

Overall, the considered framework of cooperative CR is characterized by a hierarchical structure, where one agent (the primary link) optimizes its strategy (leased time and amount of cooperation) based on the knowledge of the effects of its decision on the behavior of a second agent (the secondary network). A convenient analytical model to study this scenario is provided by Stackelberg games [36]. Related work on application of Stackelberg games to wireless communications can be found in Ref. [39], where the main focus is promoting cooperation in ad hoc networks, and in Ref. [40], where the authors are primarily interested in the optimal design of an access point in a decentralized network.

10.3.2 System Model

In the following section, we detail the model of spectrum leasing based on cooperation and the main system parameters.

10.3.2.1 Medium Access Control Layer

We consider the system sketched in Figure 10.6 where a primary (licensed) transmitter PT communicates with the intended receiver PR in slots of N_S symbols. In the same bandwidth, a secondary (unlicensed) ad hoc network $\mathcal{S}$, composed of K transmitters $\{ST_i\}_{i=1}^K$ and K receivers $\{SR_i\}_{i=1}^K$, seeks to exploit possible transmission opportunities. We assume one-to-one communication in $\mathcal{S}$, that is, the data from the secondary terminal ST_i is intended for the secondary receiver SR_i (interference channel).

The primary transmitter PT grants the use of the bandwidth to a subset $\mathcal{S}^{(k)} \subseteq \mathcal{S}$ of $k \leq K$ secondary nodes in exchange for cooperation to improve the quality of the communication link to its receiver PR. In particular, the primary decides whether to use the entire slot for direct transmission to PR or to employ cooperation. In the latter case, a fraction $1 - \alpha$ of the slot ($0 \leq \alpha \leq 1$) is used for transmission from PT to the secondary nodes in $\mathcal{S}^{(k)}$ (see Figure 10.6a). Moreover, the remaining αN_S symbols are further decomposed into two subslots according to a parameter $0 \leq \beta \leq 1$. The first subslot is of duration $\alpha\beta N_S$ and is dedicated to cooperation: the set $\mathcal{S}^{(k)}$ of active ST_i forms a distributed k-antenna array and cooperatively relays the primary codeword through DSTC toward PR [35] (Figure 10.6b). In the last subslot of duration $\alpha(1 - \beta)N_S$ symbols, the active secondary nodes are allowed to transmit their own data (Figure 10.6c). In this last subslot, the transmissions scheme amounts to an interference channel with distributed power control [26,37]. Finally, notice that the durations of different subslots are constrained to be integer numbers, which is guaranteed if αN_S and $\alpha\beta N_S$ are integers.

10.3.2.2 Physical Layer

The channels between nodes are modeled as independent proper complex Gaussian random variables, invariant within each slot, but generally varying over the slots (i.e., Rayleigh block-fading channels). We use the following notation to denote the instantaneous fading channels in each block: h_P denotes the complex channel gain between primary transmitter PT and primary receiver PR; $h_{PS,i}$ the channel gain between PT and secondary transmitter ST_i; $h_{SP,i}$ between ST_i and PR; $h_{S,ij}$ between ST_j and SR_i for any $i,j = 1, ..., K$. Moreover, we use the following notation for the average channel gains: $E[|h_P|^2] = g_P$, $E[|h_{PS,i}|^2] = g_{PS,i}$, $E[|h_{SP,i}|^2] = g_{SP,i}$, and $E[|h_{S,ij}|^2] = g_{S,ij}$. Here we are interested in a baseline scenario where the primary link is assumed to be aware of the instantaneous realizations of all the fading power gains in the system (i.e., $|h_P|^2$, $|h_{PS,i}|^2$, $|h_{SP,i}|^2$, and $|h_{S,ij}|^2$), whereas the secondary nodes are aware only of the fading power gains $|h_{S,ij}|^2$ within the secondary network. We notice that, albeit ideal, the assumption of instantaneous CSI is very common in the literature on game-theoretic applications to wireless networks (see, e.g., Refs. [26,37]) and provides a convenient framework for analysis. We remark that our focus here is to establish the role of cooperation as an enabler of spectrum leasing. For an analysis under more realistic assumptions (namely, statistical or long-term CSI available at the primary link and secondary network), the reader is referred to Ref. [41].

The transmission power of the primary is denoted as P_P. On the other hand, the ith secondary node in the active set $\mathcal{S}^{(k)}$ transmits with power $0 \leq \hat{P}_i \leq P_{\max}$, where the powers $\hat{P} = [\hat{P}_i]_{i \in \mathcal{S}^{(k)}}$ are obtained as the outcome (NE) of the power control game played between secondary nodes in the last subslot. To ensure fair behavior of the secondary nodes, one should enforce some rule on the power employed by the secondary for cooperation as a function of the power $\hat{P}_i$ that they are willing to use for their own transmission (recall the last two subslots in Figure 10.6a and b). Here we assume that each of the k activated secondary nodes in $\mathcal{S}^{(k)}$ is constrained to use the same transmission powers for both cooperation and own traffic, but the analysis could be easily extended to other models where the power for cooperation is some nondecreasing function of $\hat{P}_i$. Finally, the single-sided spectral density of the independent white Gaussian noise at the (both primary and secondary) receivers is N_0.

10.3.2.3 A Note on the Protocol

Spectrum leasing requires the primary link to be aware of the number and identity of secondary nodes available in the band of interest. In practice, this entails the need for a control channel to be used, for example, for communicating the availability of a new secondary user to the primary. This control channel can also be used for the exchange of CSI parameters among different nodes and for delivering the decision of the primary (slot allocation and cooperation parameters) to the secondary network. The need for a control channel, and the consequent reduction in spectral efficiency, is the price to be paid for the implementation of spectrum leasing. This contrasts with the commons model, where the primary is oblivious to the presence of the secondary nodes and no control channel needs to be set up.

10.3.3 COOPERATION-BASED SPECTRUM LEASING

In this section, we describe and analyze spectrum leasing based on space-time coded cooperation within a Stackelberg game framework under the assumption of instantaneous CSI. As anticipated above, here we consider information theoretic achievable rates as performance measures and thus the number of symbols N_S per blocks is assumed to be very large ($N_S \to \infty$). We first discuss the Stackelberg model of interaction between the primary link (leader) and competitive secondary network (follower). Insight on the system performance is then provided through analysis and numerical results.

10.3.3.1 Primary Link

Based on the available instantaneous CSI, the primary link selects the slot allocation parameters (α, β) and the set of cooperating secondary nodes $\mathcal{S}^{(k)} \subseteq \mathcal{S}$ toward the aim of optimizing its transmission rate $R_P(\alpha, \beta, \mathcal{S}^{(k)})$ (measured in bits/symbol or equivalently bits/s/Hz). In order to simplify the notation, we reorder the secondary links in each slot so that the first k, ST_i with $i = 1, ..., k$, belongs to the set $\mathcal{S}^{(k)}$. To start with, in the baseline case where the fraction of time leased to $\mathcal{S}^{(k)}$ is $\alpha = 0$, we set the primary transmission rate to $R_P(0, \beta, \mathcal{S}^{(k)}) = R_{\mathrm{dir}}$, where R_{dir} is the rate on the direct link between PT and PR:

$$R_P(0, \beta, \mathcal{S}^{(k)}) = R_{\mathrm{dir}} = \log_2\left(1 + \frac{|h_P|^2 P_P}{N_0}\right). \tag{10.1}$$

Instead, if the fraction of time leased to the set $\mathcal{S}^{(k)}$ of k active secondary users is $\alpha > 0$, we assume the use of a collaborative scheme based on decode-and-forward multi-hop and space-time coding, for which the achievable primary rate reads [35,42]

$$R_P(\alpha, \beta, S^{(k)}) = \min\{(1-\alpha)\,R_{PS}(S^{(k)}), \alpha\beta R_{SP}(\alpha, \beta, S^{(k)})\}, \quad \alpha > 0. \tag{10.2}$$

Notice that larger achievable rates could be achieved by more sophisticated coding/decoding schemes, for example, coding by rate splitting and decoding at the destination based on the signal received in both slots (instead of multi-hop) [42]. Here we focus on simple decode-and-forward multihop only for simplicity of presentation in order not to introduce inessential analytical complications Equation 10.2. The first term in Equation 10.2 is the rate achievable in the first subslot (Figure 10.6a) between the primary transmitter PT and all the secondary transmitters in the set $S^{(k)}$. Recalling that secondary nodes cannot cooperate among themselves for detection, this rate is easily shown to be dominated by the worst channel $|h_{PS,i}|^2$ in the set $i \in S^{(k)}$ as

$$R_{PS}(S^{(k)}) = \log_2\left(1 + \frac{\min_{i \in S^{(k)}} |h_{PS,i}|^2 P_P}{N_0}\right). \tag{10.3}$$

Notice that the factor $(1-\alpha)$ in the first term of Equation 10.2 is due to the fraction of time occupied by the first slot. The second term in Equation 10.2 is the achievable rate $\alpha\beta R_{SP}(\alpha, \beta, S^{(k)})$ between the k secondary nodes in $S^{(k)}$ and the primary receiver PR, assuming DSTC cooperation (Figure 10.6b) [35]

$$R_{SP}\left(\alpha, \beta, S^{(k)}\right) = \log_2\left(1 + \sum_{i \in S^{(k)}} \frac{|h_{SP,i}|^2 \hat{P}_i\left(\alpha, \beta, S^{(k)}\right)}{N_0}\right). \tag{10.4}$$

We remark that the rate Equation 10.4 is obtained following the ideal information-theoretic assumption of employing orthogonal STCs that are able to harness the maximum degree of diversity from cooperation (see Ref. [41] for a more practical approach). Recall that the cooperation rate Equation 10.4 depends on the powers $\hat{P}_i\left(\alpha, \beta, S^{(k)}\right)$ autonomously selected by the secondary transmitters, as the outcome (NE) of the noncooperative power control game played by the secondary terminals in the third subslot (Figure 10.6c).

To summarize, the primary link aims at solving the following rate-optimization problem:

$$\begin{aligned} &\max_{\alpha, \beta, S^{(k)}} R_P(\alpha, \beta, S^{(k)}) \\ &\text{s.t. } S^{(k)} \subseteq S,\ 0 \le \alpha, \beta \le 1 \end{aligned}, \tag{10.5}$$

where we notice that the integer constraints on the number of symbols in each subslot can be neglected at this point given the assumption $N_S \to \infty$ (see Ref. [41] for analysis of a scenario with finite N_S). This problem can be interpreted as a Stackelberg game [36]: the primary link is the Stackelberg leader, which optimizes its strategy $(\alpha, \beta, S^{(k)})$ to maximize its revenue according to Equation 10.5, being aware that its decision will affect the strategy selected by the Stackelberg follower (the secondary ad hoc network), namely the set of transmitting powers $\hat{P}\left(\alpha, \beta, S^{(k)}\right)$.

10.3.3.2 Ad Hoc Secondary Network

Any active secondary terminal ST_i, with $i = 1, ..., k$, in the set $S^{(k)}$ attempts to maximize the rate toward its own receiver SR_i discounted by the overall cost of transmission power, being aware of the parameters α, β, and $S^{(k)}$ selected by the primary and acting in a rational and selfish way. In particular, each secondary transmitter ST_i chooses its transmitting power P_i according to the NE $\hat{P}_i\left(\alpha, \beta, S^{(k)}\right)$ (we will show that it exists and is unique) of the noncooperative power control game $\langle S^{(k)}, \mathcal{P}, u_i\left(P_i, \mathbf{P}_{-i}\right)\rangle$, defined as follows. Let the set of allowed (power) strategies $\mathcal{P}$ be $\mathcal{P} = \{\mathbf{P} = [P_i]_{i \in S^{(k)}} | 0 \le P_i \le P_{\max},\ i \in S^{(k)}\}$; the utility function $u_i\left(P_i, \mathbf{P}_{-i}\right)$ of the ith secondary node (player) in the set $S^{(k)}$ is defined as the difference between the achievable transmission rate and the cost of transmitted energy, similarly to, for example, Ref. [43]. In particular, the rate achievable on the link between ST_i and SR_i is $\alpha(1 - \beta)R_i(P_i, \mathbf{P}_{-i})$ with

$$R_i(P_i, \mathbf{P}_{-i}) = \log_2\left(1 + \frac{|h_{S,ii}|^2 P_i}{N_0 + \sum_{j \in S^{(k)}, j \neq i} |h_{S,ij}|^2 P_j}\right),$$ (10.6)

where vector $\mathbf{P}_{-i}$ contains all the elements in $\mathbf{P}$ but the ith (i.e., it denotes the set of other players' strategies). Moreover, the cost for transmitted energy is $c \cdot \alpha P_i$ (recall that α is the fraction of time where the active secondary nodes are transmitting), with c being the cost per unit transmission energy. Noticing that α is a common multiplier for both rate and energy cost, the utility function equivalently reads

$$u_i(P_i, \mathbf{P}_{-i}) = (1 - \beta)R_i(P_i, \mathbf{P}_{-i}) - c \cdot P_i,$$ (10.7)

which depends on $S^{(k)}$ and β but is independent of α. We can conclude that the NE $\hat{\mathbf{P}}$ only depends on $S^{(k)}$ and β and so does the rate R_{SP} in Equation 10.4: this is made clear by using the notation $\hat{\mathbf{P}}(\beta, S^{(k)})$ for NE and $R_{SP}(\beta, S^{(k)})$ for Equation 10.4.

10.3.3.3 Analysis

In this section, we provide some insight into the performance of cooperation-based spectrum leasing under the assumption of instantaneous CSI through analysis of the Stackelberg game at hand. In particular, we are interested in determining conditions under which it is advantageous for the primary to lease the owned bandwidth for a fraction of time $\alpha > 0$ to the secondary ad hoc network. We consider at first the noncooperative power control game discussed in the previous section. It is well known that NE is a fixed point of the best responses of all the nodes in $S^{(k)}$ [36]. The best response of each user is obtained by writing the Karush–Kuhn–Tucker (KKT) conditions corresponding to the (convex) problem of maximizing the utility Equation 10.7 under the power constraint $0 \leq P_i \leq P_{\max}$.

PROPOSITION 1

The power control game $\langle S^{(k)}, \mathcal{P}, u_i(P_i, \mathbf{P}_{-i})\rangle$ has always an NE. Moreover, a necessary condition for the NE to be unique is

$$\sum_{j=1, j\neq i}^{k} \frac{|h_{S,ij}|^2}{|h_{S,ii}|^2} < 1.$$ (10.8)

Proof: It amounts to proving that the system of KKT discussed here has a unique solution; see Ref. [37] or Ref. [26].

The condition Equation 10.8 for uniqueness of the NE is intuitive because it simply imposes an upper bound on the interference: in fact, with negligible interference all the secondary links ST_i-SR_i experience no mutual interference, and the problem of optimizing the utilities Equation 10.7 decouples into k independent problems that can be easily shown to present a unique solution (for strict concavity of the utility).

We now turn to the performance of the overall Stackelberg game.

PROPOSITION 2

The optimal fraction of time $\hat{\alpha}$ leased for secondary transmission and cooperation is strictly positive if and only if there exists a subset of secondary nodes $S^{(k)} \subseteq S$ such that the following condition is satisfied

$$\frac{\hat{\beta} R_{\mathrm{SP}}(\hat{\beta}, \mathcal{S}^{(k)}) \cdot R_{\mathrm{PS}}(\mathcal{S}^{(k)})}{\hat{\beta} R_{\mathrm{SP}}(\hat{\beta}, \mathcal{S}^{(k)}) + R_{\mathrm{PS}}(\mathcal{S}^{(k)})} > R_{\mathrm{dir}}, \tag{10.9}$$

where the optimal $\hat{\beta}$ is the solution of the optimization problem

$$\hat{\beta} = \arg \max_{\beta \in [0,1]} \beta R_{\mathrm{SP}}\left(\beta, \mathcal{S}^{(k)}\right). \tag{10.10}$$

Moreover, if Equation 10.9 is verified (i.e., cooperation is employed), the optimal fraction $\hat{\alpha}$, for a given set $\mathcal{S}^{(k)}$, is given by

$$\hat{\alpha} = \frac{1}{1 + \frac{\hat{\beta} R_{\mathrm{SP}}\left(\hat{\beta}, \mathcal{S}^{(k)}\right)}{R_{\mathrm{PS}}\left(\mathcal{S}^{(k)}\right)}}. \tag{10.11}$$

Proof: See Appendix.

Equation 10.9 provided by Proposition 2 simply states that cooperation is beneficial, and thus $\hat{\alpha} > 0$, if there exists a subset of secondary transmitters such that the cooperative primary rate Equation 10.2 is larger than the direct rate R_{dir}. In fact, the left-hand side in Equation 10.9 is the cooperative primary rate Equation 10.2 with optimized $\hat{\alpha}$ Equation 10.11 (see Appendix for details). Furthermore, from Equation 10.11, it is interesting to notice that the optimal $\hat{\alpha}$ in case cooperation is selected ($\hat{\alpha} > 0$) is chosen to avoid performance bottlenecks: it decreases (more time to primary transmission) if the channels from primary to secondary transmitters are the limiting factor (i.e., for increasing $\hat{\beta} R_{\mathrm{SP}}(\hat{\beta}, k)/R_{\mathrm{PS}}(\mathcal{S}^{(k)})$), while it increases (more time to secondary transmission), if the cooperative rate limits the overall performance (i.e., for decreasing $\hat{\beta} R_{\mathrm{SP}}(\hat{\beta}, k)/R_{\mathrm{PS}}(\mathcal{S}^{(k)})$).

10.3.3.4 Numerical Examples

In this section, we consider a simple geometrical model where the set of secondary nodes is placed at approximately the same normalized distance $0 < d < 1$ from the primary transmitter PT and $1 - d$ from the primary receiver PR ($0 < d < 1$). Consequently, considering a path loss model, the average power of the channels read: $g_{\mathrm{P}} = 1$, $g_{\mathrm{PS},i} = 1/d^{\gamma}$, and $g_{\mathrm{SP},i} = 1/(1-d)^{\gamma}$, where $\gamma = 2$ is the path loss coefficient. Moreover, to further reduce the number of system parameters and get better insight into the overall performance, we set $g_{\mathrm{S},ij} = \tilde{g}_{\mathrm{S}} = 1$ and $g_{\mathrm{S},ii} = g_{\mathrm{S}}$ for $i, j = 1, ..., K$ and $i \neq j$. The primary power and the maximum secondary transmission power are $P_{\mathrm{P}} = P_{\mathrm{max}} = 1$, the cost per unit energy is $c = 0.1$, $N_0 = 1$ and, unless explicitly stated otherwise, the number of secondary transmitters is $K = 5$ and $g_{\mathrm{S}} = 20\mathrm{dB}$.

Figure 10.7 shows the achievable rate R_{P} Equation 10.2, averaged over different fading realizations, versus the normalized distance d, for α ranging from $1/8$ to $7/8$, with $\beta = 0.8$ and optimized $\mathcal{S}^{(k)}$. The rate on the direct link between PT and PR Equation 10.1, R_{dir}, is also shown as a reference. The figure reveals that spectrum leasing can significantly improve the system performance, if the parameters are chosen properly. Furthermore, as the distance d increases, activation of secondary nodes becomes demanding and it is optimal to choose a smaller value for the leased time α (recall Equation 10.11 and related discussion).

To get insight into the optimal behavior of the secondary network $\mathcal{S}$, and in particular on the optimal number of activated nodes k in $\mathcal{S}^{(k)}$, Figure 10.8 shows the primary rate versus distance d with optimized parameters $\hat{\alpha}$ and $\hat{\beta}$ (recall Proposition 2) and optimized set $\widehat{\mathcal{S}}^{(k)}$ with a constraint on $k = 1, ..., K = 5$. In other words, for each realization of fading channels, the primary rate is optimized by choosing the best subset $\widehat{\mathcal{S}}^{(k)}$ with a fixed number k of secondary users. The result of the optimization Equation 10.5 without constraining the number of secondary users is also shown.

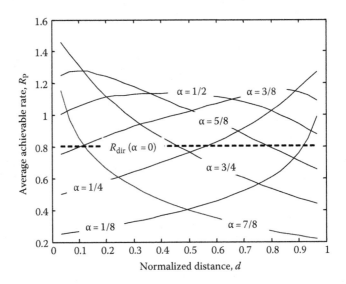

FIGURE 10.7 Primary rate $R_P(\alpha, \beta, \widehat{S}^{(k)})$ (Equation 10.2), averaged over fading, versus the normalized distance d between the primary transmitters and primary receivers, for α ranging from 1/8 to 7/8, and $\beta = 0.8$. Dashed line refers to the rate achievable through direct transmission R_{dir} ($P_P = P_{\max} = 1$, $SNR = 0$ dB, $K = 5$, and $g_S = 20$ dB).

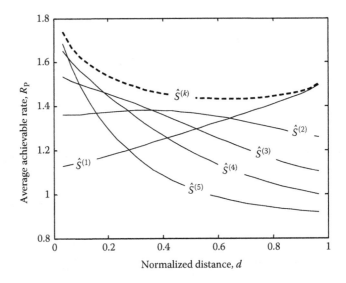

FIGURE 10.8 Primary rate $R_P(\hat{\alpha}, \hat{\beta}, \widehat{S}^{(k)})$ (Equation 10.2), averaged over fading, versus the normalized distance d, for optimal sets of secondary users $\widehat{S}^{(k)}$: $\widehat{S}^{(1)}, ..., \widehat{S}^{(5)}$ refer to the optimal sets with a constraint on the cardinality, while $\widehat{S}^{(k)}$ in the unconstrained optimal set ($P_P = P_{\max} = 1$, $SNR = 0$ dB, $K = 5$, and $g_S = 20$ dB).

It can be seen that for small distances it is better to activate (and thus cooperate with) a large number of secondary users given the large channel power gain from source to secondary network. Conversely, for large distances it is more convenient to cooperate only with the secondary users with the best instantaneous channel $|h_{PS,i}|^2$, exploiting multiuser diversity.

The average rate achieved by the (activated) secondary user, $\alpha(1 - \beta)R_i$, where R_i is given by Equation 10.6, is shown in Figure 10.9 as a function of the normalized distance d and the channel gain g_S. Because a larger distance entails a smaller optimal leased time α (see Figure 10.7), the

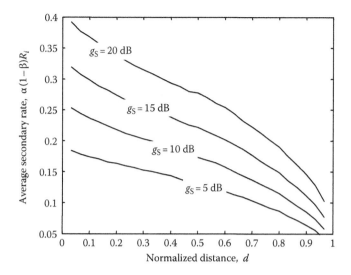

FIGURE 10.9   Secondary rate $\alpha(1-\beta)R_i$ (Equation 10.6), averaged over fading, versus the normalized distance d, in a symmetric scenario with different channel power gains g_S between secondary transmitter–receiver pairs ($g_S = 5, 10, 15, 20\,\text{dB}$, $\tilde{g}_S = 0\,\text{dB}$, $P_P = P_{max} = 1$, $SNR = 0\,\text{dB}$, and $K = 5$).

secondary rate decreases as the distance increases. Moreover, with increasing channel gain power g_S between secondary transmitter–receiver pairs, as expected, the secondary nodes are able to achieve larger rates.

10.3.4   CONCLUDING REMARKS

In this section, we have proposed and investigated a solution for the implementation of spectrum leasing in a CR scenario that complies with the property-rights model. The solution is based on the idea that secondary nodes can earn spectrum access in exchange for cooperation with the primary link. The basic premise of the proposal is that spectrum leasing may take place not necessarily on the basis of fees or charges, but also in return for an improved QoS of the incumbent primary users through cooperation with secondary terminals. By casting the problem in the framework of Stackelberg games, we have provided analytical and numerical results that have confirmed the considered model as a promising paradigm for CR networks.

10.4   CONCLUSIONS AND OPEN ISSUES

The field of CR can be still considered to be in the initial phase of development where basic definitions are subject to dispute by different communities (information theory, networking, artificial intelligence, etc.) and schools of thought. In contrast, cooperative communications is by now a well-established research area, at least from the standpoint of theoretical tools and understanding. The scope of this chapter has been a discussion on the opportunities that cooperative technology is envisioned to be able to provide in CR applications, defined according to different models. Given the current state of research in the CR field, the discussion is meant to be a preliminary effort to assess whether cooperation is a valid enabling technology for implementation of CR. Results have been derived based on the analysis of simplified communications scenarios leading to very promising conclusions for both commons and property-rights models of CR. Extending and validating these preliminary conclusions in the presence of more realistic scenarios and, possibly, of clearer directions in the future of the CR, are clearly essential steps to the actual deployment of the ideas presented in this chapter. On this note, and to conclude, here we would like to mention two particular issues that, in our opinion, deserve special attention:

- Solve the limitations of sensing in the commons model through cooperation: as discussed in Section 10.2, the major impediment to the implementation of the commons model of CR is the inherent limitation in the capability of an unlicensed node to sense the radio environment to avoid interference to the primary users. Specific issues are the well-known hidden terminal problems and the problem of detecting primary receivers. Cooperation is regarded as the key technology that might solve this problem but further research in the area is required (recall Section 10.2.2.1).
- Account for synchronization inaccuracies: the advantages of cooperative transmission are conditioned on the capabilities of radio terminals to synchronize their transmission (at the least, at the block level, but symbol synchronization is often assumed). This task is expected to be even more challenging in a CR scenario given the interference constraints on the activity of unlicensed terminals. Devising technical solutions to this problem and assessing the impact of inaccurate synchronization on the performance of cooperation for CR are thus an essential research goal.

APPENDIX

PROOF OF PROPOSITION 2

Recalling that the NE does not depend on α but only on $\mathcal{S}^{(k)}$ and β, the primary rate Equation 10.2 can be conveniently restated as $R_\mathrm{P}(\alpha, \beta, \mathcal{S}^{(k)}) = \min\{(1 - \alpha) R_\mathrm{PS}(\mathcal{S}^{(k)}), \alpha\beta R_\mathrm{SP}(\beta, \mathcal{S}^{(k)})\}$ for $\alpha > 0$, and $R_\mathrm{P}(0, \beta, \mathcal{S}^{(k)}) = R_\mathrm{dir}$ for $\alpha = 0$. Parameter β appears only in the second term of Equation 10.2 and therefore can be optimized independently by solving problem (Equation 10.10) for a given choice of $\mathcal{S}^{(k)}$. Moreover, it is easy to observe that $R_\mathrm{P}(\alpha, \beta, \mathcal{S}^{(k)})$ is the minimum of a decreasing function of α, $(1 - \alpha) R_\mathrm{PS}(\mathcal{S}^{(k)})$ and an increasing function of α, $\alpha\beta R_\mathrm{SP}(\beta, \mathcal{S}^{(k)})$, and therefore maximization is achieved when the two terms are equal: $(1 - \hat{\alpha}) R_\mathrm{PS}(\mathcal{S}^{(k)}) = \hat{\alpha}\hat{\beta} R_\mathrm{SP}(\hat{\beta}, \mathcal{S}^{(k)})$. This condition is equivalent to Equation 10.11. Finally, plugging the optimal $\hat{\alpha}$ (Equation 10.11) in $R_\mathrm{P}(\alpha, \beta, \mathcal{S}^{(k)})$, we get the result that the maximum primary rate achievable with cooperation reads

$$R_\mathrm{P}(\hat{\alpha}, \hat{\beta}, \mathcal{S}^{(k)}) = \frac{\hat{\beta} R_\mathrm{SP}(\hat{\beta}, \mathcal{S}^{(k)}) \cdot R_\mathrm{PS}(\mathcal{S}^{(k)})}{\hat{\beta} R_\mathrm{SP}(\hat{\beta}, \mathcal{S}^{(k)}) + R_\mathrm{PS}(\mathcal{S}^{(k)})}, \tag{A.1}$$

from which Equation 10.9 easily follows.

ACKNOWLEDGMENTS

The authors would like to gratefully acknowledge the very useful discussions and experimental validations by Jonathan Gambini for Section 10.2 and Igor Stanojev for Section 10.3.

REFERENCES

[1] J. Mitola, III, Cognitive radio for flexible multimedia communications, in *Proceedings of the IEEE International Workshop on Mobile Multimedia Communications (MoMuC '99)*, pp. 3–10, 1999.
[2] Federal Communications Commission, Spectrum Policy Task Force, Rep. ET Docket no. 02-135, Nov. 2002.
[3] J. O. Neel, *Analysis and Design of Cognitive Radio Networks and Distributed Radio Resource Management Algorithms*, PhD dissertation, Virginia Polytechnic Institute, Sept. 2006.
[4] S. Haykin, Cognitive radio: Brain-empowered wireless communications, *IEEE Journal on Selected Areas in Communications*, 23(2), 201–220, Feb. 2005.

[5] N. Devroye, P. Mitran, and V. Tarokh, Limits on communications in a cognitive radio channel, *IEEE Communications Magazine*, 44(6), 44–49, June 2006.

[6] Q. Zhao and B. M. Sadler, A survey of dynamic spectrum access, *IEEE Signal Processing Magazine*, 24(3), 79–89, May 2007.

[7] J. M. Peha, Approaches to spectrum sharing, *IEEE Communications Magazine*, 43(2), 10–12, Feb. 2005.

[8] G. Faulhaber and D. Farber, Spectrum management: Property rights, markets and the commons, in *Proceedings of the Telecommunications Policy Research Conference*, Oct. 2003.

[9] M. Basseville and I. Nikiforov, *Detection of Abrupt Changes: Theory and Application*, Prentice Hall, Englewood Cliffs, NJ, 1993.

[10] C. Cordeiro, K. Challapali, D. Birru, and N. Sai Shankar, IEEE 802.22: The first worldwide wireless standard based on CR, in *Proceedings of IEEE DySPAN*, pp. 328–337, 2005.

[11] http://www.ieee802.org/22/.

[12] http://www.ieee802.org/21/.

[13] Cognitive Radio, Spectrum and Radio Resource Management, Wireless World Research Forum, Working Group 6 White Paper, Dec. 2004.

[14] C. Tawil and S. Tawil, Apparatus and method for reusing satellite broadcast spectrum for terrestrially broadcast signals, Patent number: 6519446, Assignee: Northpoint Technology, Ltd, Issue date: Feb. 11, 2003.

[15] G. Kramer, I. Maric, and R. Yates, *Cooperative Communications,* Now Publishers, Inc., Hanover, MA, 2007.

[16] Y. Chen, Q. Zhao, and A. Swami, Joint design and separation principle for opportunistic spectrum access, in *Proceedings of Asilomar Conference on Signals, Systems and Computers*, 2006.

[17] A. Jovicic and P. Viswanath, Cognitive radio: An information-theoretic perspective, submitted [arXiv:cs/0604107v2].

[18] B. Wild and K. Ramchandran, Detecting primary receivers for CR applications, in *Proceedings of IEEE DySPAN*, Nov. 2005.

[19] A. Sahai, N. Hoven, and R. Tandra, Some fundamental limits on cognitive radio, in *Proceedings of Allerton Conference on Communication, Control, and Computing*, Oct. 2004.

[20] I. Maric, A. Goldsmith, G. Kramer, and S. Shamai (Shitz), On the capacity of interference channels with a partially-cognitive transmitter, in *Proceedings of IEEE International Symposium on Information Theory*, Nice, France, June 2007.

[21] P. Cheng, G. Yu, Z. Zhang, H. -H. Chen, and P. Qiu, On the achievable rate region of Gaussian cognitive multiple access channel, *IEEE Communications Letters*, 11(5), 384–386, May 2007.

[22] P. Grover and A. Sahai, What is needed to exploit knowledge of primary transmissions, in *Proceedings of IEEE DySpAN*, 2007.

[23] M. Gastpar, On capacity under receive and spatial spectrum-sharing constraints, *IEEE Transactions on Information Theory*, 53(2), 471–487, Feb. 2007.

[24] R. Etkin, A. Parekh, and D. Tse, Spectrum sharing for unlicensed bands, in *Proceedings of IEEE DySPAN*, 2005.

[25] J. Neel, J. Reed, and R. Gilles, The role of game theory in the analysis of software radio networks, in *Proceedings of SDR Forum Technical Conference*, 2002.

[26] A. Nosratinia, T. E. Hunter, and A. Hedayat, Cooperative communication in wireless network, *IEEE Communications Magazine*, 42(10), 68–73, Oct. 2004.

[27] A. Ghasemi and E. S. Sousa, Collaborative spectrum sensing for opportunistic access in fading environments, in *Proceedings of IEEE DySPAN*, Nov. 2005.

[28] G. Ganesan and Y. Li, Cooperative spectrum sensing in cognitive radio, Part I: Two user network, *IEEE Transactions on Wireless Communications*, 6(6), 2204–2213, June 2007.

[29] T. Betran, O. Simeone, and Y. Bar-Ness, Detecting primary transmitters via cooperation and memory in cognitive radio, in *Proceedings of the Conference on Information Sciences and Systems (CISS)*, 2007.

[30] O. Simeone, Y. Bar-Ness, and U. Spagnolini, Stable throughput of cognitive radios with and without relaying capabilities, *IEEE Transactions in Communications*, 55(12), 2351–2360, Dec. 2007.

[31] S. Bohacek, K. Shah, G. R. Arce, and M. Davis, Signal processing challenges in active queue management, *IEEE Signal Processing Magazine*, 21(5), 69–79, Sept. 2004.

[32] K. Eswaran, M. Gastpar, and K. Ramchandran, Bits through ARQs: Spectrum sharing with a primary packet system, *IEEE International Symposium on Information Theory (ISIT 2007)*, Nice, France, June 25–29, 2007.

[33] F. F. Digham, M. -S. Alouini, and M. K. Simon, On the energy detection of unknown signals over fading channels, in *Proceedings of the IEEE International Conference Communications (ICC)*, pp. 3575–3579, 2003.

[34] J. Gambini, O. Simeone, U. Spagnolini, and Y. Bar-Ness, Cooperative cognitive radios with optimal primary detection and packet acceptance control, in *Proceedings of the IEEE Signal Processing Advances in Wireless Communications (SPAWC), 2007.*

[35] J. N. Laneman and G. W. Wornell, Distributed space-time coded protocols for exploiting cooperative diversity in wireless networks, *IEEE Transactions on Information Theory*, 49(10), 2415–2425, Oct. 2003.

[36] M. J. Osborne and A. Rubenstein, *A Course in Game Theory*, MIT Press, 1994.

[37] G. Scutari, D. P. Palomar, and S. Barbarossa, Optimal linear precoding strategies for wideband noncooperative systems based on game theory—Part I: Nash equilibria, *IEEE Transactions on Signal Processing*, 56(3), 1230–1249, Mar. 2008.

[38] R. Etkin, A. Parekh, and D. Tse, Spectrum sharing for unlicensed bands, in *Proceedings of Allerton Conference on Communication, Control and Computing, 2005.*

[39] O. Ileri, S. C. Mau, and N. B. Mandayam, Pricing for enabling forwarding in self-configuring ad hoc networks, *IEEE Journal on selected Areas in Communications*, 23(1), 151–162, Jan. 2005.

[40] I. Stanojev, O. Simeone, and Y. Bar-Ness, Optimal design of a multi-antenna access point with decentralized power control using game theory, in *Proceedings of IEEE DySPAN*, 2007.

[41] O. Simeone, I. Stanojev, S. Savazzi, Y. Bar-Ness, U. Spagnolini, and R. Pickholtz, Spectrum leasing to cooperating ad hoc secondary networks, *IEEE Journal on Selected Areas in Communications*, 26(1), 203–213, Jan. 2008.

[42] A. Høst-Madsen and J. Zhang, Capacity bounds and power allocation for wireless relay channel, *IEEE Transactions on Information Theory*, 51(6), 2020–2040, June 2005.

[43] J. Huang, R. A. Berry, and M. L. Honig, Distributed interference compensation for wireless networks, *IEEE Journal on Selected Areas in Communications*, 24(5), 1074–1084, May 2006.

Part II

Techniques

11 Cooperative Diversity of Generalized Distributed Antenna Systems

Yifan Chen, Luoquan Hu, Chau Yuen, Yan Zhang, Zhenrong Zhang, and Predrag Rapajic

CONTENTS

Cooperative diversity in both cellular and ad hoc networks has been the subject of great research interest in the past years. This method makes use of available mobile terminals as relays that cooperate together to form a virtual antenna array. In this chapter, we consider an intuitive measure of diversity gains achievable in such networks, which is defined as the number of independent fading channels that can be averaged over to detect symbols. First, geometry-based channel models are proposed to describe the topology of a generalized distributed antenna system with cooperative users (GDAS-CU). The system architecture is comprised of M largely separated access points (APs) at one side of the link, and N geographically closed user terminals (UTs) at the other side. The UTs are operating in cooperative mode to enhance the system performance, where an idealized message sharing among the UTs is assumed. The mean cross-correlation coefficients (MCCCs) of signals received from noncollocated APs and UTs is calculated based on the network topology and the correlation models derived from the empirical data. The analysis is also extendable to more general scenarios where the APs are placed in a clustered form. Subsequently, a generalized pathloss model based on stochastic ray tracing (SRT) theory is proposed where propagation environments are suitable to be described by stochastic structures rather than being deterministically characterized. The derived model consists of two dominant terms: a logarithmic function and a power function of the distance between the transmitter and the receiver. It is shown that the proposed pathloss law which is generalized from several physically inspired SRT models, exhibits better agreement with the experimental data as compared to other existing models. Armed with the cross-correlation and pathloss model preliminaries, we investigate the diversity gains obtainable from a GDAS-CU network, which would provide critical insight into the degree of possible performance improvement when combining multiple copies of the received signal in such systems.

11.1   INTRODUCTION

Recently, there has been an explosion of interest in cooperative wireless communications due to the fact that these mechanisms could substantially enhance system performance with respect to much less power consumption, higher system capacity, smaller packet loss rate, and higher network resilience. The concept of cooperation can be applied in a variety of wireless systems and networks. From the physical point of view, cooperative communication explores the benefits of a multiuser environment by creating a virtual multiple-input-multiple-output (MIMO) system. One of the feasible cooperative schemes is the generalized distributed antenna system, which refers to a virtual MIMO comprising multiple antennas collocated at the user terminal (UT) and several geographically scattered access points (APs), each with multiple antennas collocated within an AP at the other side [1,2]. Such a system has the advantage of macrodiversity that is inherent to the widely spaced antenna, and therefore, offers the capability to enhance signal quality and improve coverage. A sectorized architecture is employed where each AP has a separate feeder to a central base station (BS), which ensures that various benefits of a MIMO system such as diversity combining and interference reduction can be realized through cooperation among APs [3,4]. In this chapter, we consider a more general scenario where multiple spatially scattered UTs may cooperate with one another to transmit or receive signals from APs, which will be referred to as the generalized distributed antenna system with cooperative users (GDAS-CU) [5,6]. It makes use of available UTs as relays to work together to form a virtual antenna array. It has been shown that full diversity can be achieved by both amplify-and-forward and decode-and-forward techniques in the relays [7].

As the system diversity is dependent on fading correlation of the multipaths in the GDAS-CU, it is thus important to investigate the cross-correlation between two links from two APs at one side of the link to two UTs at the other side. There is currently no well-established model for the cross-correlation coefficient. Two key parameters have been considered in the past studies: (1) the angle-of-arrival difference (AAD) and (2) the relative distance difference (RDD) between two AP–UT paths. In most cases, the AAD has the strongest influence on the measured cross-correlation [8,9]. However, models considering both AAD and RDD also exist [10,11]. With geometry-based channel models, network-level synthesization of GDAS-CU could be facilitated by generating location-dependent correlation values derived from the empirical data in Refs. [8,11]. The theoretical result of the average correlation can be derived, which roughly reflects the overall degree of correlation and provides critical insight into the quality of fading wireless channels.

The diversity level is also related to the received signal strength. Conventionally, pathloss is simply modeled as a linear function of $\log_{10} r$, where r is the distance of separation between the transmitter and the receiver [12]. Nevertheless, this simplified power decaying model lacks physical reasoning and does not always yield accurate curve fitting of measurement results. In recent years, an alternative approach based on stochastic ray tracing (SRT) is used to systematically study the wireless propagation channels, which overcomes the aforementioned limitation by providing sound physical understanding to the underlying propagation phenomenon [13–21]. The radio channels are described as random media, such as percolation lattices and the electromagnetic waves are considered as stochastic rays. An appealing observation made from these SRT models is that an additional pathloss factor (apart from the logarithmic term) appears in all the models. We take into account this phenomenon and propose a generalized power decaying model that is composed of two dominant items: $10n \log_{10} r$ and αr^{δ}, where n, α, and δ are model parameters depending on the scattering conditions [14]. It is shown that this model provides a better fit to the measured data than other existing models. Therefore, it would be more relevant to apply the revisited signal attenuation formula for the analysis of cooperative diversity in GDAS-CU networks.

Armed with the model preliminaries, a physically inspired metric is employed to quantify the diversity gains of a GDAS-CU, which is defined as the number of independent fading paths that can be averaged over to detect symbols. From information-theory point of view, it also corresponds to the signal-to-noise ratio (SNR) exponent of the error probability at high SNR [22,23]. The cardinality of the overall "diversity-paths" set offered by the GDAS-CU is calculated. As this measure provides the largest number of redundant copies of transmitted information, it can be used as a benchmark to access the performances bound of various diversity-based cooperative schemes that have been extensively pursued in recent years including relay protocols and coordination, cooperative space-time transmit diversity, spatial multiplexing, and so on [7,24–31].

The chapter is organized as follows. In Section 11.2, we briefly discuss the topology of GDAS-CU networks, where both single- and multicell architectures are introduced. In Section 11.3, the mean cross-correlation coefficients (MCCCs) in such systems are derived based on either AAD-only or AAD/RDD-dependent models. Subsequently, a comprehensive discussion on power decaying patterns in wireless networks is presented in Section 11.4. In Section 11.5, the analysis of cooperative diversity gains defined as the number of independent fading paths is provided. We also introduce a new channel model parameter called generalized cross-correlation coefficient (GCCC) in this section. Numerical examples are given in Section 11.6, where the characteristics of the MCCCs, mean received power, and GCCC are investigated. Subsequently, the influence of the mean received power and the GCCC on multipath diversity is studied. Finally, some open issues and concluding remarks are drawn in Section 11.7.

List of Abbreviations

MIMO	multiple-input-multiple-output
UT	user terminal
AP	access point
BS	base station
GDAS-CU	generalized distributed antenna system with cooperative users
AAD	angle-of-arrival difference
RDD	relative distance difference
SRT	stochastic ray tracing
SNR	signal-to-noise ratio
MCCC	mean cross-correlation coefficient
GCCC	generalized cross-correlation coefficient
pdf	probability density function
AOA	angle-of-arrival

2D	two-dimensional
RMS	root-mean-square
DIV	diversity
AOD	angle-of-departure

11.2 GDAS-CU SYSTEM TOPOLOGY

11.2.1 SINGLE-CELL NETWORK

The simplest system topology is a single-cell network where the APs are placed in a single service region such as an outdoor street canyon or an indoor conference room. Figure 11.1a depicts the geometrical structure of a single-cell GDAS-CU. A number of distributed APs spaced apart by a large distance are placed in the deployment cell. Each AP itself is an antenna array and is connected to the central BS through optical fibers, coaxial cable, or radio link [1,2]. To synthesize the randomness in the AP placement owing to the complex landform of real environments, it is assumed that these APs are uniformly distributed in a circular cell with radius R. This topology model is similar to the random antenna layout in Ref. [32]. However, the UTs may not be located at the center of the circular region, which is different from the structure in Ref. [32]. At the UT site, it is supposed that the message knowledge can be ideally shared among the devices if they are geographically close to each other [33,34]. Eventually, they can fully cooperate with one another during the process of transmission and reception.

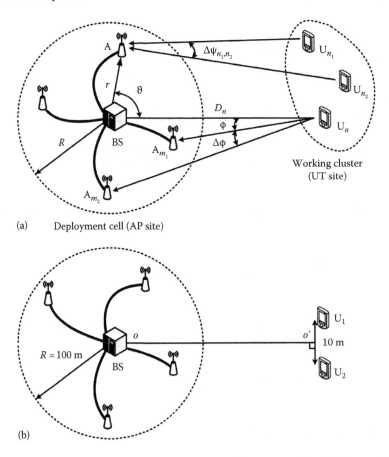

FIGURE 11.1 (a) Pictorial illustration of a single-cell GDAS-CU network and (b) a possible single-cell network architecture in urban environment.

In other words, we are considering a wireless network with M transmitting UTs denoted $\mathcal{S}_U = \{1, 2, \ldots, M\}$. A source UT, $m \in \mathcal{S}_U$, has information to transmit to the central BS through the N APs, potentially using terminals $\mathcal{S}_U - \{m\}$ as relays to perform cooperative diversity. It is assumed that there is no information loss between the source terminal and the relays and between the central BS and the APs. In the following section, the antenna layout of the UTs will be analyzed using a deterministic approach as the number of cooperative UTs is usually small. A two-UT network in a typical outdoor microcellular environment is illustrated in Figure 11.1b.

11.2.2　Multicell Network

In real-life deployment, APs could be distributed in multiple cells to serve the region of interest. For example, the candidate locations of APs may be constrained by the street layout or building structure, which causes clustered placement of APs. To assess the effect of multiple deployment cells on the average cross-correlation, a simplified hypothetical architecture is considered in this chapter, as shown in Figure 11.2a. The APs are homogeneously placed in a number of circular clusters, and each AP is connected to a central BS through a separate feeder. Figure 11.2b illustrates some possible network topologies, where the vertices of each regular polygon (equilateral triangle, square, regular pentagon, etc.) correspond to the centers of circular cells. It is worth noting that the geometries in Figures 11.1 and 11.2 are not meant to be an exact representation of the actual AP placement. Rather, they capture the main features of network structure through a set of condensed geometrical parameters, made clear in the following discussions.

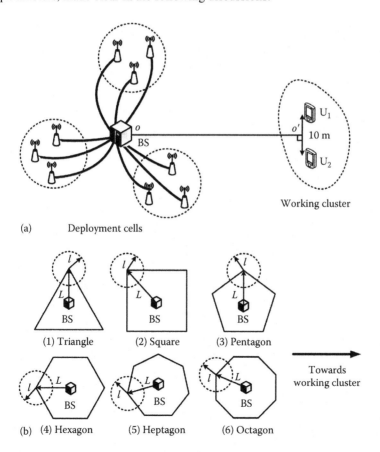

FIGURE 11.2 (a) Pictorial illustration of a multicell GDAS-CU network and (b) some possible multicell network architectures.

11.3   MEAN CROSS-CORRELATION IN GDAS-CU NETWORK

11.3.1   CORRELATION DEPENDING ON AAD ONLY

Consider the system geometry pictured in Figure 11.1a. First of all, we want to determine the average value of the cross-correlation coefficient at APs. Let us define the spatial locations of APs as $\mathbf{A}_1, \mathbf{A}_2, \ldots, \mathbf{A}_M \in \mathcal{C}_A$ and the locations of UTs as $\mathbf{U}_1, \mathbf{U}_2, \ldots, \mathbf{U}_N \in \mathcal{C}_U$, where $\mathcal{C}_A$ and $\mathcal{C}_U$ represent the deployment cell and the working cluster, respectively, as shown in Figure 11.1a. The first step is to derive the probability density function (pdf) of the AAD between two paths from one UT, $\mathbf{U}_n$ belonging to $\mathcal{C}_U$, to any two APs, $\mathbf{A}_{m_1}$ and $\mathbf{A}_{m_2}$ belonging to $\mathcal{C}_A$.

The piecewise continuous model proposed by Sørensen is based on the 900 MHz measurements conducted in Aarhus, Denmark [8]. A vehicle with a roof-mounted omnidirectional antenna provided simultaneous measurements of received signals from three APs located in the downtown area. The measurement site was characterized by almost uniform building height of about 4 to 5 storeys and a gently rolling terrain. A number of measurement routes with an average length of 400 m were defined by driving around a single or a group of apartment blocks. The independent nonparametric data bootstrap's procedure has been applied to estimate a confidence interval for the correlation coefficient. The shape of cross-correlation against AAD is summarized as [8]

$$\rho(\Delta\phi) = \begin{cases} 0.78 - 0.0056|\Delta\phi|, & \text{if } 0° \leq |\Delta\phi| \leq 15° \\ 0.48 - 0.0056|\Delta\phi|, & \text{if } 15° \leq |\Delta\phi| \leq 60° \\ 0, & \text{if } |\Delta\phi| \geq 60° \end{cases} \tag{11.1}$$

where $\Delta\phi$ is the AAD. This model describes the main characteristics that are likely to be observed in urban areas. Note that though the actual ρ against $\Delta\phi$ may change for a different environment, the general methodology presented here is still applicable.

As the APs are uniformly distributed within a circle, the pdf of AOA viewed from the nth UT ($n = 1, 2, \ldots, N$) when the UT locates outside the deployment cell is given by [35]:

$$f_{\phi,n}(\phi) = \begin{cases} \dfrac{2D_n \cos\phi \sqrt{D_n^2 \cos^2\phi - D_n^2 + R^2}}{\pi R^2}, & -\sin^{-1}\left(\dfrac{R}{D_n}\right) \leq \phi \leq \sin^{-1}\left(\dfrac{R}{D_n}\right) \\ 0, & \text{otherwise} \end{cases} \tag{11.2}$$

where D_n is the distance of separation between the central BS and the nth UT as depicted in Figure 11.1a. When the UT lies within the cell, $(D_n/R) < 1$. The pdf of AOA is (the derivation is given in Appendix)

$$f_{\phi,n}(\phi) = \frac{R^2 + D_n^2 + 2RD_n \cos\left(\sin^{-1}\frac{D_n \sin\phi}{R} + \phi\right)}{2\pi R^2}, \quad 0 \leq \phi \leq 2\pi \tag{11.3}$$

Subsequently, the pdf of AAD is obtained as [36]

$$f_{\Delta\phi,n}(\Delta\phi) = \begin{cases} \displaystyle\int_0^\infty f_{\phi,n}(\Delta\phi + \phi) f_{\phi,n}(\phi) d\phi, & \Delta\phi \geq 0 \\ \displaystyle\int_{-\Delta\phi}^\infty f_{\phi,n}(\Delta\phi + \phi) f_{\phi,n}(\phi) d\phi, & \Delta\phi < 0 \end{cases} \tag{11.4}$$

and the MCCC at any two APs in the deployment cell is

$$\rho_A = \mathbb{E}(\rho) = \frac{1}{N} \sum_{n=1}^{N} \int_{\Delta\phi} \rho(\Delta\phi) f_{\Delta\phi,n}(\Delta\phi) d(\Delta\phi) \tag{11.5}$$

where the subscript A denotes that the correlation is observed at the AP site and $\mathbb{E}$ represents the expectation operator.

The next step is to derive the average cross-correlation among UTs. The AAD between two propagation paths from any AP, $\mathbf{A} \in \mathcal{C}_A$, to two UTs $\mathbf{U}_{n_1}$ and $\mathbf{U}_{n_2}$ can be obtained from the elementary geometry shown in Figure 11.1a

$$\Delta \psi_{n_1,n_2}(\mathbf{A}) = \cos^{-1} \left(\frac{\left| \mathbf{U}_{n_1} \mathbf{A} \right|^2 + \left| \mathbf{U}_{n_2} \mathbf{A} \right|^2 - \left| \mathbf{U}_{n_1} \mathbf{U}_{n_2} \right|^2}{2 \left| \mathbf{U}_{n_1} \mathbf{A} \right| \left| \mathbf{U}_{n_2} \mathbf{A} \right|} \right) \tag{11.6}$$

The MCCC at any two UTs is given by

$$\rho_U = \int_{\mathcal{C}_A} \rho(\Delta \psi_{n_1,n_2}(\mathbf{A})) f_\mathbf{A}(\mathbf{A}) d\mathbf{A} \tag{11.7}$$

where the subscript U denotes that ρ is observed at the UT site. $f_\mathbf{A}(\mathbf{A})$ is the distribution function of APs, which can be more conveniently expressed in polar coordinates (r, ϑ). Equation 11.7 can thus be reformulated as

$$\rho_U = \int_0^R \int_0^{2\pi} \rho \left(\Delta \psi_{n_1,n_2}(r, \vartheta) \right) \frac{r}{\pi R^2} d\vartheta dr$$

$$= \int_0^R \int_0^{2\pi} \rho \left(\cos^{-1} \frac{2r^2 - 2rr_{n_1} \cos \left(\vartheta - \vartheta_{n_1} \right) - 2rr_{n_2} \cos \left(\vartheta - \vartheta_{n_2} \right) + 2r_{n_1} r_{n_2} \cos \left(\vartheta_{n_1} - \vartheta_{n_2} \right)}{2 \sqrt{r^2 + r_{n_1}^2 - 2rr_{n_1} \cos \left(\vartheta - \vartheta_{n_1} \right)} \sqrt{r^2 + r_{n_2}^2 - 2rr_{n_2} \cos \left(\vartheta - \vartheta_{n_2} \right)}} \right)$$

$$\times \frac{r}{\pi R^2} d\vartheta dr \tag{11.8}$$

where $(r_{n_1}, \vartheta_{n_1})$ and $(r_{n_2}, \vartheta_{n_2})$ are the polar coordinates of $\mathbf{U}_{n_1}$ and $\mathbf{U}_{n_2}$, respectively.

In the special case where the channel correlations observed at the AP and UT sites are decoupled (i.e., the widely used "Kronecker" model in conventional MIMO systems, [37,38]), the mean correlation between paths $\mathbf{A}_{m_1} \rightarrow \mathbf{U}_{n_1}$ and $\mathbf{A}_{m_2} \rightarrow \mathbf{U}_{n_2}$ can be calculated as

$$\rho_{AU} = \rho_A \times \rho_U \tag{11.9}$$

Nevertheless, real-life measurements show that the "Kronecker" model may yield systematic prediction error for the analysis of small-scale fading [39]. It is plausible that the model inaccuracy also exists in the case of large-scale fading. Therefore, Equation 11.9 serves only as a lower bound of the actual overall cross-correlation, and more experimental data are required to provide further insight into the formulation of ρ_{AU}.

Finally, to gauge the access qualities of the GDAS-CU system, we calculate the mean received power as

$$\overline{P_\mathrm{R}} = \frac{1}{N} \sum_{n=1}^N \mathbb{E}(P_R (\mathbf{U}_n, \mathbf{A}))$$

$$= \frac{1}{N} \sum_{n=1}^N \int_{\mathcal{C}_A} P_R (\mathbf{U}_n, \mathbf{A}) f_\mathbf{A}(\mathbf{A}) d\mathbf{A}$$

$$= \frac{1}{N} \sum_{n=1}^N \int_0^R \int_0^{2\pi} P_\mathrm{R} (\mathbf{U}_n, \mathbf{A}) \frac{r}{\pi R^2} d\vartheta dr \tag{11.10}$$

where P_R is dependent on the propagation loss from UT ($\mathbf{U}_n$) to AP ($\mathbf{A}$). More details on the modeling of P_R are provided in Section 11.4. As shown in Sections 11.5 and 11.6, $\overline{P_R}$ is one of the key parameters that determine the system diversity.

11.3.2 CORRELATION DEPENDING ON BOTH AAD AND RDD

In this section, we consider the scenario in which the cross-correlation is AAD/RDD dependent. Similar to Refs. [10,11], we introduce the following variable χ to denote the distance ratio between two links with propagation distances d_1 and d_2, respectively

$$\chi = \left| 10 \log_{10} \left(\frac{d_1}{d_2} \right) \right| \quad \text{[dB]} \tag{11.11}$$

A cross-correlation prediction table is presented in Ref. [11], which is based on the 900 MHz measurements performed in a small-cell urban environment in Mulhouse, France. Two APs with omnidirectional antennas were set up, where the antennas were about 30 m above ground level and 10 m over the skyline of surrounding buildings. A vehicle equipped with roof-mounted quarter-wavelength antenna measured the signals from both APs simultaneously. The measurements were performed in the surrounding area, up to 3 km away from the APs. A contour plot of the predicted correlation values in Ref. [11] with respect to AAD and RDD are shown in Figure 11.3a for illustration purpose.

To ensure a backward compatibility of the AAD/RDD-dependent model with the AAD-only model in Equation 11.1, a general expression that includes Equation 11.1 as a special case needs to be defined. Two collocated APs ($\chi = 0$ dB) have fully correlated shadow fading, and the correlation level decreases with increasing χ. At some threshold point $\chi = \chi_{th}$ and the correlation coefficient reaches the minimum, which may not be zero due to the common local scattering around the UT. The following model is thus introduced

$$\rho(\Delta\phi, \chi) = \begin{cases} g(\chi, \chi_{th}, \alpha)(0.78 - 0.0056|\Delta\phi| + \sigma_1) + \sigma_2, & \text{if } 0° \leq |\Delta\phi| \leq 15° \\ g(\chi, \chi_{th}, \alpha)(0.48 - 0.0056|\Delta\phi| + \sigma_1) + \sigma_2, & \text{if } 15° \leq |\Delta\phi| \leq 60° \\ g(\chi, \chi_{th}, \alpha)\sigma_1 + \sigma_2, & \text{if } |\Delta\phi| \geq 60° \end{cases} \tag{11.12}$$

where

$$g(\chi, \chi_{th}, \alpha) = \begin{cases} \left(1 - \frac{\chi}{\chi_{th}}\right)^{\alpha}, & \text{if } \chi \leq \chi_{th} \\ 0, & \text{if } \chi > \chi_{th} \end{cases} \tag{11.13}$$

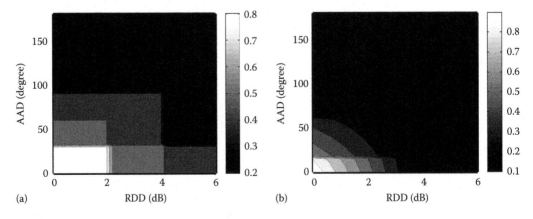

FIGURE 11.3 Contour plots of the correlation models derived from (a) the prediction table in Ref. [11] and (b) Equation 11.12.

and χ_{th} denotes the threshold distance ratio where the RDD-dependent correlation reaches its minimum value. α controls the shape of the distance-dependent function. σ_1 is related to the standard deviation of the correlation at each AAD, which was found to be approximately 0.1 on average in Ref. [8]. σ_2 quantifies the minimum correlation coefficient due to the common local scattering around the UTs. A contour plot of the function in Equation 11.12 is depicted in Figure 11.3b for $\alpha = 1$, $\chi_{th} = 4$ dB, $\sigma_1 = 0.1$, and $\sigma_2 = 0.1$. A parallel trend to the pattern shown in Figure 11.3a can be observed from Figure 11.3b. To further verify the applicability of the proposed model, the authors have performed simulation studies by considering a similar cell layout with the measurement area in Ref. [8] (see also Figure 11.4a and c). The following model parameters have been assumed: $\alpha = 0.5$, $\chi_{th} = 7$ dB, $\sigma_1 = 0.1$, and $\sigma_2 = 0$. Comparing Figure 11.4b to d shows that, in general the model in Equation 11.12 can capture the main characteristics of scattered correlations at each AAD. Finally, it is worth noting that Equation 11.12 reduces to Equation 11.1 for $\alpha = 0$, $\sigma_1 = 0$, and $\sigma_2 = 0$.

Subsequently, the average correlation between two paths from $\mathbf{U}_n$ to any two APs, $\mathbf{A}_{m_1}$ and $\mathbf{A}_{m_2}$, can be derived as

$$\rho_A = \frac{1}{N} \sum_{n=1}^{N} \iint \rho\left(\Delta\phi\left(\mathbf{A}_{m_1}, \mathbf{A}_{m_2}, \mathbf{U}_n\right), \chi\left(\mathbf{A}_{m_1}, \mathbf{A}_{m_2}, \mathbf{U}_n\right)\right) f_{\mathbf{A}_{m_1}}\left(\mathbf{A}_{m_1}\right) f_{\mathbf{A}_{m_2}}\left(\mathbf{A}_{m_2}\right) d\mathbf{A}_{m_1} d\mathbf{A}_{m_2}$$

$$= \frac{1}{N} \sum_{n=1}^{N} \frac{1}{(\pi R^2)^2} \iiiint \rho(\Delta\phi(x_1, y_1, x_2, y_2, x_n, y_n), \chi(x_1, y_1, x_2, y_2, x_n, y_n)) dx_1 dy_1 dx_2 dy_2 \quad (11.14)$$

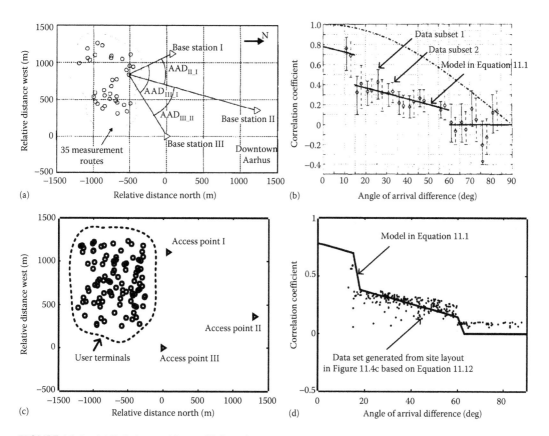

FIGURE 11.4 (a) Relative positions of BSs and measurement area and (b) measured correlation coefficients in Ref. [8]; (c) hypothetical measurement site and (d) the corresponding correlation coefficients generated from Equation 11.12 in the simulation study.

where (x_1, y_1), (x_2, y_2), and (x_n, y_n) are the Cartesian coordinates of $\mathbf{A}_{m_1}$, $\mathbf{A}_{m_2}$, and $\mathbf{U}_n$, respectively.

The MCCC among UTs is obtained in the same manner as Equation 11.8 with inclusion of RDD-dependent cross-correlation:

$$
\begin{aligned}
\rho_U &= \int_0^R \int_0^{2\pi} \rho\left(\Delta\psi_{n_1,n_2}(r,\vartheta), \chi_{n_1,n_2}(r,\vartheta)\right) f_{r,\vartheta}(r,\vartheta) d\vartheta dr \\
&= \int_0^R \int_0^{2\pi} \rho\left(\Delta\psi_{n_1,n_2}(r,\vartheta), \chi_{n_1,n_2}(r,\vartheta)\right) \frac{r}{\pi R^2} d\vartheta dr
\end{aligned}
\tag{11.15}
$$

with χ_{n_1,n_2} being the distance ratio between the paths $\mathbf{A} \to \mathbf{U}_{n_1}$ and $\mathbf{A} \to \mathbf{U}_{n_2}$ as defined in Equation 11.11.

Finally, to derive the values of ρ_A, ρ_U, and ρ_{AU} for multicell networks, we simply change the circular integration region in Equations 11.5, 11.8, 11.14, and 11.15 to the multiple circular regions in Figure 11.2.

11.4 POWER DECAYING IN WIRELESS NETWORKS

11.4.1 EXISTING PATHLOSS MODELS

The power decaying or pathloss model in wireless channels plays an important role in wireless linkage reliability. The pathloss representing signal attenuation measured in dB is defined as the difference between the effective transmitted power and the received power:

$$
PL = 10 \log_{10} \frac{P_T}{P_R}
\tag{11.16}
$$

where P_T and P_R represent the transmitted and received signal strength, respectively. Following from Equation 11.16, we may describe the pathloss at distance r as

$$
PL(r) = -P_R(r)
\tag{11.17}
$$

where $P_R(r)$ measured in dBW is the mean received power in channels when $P_T = 1$ dBW.

Empirical pathloss models are mainly derived based on the curve fitting of measurement data. Conventionally, the power decaying relationship is formulated as

$$
PL(r) = 10 n_1 \log_{10} r + c_{01}
\tag{11.18}
$$

or

$$
PL(r) = \begin{cases} 10 n_1 \log_{10} r + c_{01}, & 1 < r \le r_{\text{break}1} \\ 10 n_2 \log_{10} r + c_{02}, & r_{\text{break}1} < r \le r_{\text{break}2} \\ 10 n_3 \log_{10} r + c_{03}, & r > r_{\text{break}2} \end{cases}
\tag{11.19}
$$

where n_i and c_{0i} ($i = 1, 2, 3$) are model parameters to be identified. The breakpoints $r_{\text{break}i}$ define the threshold locations where the slope of the power decaying curve varies, which are obtained through measurement campaigns. Equation 11.18 is commonly referred to as the linear model of $\log_{10} r$ while Equation 11.19 is called the multi breakpoint pathloss model [40]. Dual slope model [40] is a special case of Equation 11.19, for which only one breakpoint appears.

11.4.2 Pathloss Derived from Stochastic Ray Tracing

The idea of modeling the urban cell as a percolating lattice was first proposed in Refs. [18,19]. The percolation theory [41] was then applied to analyze the signal propagation in such a channel formulation. The authors of Refs. [15,20,21] extended the work of Ref. [18] by treating the radiating wave as a stream of photons propagating in the environment; the ideas of photon particles and random walks were also used in Ref. [16]. A pathloss model based on such an approach can be expressed as [15]

$$PL(r) = 20 \log_{10} r + \alpha_1 r + c_1 \tag{11.20}$$

On the other hand, the pathloss formula derived from the random walk model in our previous work [13] is given by

$$PL(r) = 10 \log_{10} r + \alpha_1' r + c_1' \tag{11.21}$$

where α_1, α_1', c_1, and c_1' are constants.

In Refs. [12,42], a modified pathloss model that introduces an extra attenuation factor has the same mathematical structure as Equations 11.20 and 11.21. This model is composed of a free space pathloss and an additional loss term that increases exponentially with r. Nevertheless, a close examination of Equations 11.20 and 11.21 reveals that the additional loss would increase with r at an overwhelming speed, which is certainly implausible and is revisited in a later section.

The authors of Ref. [16] applied the Boltzmann and diffusion equations to describe the propagation process of radio waves. This method is based on the concepts of statistical physics and transport theory. The corresponding pathloss model can be summarized as [16]

$$PL(r) = 5 \log_{10} r + \alpha_2 r + c_2 \tag{11.22}$$

Moreover, the transport theory was also applied to analyze the pathloss in indoor channels [17], where the power decaying is given by

$$PL(r) = 15 \log_{10} r + \alpha_3 r + c_3 \tag{11.23}$$

α_i and c_i $(i = 2, 3)$ are constants. An interesting observation made from Equations 11.20 through 11.23 is that all the SRT-based pathloss models share a common structure composed of two dominant items: $\log_{10} r$ and $\alpha_i r$.

11.4.3 Generalized Pathloss Models

11.4.3.1 Review of a Novel SRT Model Based on Random Lattice Theories

Dense urban environments are populated with rectangular buildings, straight streets, and occasionally mega-scale plazas with large areas. The authors of Ref. [18] introduced a random lattice to model this physical propagation scenario, where percolation theory [41] was applied in the domain of channel modeling. In our previous paper [13], we extended the work of Refs. [15,18–21] to obtain several pathloss models based on the methodology of random lattice. This novel SRT model is briefly reviewed in this section.

To form site percolation, each cell in a lattice may be either open with probability p or closed otherwise. The structure of a percolation lattice exhibits high visual similarity with the urban built-up areas as indicated in Ref. [18]. Each closed cell may represent a dominant reflector in the channel such as building walls. Each multipath component propagating in random lattice channels is considered as a stochastic ray. Without loss of generality, we assume that the Tx is located at the origin of a 2D propagation plane. By using the maximum entropy method [43], the distribution function of stochastic rays that undergo a certain number of collisions at the location with polar coordinate (r, θ) is derived as [44]

$$Q_i(r, \theta) = \frac{2}{\pi D_i^2} \exp\left(-\frac{2r}{D_i}\right), \quad i \geq 1 \tag{11.24}$$

where D_i is the average distance traveled by stochastic rays that undergo i collisions when reaching (r, θ). Because Equation 11.24 is independent of the abscissa component θ, it is omitted in the following discussion for simplicity. In general, there is no closed-form solution for D_i. However, in the special case of a random-walk process where each movement of the propagating particle should exhibit equal probability and equal length in all possible directions, an analytical solution $D_i = \bar{d}\sqrt{i}$ exists [45]. Here $\bar{d}$ is the distance traveled by a particle at each step of the 2D random walk. In Ref. [13] we have also derived $\bar{d}$ in a percolation lattice as follows:

$$\bar{d} = \frac{a}{\sqrt{1-p}} \tag{11.25}$$

where
   a is the cell-side length in the lattice
   p is the probability that a cell is not occupied

For a more anomalous propagation scenario, the relationship between D_i and $\bar{d}$ is conjectured to be [13]

$$D_i = \bar{d}i, \quad i \geq 1 \tag{11.26}$$

A more detailed discussion on Equation 11.26 and its further generalization have been presented in Ref. [13]. Substituting Equations 11.25 and 11.26 into Equation 11.24 yields

$$Q_i(r) = \frac{2(1-p)}{\pi a^2 i^2} \exp\left(-\frac{2r\sqrt{1-p}}{ai}\right), \quad i \geq 1 \tag{11.27}$$

It can be readily inferred from Equation 11.27 that the probability of occurrence for stochastic rays undergoing fewer reflections is larger than those undergoing more reflections. This is consistent with the well-known fact that the electromagnetic waves experiencing less reradiations play the most dominant role in wireless communications.

11.4.3.2 Pathloss Model Derived from the Aforementioned Method

Suppose that both the Tx and Rx antennas are isotropic and $P_T = 1$ dBW. Subsequently, the mean received power in random lattice channels derived from Equation 11.24 is given by [13,44]

$$P_R(r) = \sum_{i=1}^{\infty} \exp(-\xi i) Q_i(r) \tag{11.28}$$

where
   $\xi = L \ln 10/10$
   L is the average reflection loss during every collision

Meanwhile L also takes into account the frequency dependence of channels. Note that Equation 11.28 corresponds to a non-line-of-sight propagation condition.

PROPOSITION 1

The pathloss of stochastic rays in a random lattice with the level of analomousness defined by Equation 11.26 is

$$PL(r) \approx 7.5 \log_{10} r + 12.28 \sqrt{\frac{\sqrt{1-p}\xi}{a}} r^{\frac{1}{2}} - 10 \log_{10} \left[\frac{2^{\frac{3}{4}}}{\sqrt{\pi} a^{\frac{3}{4}}} \frac{(1-p)^{\frac{3}{8}}}{\xi^{\frac{3}{4}}} \right] \tag{11.29}$$

Proof: Substituting Equation 11.27 into Equation 11.28 and replacing the summation operator in Equation 11.28 with integration yields

$$P_R(r) = \sum_{i=1}^{\infty} \exp(-\xi i) Q_i(r)$$

$$\approx \int_1^{\infty} \frac{2(1-p)}{\pi a^2 y^2} \exp\left[-\xi y - \frac{2r}{a/\sqrt{1-py}} \right] dy$$

$$\approx \frac{2(1-p)}{\pi a^2} \int_0^{\infty} y^{-2} \exp\left[-\xi y - \frac{2r}{a/\sqrt{1-py}} \right] dy \tag{11.30}$$

The approximation identity in the last row of Equation 11.30 holds if the integrand approaches zero in the domain of $(0, 1]$, which is satisfied because the probability of stochastic rays is small when the reflection number $i = 1$. Using [[46], Eq. (3.478 4)], we have

$$P_R(r) \approx \frac{4}{\pi a^{\frac{3}{2}}} \sqrt{\frac{1-p}{\xi}} r^{-\frac{1}{2}} K_{-1} \left[2\sqrt{\frac{2\sqrt{1-p}\xi r}{a}} \right] \tag{11.31}$$

Because $K_{-1}(z) = K_1(z)$ [[46], Eq. (8.486 16)], we further apply [[46], Eq. (8.451 6)], where the asymptotic expansion of Bessel functions is used for a large value of variables. Taking the first term of [[46], Eq. (8.451 6)], we have

$$K_1(z) \approx \sqrt{\frac{\pi}{2z}} \exp(-z) \tag{11.32}$$

Subsequently, the following can be derived from Equation 11.31

$$P_R(r) \approx \frac{2^{\frac{3}{4}}}{\sqrt{\pi} a^{\frac{5}{4}}} \frac{(1-p)^{\frac{3}{8}}}{\xi^{\frac{3}{4}}} r^{-\frac{3}{4}} \exp\left(-2\sqrt{\frac{2\sqrt{1-p}\xi r}{a}} \right) \tag{11.33}$$

Taking logarithmic operation on both sides of Equation 11.33, the pathloss is given by

$$PL(r) \approx 7.5 \log_{10} r + 12.28 \sqrt{\frac{\sqrt{1-p}\xi}{a}} r^{\frac{1}{2}} - 10 \log_{10} \left[\frac{2^{\frac{3}{4}}}{\sqrt{\pi} a^{\frac{3}{4}}} \frac{(1-p)^{\frac{3}{8}}}{\xi^{\frac{3}{4}}} \right] \tag{11.34}$$

This completes the proof.

Let $\alpha_4 = 12.28\sqrt{\sqrt{1-p}\xi/a}$ and c_4 represent the last term of Equation 11.29; we have

$$PL(r) = 7.5\log_{10}r + \alpha_4 r^{\frac{1}{2}} + c_4 \tag{11.35}$$

Equation 11.35 indicates that the pathloss model based on the stochastic ray method in Ref. [13] is a combination of the traditional linear model with pathloss exponent $n = 0.75$ and a correction factor represented by the square root of r. It is worth emphasizing that the mathematical structure of Equation 11.35 is similar to Equations 11.20 through 11.23.

To determine the input parameters in Equation 11.35, some general guidelines are described as follows. $1 - p$ is the ratio of the occupation area of obstacles to the total area of service region. a can be obtained from Equation 11.25 if $\bar{d}$ is known. Otherwise, it can be directly derived from the distribution of obstacles. To determine the value of L is a more difficult task. It is unlikely to obtain an exact value of L when the electrical properties of various scattering materials and the incident angles of multipath components are not known. Furthermore, Because the SRT is used to characterize an average propagation condition, the value of L can only represent a mean reflection loss. In general, a reasonable range of L is $6 \le L \le 15$ (in dB). For example, $L = 6$ dB indicates that three-quarters of the energy is lost after each collision. Following from the above discussions on the practical values of a, p, and L, typically α_4 is in the range of $1 \sim 3$.

11.4.3.3 Generalized Pathloss Model

Figure 11.5 illustrates the predicted pathloss versus r for the models expressed in Equations 11.18, 11.20, 11.22, and 11.35. It is apparent that the signal attenuation depicted by Equations 11.20 and 11.22 increases rapidly with r. The main reason is that the additional attenuation term $\alpha_i r$ in Equations 11.20 and 11.22 is a linear function of r. On the other hand, Equation 11.35 predicts a more moderate increase in pathloss because the additional term is a fractional power function of r.

Subsequently, an interesting observation is that a generalized pathloss model can be formulated as follows:

$$PL(r) = 10n\log_{10}r + \alpha r^{\delta} + C \tag{11.36}$$

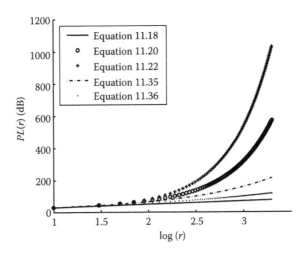

FIGURE 11.5 Comparison among different pathloss models where $n_1 = 2$, $c_{01} = 10$ in Equation 11.18; $\alpha_1 = 0.25$, $c_1 = 7.5$ in Equation 11.20; $\alpha_2 = 0.5$, $c_2 = 20$ in Equation 11.22; $\alpha_4 = 4$, $c_4 = 10$ in Equation 11.35; and $n = 1.5$, $\alpha = 5$, $\delta = \frac{1}{3}$, $C = 4.2$ in Equation 11.36.

where n, α, δ, and C are constants. In general, the parameters n, α, and δ are positive values. From Equation 11.36 we have

$$\frac{\partial PL(r)}{\partial r} = \frac{10n}{\ln 10}\frac{1}{r} + \alpha\delta r^{\delta-1} \qquad (11.37)$$

Equation 11.37 can be analyzed in the limiting case where the Rx is extremely far away from the Tx (e.g., $r = 10^4$ m). Suppose a hypothetical scenario where there is an incremental change of distance from r to $r + 1$. An intuitive physical appeal tells us that the additional pathloss, $\partial PL(r)/\partial r$, should approach zero. Therefore,

$$\lim_{r\to\infty} \frac{\partial PL(r)}{\partial r} = 0 \qquad (11.38)$$

Equation 11.38 requires $\delta < 1$ in Equation 11.37. We can thus infer that δ should be less than 1 in order to satisfy the condition that the pathloss does not grow radically with r. Consequently, the existing models including Equations 11.20 through 11.22 and the modified signal attenuation model in Ref. [42] are appropriate only for indoor short-range channels. For outdoor macro- and mega-cells with large propagation distance, these models tend to overestimate the increase in $PL(r)$ with r. Figure 11.5 also illustrates the generalized pathloss model in Equation 11.36 compared with the other models.

11.4.4 Experimental Verifications and Discussions

11.4.4.1 Verification of the Proposed Model through Experimental Data

The empirical results shown in Ref. [15] have been reproduced in Figure 11.6. The measurements were taken in the "Prati" district of Rome, which is a classical dense and homogeneous urban environment. A 900 MHz sinusoidal wave is used to probe the channel. We assume that the interval distance $a = 20$ m, which closely estimates the average width of boulevards, and the inoccupation probability $p = 0.7$. It is further assumed that $L = 12$ dB. The pathloss calculated from Equation 11.35 is plotted in Figure 11.6. It can be seen that the theoretical results exhibit good

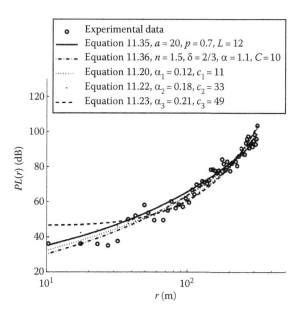

FIGURE 11.6 Pathloss versus distance for different models. The experimental data obtained in Ref. [15] are marked with circles.

agreement with the experimental data. Furthermore, the root-mean-square (RMS) error can be used to evaluate the curve-fitting accuracy

$$\sigma = \sqrt{\frac{1}{J} \sum_{j=1}^{J} \left[PL_m(r_j) - PL_t(r_j) \right]^2} \tag{11.39}$$

where
  J is the total number of recorded data points
  the subscripts m and t denote the measured data and the theoretical results, respectively

The RMS error corresponding to Equation 11.35 is 4.13 dB. It is worth mentioning that if the values of a, p, and L vary in a small range, the RMS error will also change within a small interval. Therefore, the prediction accuracy is rather robust to the model parameter selection process.

Next, we identify the parameters of Equation 11.36 by directly fitting Equation 11.36 with the experimental data, where we set $n = 1.5$ and $\delta = 2/3$. The RMS error is still used to evaluate the curve-fitting accuracy. If $\alpha = 1.1$ and $C = 10$ in Equation 11.36, the RMS error of the model is 3.54 dB, which is rather satisfactory and smaller than the results in Ref. [13]. Obviously, the model expressed in Equation 11.36 could be used as a benchmark to provide a lower bound of the RMS error. To further exemplify the usefulness of the proposed model, we compare 11.36 to other pathloss models (e.g., [15–17]) in terms of the prediction accuracy. Figure 11.6 illustrates the fitting results of all the models with reference to the same set of experimental data, where the best-fit parameters are shown in the figure. We further calculate the RMS error of different models as follows: $\sigma = 3.70$ dB in 11.20, $\sigma = 4.49$ dB in 11.22, and $\sigma = 5.35$ dB in 11.23. All of them are larger than the result obtained with the proposed generalized model.

11.4.4.2 Further Remarks

A large amount of measurement data are required to provide critical insight into the selection of proper model parameters in Equation 11.36. Due to the lack of sufficient experimental results, we are not able to provide a comprehensive description of the model parameters in this work. However, it is reasonable to assume that the parameters n and δ are less dependent on the propagation scenarios, as the geometrical and electrical properties of obstacles in wireless channels mainly affect the parameters α and C. As a result, it is of practical interest to determine the values of α and C through measurement campaigns. The empirical pathloss models indicate that the loss exponent in Equation 11.36 should be around 1.6 [12]. Considering that there is an additional attenuation term in Equation 11.36, we propose to choose $n = 1.5$. There is no strict constraint on δ. A general requirement is to select $\delta < 1$ as discussed above.

Finally, we can rewrite Equation 11.36 as

$$P_R(r)(W) = cr^{-n} 10^{-\frac{\alpha r^\delta}{10}} \tag{11.40}$$

where c is a constant. If $n = 2$ and $\alpha = 0$, Equation 11.40 reduces to the free space propagation model [12]. A degenerate case of Equation 11.40 where $\delta = 1$ has been applied to model the signal attenuation regime in wireless networks [47,48]. Apparently, a more accurate description of the radio wave propagation in wireless networks could be obtained when more appropriate values of δ are chosen.

11.5 COOPERATIVE DIVERSITY ANALYSIS

Diversity is commonly known as the redundancy of the transmitted signal in a particular communication system. In this work, we focus on an intuitive measure of diversity level achievable in wireless

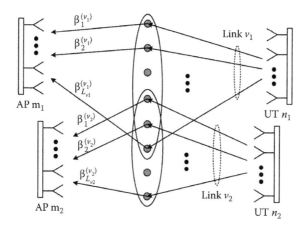

FIGURE 11.7 Double-directional channel impulse responses for two wireless links υ_1 and υ_2. Some scattering elements are common to both links, which cause the cross-correlation between υ_1 and υ_2.

networks, which is defined as the number of independent propagation paths that can be averaged over to detect symbols and is abbreviated as diversity (DIV) in the following discussions. The main objective is to estimate the diversity level in a GDAS-CU network, which should be simple and yet have sound physical basis.

Figure 11.7 shows two communication links υ_1 and υ_2 connecting UT n_1 to AP m_1 and UT n_2 to AP m_2, respectively. A double-directional channel impulse response for path υ_1 is taken to be

$$h_{\upsilon_1}(\theta_T, \theta_R) = \sum_{l=1}^{L_{\upsilon_1}} \beta_l^{(\upsilon_1)} \delta\left(\theta_T - \theta_{T,l}^{(\upsilon_1)}\right) \delta\left(\theta_R - \theta_{R,l}^{(\upsilon_1)}\right) \tag{11.41}$$

where $\{\theta_{T,l}^{(\upsilon_1)}\}$ and $\{\theta_{R,l}^{(\upsilon_1)}\}$ are the AODs and AOAs. The channel is characterized by the coefficients $\beta_l^{(\upsilon_1)}$ that couple the multiple transmit angles with multiple receive angles. We will assume that the discrete model in Equation 11.41 has considered the finite dimensionality of the spatial signal space due to a finite number of antenna elements and a finite array aperture (i.e., the propagation paths have been partitioned into resolvable AOD/AOA bins [49]). Therefore, L_{υ_1} defines the diversity level of link υ_1, and $\{\beta_l^{(\upsilon_1)}\}$ are independently distributed random variables. Nonetheless, in real-life scenarios, a diversity link is broken if the received signal power fails to achieve an appropriately specified threshold Υ. We assume that $\left(\beta_l^{(\upsilon_1)}\right)^2$ are identical lognormal fading signals with pdf:

$$f\left(\left(\beta_l^{(\upsilon_1)}\right)^2; \mu_{\upsilon_1}, s_{\upsilon_1}\right) = \frac{1}{\left(\beta_l^{(\upsilon_1)}\right)^2 s_{\upsilon_1} \sqrt{2\pi}} e^{-\left[\ln\left(\beta_l^{(\upsilon_1)}\right)^2 - \mu_{\upsilon_1}\right]^2 / 2 s_{\upsilon_1}^2} \tag{11.42}$$

where μ_{υ_1} and s_{υ_1} are the location and shape parameters, respectively. Note that the power received through link υ_1, $P_R^{(\upsilon_1)}$, is related to the two parameters through the following relationship $P_R^{(\upsilon_1)} = e^{\mu_{\upsilon_1} + s_{\upsilon_1}^2/2}$. The outage probability is basically equal to the cumulative density function of the output signal

$$P_{\text{out}}^{(\upsilon_1)} = \Pr\left(\left(\beta_l^{(\upsilon_1)}\right)^2 \le \Upsilon\right) = \frac{1}{2} + \frac{1}{2}\text{erf}\left[\frac{\ln \Upsilon - \mu_{\upsilon_1}}{s_{\upsilon_1}\sqrt{2}}\right] \tag{11.43}$$

where $\Pr(\cdot)$ represents probability. Effectively, the number of diversity paths for υ_1 should be scaled down by $\left(1 - P_{\text{out}}^{(\upsilon_1)}\right)$ to account for the outage loss.

Consider another radio link v_2 illustrated in Figure 11.7. The double-directional channel impulse response can be written by following the similar procedure with Equation 11.41. When propagation paths v_1 and v_2 are correlated, a number of scattering elements will be common to the two paths. The discrete channel models for both v_1 and v_2 can thus be rewritten as

$$h_{v_1}(\theta_T, \theta_R) = \sum_{l=1}^{L_{v_1,v_2}} \beta_l^{(v_1,v_2)} \delta\left(\theta_T - \theta_{T,l}^{(v_1)}\right) \delta\left(\theta_R - \theta_{R,l}^{(v_1)}\right)$$

$$+ \sum_{l=L_{v_1,v_2}+1}^{L_{v_1}} \beta_l^{(v_1)} \delta\left(\theta_T - \theta_{T,l}^{(v_1)}\right) \delta\left(\theta_R - \theta_{R,l}^{(v_1)}\right) \tag{11.44}$$

$$h_{v_2}(\theta_T, \theta_R) = \sum_{l=1}^{L_{v_1,v_2}} \beta_l^{(v_1,v_2)} \delta\left(\theta_T - \theta_{T,l}^{(v_2)}\right) \delta\left(\theta_R - \theta_{R,l}^{(v_2)}\right)$$

$$+ \sum_{l=L_{v_1,v_2}+1}^{L_{v_2}} \beta_l^{(v_2)} \delta\left(\theta_T - \theta_{T,l}^{(v_2)}\right) \delta\left(\theta_R - \theta_{R,l}^{(v_2)}\right) \tag{11.45}$$

The scattering coefficients associated with the first terms of Equations 11.44 and 11.45 are identical to both paths, while those with the second terms are distinct and unique to the relevant path. We can now express the cross-correlation between the signal fading on the two paths as

$$\rho_{v_1,v_2} = \frac{\mathbb{E}\left(\sum_{l=1}^{L_{v_1}} \chi_l^{(v_1)} \times \sum_{l=1}^{L_{v_2}} \chi_l^{(v_2)}\right)}{\sqrt{\mathbb{E}\left[\left(\sum_{l=1}^{L_{v_1}} \chi_l^{(v_1)}\right)^2\right]}\sqrt{\mathbb{E}\left[\left(\sum_{l=1}^{L_{v_2}} \chi_l^{(v_2)}\right)^2\right]}} \tag{11.46}$$

where $\chi_l = \beta_l^2 - \mathbb{E}\left(\beta_l^2\right)$. Substituting Equations 11.44 and 11.45 into Equation 11.46 yields

$$\rho_{v_1,v_2} = \frac{\sum_{l=1}^{L_{v_1,v_2}} \left(\sigma_l^{(v_1,v_2)}\right)^2}{\sqrt{\sum_{l=1}^{L_{v_1}} \left(\sigma_l^{(v_1)}\right)^2} \times \sqrt{\sum_{l=1}^{L_{v_2}} \left(\sigma_l^{(v_2)}\right)^2}} \tag{11.47}$$

where $\{\sigma_l^{(v_1)}\}$ and $\{\sigma_l^{(v_2)}\}$ are the standard deviations of $\{\chi_l^{(v_1)}\}$ and $\{\chi_l^{(v_2)}\}$, respectively. $\{\sigma_l^{(v_1,v_2)}\}$ are the standard deviations for those scatterers common to both wireless links. If $\sigma_l^{(v_1)} = \sigma_l^{(v_2)} = \sigma_l$ (i.e., identical and independent faders), Equation 11.47 can be simplified as

$$\rho_{v_1,v_2} = \frac{L_{v_1,v_2}}{\sqrt{L_{v_1}L_{v_2}}} \tag{11.48}$$

which leads to $L_{v_1,v_2} = \rho_{v_1,v_2}\sqrt{L_{v_1}L_{v_2}}$. Along the lines of derivation for Equations 11.46 through 11.48, we introduce GCCC as a new cross-correlation measure among three radio links v_1, v_2, and v_3

$$\varrho_{v_1,v_2,v_3} = \frac{\mathbb{E}\left(\sum_{l=1}^{L_{v_1}} \chi_l^{(v_1)} \times \sum_{l=1}^{L_{v_2}} \chi_l^{(v_2)} \times \sum_{l=1}^{L_{v_3}} \chi_l^{(v_3)}\right)}{\sqrt[3]{\mathbb{E}\left[\left(\sum_{l=1}^{L_{v_1}} \chi_l^{(v_1)}\right)^3\right]}\sqrt[3]{\mathbb{E}\left[\left(\sum_{l=1}^{L_{v_2}} \chi_l^{(v_2)}\right)^3\right]}\sqrt[3]{\mathbb{E}\left[\left(\sum_{l=1}^{L_{v_3}} \chi_l^{(v_3)}\right)^3\right]}}$$

$$= \frac{\sum_{l=1}^{L_{v_1,v_2,v_3}} \gamma_l^{(v_1,v_2,v_3)}}{\sqrt[3]{\sum_{l=1}^{L_{v_1}} \gamma_l^{(v_1)}} \times \sqrt[3]{\sum_{l=1}^{L_{v_2}} \gamma_l^{(v_2)}} \times \sqrt[3]{\sum_{l=1}^{L_{v_3}} \gamma_l^{(v_3)}}} \tag{11.49}$$

where $\{\gamma_l^{(v_1)}\}$, $\{\gamma_l^{(v_2)}\}$, and $\{\gamma_l^{(v_3)}\}$ are the third moment of the faders associated with the relevant paths. $\{\gamma_l^{(v_1,v_2,v_3)}\}$ are the third moment for those scatterers common to all three links. If $\gamma_l^{(v_1)} = \gamma_l^{(v_2)} = \gamma_l^{(v_3)} = \gamma_l$, Equation 11.49 can be simplified as

$$\varrho_{v_1,v_2,v_3} = \frac{L_{v_1,v_2,v_3}}{\sqrt[3]{L_{v_1}L_{v_2}L_{v_3}}} \tag{11.50}$$

Thus, we have

$$L_{v_1,v_2,v_3} = \varrho_{v_1,v_2,v_3}\sqrt[3]{L_{v_1}L_{v_2}L_{v_3}} \tag{11.51}$$

In general, further extension of Equations 11.49 through 11.51 to the scenario of K ($K > 3$) wireless links is very difficult. However, it is reasonable to conjecture that for $K > 3$, the number of scattering elements universal to all the links would be small and thus can be ignored. We note that no empirical data have been reported in the literature to characterize the values of ϱ_{v_1,v_2,v_3}. Nevertheless, this parameter is critical to the attainable DIV in a GDAS-CU network, as made clear in Section 11.6.3.

We now represent the redundant copies of received signals through paths v_1, v_2, and v_3 as sets $\mathcal{D}_{v_1}$, $\mathcal{D}_{v_2}$, and $\mathcal{D}_{v_3}$, respectively. Apparently, the DIV is given by the number of elements in the union of the three sets

$$\begin{aligned}
\text{DIV} &= \left|\mathcal{D}_{v_1} \cup \mathcal{D}_{v_2} \cup \mathcal{D}_{v_3}\right| \\
&= \left|\mathcal{D}_{v_1}\right| + \left|\mathcal{D}_{v_2}\right| + \left|\mathcal{D}_{v_3}\right| - \left|\mathcal{D}_{v_1} \cap \mathcal{D}_{v_2}\right| - \left|\mathcal{D}_{v_1} \cap \mathcal{D}_{v_3}\right| - \left|\mathcal{D}_{v_2} \cap \mathcal{D}_{v_3}\right| + \left|\mathcal{D}_{v_1} \cap \mathcal{D}_{v_2} \cap \mathcal{D}_{v_3}\right|
\end{aligned} \tag{11.52}$$

Subsequently, the cardinality of each "diversity-paths" set can be computed as

$$\begin{cases}
\left|\mathcal{D}_v\right| = \left(1 - P_{\text{out}}^{(v)}\right) L_v & v \in \{v_1, v_2, v_3\} \\
\left|\mathcal{D}_{v_a} \cap \mathcal{D}_{v_b}\right| = \left(1 - \max\{P_{\text{out}}^{(v_a)}, P_{\text{out}}^{(v_b)}\}\right) \times L_{v_a,v_b} & \{v_a, v_b\} \subset \{v_1, v_2, v_3\} \\
\left|\mathcal{D}_{v_1} \cap \mathcal{D}_{v_2} \cap \mathcal{D}_{v_3}\right| = \left(1 - \max\{P_{\text{out}}^{(v_1)}, P_{\text{out}}^{(v_2)}, P_{\text{out}}^{(v_3)}\}\right) \times L_{v_1,v_2,v_3}
\end{cases} \tag{11.53}$$

The final result of DIV can be obtained by substituting Equation 11.53 into Equation 11.52.

Extending Equation 11.52 to the general situation with MN links yields

$$\begin{aligned}
\text{DIV} &= \left|\bigcup_{1 \leq v \leq MN} \mathcal{D}_v\right| \\
&\approx \sum_{v=1}^{MN} \left|\mathcal{D}_v\right| - \sum_{1 \leq (v_1,v_2) \leq MN} \left|\mathcal{D}_{v_1} \cap \mathcal{D}_{v_2}\right| + \sum_{1 \leq (v_1,v_2,v_3) \leq MN} \left|\mathcal{D}_{v_1} \cap \mathcal{D}_{v_2} \cap \mathcal{D}_{v_3}\right| \\
&\approx \sum_{v=1}^{MN} \left(1 - P_{\text{out}}^{(v)}\right) L_v - \sum_{1 \leq (v_1,v_2) \leq MN} \left(1 - \max\{P_{\text{out}}^{(v_1)}, P_{\text{out}}^{(v_2)}\}\right) \times \varrho_{v_1,v_2}\sqrt{L_{v_1}L_{v_2}} \\
&\quad + \sum_{1 \leq (v_1,v_2,v_3) \leq MN} \left(1 - \max\{P_{\text{out}}^{(v_1)}, P_{\text{out}}^{(v_2)}, P_{\text{out}}^{v_3}\}\right) \times \varrho_{v_1,v_2,v_3}\sqrt[3]{L_{v_1}L_{v_2}L_{v_3}}
\end{aligned} \tag{11.54}$$

It is worth emphasizing that though Equation 11.54 correctly counts the available DIV and yields a substantially correct intuitive picture, a large number of measurement data are required to facilitate a realistic implementation of the formulation (e.g., parameterizations of cross-correlations and pathloss laws, estimation of diversity level at individual links, etc.).

11.6 NUMERICAL EXAMPLES

11.6.1 CHARACTERISTICS OF MCCCs

In this section, we study the characteristics of MCCCs in a GDAS-CU network. In the following example, two UTs, U_1 and U_2, cooperate together to transmit and receive signals as illustrated in Figure 11.1b. The radius of the deployment cell is 100 m and the distance of separation between U_1 and U_2 is 10 m. We represent the centers of the circular cell and U_1U_2 as o and o', respectively (see also Figure 11.1b). We further assume that $oo' \perp U_1U_2$. As it is difficult to obtain closed-form solutions for the integrals in Equations 11.5, 11.8, 11.14, and 11.15, numerical approaches are applied to calculate the integrations. Figure 11.8a plots the MCCCs against $|oo'|/R$ for both the AAD-only and AAD/RDD-dependent models. As $|oo'|/R$ increases, the UTs move away from the center of the cell. The angular range of the AOA at the UT decreases. Therefore, ρ_A, ρ_U, and ρ_{AU} increase. Furthermore, ρ_U is much larger than ρ_A due to the smaller spacing between the UT ports. It is also observed that both ρ_A and ρ_{AU} feature a steep transition slope when the UTs are near the boundary of the deployment cell (i.e., when $(|oo'|/R) = 1$). However, the transition is more gradual for ρ_U and indeed, there is no remarkable change in the value of ρ_U throughout the entire range of $|oo'|/R$. Another important

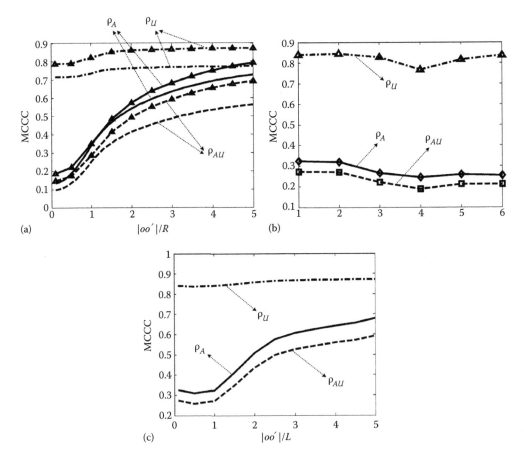

FIGURE 11.8 (a) MCCCs against the ratio of $|oo'|$ to R. Curves marked with triangles correspond to the mean cross-correlation derived from the AAD/RDD-dependent model in Equation 11.12; (b) influence of various multicell structures where the horizontal axis indicates the index of different topological models shown in Figure 11.2b: 1. Triangle, 2. Square, 3. Pentagon, 4. Hexagon, 5. Heptagon, 6. Octagon ($|oo'|/L = 1$, see also Figure 11.2b); and (c) MCCCs against the ratio of $|oo'|$ to L for triangle network architecture.

observation is that, with inclusion of the RDD dependency, the mean cross-correlation increases for both ρ_A and ρ_U. Finally, it can be seen that as $|oo'|/R$ increases further, the MCCCs eventually approaches an upper bound.

We now investigate the influence of multiple cells on the system correlation level. The network topologies considered in the numerical example are illustrated in Figure 11.2b with $L = 100$ m and $l = 20$ m. Note that L quantifies the overall spread of the AP placement and l defines the size of each cluster. As can be seen from Figure 11.8b, the impact of various topological structures on the mean correlation is insignificant. This phenomenon is also observed for other values of $|oo'|$. Subsequently, Figure 11.8c plots the MCCCs as $|oo'|/L$ increases from 0.1 to 5, for which a parallel trend to Figure 11.8a is observed. Following from the results in Figure 11.8b and c, it is expected that $|oo'|/L$ would be the most dominant design parameter compared to others such as the number and size of cells.

11.6.2 CHARACTERISTICS OF MEAN RECEIVED POWER AND OUTAGE PROBABILITY

In this section, the properties of the mean received power and the corresponding outage probability for various pathloss models presented in Section 11.4 are investigated. The same system structure with the example in Section 11.6.1 is used. Figure 11.9a shows the values of $P_R(oo')/\overline{P_R}$, which compares the average signal strength in a GDAS-CU network to the conventional collocated antenna system. Obviously, the ratio is smaller than 1 for most of the values of $|oo'|/R$, which implies that GDAS-CU would achieve higher received signal strength than the collocated antenna system. Furthermore, $P_R(oo')/\overline{P_R}$ reaches the minimum value in the proximity of $(|oo'|/R) = 1$ and eventually approaches to 1 as $|oo'|/R$ increases further.

Figure 11.9b depicts the outage probability for different power decaying laws. For the lognormal fading at each link, $s = 4$ dB (cf. Equation 11.42), and the outage threshold is taken to be 20 dB below the mean received signal at $(|oo'|/R) = 5$. As can be seen from Figure 11.9b, P_{out} increases with $|oo'|/R$ and overall there is a radical increase in P_{out} when the UTs are near the boundary of the deployment cell.

Finally, it is made clear from Figure 11.9a and b that different pathloss models yield considerably different values of $P_R(oo')/\overline{P_R}$ and P_{out}, which demonstrates the significance of applying an accurate propagation model for diversity analysis.

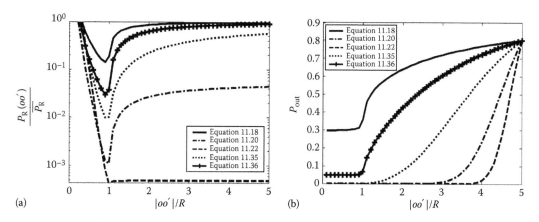

(a)

(b)

FIGURE 11.9 (a) $P_R(oo')/\overline{P_R}$ and (b) P_{out} against the ratio of $|oo'|$ to R for different pathloss models where $n_1 = 2$, $c_{01} = 10$ in Equation 11.18; $\alpha_1 = 0.25$, $c_1 = 7.5$ in Equation 11.20; $\alpha_2 = 0.5$, $c_2 = 20$ in Equation 11.22; $\alpha_4 = 4$, $c_4 = 10$ in Equation 11.35; and $n = 1.5$, $\alpha = 5$, $\delta = 1/3$, $C = 4.2$ in Equation 11.36.

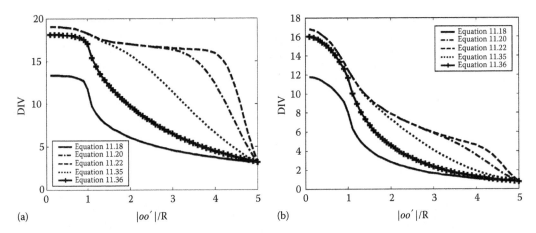

FIGURE 11.10 DIV estimates where the GCCC is derived by scaling down the values of ρ_A (AAD-only model) in Figure 11.8a by (a) 1/2 and (b) 1/3, respectively. The model parameters for various power decaying laws are $n_1 = 2$, $c_{01} = 10$ in Equation 11.18; $\alpha_1 = 0.25$, $c_1 = 7.5$ in Equation 11.20; $\alpha_2 = 0.5$, $c_2 = 20$ in Equation 11.22; $\alpha_4 = 4$, $c_4 = 10$ in Equation 11.35; and $n = 1.5$, $\alpha = 5$, $\delta = 1/3$, $C = 4.2$ in Equation 11.36.

11.6.3 INFLUENCE OF MEAN RECEIVED POWER AND GCCC ON MULTIPATH DIVERSITY GAINS

In this section, we explore the effect of mean received power and GCCC on the total number of diversity paths. The same system architecture with the examples in Sections 11.6.1 and 11.6.2 is employed. There are three APs and two UTs in the wireless network. For simplicity, the number of diversity paths associated with each individual link is assumed to be five. Similar to Section 11.6.2, $s = 4$ dB and the outage threshold is taken to be 20 dB below the mean received signal at $(|oo'|/R) = 5$ for the large-scale fading. The MCCCs (AAD-only model) can be obtained from Figure 11.8a for different values of $|oo'|/R$. Though there is no established model for the GCCC ϱ, it is reasonable to conjecture that the GCCC will increase with $|oo'|/R$ and its value would be less than the MCCCs predicted in Figure 11.8a. For illustration purpose, we consider two different models of ϱ against $|oo'|/R$, which are derived by scaling down the values of ρ_A (AAD-only model) in Figure 11.8a by 1/2 and 1/3, respectively. The corresponding DIV estimates are shown in Figure 11.10 for various pathloss models. It can be seen that on one hand, the diversity level is sensitive to the shape of ϱ against $|oo'|/R$. Smaller values of ϱ yield larger overall diversity gains as can be observed by comparing Figure 11.10a to b, provided that the MCCCs remain unchanged. This phenomenon is expected as can be seen from the expressions of Equations 11.53 and 11.54. On the other hand, the variation in DIV estimates is remarkable for different pathloss models. Moreover, the down trend of the multipath diversity with respect to $|oo'|/R$ is present in both Figure 11.10a and b. Similar to Figures 11.8a, c, and 11.9, most of the curves in Figure 11.10 exhibit a "transition zone" in the proximity of $(|oo'|/R) = 1$.

11.7 SUMMARY AND OPEN ISSUES

The diversity analysis for a GDAS-CU network has been carried out in this chapter. Starting with a simple network topology and system architecture, the average correlation coefficients have been derived. We have also addressed more general scenarios where the APs are grouped into multiple clusters and the cross-correlation is both AAD- and RDD-dependent. Subsequently, a generalized pathloss model based on SRT theory has been proposed where propagation environments are suitable to be described by stochastic structures rather than being deterministically characterized. It is shown that the proposed pathloss law, which is generalized from several physically inspired SRT models, exhibits better agreement with the experimental data than other existing models. It is found that the received signal power has a significant impact on the number of redundant links. We have also provided an intuitive method for calculating the system diversity levels. To facilitate the DIV computation, a new parameter called GCCC has been introduced, which is shown to have an important influence on the attainable system diversity.

The following open issues should be addressed in future to facilitate a realistic implementation of the proposed analytical framework:

- More empirical data are required to describe the cross-correlation against AAD and RDD in various service environments
- Model parameters of the generalized power decaying law are to be determined for different environments through comprehensive measurement campaigns
- Characterization of the generalized cross-correlation would be of significant practical interest

APPENDIX: DERIVATION OF EQUATION 11.3

Figure A.1 illustrates the geometry when the UT lies within the deployment cell. Let us consider the differential strip between $\alpha = \phi$ and $\alpha = \phi + \Delta\phi$. Because the APs are assumed to be uniformly distributed within the circular cell, the area within the strip is proportional to the probability of AOA at the UT. With some manipulations, the area can be expressed as

$$A(\phi \leq \alpha \leq \phi + \Delta\phi) = \frac{1}{2}\sqrt{R^2 + D^2 + 2RD\cos\left(\sin^{-1}(D\sin\phi/R) + \phi\right)}$$
$$\times \sqrt{R^2 + D^2 + 2RD\cos\left(\sin^{-1}(D\sin(\phi + \Delta\phi)/R) + \phi + \Delta\phi\right)} \sin\Delta\phi$$

$$(A.1)$$

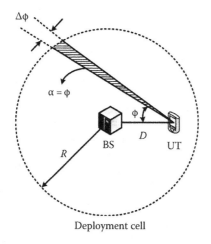

Deployment cell

FIGURE A.1 Illustration of the geometry for deriving the pdf of AOA at the UT when UT lies within the deployment cell.

Subsequently, the probability of AOA at UT is derived as

$$\Pr(\phi \leq \alpha \leq \phi + \Delta\phi) = \frac{A(\phi \leq \alpha \leq \phi + \Delta\phi)}{\pi R^2} \tag{A.2}$$

and the pdf of AOA is given by

$$f_\phi(\phi) = \lim_{\Delta\phi \to 0} \frac{\Pr(\phi \leq \alpha \leq \phi + \Delta\phi)}{\Delta\phi} = \frac{R^2 + D^2 + 2RD\cos\left(\sin^{-1}(D\sin\phi/R) + \phi\right)}{2\pi R^2} \tag{A.3}$$

This completes the derivation.

REFERENCES

[1] W. Roh and A. Paulraj, Outage performance of the distributed antenna systems in a composite fading channel, in *Proc. IEEE Veh. Technol. Conf. '02 Fall*, Vancouver, British Columbia, Canada, pp. 1520–1524, 2002.

[2] N. H. Dawod, I. D. Marsland, and R. H. M. Hafez, Improved transmit null steering for MIMO–OFDM downlinks with distributed base station antenna arrays, *IEEE J. Select. Areas Commun.*, 24, 419–426, 2006.

[3] V. Roy and C. L. Despins, Planning of GSM-based wireless local access via simulcast distributed antennas over hybrid fiber-coax, in *Proc. IEEE GLOBECOM'99*, Rio de Janeireo, Brazil, pp. 1116–1120, 1999.

[4] H. Yanikomeroglu and E. S. Sousa, Power control and number of antenna elements in CDMA distributed antenna systems, in *Proc. IEEE ICC'98*, Atlanta, GA, pp. 1040–1045, 1998.

[5] Y. Chen, C. Yuen, Y. Zhang, and Z. Zhang, Diversity gains of generalized distributed antenna systems with cooperative users, in *Proc. IEEE WCNC 2007*, Hong Kong, China, 2007.

[6] Y. Chen, C. Yuen, Y. Zhang, and Z. Zhang, Cross-correlation analysis of generalized distributed antenna systems with cooperative diversity, in *Proc. IEEE VTC 2007 Spring*, Dublin, Ireland, 2007.

[7] J. N. Laneman and G. W. Wornell, Distributed space-time-coded protocols for exploiting cooperative diversity in wireless networks, *IEEE Trans. Inform. Theory*, 49, 2415–2425, 2003.

[8] T. B. Sørensen, Slow fading cross-correlation against azimuth separation of base stations, *IEE Electron. Lett.*, 35, 127–129, 1999.

[9] F. Graziosi, M. Pratesi, M. Ruggieri, and F. Santucci, Joint characterization of outage probability and handover performance in cellular mobile networks, in *Proc. IEEE Veh. Technol. Conf. (VTC98)*, Ontario, Ottawa, Canada, pp. 1054–1058, 1998.

[10] T. Klingenbrunn and P. Mogensen, Modelling cross-correlated shadowing in network simulations, in *Proc. IEEE VTC'99*, Amsterdam, Netherlands, pp. 1407–1411, 1999.

[11] K. Zayana and B. Guisnet, Measurements and modelisation of shadowing cross-correlations between two base-stations, in *Proc. IEEE ICUPC'98*, pp. 101–105, 1998.

[12] T. S. Rappaport, *Wireless Communications Principles and Practice*, Chapter 3, Englewood Cliffs, NJ: Prentice-Hall, 1996.

[13] L. Q. Hu, H. Yu, and Y. Chen, Path loss models based on stochastic rays, *IET Microw. Antennas Propag.*, 1(3), 602–608, 2007.

[14] L. Q. Hu, H. Zhu, and Y. Chen, Generalized path loss model for wireless channels in homogenous propagation environments, in *Proc. 2007 Asia-Pacific Microwave Conference*, pp. 1–4, Bangkok, Thailand, 2007.

[15] M. Franceschetti, J. Bruck, and L. Schulman, A random walk model of wave propagation, *IEEE Trans. Antennas Propag.*, 52(5), 1304–1317, 2004.

[16] D. Ullmo and H. U. Baranger, Wireless propagation in buildings: A statistical scattering approach, *IEEE Trans. Veh. Technol.*, 47(9), 947–955, 1999.

[17] R. Janaswamy, An indoor pathloss model at 60 GHz based on transport theory, *IEEE Antennas Wireless Propag. Lett.*, 5, 58–60, 2006.

[18] G. Franceschetti, S. Marano, and F. Palmieri, Propagation without wave equation toward an urban area model, *IEEE Trans. Antennas Propag.*, 47(9), 1393–1404, 1999.

[19] S. Marano, F. Palmieri, and G. Franceschetti, Statistical characterization of ray propagation in a random lattice, *J. Opt. Soc. Am. A*, 16(10), 2459–2464, 1999.

[20] M. Franceschetti, Stochastic rays pulse propagation, *IEEE Trans. Antennas Propag.*, 52(19), 2742–2752, 2004.

[21] S. Marano and M. Franceschetti, Ray propagation in a random lattice: A maximum entropy, anomalous diffusion process, *IEEE Trans. Antennas Propag.*, 53(6), 1888–1896, 2005.

[22] L. Zheng and D. N. C. Tse, Diversity and multiplexing: A fundamental tradeoff in multiple antennas channels, *IEEE Trans. Inform. Theory*, 49, 1073–1096, 2003.

[23] R. S. Kennedy, *Fading Dispersive Communication Channels*, London: John Wiley and Sons, 1968.

[24] A. Sendonaris, E. Erkip, and B. Aazhang, User cooperation diversity—Part I: System description, *IEEE Trans. Commun.*, 51, 1927–1938, 2003.

[25] A. Sendonaris, E. Erkip, and B. Aazhang, User cooperation diversity—Part II: Implementation aspects and performance analysis, *IEEE Trans. Commun.*, 51, 1939–1948, 2003.

[26] J. N. Laneman, D. N. C. Tse, and G. W. Wornell, Cooperative diversity in wireless networks: Efficient protocols and outage behavior, *IEEE Trans. Inform. Theory*, 50, 3062–3080, 2004.

[27] J. Boyer, D. D. Falconer, and H. Yanikomeroglu, Multihop diversity in wireless relaying channels, *IEEE Trans. Commun.*, 52, 1820–1830, 2004.

[28] R. U. Nabar, H. Bolcskei, and F. W. Kneubuhler, Fading relay channels: Performance limits and space-time signal design, *IEEE J. Sel. Areas Commun.*, 22, 1099–1109, 2004.

[29] A. J. Jardine, J. S. Thompson, S. McLaughlin, and P. M. Grant, Dual antenna cooperative diversity techniques, *IET Proc. Commun.*, 153(4), 556–564, 2006.

[30] K. J. R. Liu, W. Su, and A. Wittneben (Eds.), Special issue on cooperative communications and networking, *IEEE J. Sel. Areas Commun.*, 25(2), 2007.

[31] G. K. Kramer, R. Berry, A. El Gamal, H. El Gamal, M. Franceschetti, M. Gastpar, and J. Laneman (Eds.), Special issue on models, theory, and codes for relaying and cooperation in communication networks, *IEEE Trans. Inform. Theory*, 53(10), 2007.

[32] H. Zhuang, L. Dai, L. Xiao, and Y. Yan, Spectral efficiency of distributed antenna system with random antenna layout, *IEE Electron. Lett.*, 39, 495–496, 2003.

[33] N. Devroye, P. Mitran, and V. Tarokh, Achievable rates in cognitive radio channels, *IEEE Trans. Inform. Theory*, 52, 1813–1827, 2006.

[34] N. Devroye, P. Mitran, and V. Tarokh, Limits on communcation in a cognitive radio channel, *IEEE Commun. Mag.*, 44, 44–49, 2006.

[35] P. Petrus, J. H. Reed, and T. S. Rappaport, Geometrical-based statistical macrocell channel model for mobile environments, *IEEE Trans. Commun.*, 50, 495–502, 2002.

[36] A. Papoulis and S. U. Pillai, *Probability, Random Variables and Stochastic Processes*, 4th edn., New York: McGraw-Hill, 2002.

[37] K. Yu, M. Bengtsson, B. Ottersten, D. McNamara, P. Karlsson, and M. Beach, Modeling of wideband MIMO radio channels based on NLOS indoor measurements, *IEEE Trans. Veh. Technol.*, 53, 655–665, May 2004.

[38] J. P. Kermoal, L. Schumacher, K. I. Pedersen, P. E. Mogensen, and F. Frederiksen, A stochastic MIMO radio channel model with experimental validation, *IEEE J. Sel. Areas Commun.*, 20, 1211–1226, 2002.

[39] H. Ozcelik, M. Herdin, W. Weichselberger, J. Wallace, and E. Bonek, Deficiencies of "Kronecker" MIMO radio channel model, *IEE Electron. Lett.*, 39(16), 1209–1210, 2003.

[40] H. L. Bertoni, *Radio Propagation for Modern Wireless Systems*, Chapter 2, Englewood Cliffs, NJ: Prentice-Hall, 1999.

[41] G. Grimmett, *Percolation*, Chapter 5, New York: Springer-Verlag, 1989.

[42] D. M. J. Devasirvatham, C. Banerjee, M. J. Krain, and D. A. Rappaport, Multi-frequency radiowave propagation measurements in the portable radio environment, in *Proc. IEEE ICC'90*, Atlanta, GA, pp. 1334–1340, 1990.

[43] T. M. Cover and J. A. Thomas, *Elements of Information Theory*, Chapter 11, New York: Wiley, 1991.

[44] L. Q. Hu and H. Zhu, No-wave approaches and its application to received power of radio wave propagation, in *Proc. 2006 China–Japan Joint Microwave Conference*, Chengdu, China, pp. 605–609, 2006.

[45] B. K. Holland, *What are the Chances?: Voodoo Deaths, Office Gossip, and Other Adventures in Probability*, Chapter 6, Baltimore: The Johns Hopkins University Press, 2002.

[46] I. S. Gradshteyn and I. M. Ryzhik, *Table of Integrals, Series, and Products*, 6th edn., Chapters 3 and 8, New York: Academic, 2000.

[47] L. L. Xie and P. R. Kumar, A network information theory for wireless communication: Scaling laws and optimal operation, *IEEE Trans. Inform. Theory*, 50(5), 748–767, 2004.

[48] L. L. Xie and P. R. Kumar, On the path-loss attenuation regime for positive cost and linear scaling of transport capacity in wireless networks, *IEEE Trans. Inform. Theory*, 52(6), 2313–2328, 2006.

[49] A. M. Sayeed, Deconstrcting multi-antenna fading channels, *IEEE Trans. Signal Processing*, 50, 2563–2579, 2002.

12 Cooperative ARQ Protocols

*Julián Morillo-Pozo, David Fusté-Vilella,
and Jorge García-Vidal*

CONTENTS

This chapter reviews first the main ideas that motivate and serve as background for the research field of cooperative automatic repeat request (C-ARQ) protocols. It revisits some of the classical results on ARQ protocols, such as basic throughput versus reliability trade-offs, and reviews well-known concepts such as hybrid ARQ (H-ARQ) type I and II or packet combining methods. Regarding the cooperative networking field, its historical development is also reviewed, from the classical relay channel and the cooperative diversity channels to the more recent proposals.

The more important results on setting the potentials and the limits on what can be achieved by means of cooperation are also presented, to introduce next the main C-ARQ proposals made to date. These proposals differ on whether they are devised as a cross-layer mechanism between layers 1 and 2, they are pure layer 2 mechanisms, or they are a cross-layer mechanism between layers 2 and 3; on whether they are devised for cellular, sensor, infrastructured wireless local area network (WLAN), ad hoc, or mesh networks; or on whether they use or not frame combining.

The challenges and opportunities presented by the C-ARQ protocols, and its impact on the architecture of the future next generation wireless networks are also discussed as a conclusion of the chapter.

12.1   INTRODUCTION

The proliferation of mobile wireless multifunctional devices is an important phenomenon that will shape the future Internet. Up to now, most of these devices are connected using legacy protocols, which use a point-to-point abstraction for links, ignoring some fundamental properties of wireless media, such as the broadcast nature of wireless transmission or the various forms of diversity. During recent years, the research community has been increasingly interested in cooperative protocols, which allow nodes to virtually increase their communication and processing capacities, and is aware of the special characteristics of wireless transmission for reducing the cooperation overhead or for obtaining diversity gains.

An important example of these cooperative mechanisms is the C-ARQ. ARQ is a widely used method for achieving reliability by means of retransmission. In its cooperative version, nodes other than the sender or the destination play a role in the packet retransmission process, thus exploiting the broadcast nature of wireless transmission and space diversity.

In this chapter, we review the main results obtained to date in the field of C-ARQ protocols research. Our goal is to present a comprehensible perspective on what is known and on what the current challenges are in this new and interesting area of networking. The chapter is organized as follows: In Section 12.2, we briefly describe the classical ARQ mechanisms used extensively in current communication systems. In Section 12.3, we address the specific case of the general C-ARQ protocol, with a discussion of some of its distinctive characteristics. Section 12.4 is devoted to exploring the potential and limits of this approach. In Section 12.5, we describe the main proposals made to date. Section 12.6 presents experimental results, which show the improvements in terms of link reliability and throughput that can be achieved. Finally, Section 12.7 is devoted to conclusions and open issues.

12.2   AUTOMATIC REPEAT REQUEST PROTOCOLS

Given the error-prone nature of wireless networks, error-control mechanisms are a critical aspect of their design. It is therefore not surprising that although they have been the subject of study for many years, they still receive considerable attention. In this section, we briefly review some of the classical error-control mechanisms. Many excellent surveys and books are available on this subject. The interested reader is referred, for instance, to Refs. [1–3], or Ref. [4] for more in-depth discussions.

Basically, error control is usually achieved through two distinct approaches: ARQ and forward error correction (FEC).

In ARQ systems, before transmission, information is split into messages to which parity-check bits are added, to form codewords of an error-detecting code, which are then transmitted through the communication channel. In the receiver, the code is used to identify erroneously received codewords. If the receiver does not detect any error, the received message is assumed to be error free and is delivered to the user. There is always a certain probability of an undetected error, but if enough redundant bits are used, this probability can be made very small. If the receiver detects an error, it discards the erroneously received codeword, and a request for a retransmission is sent through a return channel. Depending on the number of packets that can be sent without a positive acknowledgment, and on how the packet retransmissions are managed, we have three retransmission strategies: stop-and-wait (SW), go-back-N (GBN), and selective-repeat (SR); see for instance Ref. [5] for a detailed discussion.

ARQ is a highly reliable, simple, and well-understood mechanism. Its main drawback is that throughput is not constant, because it depends on the number of transmissions needed to receive a correct codeword. Because of these characteristics, ARQ has become very popular in data communication systems, where keeping a constant throughput is not an essential requirement.

In FEC systems, an error-correcting code is used to detect the presence of erroneous bits in the received data and to correct them if possible. The receiver uses this code to determine the exact error locations, thus delivering the correct information to the user. However, unless very long powerful codes are used, there is a nonnegligible probability that the correction process will fail, delivering incorrect data, which means that it is not easy to achieve high reliability. Nevertheless, because no retransmissions are required, FEC systems maintain a constant throughput regardless of the bit error rate. FEC is especially appropriate in communication systems, where return channels are not available or retransmission is not possible.

ARQ and FEC drawbacks can be overcome by using combinations called H-ARQ schemes, which essentially consist in embedding the FEC into an ARQ system. Usually, these mechanisms are classified into type-I and type-II schemes.

Type-I H-ARQ schemes use a code that is designed for simultaneous error detection and correction. When the reception errors can be corrected, the system behaves simply as an FEC. When the code detects an uncorrectable error, a retransmission is requested. Because a code is used for both error detection and correction, it requires more parity-check bits, leading to an increased overhead. These error control schemes work quite well on channels that are essentially stationary, because the coding scheme can be selected to make retransmission relatively infrequent while maintaining an acceptable coding overhead. In nonstationary channels, however, and unless some adaptative strategy is used (see for instance Ref. [6]), the protocol can become quite inefficient, as for increasing signal-to-noise ratio (SNR) the error detection coding can lead to a useless overhead, while for decreasing SNR, the probability that each received packet contains uncorrectable error patterns increases.

Type-II H-ARQ schemes naturally include a rate adaptation strategy, which make them attractive for nonstationary channels. In these schemes, the transmitter first sends the codeword obtained and adds the redundancy bits of an error-detecting code to the data message. If this first transmission is correct, the process ends. Otherwise, the receiver stores the erroneously received codeword and requests a retransmission. The retransmission, however, does not consist in sending again the codeword. Instead, a packet containing redundancy bits to be used in a packet recovery process is sent. The receiver thus performs an error-correction operation, using the packet containing the data bits and the additional redundancy bits received in the second packet. If this operation is not successful, the received redundancy bits are stored and the process is repeated, requesting transmission of additional redundancy bits.

Different variants of type-II H-ARQ schemes have appeared in the literature. In Ref. [7], the additional redundancy bits are obtained using an invertible error-correcting code. Given the invertibility property of the error-correcting code, if this packet is correctly received, the receiver can recover the original data message, and the process ends. Otherwise, it can append the received redundancy bits with the previously stored codeword to form a 1/2 rate code, performing an error-correction operation. If this transmission is unsuccessful, the packet is combined with the previously stored redundancy bits, and the process continues until the packet is correctly received or reconstructed. Other authors have proposed the use of punctured codes: Transmission starts with the highest rate code. If the first transmission fails, more redundancy bits are added in retransmitted packets; see for instance Ref. [8] (rate compatible punctured convolucional codes [RCPC]), [9] (punctured convolutional coding), [10] (punctured Reed–Solomon), or [11] (punctured maximum distance codes).

As we have seen, in type-II H-ARQ schemes, erroneously received packets are not discarded, but they are used to improve the packet reconstruction process. This idea is extensively exploited in packet-combining schemes, first proposed in Ref. [12]; see for instance Refs. [13–15], or Ref. [16]. Hard decision is used in Ref. [12]: the erroneous copies are XORed to locate the errors in a combined copy, using an exhaustive search method. This simple concept has the advantage that it can be used in existing transmitters that only append frame check sequences (FCS) and that only software or firmware modifications are needed. In soft-decision schemes (see [13]), the receiver combines noisy packets to obtain a code rate that is low enough to achieve reliable communication.

A common characteristic of all the schemes discussed here is that the retransmitting node is always the node that performs the transmission of the original packet, which clearly has a correct copy of it. C-ARQ introduces new dimensions into the design space, allowing nodes other than the sender or the destination to play a role in the packet retransmission process, thus exploiting the broadcast nature of wireless transmissions and space diversity.

12.3 C-ARQ PROTOCOLS

In this section, we describe a simple canonical C-ARQ mechanism and discuss the novel aspects that appear in its design. We focus on the problem of transmitting packets over a wireless link between two nodes, T and R. Nodes T and R could be the base station (BS)/access point (AP) and the client in a cellular network/infrastructured wireless local area network (WLAN), two nodes of an ad hoc or mesh wireless network, or two networked sensors. A medium access control (MAC) ensures that nodes transmit on essentially orthogonal channels as well as integrating the required signaling.

We assume that there is a set of nodes, which we call cooperators, whose conditions for transmission to R are especially favorable, probably because they are in the vicinity of R, and which are also willing to cooperate with R. We call this set, in which node R is included, $C(R)$, while the number of elements of this set $|C(R)| = M$ (see Figure 12.1).

Let T transmit a frame addressed to R. We assume that nodes of $C(R)$ can identify that the final destination of the frame is R even in the presence of transmission errors. After receiving the frame, every node checks for its correctness, using for instance a CRC. If node R finds that the frame has suffered errors, it sends a signaling packet to its cooperators requesting a retransmission of their frame copies. If any of its cooperators has the correct frame, it will retransmit it. Because we assume that nodes from $C(R)$ have a better transmission channel to R than node T, the cooperator with the correct copy of the frame will keep retransmitting the packet until it is correctly received by R. If there are no such correct copies, R will ask for a packet retransmission from node T.

In the previous description we assume that the cooperators are close to R. However, we could also consider the case in which they are close to T, meaning that we can take for granted that they receive correctly the packet sent by T, and that the retransmission from the cooperator to R exploits the fact that its channel to R suffers independent fading. Of course, intermediate situations are also possible.

This general description leaves many open issues and gives room for many possible variants. For instance, we have not described any mechanism for determining the cooperating neighbors, for requesting the copies of the packets, for increasing the probability of receiving a correct destination address in a packet, or for reducing the exchange overhead between node R and its cooperators.

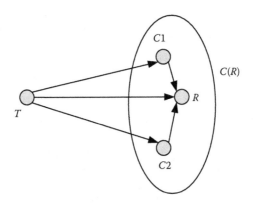

FIGURE 12.1 C-ARQ scenario. In this case $|C(R)| = M = 3$.

Moreover, we could exploit some ideas borrowed from H-ARQ or packet-combining schemes. For instance, if any of the nodes in $C(R)$ has a correct copy of the packet, it can send redundancy bits to R to help in the packet recovery process, thus introducing a type II H-ARQ scheme. Another possibility is that even if the frame was received with errors by the nodes in $C(R)$, these nodes can keep a copy of it. If node R finds that the frame has suffered errors, and there are no cooperators with correct copies of it, R will ask for incorrect copies of the frame. Once R has received the information sent by its cooperators (assuming that no received copy of the packet has a correct CRC), R produces a possible correct copy of the frame using a suitable algorithm. This possible correct copy is checked against the CRC. If this reconstructed packet is still incorrect, a packet retransmission from node T is requested.

Finally, some authors propose that several nodes retransmit simultaneously using a suitable space-time code for achieving transmit diversity, as in the Alamouti scheme briefly discussed later.

Although C-ARQ is essentially a variant of the classical ARQ mechanisms, it introduces two new dimensions, namely space diversity, as packets are received at different points of the space suffering different shadowing and multipath fading effects, and cooperation, because a set of nodes must be coordinated for a cooperative frame recovery process. Some consequences of this are as follows:

- Like other cooperative mechanisms, C-ARQ basically exploits broadcast transmission and space diversity. Broadcast transmission introduces a natural repetition code, in the sense that many copies of the same packet are received by different nodes. Space diversity allows a diversity gain, because the packet copies received by nodes of $C(R)$ are sent through statistically independent channels.
- Like any other cooperative protocol, C-ARQ depends on the willingness of nodes to cooperate. In some applications, in which all nodes belong to a single organization and work toward a common goal, (for instance, a public-safety communication system or in a mesh network backbone) this poses no problem. In networks where users do not have any relationship with each other, apart from sharing the same network, cooperation incentives can be necessary; see for instance Chapter 1 of Ref. [17].
- C-ARQ is an opportunistic mechanism, meaning that its effectiveness depends on the fact that the receiving node has nearby nodes willing to cooperate. In general, this cannot be guaranteed. Designers of C-ARQ mechanisms should aim at having an increased performance when cooperation is possible or otherwise, a performance equivalent to noncooperative mechanisms.
- Cooperator selection policy is a key factor influencing the overall performance of C-ARQ. Provided that enough potential cooperators are present, the node must choose the optimal set of cooperator nodes, defining the size of this set and the characteristics of the nodes. Note that the receiving node usually has only a partial knowledge of the potential cooperators, and depending on the degree of this knowledge, the decision policy should be chosen accordingly. Cooperator selection is largely an open research issue. Studies dealing with the performance of cooperative schemes depending on the cooperators position can be found in Ref. [18] or Ref. [19].
- In C-ARQ, some message exchange is required to coordinate the operation of the cooperators, which can introduce a considerable overhead, having a negative impact on performance. Defining mechanisms with low coordination overhead is critical for an optimal operation of the cooperative protocols; see for instance Ref. [20]. Moreover, exploiting the broadcast transmission means that nodes have to process and buffer more packets than in noncooperative networks. In some cases, for instance in sensor networks, this could lead to node resource exhaustion or to livelock situations.
- During cooperation phases, cooperator nodes cannot be in sleep state. It is well known that in low-power transmission systems, reception power and transmission power are not

drastically different, meaning that the power consumption due to many receptions of the same packet is nonnegligible. As an example, in 802.11 cards, network interface card (NIC) power consumption during transmission, reception, and idle states are similar: for instance, transmitting state 1675 mW, receiving state 1425 mW, and idle state 1319 mW; see Ref. [21]. Only when the NIC is put into sleep state, the power consumption is significatively reduced (e.g., 177 mW in Ref. [21]). On the other hand, cooperation can be also used to balance energy consumption among nodes; see Refs. [18,22].

- When using cooperative protocols, the MAC layer should be designed to allow nonstandard signaling message exchanges. As an example, in several cooperative protocols, acknowledgment messages can be sent by more than one node or by a node that is not the one to which the frame was addressed, which differs considerably from current MAC signaling, such as the IEEE 802.11 signaling; see for instance Refs. [20,23,24].

- Most communication stacks deal with objects such as users, services, nodes, network attachment points, and/or paths; see Ref. [25]. A stack with cooperative protocols could include new objects, such as the object cloud suggested in Ref. [26] which includes a given node and its cooperators. New objects lead to new addressing schemes required to designate the new objects.

- In multi-hop networks, there is a strong interaction between reliability mechanisms and routing. It is well known that many ad hoc routing protocols suffer from electing poor links while building routing paths. In this context, increasing link reliability can have a large impact on the network performance. Furthermore, it is well known that transmission control protocol (TCP) over wireless networks can have a poor performance, due to frequent packet losses. Once again, the study of the interaction between layer-2 reliability mechanisms, such as C-ARQ, and TCP is of the utmost importance. In both cases we are calling for a cross-layer design of different protocols.

12.4   COOPERATION POTENTIALS AND LIMITS

Most papers set the potential gains of C-ARQ schemes referring to two well-known models: antenna array and cooperative diversity systems.

Systems that use arrays of multiple transmit and receive antennas are a way of achieving independent paths in wireless systems, exploiting what is known as space diversity. This independency appears when there is a minimal separation between the antennas in the array. This minimal separation depends on the scattering environment, antenna directivity, and so on. Antenna array systems can be single-input multiple-output (SIMO), multiple-input single-output (MISO), or multiple-input multiple-output (MIMO). We briefly discuss the two first types of systems. The reader can find a more comprehensive review in, for instance, Chapter 7 of Ref. [27].

In SIMO systems, we have a single transmitter antenna and an array of multiple receiver antennas. Signals of each receiver antenna are combined to obtain a resultant signal that is passed through a standard demodulator. The process can obtain two types of performance gains: array gain and diversity gain. Array gain results from coherent combining of the multiple receive signals, leading to an increased equivalent SNR. Diversity gain occurs because the combining of multiple independent fading paths leads to a more favorable fading distribution for the resulting signal.

In MISO systems, there are multiple transmitter antennas, with the transmit power divided among them, and a single receiver antenna. The benefits of a MISO system depend on the channel knowledge of the transmitter. In the more usual case, in which the channel is not known at the transmitter, and when the transmitter has two antennas, the system can still achieve a diversity gain, although not an array gain, by using what is known as the Alamouti scheme [28].

In array systems, there is no penalty cost in putting together the signals of the different branches, while in C-ARQ, there is a nonnegligible overhead in the coordination and communication between cooperator nodes and the receiving node. However, array systems are an interesting conceptual model for understanding C-ARQ performance, because they introduce the concept of diversity gain. Diversity gain can be quantified by using the diversity order, d, defined as

$$d \equiv \lim_{\text{SNR} \to \infty} \frac{-\log(\text{BER}(\text{SNR}))}{\log(\text{SNR})}, \qquad (12.1)$$

where
   SNR is the signal-to-noise ratio
   BER(SNR) is the bit error rate in the function of the SNR of the received signal

It can be shown that in Rayleigh fading channels, an array of M receive antennas can achieve a diversity order of M when the optimal signal combining policy is used; see for instance Chapter 7 of Ref. [27]. This can lead to a remarkable reduction in BER. Cooperative communications can achieve similar reductions, but using single-antenna nodes.

Cooperative diversity is another model frequently used for assessing the potentials of cooperative communications; see Refs. [29,30]. The cooperative diversity model is a recently proposed variant of the classical relay channel; see Ref. [31]. The relay channel is a three-terminal communication channel. The source and destination establish a communication, and there is a third node, the relay, which overhears the transmission of the source and which cooperates with the aim of improving the communication between source and destination. In the cooperative diversity variation, we have two nodes with information to communicate to a receiver node. Both source nodes also act as a relay of the other source node.

For describing the cooperative diversity model we use the overview in Ref. [32]. The reader is referred to this work and references therein for a more detailed discussion. Assume that radios $0, \ldots, T-1$ cooperate to communicate a message produced by radio 0 to a given destination radio T. For sending a message $\underline{s} = (s[0], \ldots, s[M-1])^T$ of M symbols taken from an alphabet of size S, the system makes K uses of the channel, thus achieving a spectral efficiency of $M \times \log(S)/K$. We assume a narrowband frequency flat fading channel, meaning that if $\mathbf{X}$ is a $K \times T$ matrix with column r given by the symbols transmitted by radio r during the time slots $0, \ldots, K-1$, the received signal in radio i can be characterized as

$$\underline{y_i} = \mathbf{X}\underline{h_i} + \underline{w_i} \qquad (12.2)$$

where $\underline{h_i}$ is a vector of the relay fading coefficients, which is assumed to be known at radio i, and the components of vector $\underline{w_i}$ are mutually independent random variables, which model the thermal noise and other interference received at radio i during each time slot. Cooperative diversity schemes can be characterized by a set of $K \times T$ matrices $\mathbf{G}(\underline{s})$ associated with each message $\underline{s}$. This matrix defines the transmitted signals of radios $0, \ldots, T-1$ when message $\underline{s}$ must be communicated from radio 0 to radio T.

The two simplest forms of cooperative diversity are amplify and forward (AF) and decode and forward (DF). Assume now $M = 1$ and $K = T$, leading to a system with spectral efficiency of $1/T$. In AF, radio 0 transmits a symbol associated to message $s \in \{0, 1\}$. Radio $r \in \{1, \ldots, T-1\}$ receives symbol $\underline{y_r}$ and transmits at time slot r the signal $\beta_r Z_r$, where β_r is a constant that guarantees that node r transmits at power P_r and $Z_r = h_r^H \underline{y_r}$ (recall that we assume $\underline{h_r}$ is known at radio r). In other words, in AF each node retransmits a scaled version of the samples received through an orthogonal channel and is defined by matrix $\mathbf{G}(s) = \text{diag}(\beta_0 Z_0, \ldots, \beta_{T-1} Z_{T-1})$.

In DF each radio first estimates the symbol transmitted by radio 0 and retransmits it in orthogonal channels scaling the transmission power to guarantee that radio r transmits with power constraint P_r, leading to a matrix $\mathbf{G}(s) = \text{diag}(\sqrt{P_0}\widehat{s_0}, \ldots, \sqrt{P_{T-1}}\widehat{s_{T-1}})$.

As one may observe, the spectral efficiency of these schemes decreases with the number of relays. Greater spectral efficiency can be achieved using space-time codes, leading to a $K \times T$ matrix $\mathbf{G}(s)$ with $K < T$.

Cooperative transmission is equivalent to a MISO system with a per antenna power constraint; see Ref. [32]. In fading channels, one can find the diversity order of these schemes by using the concept of outage probability. Assuming that we chose a fixed rate R to encode a message, and that during the message transmission the fading coefficients are quasistatic, then the channel mutual information becomes a random variable as a function of the fading coefficients. The outage probability is defined as the probability of the mutual information falling below the rate R. We can thus define a diversity order in similar terms as before, substituting the BER(SNR) by the outage probability P_{outage}(SNR). In Ref. [33] it is shown that for $T = 2$, both AF and DF achieve a full diversity order of 2.

The cooperative diversity systems described above assume that a fixed time division multiple access (TDMA) frame structure is used, meaning that relays always cooperate. This lack of flexibility can decrease system throughput, which can be overcome using the ARQ concept: nodes cooperate only if the received frame is incorrect. Reference [34] analyses the achievable throughput in such a system for two different schemes: H-ARQ type I (HARQ-TI) and H-ARQ with chase combining (HARQ-CC).

Using the previous notation, radio 0 sends a message to radio T, while radios $r \in \{1, \ldots, T-1\}$ act as potential cooperators. Block Rayleigh fading is assumed, meaning that during time slot n, the vector of relay fading coefficients $h_r^{(n)}$ remains constant, although it changes independently with each transmission, while the coefficients of these vectors are mutually independent, circularly symmetric, complex Gaussian variables with power normalized to unity. In the analysis it is also assumed that the average power received by the relay from the source node and the other relays is larger than the average power received by radio T. All nodes have a transmit power of P and are impaired by Gaussian noise with one-sided power spectral density of N_0.

In HARQ-TI, during time slot $N = 1$, the source ($r = 0$) sends a packet containing a codeword with a rate C_0. The packet is correctly received by radio $r \in 1, \ldots, T$ if C_0 is below the instantaneous achievable rate, given by $\log_2(1 + |h_{r0}^{(1)}|^2 \frac{P}{N_0})$. If the packet is incorrectly received at the destination, the cooperators that have received the packet correctly and the sender node retransmit the packet in the next time slot, using space-time-block-coding. If $\widetilde{k}^n$ is the number of nodes that perform the retransmission at time slot n, the instantaneous achievable rate at node T is given by

$$ log_2\left(1 + \left(\left|h_{T0}^{(1)}\right|^2 + \sum_{i=1}^{\widetilde{k}^n}\left|h_{Ti}^{(1)}\right|^2\right)\frac{P}{N_0}\right), \tag{12.3} $$

where $i \in 1, \ldots, \widetilde{k}^n$ denote the radios that have correctly received the packet at any of the time slots $1, \ldots, n$.

Should the second retransmission be incorrect, the process of transmission and retransmission continues, incorporating at each retransmission new relays that correctly decode the packet. Given that the previous $n-1$ transmissions were unsuccessful, the expression of the probability of unsuccessful decoding at the nth transmission is obtained, thus deriving the expected throughput of the system.

In HARQ-CC, previously received packets are soft-combined before detection, using the already discussed scheme of Ref. [13]. The derivation follows analogous steps, with the difference that in this case the achievable rates depend not only on $\widetilde{k}^n$ but also on $\widetilde{k}^1$, $i \in \{1, \ldots, n-1\}$.

The numerical results show a remarkable increase in throughput with respect to a SISO system, especially for low values of SNR, where performance becomes closer to that obtained by a MISO system with T transmitting antennas.

A further interesting analysis of C-ARQ appears in Ref. [35]. Once again, it is assumed that all channels are subject to flat frequency nonselective Rayleigh fading. The performance of traditional ARQ and three variants of C-ARQ are studied. In all cases, a single relay is assumed. In the first variant of C-ARQ, the relay always retransmits provided it has received a correct copy of the frame. In the second variant, the retransmission is performed by either the source or the relay, depending on which node has a better channel to destination. In the third variant, both the source and the channel retransmit simultaneously, using the Alamouti scheme. Employing an approximate expression for the packet error rate (PER), and assuming a truncated ARQ in which the number of retransmission is limited to one, expressions for PER after retransmission are obtained. The authors show that the three schemes are capable of extracting full diversity order.

An alternative technique of analysis for C-ARQ based on Markov chain modeling is presented in Ref. [36]. The channels between source, destination, and relay are modeled using two-state Markov chains. If the channel is in the good state, all frames are received, while in the bad state all frames are lost. Transition probabilities are fixed to model a quasistatic Rayleigh fading channel. Depending on the state of the different channels, one can derive another Markov chain that indicates the status of the relay node, that is, whether it will decode the packet correctly and whether it will retransmit it successfully if requested. For the case in which we have more than one relay, an equivalent super neighbor node is obtained. Finally, a Markov chain that captures the state of the channels between source and destination and the status of the relay node are derived, and the throughput and the induced delay are then obtained in a subsequent analysis.

12.5 PRACTICAL C-ARQ SCHEMES

In this section, we present the main C-ARQ proposals made to date. These proposals differ as to whether they are devised as a cross-layer mechanism between layers 1 and 2, whether they are pure layer 2 mechanisms, whether they are a cross-layer mechanism between layers 2 and 3; whether they are devised for cellular, sensor, infrastructured WLAN, ad hoc, or mesh networks; or whether they use frame-combining. An interesting example of a practical integration of C-ARQ in a more complete protocol structure are the proposals oriented to the 802.16j protocol.

12.5.1 NODE COOPERATIVE STOP AND WAIT (NCSW)

The NCSW is a C-ARQ scheme proposed in Refs. [36,37]. The protocol operates essentially according to the description given in Section 12.2. The motivation of this proposal is to combat the effects of channel fading. Due to the inherent characteristics of the fading process in wireless channels, frame errors appear in bursts rather than randomly. When the link between two communicating nodes is experiencing frame errors, there is a high probability that the bad channel condition will persist for as long as the transmission time of multiple data frames. Conventional retransmission schemes are not very effective in such environments with bursty frame errors, because the sender retransmits data frames by remaining oblivious to the fact that even the retransmitted packets will encounter bit errors with very high probability due to the persistence of the bad channel condition.

The NCSW protocol either does not focus or does not define how the relationship or coordination between cooperators, source, and destination is established. However, it does introduce the concept of cooperation group defined as a subset of nodes that can reach one another with a single hop. Those groups may be set up during connection stage or link-level handshaking (e.g., RTS/CTS in 802.11). For example, in the case of establishment during link-level handshaking, a good cooperator would be a node that overhears both RTS and CTS packets. For the case of establishment during connection stage, one may consider, for example, a routing protocol such as ad hoc on-demand distance vector AODV [38] in which a route request (RREQ) and a route reply (RREP) are exchanged during the establishment of a path: a node that overhears both the RREQ and the RREP for a given hop would be a good candidate for forming part of the cooperation group for that hop.

Once the cooperation group has been defined, three different channels can also be defined:

- Primary channel: channel between the source and the destination nodes
- Interim channel: channel between the source and a neighbor in the cooperation group
- Relay channel: channel between a neighbor in the cooperation group and the destination

Neighbor nodes keep listening to the shared channel during the transmissions from the source and assist the sender and the receiver if errors occur.

Under these conditions the NCSW is compared with the conventional retransmission scheme in Ref. [37]. In this work, the NCSW is studied by simulation using a Rayleigh fading model for the generation of channels. The main conclusions of the study can be summarized as follows:

- If the qualities of the interim and relay channels are good, having even one or two neighbor nodes can significantly improve the throughput
- However, when the qualities of the interim and relay channels are poor, more neighbor nodes are required to achieve the same level of performance gain
- Individual frame errors cannot be prevented, but cooperation of the neighbor nodes can reduce the impact of bursty errors

12.5.2 ARQ-C PROTOCOL

We now focus on a proposal for sensor networks and more specifically for microwave recharged sensor nodes [22]. The ARQ protocols described in Ref. [22] (one of them being the ARQ-C, which is the title of this section) are intended for a reliable and fair data transmission from the sensor nodes to a BS (uplink). On analysis, the RF transmission channel (uplink) is considered to suffer from both path loss and fading. The downlink channel is a very special case due to the fact that it is a microwave (MW) channel used by the BS both to communicate with the sensor nodes and to recharge them. Only path loss is considered for this channel. Three different ARQ protocols are compared:

- Conventional ARQ
- ARQ-C
- ARQ-C^N

In Ref. [22], a performance comparison is carried out among the three protocols in terms of the parameter that the authors call achievable saturation throughput. Results indicate that higher throughput, or equivalently lower power consumption for a target throughput, is achievable by the two cooperative protocols, when compared with the conventional ARQ protocol.

The comparison is made for a very specific system in which sensor nodes do not normally need to receive data from other sensors. An optional module may be added to the sensor nodes to provide the RF reception capability that is required by the two cooperative data link protocols discussed. The BS is responsible for collection of data for the entire set of sensor nodes, and for this purpose, it is necessary to design a data link protocol that makes the RF channel reliable and equally available to the sensor nodes.

As in many other works, it is assumed that transmission errors may occur only on the uplink RF channel, as the sensor node power budget limits the SNR. Transmission errors on the downlink MW channel are negligible due to the relatively high power of the MW signal and can be easily overcome with a timeout-triggered retransmission at the BS.

On this specific system, all decisions are made at the BS. Sensor nodes obey the control frames received from the base station. The BS is responsible for choosing of the cooperating sensor nodes and for scheduling collision-free transmission and retransmissions of the sensor nodes. As long as the BS controls everything, problems such as coordination between nodes, distributed management of the cooperators, distributed election of cooperators, and so on, do not appear.

One of the two proposals made in Ref. [22] is the ARQ-C protocol. If the BS does not successfully receive a transmission from a source node, it requests the frame retransmission from another node (the relay). If chosen wisely, the relay may increase the probability of delivering the data frame successfully without requiring any further retransmission. The relay node can help only if it has correctly received the data frame transmitted by the source, so there is no type of frame combining on this mechanism.

For each transmission attempt, only one of the multiple potential cooperating nodes is chosen to be the relay. The BS makes such a choice, thus effectively creating a situation of load (and power) balancing among the sensor nodes. The BS may choose in a probabilistic way according to some predefined distribution values. Note that the required intelligence resides entirely at the BS. Sensor nodes are ordered to overhear and transmit by the BS via the recharging MW channel.

The ARQ-C^N protocol is a recursive version of the ARQ-C protocol. The rationale behind this recursive protocol is the assumption that the relay node is chosen to have (a) a higher probability of successful (re)transmission than the source (or previous relay) and (b) to require a lower power to transmit the data frame to the BS.

As stated previously, the three protocols are compared in terms of the saturation throughput. The saturation throughput (S) defines the throughput values ($\leq S$) that can be sustained by the system under two constraints: the average power consumption at each sensor node cannot exceed the power recharge rate, and fair access is granted to all sensor nodes in the area surrounding the BS. This criterion for comparison purposes is also used for the election of cooperators, so the cooperating nodes are chosen to maximize the saturation throughput. The problem is formulated mathematically in Ref. [22] using queuing models for each of the three protocols. The formulation consists of two equations representing the two constraints for each case.

For the evaluation and comparison, both path loss and fading are taken into account in the RF uplink transmission, while only path loss is taken into account in the MW downlink recharging signal (only for recharging purposes—recall that the downlink is supposed to be error-free). Fading is assumed to be Rayleigh slow and flat; that is, the fading coefficients are considered constant over a single frame transmission. This assumption combined with the employment of BPSK is very common throughout all C-ARQ analysis and proposals. The last assumption, which is not usually true, is that the energy consumption at the sensor node is due to transmissions, so it is assumed that the energy per bit necessary for overhearing a transmission is negligible.

The two C-ARQ protocols always yield higher saturation throughput than that achieved by the non-C-ARQ. In fact, they may even more than double the saturation throughput of the noncooperative case; or equivalently, the required power to operate the system is half when compared with the non-C-ARQ protocol.

12.5.3 MULTI-RADIO DIVERSITY (MRD)

The MRD wireless system, proposed in Ref. [23], uses path diversity to improve loss resilience in WLANs. MRD coordinates wireless receptions among multiple radios to improve loss resilience in the face of path-dependent frame corruptions over the radio. MRD incorporates two techniques to recover from bit errors and lower the loss rates observed by higher layers, without consuming much extra bandwidth:

- Frame-combining, in which multiple, possibly erroneous, copies of a given frame are combined together in an attempt to recover the frame without retransmission. This technique is also used in Refs. [20,39] although the way in which the frames are combined presents some differences among the different proposals.
- Request-for-acknowledgment (RFA), a low-overhead retransmission scheme that operates above the link layer and below the network layer to attempt to recover from frame-combining failures. Ways of reducing the overhead incurred by these types of protocols are also devised in Ref. [20] although with a totally different approach.

A point of great interest in this proposal is that it has really been designed and implemented as a fully functional WLAN infrastructure based on 802.11a, and throughput gains up to 2.3× have been measured over single radio communications employing 802.11's autorate adaptation scheme.

The approach uses path diversity, relying on multiple APs covering a given area (for uplink diversity) and multiple radios on the user's device (for downlink diversity). The hypothesis underlying this system is one that appears throughout this chapter: because frame losses are often path-dependent (e.g., due to multipath fading), location-dependent (e.g., due to noise), and statistically independent between different receiving radios, multiple radios that all receive versions of the same transmission may together be able to correctly recover a frame, even when any given individual radio cannot. This same assumption is the main hypothesis in all frame-combining proposals, as is the case of Ref. [20].

In MRD, the entity that performs this frame-combining task is the MRD combiner (MRDC).

The MRD frame-combining algorithm divides each frame into blocks. For each block, the algorithm assumes that at least one of the received copies of a frame (including any possible retransmissions) contains the correct bit values for that block. The algorithm then attempts to reconstruct the correct frame by trying every version received for each block. The process succeeds if a particular block combination passes the checksum embedded in the data frame and fails once the search exhausts all possible block choices for each block. In that final case, MRD uses a lightweight retransmission scheme running above the WLAN link layer to further improve error recovery. At the sender, the MRD sender (MRDS) buffers all frames that have not yet been acknowledged and retransmits any frame that it believes has not been successfully received by the MRDC (after frame combining). To prevent adverse interactions caused by ARQ schemes at two different layers, MRD turns off link-layer retransmission altogether. The proposed ARQ protocol is designed to have low overhead, using a request for ACK (RFA) technique rather than the traditional ACKs or NACKs. With RFA, the MRDS explicitly requests an ACK from the MRDC for certain frames and decides whether and when to retransmit frames based on this feedback.

All these techniques and schemes arise from the fact that traditional ARQ-based retransmission works well when the duration of channel degradation is short. But when the channel quality deteriorates for a long period, link-layer ACK becomes ineffective and wasteful.

To improve this traditional ARQ, in the MRD architecture the APs forward all frames—including those that are corrupted—to the MRDC, which filters redundant data frames received by multiple radios and invokes the frame-combining procedure when needed. The MRDS uses the RFA protocol to obtain the results of the frame-combining procedure from the MRDC.

12.5.4 C-ARQ/FC

This section is devoted to the C-ARQ proposal reported in Ref. [20]. This paper proposes a low coordination overhead C-ARQ scheme with an integrated frame combiner, which exploits space diversity and cooperation between neighboring nodes. This scheme is a variant of the basic operation of C-ARQ, called C-ARQ with frame combiner (C-ARQ/FC). In this variant of C-ARQ, even when no cooperator node has a correct copy of the frame, the cooperators send their incorrect frames to the destination node (let us say y) that, using this information, tries to reconstruct the original frame. The frame reconstruction is performed by means of a so-called FC. Only if the frame reconstruction is unsuccessful will y request a retransmission from the original sender (let us say x).

The study shows how both C-ARQ (in its basic operation version) and C-ARQ/FC lead to a considerably low equivalent frame error rate (FER), defined as the probability that a frame requires a retransmission from node x. However, both mechanisms introduce an extra coordination overhead, due to the required node-to-cooperators and cooperators-to-node communication. This is exacerbated in the latter case, as the cooperator nodes must send copies of erroneous frames to y. Decreasing the equivalent FER and increasing the coordination overhead have opposite effects on the efficiency of the protocol. It is thus necessary to study under which conditions these techniques lead to practical benefits.

The main findings derived from Ref. [20] are

1. Both C-ARQ and C-ARQ/FC reduce the equivalent FER considerably compared with the classical ARQ for all the studied scenarios. In channels with a strong LOS component, the equivalent FER for C-ARQ/FC can be several orders of magnitude lower than for C-ARQ. This difference in equivalent FER values is considerably less in the case of NLOS scenarios.
2. In channels with strong LOS component and low SNR, the maximum achievable throughput of C-ARQ/FC is much higher than for C-ARQ and classic ARQ. The difference reduces when the LOS component becomes weaker, when C-ARQ becomes the best option. In Rayleigh channels, for instance, C-ARQ and even classic ARQ achieve a higher efficiency than C-ARQ/FC.

The low coordination overhead C-ARQ/FC protocol proposed in Ref. [20] is motivated by examining the coordination overheads of C-ARQ.

12.5.4.1 Overhead of C-ARQ

The C-ARQ protocol operates in two phases; see Figure 12.2a. Phase 1 starts when node x sends a frame to node y. If the frame is correctly received by y, node y will send back a short signaling message with an acknowledgment, and provided that this short message is received correctly by x, the frame transmission process will be finished. If y receives an erroneous copy of the frame sent by x, Phase 2 of the protocol starts: Node y will broadcast a short signaling message asking their cooperators for a correct copy of the frame. The cooperator nodes start a round robin in which they send either (1) a short signaling message indicating that they do not have a correct copy of the frame or (2) a message with the copy of the received frame, if they have received the frame correctly.

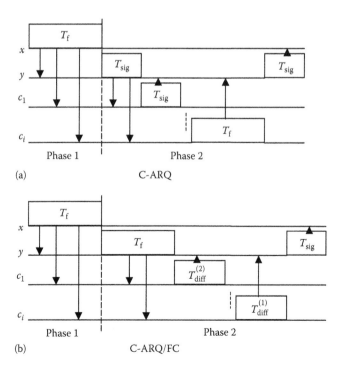

FIGURE 12.2 C-ARQ and C-ARQ/FC operation. We show the case in which cooperator c_i is the first to have the correct copy of the frame.

In the latter case, the round robin is stopped after the correct frame is sent to y. Phase 2 will have different average durations depending on two possible conditions:

1. If no cooperator has a correct copy of the packet, the duration of Phase 2 is given by $A_{\text{Ph2}}^{(1)} = (M + 1) \times T_{\text{sig}}$, where T_{sig} is the transmission time of the signaling messages and $M - 1$ is the number of cooperators of y.
2. Otherwise, the average duration of Phase 2 is $A_{\text{Ph2}}^{(2)} = T_{\text{f}} + ((M/2) + 1) \times T_{\text{sig}}$, where T_{f} is the transmission time of a data frame.

12.5.4.2 Low Coordination Overhead C-ARQ/FC

In C-ARQ/FC protocol, even when any cooperator node has a correct copy of the frame, the cooperators send their incorrect frames to y, which uses this information to try to reconstruct the original frame. However, a direct implementation would lead, however, to a too high coordination overhead. To reduce the signaling overhead, the strong correlation between erroneous frames received by different nodes is exploited: the cooperators do not transmit to node y the entire erroneous frames but the differences (bits that are not equal) with respect to the frame received by y. For reasonable values of SNR, the number of erroneous bits per frame is expected to be small, which means that if the cooperators indicate in some compressed format the bits that differ from the frame received by y, they will need much shorter packets than sending the whole frame.

The C-ARQ/FC operation can be described as follows: Phase 1 is identical to that in C-ARQ. Phase 2 starts when node y receives an erroneous frame sent by x. Node y broadcasts to its cooperators a message that contains its erroneously received frame; see Figure 12.2b. The cooperators will XOR bit-by-bit the frame they received from x and the frame sent by y. They code in a compressed format the position of the bits where the cooperator's frame and the y node's frame differ and establish a round robin for sending this information to node y. If a cooperator has the correct frame, it will indicate so with a flag in the message that carries the differences. In this case the other cooperators will also receive this message with the flag set, and the round robin will be stopped. If no cooperator has the correct frame, the round robin will be completed and node y will use this information for building the correct copy, using a frame combiner algorithm. If it does not succeed, y will ask for a retransmission to node x. In Ref. [20] a simple majority-voting (MV) frame combiner is considered.

The main conclusions drawn from the performance study presented in Ref. [20] are

1. Both C-ARQ and C-ARQ/FC achieve great improvements in terms of equivalent FER over conventional ARQ, even for $M = 3$. This improvement can be of several orders of magnitude for the three different channels considered.
2. C-ARQ/FC improves equivalent FER considerably over C-ARQ for the AWGN channel (i.e., $K \rightarrow \infty$). For Rice $K = 5$ the improvement is lower. For the Rayleigh channel ($K = 0$) the difference is negligible.

12.5.4.3 Protocol Efficiency

Reducing the equivalent FER is usually not the most important effect of an ARQ system. It is expected that the use of an ARQ system will also have a positive impact on the system efficiency, defined as the maximum achievable throughput.

For AWGN channel and low SNR values, when C-ARQ/FC is used the system efficiency increases many times in comparison with that of C-ARQ or classical ARQ. For instance, if SNR = 5, the efficiency achieved with C-ARQ/FC is 0.46, whereas for C-ARQ the efficiency drops to 0.05, and for the classical ARQ the efficiency is 0.02. Note that for SNR = 5, the equivalent FER for C-ARQ/FC is several orders of magnitude lower than for C-ARQ or classical ARQ (see Figure 2 of Ref. [20]), while the signaling overhead is reasonably low. The combined effect leads to a spectacular increase in efficiency. The difference decreases as the SNR increases, and it is negligible for SNR values higher than 8. When the NLOS components become stronger, C-ARQ becomes the more efficient mechanism. For Rayleigh channels, even classical ARQ is more efficient than C-ARQ/FC.

12.5.5   Hybrid Type-II C-ARQ

In Ref. [19], a proposal of a C-ARQ employing FEC in a manner equivalent to the H-ARQ is presented. A related mechanism is also discussed in Ref. [34].

12.5.5.1   Protocol's Operation

1. The source encodes its message with a convolutional code of code rate $R_c = 1/2$ and transmits the odd-numbered bits of the codeword in a broadcast transmission to relay and destination. Note here the introduction of a convolutional code to work in conjunction with the C-ARQ (this is not the case, for example, of NCSW). If the destination decodes successfully, transmission continues with the next block.
2. If decoding fails, the respective mobile station(s) (relay and destination) send out a broadcast message to inform the source (and relay) about this event. The feedback signal can be sent at very low code rate and therefore be assumed to be successfully decodable (once again we have the assumption about the error-free condition of the reverse channel, as in the case, for example, of the analysis of NCSW). If the destination is not able to decode the source message, but the relay successfully decodes it, the relay reencodes the source message to obtain the even-numbered bits of the codeword and transmits them to the destination.
3. If the destination transmits a repeat query and the relay is able to send the even-numbered bits of the encoded source message, the destination assembles the two messages into the complete (received) codeword and repeats the decoding process. If it is not able to decode, a packet error is declared.

This protocol constitutes a cooperative extension to conventional H-ARQ, just as cooperative relaying in general can be seen as an extension of conventional relaying [40]. Its advantage over conventional protocols is twofold:

1. First, the benefits of path loss reduction can be exploited in asymmetric networks because the relay normally needs less power than the source to provide the destination with additional redundancy
2. Second, the relay path provides a statistically independent channel and therefore the diversity required by FEC techniques to code over different fading realizations in environments where the channel coherence time is large and no temporal diversity is available

The performance of the different protocols is determined in terms of packet error rate versus SNR. For direct transmission without feedback, the destination declares a packet error whenever it is not able to decode the source message relying solely on the transmission from the source, which implies that the protocol has only one opportunity. In ARQ-type protocols, the destination can request additional redundancy (from the source for standard H-ARQ and from the relay in the cooperative case), which implies that these protocols have two decoding opportunities.

The SNR required by any protocol to attain a specific target packet error rate (typically, 10^{-2}) is calculated in Ref. [19]. From this the authors calculate the SNR gain of a protocol for any given channel coherence time as the difference between the SNRs required by uncoded direct transmission and the investigated protocol to achieve this specific PER. The SNR gain depends mainly on the available degree of temporal diversity and, for cooperative H-ARQ, the network geometry. Another factor affecting the SNR gain, and not explicitly mentioned by the authors of Ref. [19], is the spatial diversity achieved by C-ARQ.

The simulations show the dependency of the protocols on the presence of temporal diversity.

The results also make clearly visible the dependence of the performance of cooperative H-ARQ schemes on the network geometry. For positions close to the source, the relay experiences better reception conditions, while placing the relay toward the destination results in larger transmit power savings. In fast fading environments, the relay should be placed close to the source, while in slow fading regimes a position roughly in the middle between source and destination is favorable. The results also show how the cooperative approach outperforms conventional H-ARQ for a wide range of network geometries when temporal diversity becomes scarce. This result very nicely illustrates the fact that the cooperative approach provides a second source of diversity. Although not explicitly mentioned by the authors, we can call this second source of diversity the spatial diversity provided by cooperative protocols. In a slow fading environment, the cooperative scheme yields an SNR gain of 2.5 dB over H-ARQ and 12.5 dB over uncoded direct transmission.

One of the most relevant conclusions that can be obtained from Ref. [19] is the importance of the relay position for the performance of cooperative H-ARQ. The performance of the cooperative H-ARQ depends mainly on the reception conditions of the relay, that is, upon the relay's distance from the source. A maximum SNR gain of 2.75 dB over conventional H-ARQ is attained if the relay is positioned at $0.4d$ from the source in the line between source and destination, provided that d is the distance between source and destination.

12.5.6 HARBINGER

We now discuss the protocol proposed in Ref. [41]. In this work, a practical approach to networks comprising multiple relays operating over orthogonal time slots is proposed, based on a generalization of H-ARQ. As in all other C-ARQ protocols seen in this chapter, and in contrast with conventional H-ARQ, retransmitted packets do not need to come from the original source radio, but can instead be sent by relays that overhear the transmission. Assuming a block fading environment, the results obtained in Ref. [41] indicate a significant improvement in the energy-latency trade-off when compared with conventional multi-hop protocols implemented as a cascade of point-to-point links.

Note that when multiple relays are considered, the scheduling of the relays becomes a fundamental issue. The relays must know if and when to transmit and ideally should be able to make these decisions in a distributed fashion. To treat this problem, Ref. [41] proposes a baseline protocol that the authors call Hybrid-ARq-Based INtracluster GEographic Relaying (HARBINGER) and compare it with some other candidate protocols.

12.5.6.1 System Model

In the system model in Ref. [41] a network is divided into clusters of nodes, each cluster $N = \{Z_k : 1 \leq k \leq K\}$ consisting of a source $Z_s = Z_1$, a destination $Z_d = Z_K$, and $K_r = K - 2$ relays. Each node has a single half-duplex radio and a single antenna. When any node in N transmits, all nodes also in N may receive the signal over a block fading channel.

Messages that must travel outside the cluster are routed from cluster to cluster, and a higher level networking protocol is still needed to handle this routing. However, the networking protocol only has to route at the cluster level rather than at the node level. Although this concept is similar to other hierarchical routing protocols, the key difference is that routing within the cluster is now handled implicitly by the retransmission process of the ARQ protocol rather than explicitly by a network layer routing algorithm.

Continuing with the system model, time is divided into slots, which are of equal duration. During slot s, a node may either transmit or receive, but cannot do both.

The source begins by encoding a b bit message into a codeword of length n symbols. The codeword is broken into M blocks, each of length $L = n/M$ and rate $R = b/L$. The code itself may simply be a repetition code, in which case all M blocks are identical and each node will diversity combine all blocks that it has received. More generally, incremental redundancy could be used, whereby each block is obtained by puncturing a rate $r_M = R/M$ mother code. With incremental redundancy, a different part of the codeword is transmitted each time, and after the mth block, a receiver passes the rate $r_m = R/m$ code that it has received until then through its decoder (code combining).

The system model also treats the relationship between routing, relaying, forwarding, and ARQ. With conventional multi-hop, messages must flow through the cluster as a series of direct transmissions determined a priori by a routing algorithm. The destination may not decode the source's direct transmission, even if the instantaneous source–destination SNR is sufficiently high to do so.

If the destination is also allowed to "hear" the source, then several other options are possible, but the problem arises of determining which node should transmit at each time. The solution the authors in Ref. [41] advocate for selecting which node in a multiple relay network transmits a particular block is to embed the selection process into the H-ARQ protocol.

One important characteristic of HARBINGER is that a node is supposed to have an accurate estimate of its own position as well as the position of the source and destination. It can measure its own position with an onboard global positioning system (GPS) receiver, and the header of each message could contain the location of the source and destination.

The protocol is designed so that the node closest to the destination, between the nodes that have decoded the message, will transmit the next block of the message. In practice, the protocol could begin with the source sending out block $m = 1$. Following the transmission of this block, the network enters a contention period. The contention period interval is divided into K_r subintervals, one for each relay. During the first interval, relay Z_{K-1}, which is closest to the destination, sends an ACK if it has decoded the packet; otherwise, it will remain silent. This process continues for all relays. Once a node has sent an ACK signal, the network will then know which node is closest to the destination and that node is free to send the second block. If no node sends an ACK during the contention period, then the second block will simply be sent by the source.

Two other options presented in Ref. [41] for the relay election would be to take into account the instantaneous SNR instead of the distance or to give channel access to the relays randomly.

The three options in the previous section and the traditional multi-hop relaying are compared in Ref. [41]. The different protocols are compared in terms of throughput, energy-delay trade-off, effect of network topology, and total energy consumption, in Ref. [41]. A comparison between diversity combining versus code combining is also shown. The reader is referred to Ref. [41] for more in-depth details.

12.5.7   C-ARQ IN 802.16

The purpose of the IEEE 802.16j standard [42] is to enhance coverage, throughput, and system capacity of 802.16 networks by specifying multi-hop relay capabilities and functionalities of interoperable relay stations and base stations. The multi-hop relay is a promising solution for expanding coverage and for enhancing throughput and system capacity. To allow for the use of multi-hop relay, the standard 802.16j defines three types of stations:

- MMR-BS: Mobile multi-hop relay-enabled base station
- RS: Relay station providing extended cell coverage
- MS: Mobile station that can attach to either 802.16 BS, MMR-BS, or RS

One of the requirements of the 802.16j standard is that quality of service (QoS) and H-ARQ should be supported by relay as defined in 16e systems; moreover, the MS and BS solutions should be 802.16e compliant and the existing 802.16 BS should be able to software upgrade to MMR-BS. The goal of multi-hop relay networks, as is the case for the proposed 802.16j standard, is to allow higher capacities and coverage, but not the absence of infrastructure. The throughput enhancement can be achieved through the idea of virtual antenna arrays in which multiple cooperating relays act as distributed MIMO. The main challenges to be met for this to work are the synchronization and sharing of channel state information. Note that, for the case of 802.16j, routing is not an issue as long as relays are fixed. Relays can help overcome obstacles and improve the capacity by decreasing the distance of transmissions, but at the price of increasing delays. For this reason, the number of hops must be limited to two or three.

There have been several proposals for cooperative mechanisms, called cooperative relaying in this context; see Refs. [43–46]. In Refs. [43–45] the use of cooperative diversity systems is proposed, while Ref. [46] describes the use of a C-ARQ mechanism. Next, we focus on the latter proposal.

Cooperative relay technology provides reliable downlink frame transmission by either transmitting synchronous frames simultaneously or by adopting channel coding mechanisms. Both schemes are applied to PHYsical layer (PHY). In spite of the fact that they can support reliable downlink frame transmission, the technology is only used for downlink frame transmission without MAC support. However, if MAC uses cooperative relays, more reliable frame transmission is achieved. The proposal [46] introduces an ARQ scheme with multiple transmission paths to provide reliable frame transmission by using cooperative relays.

It can be assumed that a MS transmits an uplink frame to at least two cooperative relays. When a RS receives an erroneous frame where only a single bit is incorrect, all bits of the frame are useless because it is impossible to recover the frame. However, another RS may receive it successfully, and thus relay it toward its mobile relay-base station (MR-BS). An MR-BS has a higher probability of receiving a frame successfully compared with utilizing a single relay.

ARQ is one of the reliable transmission schemes provided by MAC. ARQ can be applied to reliable frame transmission by using cooperative relays at the MAC layer. Both MR-BS and MS are the terminal entities each ARQ path begins or ends.

Figure 12.3 shows an example where an ARQ is used by supporting multiple paths when cooperative relays are deployed. In this figure, an MS transmits a frame to both MRS simultaneously, and then each MRS filters out erroneous frames by successfully relaying received frames toward MR-BS. To make this possible, MR-BS is required to configure each MRS to provide the ARQ scheme.

An MR-BS may establish multiple paths for single ARQ-enabled connections between itself and an MS via several MRSs. An MRS on each path should drop incorrectly received frames with corrupt CRC. Otherwise, it will forward the successfully received frames to the next destination on each path.

12.6 EXPERIMENTAL RESULTS

We implement a prototype of C-ARQ as a Linux user space program that interposes itself between the IP stack and the wireless device driver. Click Modular Router [47] and MadWifi [48] are used to implement our prototype.

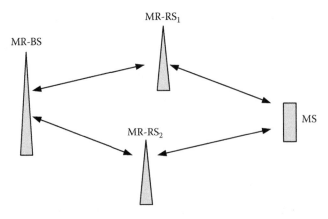

FIGURE 12.3 Multiple-path-establishment for ARQ using cooperative relays.

FIGURE 12.4 Test bed.

The test bed consists of one receiver node, one transmitter node, and optionally one cooperator node, and it is located inside one of the buildings on the UPC campus (see Figure 12.4). The tests are divided into two different types: delivery ratio tests using User Datagram Protocol (UDP) traffic and throughput tests using TCP traffic. To compare the effect of using C-ARQ, we perform our tests twice; first, without the cooperator node, and second with the cooperator node located in the static point of the map P_{Coop}. Moreover, to obtain statistical results, we perform every UDP test ten times per point and every TCP test five times per point, both with the transmitter node always located at point P_{Tx} and with the receiver node located through the ten points P_{Rx}. In all tests we disable Wi-Fi retransmissions to emphasize retransmissions of C-ARQ. We use 802.11a to prevent interferences with the campus wireless network, and the bit rate was fixed at 6 Mbps.

Delivery ratio tests consist in using the transmitter node to send UDP packets to the receiver node and counting how many packets reach the receiver node. At every point P_{Rx}, 100 UDP packets are sent every 100 milliseconds and ten times per point. Figure 12.5 gives the mean delivery ratio at every point of the map with and without C-ARQ.

Results show that C-ARQ drastically improves the delivery ratio of UDP traffic, especially at the points where connectivity between transmitter and receiver nodes starts to become bad (i.e., P_{Rx5}, P_{Rx7}, and P_{Rx8}). Obviously, at the points where there is good connectivity (i.e., P_{Rx1}, P_{Rx2}, P_{Rx3}, P_{Rx4}, and P_{Rx6}), the improvement is lower or null, and at the points where there is no connectivity (i.e., P_{Rx9} and P_{Rx10}), the improvement is maximum. However, many experiments show great variations between some near points, mainly due to the dynamic behavior of the wireless channel. Therefore, the level of improvement achieved by C-ARQ is strictly related to the position of the cooperator nodes and to the properties of the environment (such as doors, walls, people, and all objects involved in the scenario).

Furthermore, throughput tests consist in sending to the receiver node a 13 MB file from the transmitter node via a secure FTP connection with TCP traffic. As previously stated, the file is downloaded five times per point. Figure 12.6 gives the mean throughput achieved at every point of the map with and without the use of C-ARQ. Results show that C-ARQ improves the throughput of TCP traffic, although not as visibly as in the UDP case. The bad behavior of TCP over wireless networks is well known. This is especially true at the points where there is not good connectivity between the transmitter and the receiver nodes (i.e., from P_{Rx3} to P_{Rx10}). In this case the majority of TCP traffic passes through the cooperator node. These cases could be considered as a two-hop

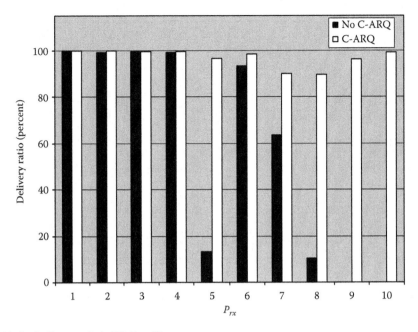

FIGURE 12.5 Delivery ratio in UDP traffic.

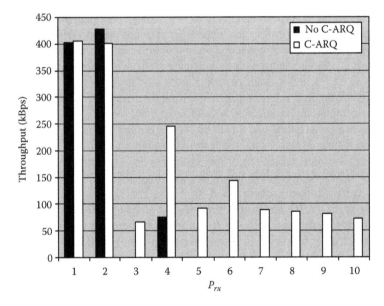

FIGURE 12.6 Throughput in TCP traffic.

wireless network, and due to this increment in the number of hops and to the increment in the end-to-end delay, TCP is not able to manage higher throughputs. However, C-ARQ expands the range of the receiver node without stalling the secure FTP connection.

It is important to bear in mind that all these tests were executed with a real prototype and consequently, with a real wireless channel. No assumptions or simulations were made. These facts, together with the optimistic results achieved by the experiments, provide C-ARQ with sufficient importance to be considered for integration into our daily wireless networks.

12.7   CONCLUSIONS AND OPEN RESEARCH ISSUES

ARQ is a highly reliable, simple, well understood, and extensively used error-control mechanism, which achieves reliability through packet retransmission. Due to the inherent characteristics of the fading process in wireless channels, errors appear in bursts rather than randomly. Conventional ARQ is not very effective in such environments and will cause significant degradation in the performance of link layer. C-ARQ is a recently proposed variant of ARQ in which nodes other than the sender or the destination can play a role in the packet retransmission process, thus exploiting the broadcast nature of wireless transmissions and space diversity.

The potential gains of C-ARQ schemes can be obtained by referring to two well-known models: antenna array and cooperative diversity systems. The antenna array model consists of the use of arrays of multiple transmit or receiving antennas as a way of achieving independent paths in wireless systems, exploiting the already mentioned space diversity. The cooperative diversity model is a variant of the classical relay channel in which there are two cooperating nodes, with information to communicate to a receiver node. Using these framework models, it has been shown for instance that when a relay is used, even the simplest forms of cooperative diversity achieve full diversity order of 2, as is also the case for C-ARQ schemes. Thus, it is possible to state that these models, and more precisely the antenna array, represent the limits on the potentials of C-ARQ protocols.

Furthermore, cooperative diversity systems assume that a fixed TDMA frame structure is used, meaning that relays always cooperate. This lack of flexibility can decrease system throughput, which can be overcome by using the ARQ concept: nodes cooperate only if the received frame is incorrect. Although in networks where adaptive coding and retransmission mechanism are used, the achievable throughput depends on many factors, studies nevertheless show that C-ARQ can also achieve major throughput gains.

The experimental results presented clearly support these conclusions. Improvement in terms of FER of C-ARQ is beyond dispute, while the improvement in terms of throughput is also shown.

A further important performance figure is power consumption. C-ARQ can effectively reduce the power consumption of nodes in a wireless network by either preventing wasteful retransmissions from a sender while the sender–receiver channel is in a deep fade or by preventing retransmissions from far away nodes, which need more power than a node closer to the receiver that has probably received the frame correctly and can relay it to the final destination. This performance metric is specially interesting in the field of sensor networks, where several proposals show that reductions in power consumption can be achieved when these mechanisms are used.

Thus, the main performance characteristics of C-ARQ are now well understood. However, some important research issues remain open; some of them have been identified in this chapter:

- Practical experience with prototypes and implementations. There are many performance evaluations by means of analytical models and simulations, but we have observed a lack of experience with practical implementations. Although in this chapter some experimental results obtained from a test bed using a prototype are presented, more work still needs to be done in this direction. Note that not only experimental results are important, but also the practical issues that may arise, when facing with a real implementation of these kinds of protocols.

- How cooperation can be induced so that all nodes have a positive reward. Acting as a cooperator has, for example, a cost in terms of power consumption, as long as a node must receive or transmit frames on behalf of another node. Although many wireless networks may be considered as naturally cooperative, one can think of many examples where this is not the case. In these situations, incentives for each user to allow cooperation need to be studied in greater detail.

- Cooperator selection policy is another key factor influencing the overall performance of C-ARQ and is largely an open research issue. Although studies exist, some of which are presented here, on the impact of the position of the cooperators with regard to the sender and receiver of a transmission, they lack practical selection policies that can be implemented in a distributed manner.

- Defining mechanisms with low coordination overhead is critical for an optimal operation of the cooperative protocols. There is no discussion on the potential benefits that a C-ARQ protocol can bring, but to exploit them extensively, low coordination overhead mechanisms need to be designed and evaluated. Moreover, the MAC layer should be designed to allow nonstandard signaling message exchanges and to allow for cooperation selection, coordination among cooperators, and acknowledgments coming from nodes other than the receiver, and so on.

- Many cross-layer design issues appear in C-ARQ design. Interaction with routing protocols in multi-hop networks or interaction with TCP are two clear examples. Once again, this is a largely unexplored research field.

REFERENCES

[1] S. Lin, D. J. Costello, and M. J. Miller, Automatic-repeat request error control schemes, *IEEE Communications Magazine*, 22(12), 5–17, 1984.

[2] H. Liu, H. Ma, M. El Zarki, and S. Gupta, Error control schemes for networks: An overview, *Mobile Networks and Applications*, 2, 167–182, 1997.

[3] D. J. Costello, J. Hagenauer, H. Imai, and S. B. Wicker, Applications of error-control coding, *IEEE Transactions on Information Theory*, 44(6), 1998.

[4] S. Lin and D. J. Costello, *Error Control Coding*, Prentice-Hall, Englewood Cliffs, NJ, ISBN 0130426725, 2nd edn., 2004.

[5] D. Bertsekas and R. Gallager, *Data Networks*, Prentice-Hall, Englewood Cliffs, NJ, ISBN 0132016745, 2nd edn., 1992.

[6] M Rice and S. B. Wicker, Adaptative error control for slowly varying channels, *IEEE Transactions on Communications*, 42(3), 917–926, Mar. 1994.

[7] S. Lin and P. S. Yu, An hybrid ARQ scheme with parity retransmission for error control of satellite channels, *IEEE Transactions on Communications*, 30(7), Part 2, 1701–1719, Jul. 1982.

[8] J. Hagenauer, Rate-compatible punctured convolutional codes (RCPC codes) and their applications, *IEEE Transactions on Communications*, 36(4), 389–400, 1988.

[9] S. Kallel and D. Haccoun, Generalized type II hybrid ARQ scheme using punctured convolutional coding, *IEEE Transactions on Communications*, 38(11), 1938–1946, 1990.

[10] S. B. Wicker, Type II hybrid-ARQ protocols using punctured Reed-Sollomon codes, *IEEE Military Communications Conference*, MILCOM 1991.

[11] S. B. Wicker and B. A. Harvey, Type-II hybrid-ARQ protocols using punctured MDS codes, *IEEE Transactions on Communications*, 42(2–4), 1431–1440, 1994.

[12] P. S. Sindhu, Retransmission error control with memory, *IEEE Transactions on Communications*, Com-25(5), 473–479, 1977.

[13] D. Chase, Code combining—a maximum-likelihood decoding approach for combining an arbitrary number of noisy packets, *IEEE Transactions on Communications*, Com-33(5), 385–393, 1985.

[14] B. A. Harvey and S. B. Wicker, Packet combining systems based on the viterbi encoder, *IEEE Transactions on Communications*, 42(2–4), 1544–1557, 1994.

[15] S. S. Chakraborty, E. Yli-Juuti, and M. Liinaharja, An ARQ scheme with packet combining, *IEEE Communications Letters*, 2(7), 200–202, 1998.

[16] Q. Zhang and S. A. Kassam, Hybrid ARQ with selective combining for fading channels, *IEEE Journal on Selected Areas in Communications*, 17(5), 867–880, 1999.

[17] F. P. Fitzek and M. D. Katz, Ed., *Cooperation in Wireless Networks: Principles and Applications*, Springer, New York. ISBN 140204710X, 2006.

[18] H. Ochiai, P. Miran, and V. Tarokh, Collaborative beamforming for distributed wireless ad-hoc wireless sensor networks, *IEEE Transactions on Signal Processing*, 53(11), 4110–4124, 2005.

[19] E. Zimmermann, P. Herhold, and G. Fettweis, The impact of cooperation on diversity-exploiting protocols, *IEEE 59th Vehicular Technology Conference*, VTC 2004-Spring., Vol. 1, pp. 410–414, 2004.

[20] J. Morillo and J. García-Vidal, A low coordination overhead C-ARQ protocol with frame-combining, *18th Annual IEEE International Symposium on Personal, Indoor and Mobile Radio Communications*, IEEE PIMRC, Athens, Greece, 2007.

[21] S. Chandra, Wireless network interface energy consumption, *Multimedia Systems*, 9, 185–201, 2003.

[22] M. Tacca, P. Monti, and A. Fumagalli, Cooperative and non-cooperative ARQ protocols for microwave recharged sensor nodes, *Proceedings of the Second European Workshop on Wireless Sensor Networks*, pp. 45–46, 2005.

[23] A. Miu, H. Balakrishnan, and C. E. Koksal, Improving loss resilience with multi-radio diversity in wireless networks, *The Eleventh Annual International Conference on Mobile Computing and Networking*, ACM MobiCom, Cologne, 2005.

[24] P. Liu, Z. Tao, and S. Panwar, A cooperative MAC protocol for wireless local area networks, *IEEE International Conference on Communications*, IEEE ICC, Seoul, 2005.

[25] J. Saltzer, On the naming and binding of network destinations, *RFC 1498*, 1993.

[26] J. Garcia-Vidal, Addressing and forwarding in cooperative wireless networks, *UPC-DAC-RR-XCSD-2005-8*, DAC-UPC Research Report, December 2005.

[27] A. Goldsmith, *Wireless Communications*, Cambridge University Press, Cambridge, U.K., 2005.

[28] S. Alamouti, A simple transmit diversity technique for wireless communications, *IEEE Journal on Selected Areas in Communications*, 1451–1458, Oct. 1998.

[29] A. Sendonaris, E. Erkip, and B. Aazhang, User cooperation diversity—Part I & II, *IEEE Transactions on Communications*, 51(11), 1927–1948, 2003.

[30] J. N. Laneman and G. W. Wornell, Distributed space-time coded protocols for exploiting cooperative diversity in wireless networks, *IEEE Transactions on Information Theory*, 49(10), 2415–2425, Oct. 2003.

[31] T. M. Cover and A. A. El Gamal, Capacity theorems for the relay channel, *IEEE Transactions on Information Theory*, 25, 572–584, 1979.

[32] A. Scaglione, D. L. Goeckel, and J. N. Laneman, Cooperative communications in mobile ad hoc networks, *IEEE Signal Processing Magazine*, 23(5), 18–29, 2006.

[33] J. N. Laneman, D. Tse, and G. Wornell, Cooperative diversity in wireless networks: Efficient protocols and outage behaviour, *IEEE Transactions on Information Theory*, 50(12), 3062–3080, 2004.

[34] I. Stanojev, O. Simeone, Y. Bar-Ness, and C. You, Performance of multiple-relay collaborative hybrid-ARQ protocol over fading channels, *IEEE Communication Letters*, 10(7), 522–524, 2006.

[35] G. Yu, Z. Zhang, and P. Qiu, Cooperative ARQ in wireless networks: Protocols description and performance analysis, *IEEE International Conference on Communications*, (ICC 2006), Vol. 8, pp. 3608–3614, Istanbul, 2006.

[36] M. Dianati, X. Ling, K. Naik, and X. Shen, A node cooperative ARQ scheme for wireless ad-hoc networks, *IEEE Wireless Communications and Networking Conference*, New Orleans, 2005.

[37] M. Dianati, X. Ling, S. Naik, and X. Shen, A node cooperative ARQ scheme for wireless ad-hoc networks, *IFIP Networking conference, Waterloo, 2005.*

[38] C. Perkins, E. Belding-Royer, and S. Das, Ad hoc on-demand distance vector (AODV) routing, *RFC 3561*, 2003. http://www.ietf.org/rfc/rfc3561.txt

[39] J. Morillo, J. García-Vidal, and A. Pérez-Neira, Collaborative ARQ in wireless energy-constrained networks, *Proceedings of the 2005 Joint Workshop on Foundations of Mobile Computing*, ACM-SIGMOBILE DIAL-M-POMC, Cologne, 2005.

[40] P. Herhold, E. Zimmermann, and G. Fettweis, A simple cooperative extension to wireless relaying, *2004 International Zurich Seminar on Communications*, Zurich, 2004.

[41] Bin Zhao and M. C. Valenti, Practical relay networks: A generalization of hybrid-ARQ, *IEEE Journal on Selected Areas in Communications*, 23(1), 7–18, 2005.

[42] IEEE 802.16's Relay Task Group, http://ieee802.org/16/relay/

[43] W. Ni, L. Cai, G. Shen, and S. Jin, Cooperative relay approaches in IEEE 802.16j, IEEE C802.16j-07/258.

[44] J. Chui and A. Chindapol, Clarifications on cooperative relaying, IEEE C802.16j-07/242r2.

[45] J. Chui and A. Chindapol, Cooperative relaying in downlink for IEEE 802.16j, IEEE S802.16j-07/124.

[46] S. Jin, C. Yoon, and Young-il Kim, An ARQ with cooperative relays in IEEE 802.16j, IEEE C802.16j-07/250rl.

[47] Click modular router. http://www.read.cs.ucla.edu/click/

[48] Madwifi: Multiband Atheros driver for WiFi. http://madwifi.org/

13 Impact of Cooperative Transmission on Network Routing

Zhu Han and Vincent H. Poor

CONTENTS

Cooperative communications has attracted significant recent attention as a transmission strategy for future wireless networks. Cooperative communications efficiently take advantage of the broadcast nature of wireless networks to allow network nodes to share their messages and transmit cooperatively as a virtual antenna array, thus providing diversity that can significantly improve system performance. Cooperative communications can be applied in a variety of wireless systems and networks. In the research community, a considerable amount of work has been done in this area for networks such as cellular, WiFi, ad hoc/sensor networks, and ultrawideband (UWB). These ideas are also working their way into standards; e.g., the IEEE 802.16 [1] (WiMAX) standards body for future broadband wireless access has established the 802.16j Relay Task Group to incorporate cooperative relaying mechanisms into this technology. Most existing work on cooperative communications concentrates on the physical (PHY) and medium access control (MAC) layers of wireless networks, examining issues such as capacity improvement, power control, and relay selection. The impact on the higher layers, such as routing in the network layer, has not been fully investigated. With cooperative communications, link-level performance can be improved by using relays. So, optimal route selection depends not only on the nodes of the direct links but also on the relays as well. In this chapter, we examine methods for improving traditional routing schemes by considering the use of cooperative diversity over a variety of wireless systems such as ad hoc, sensor, and general wireless networks.

13.1   INTRODUCTION

This chapter considers the impact of cooperative communications on the network layer over a range of wireless network scenarios. In addition, the chapter provides insights into the vertical integration of wireless networks by cross-layer optimization. The goals of this chapter are to provide the reader with an understanding of the impact of cooperative communications on network routing and to provide new perspectives on system optimization based on the latest research in this area. To achieve these goals, this chapter is organized as follows.

An introduction is given in the remainder of the current section. Some basic cooperative communication protocols are examined in Section 13.2, followed by a literature review and brief discussion of the impact of cooperative communications on the various layers of wireless networks. Then for several specific types of wireless networks, three case studies are presented in Sections 13.3 through 13.5. These case studies are summarized by the items listed below. Finally, conclusions are drawn in Section 13.6.

Before turning to some basic protocols in the following section, we first discuss some general problems associated with routing in wireless networks and how we approach these problems in this chapter.

1. *Routing problems in selfish ad hoc networks*: In ad hoc networks consisting of autonomous nodes without centralized control, the individual nodes may not be willing to fully cooperate to accomplish the overall network goals. Specifically for the packet forwarding problem, forwarding of other nodes' packets consumes a node's limited battery resources. Therefore, it may not be in a node's immediate best interest to forward the packets of other nodes. However, failure of a node to cooperate by forwarding other nodes' packets will severely affect the network functionality and ultimately will impair the node's own long-term interests even if such a node is able to transmit its own packets. Hence, it is crucial to design a mechanism to enforce cooperation among selfish nodes.

   To overcome this problem, we consider the application of a repeated game that can induce all nodes to forward each other's packets. The essence of the repeated game is to impose enough of a threat of punishment in the future for current noncooperation, so as to limit inefficient competition. Therefore, no node has an incentive to be uncooperative, and cooperation among selfish nodes for packet forwarding is enforced. However, the nodes on the boundary of the network cannot benefit from this strategy, as the other nodes do not depend on them and can drop their packets without worrying about future punishment. This problem is sometimes known as the curse of the boundary nodes. To overcome this problem, an approach based on coalition games is proposed here, in which the boundary nodes can use cooperative communications to help the backbone nodes in the middle of the network. In return, the backbone nodes are willing to forward the boundary nodes' packets. We study the stability of the coalitions using the concept of a core. Then we investigate two types of fairness, namely, the min–max fairness using nucleolus and the average fairness using the Shapley function. Finally, a protocol is designed using both repeated games and coalition games, and simulation results demonstrate the effectiveness of the proposed protocol.

2. *Routing problems in sensor networks*: In wireless sensor networks, extending the lifetimes of battery-operated devices is considered a key design issue, which increases the capability of uninterrupted information exchange and alleviates the burden of replenishing batteries. Due to the burden of forwarding others' packets, the nodes around the sink in a sensor network tend to run out of battery energy first. So even if the other nodes still have energy, the whole network cannot function.

   To overcome this problem, we consider the situation in which closely located nodes use cooperative communications to reduce the load or even to avoid packet-forwarding requests to the nodes that have critical battery lives. The performance improvement obtained by this technique is analyzed for a special two-dimensional (2D) disk case. Further, for general networks in which information-generation rates are fixed, a new routing problem is formulated as a linear programming problem. For general networks with random information-generation rates, the cost for routing is dynamically adjusted according to the amount of energy remaining and potential cooperation partners. From the analysis and the simulation results, the proposed method can reduce the payloads of energy-depleting nodes by about 90 percent in the 2D disk case network with a sink in the center and can improve the network lifetimes of general networks by about 10 percent, compared with existing techniques.

3. *Routing problems in general multi-hop networks*: For general multi-hop networks, routing protocols, such as the shortest path algorithm, depend on the cost for each hop. Most current cooperative routing schemes try to improve the shortest path solution using cooperative communications. However, this type of approach produces suboptimal solutions, because the optimal route with cooperative communications might not be the same as the shortest path. On the other hand, to look for the optimal cooperative route directly is a computationally intensive problem. Another approach is to seek ways to quantify the cost of each hop, incorporating the effects of cooperative communications. Then, currently available routing schemes can be directly applied to find the optimal route. We discuss a formal design philosophy based on this idea and describe some challenges for cooperative routing algorithms. Finally, we also briefly discuss recent standards developed in this area, notably the IEEE 802.16j standard, to provide the reader with perspectives from an industrial point of view.

13.2 COOPERATIVE TRANSMISSION—A NEW COMMUNICATION PARADIGM

In this section, we first classify the currently known cooperative transmission (CT) protocols [2,3]. Then, we review briefly the current state of the art in cooperative communications, including its impact on different layers. Then, in the following sections, we consider the impact on the network layer in more depth.

13.2.1 COOPERATIVE TRANSMISSION PROTOCOLS

To illustrate the basic idea of CT, a highly simplified topology with one source node, two relay nodes, and one destination node is shown at the bottom of Figure 13.1. CT is conducted in two

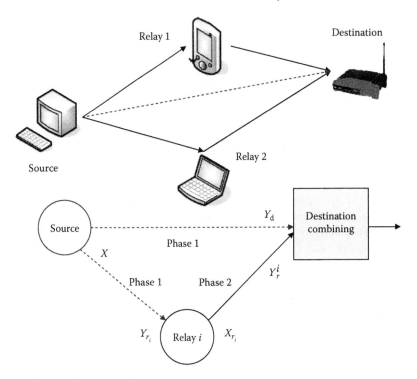

FIGURE 13.1 Cooperative communication system model.

phases. In Phase 1, the source broadcasts a message to the destination and relay nodes. In Phase 2, relay nodes send information to the destination (in different time slots or on different orthogonal channels), and the destination combines messages from the source and relays. It has been shown that the capacity region of this communication channel can be significantly increased by such techniques, and that the performance gain of CT is proportional to the number of relays in the Rayleigh fading case [4]. Here, we denote the source node as s, the several relays nodes as r_i, and the destination nodes as d. Now we can describe the process more precisely. In Phase 1, the received signals Y_d and Y_{r_i} at the destination d and relay r_i, respectively, can be expressed as

$$Y_d = \sqrt{P_s G_{s,d}}X + n_d, \tag{13.1}$$

and

$$Y_{r_i} = \sqrt{P_s G_{s,r_i}}X + n_{r_i}, \tag{13.2}$$

where

P_s represents the transmit power to the destination from the source
X is the unit-energy information symbol transmitted by the source in Phase 1
$G_{s,d}$ and G_{s,r_i} are the channel gains from s to d and, r_i respectively
n_d and n_{r_i} are samples from independent (discrete-time) additive white Gaussian noise (AWGN) processes, independent of X

Without loss of generality, we assume that the noise power is the same for all the links and is denoted by σ^2.

A number of different types of cooperative communication protocols have been developed in the literature. Here, we describe several of the most relevant of these briefly as follows:

1. Direct transmission: Without the relay nodes' help, the signal-to-noise ratio (SNR) that results from s to d can be expressed as

$$\Gamma_{s,d}^{DT} = \frac{P_s G_{s,d}}{\sigma^2}, \tag{13.3}$$

and the capacity of the direct transmission channel is

$$R_{s,d} = W \log_2 \left(1 + \Gamma_{s,d}^{DT}\right), \tag{13.4}$$

where W is the bandwidth used for information transmission.

2. Amplify-and-forward (AF) CT: In Phase 2, relay i amplifies Y_{r_i} and forwards it to the destination with transmitted power P_{r_i}. The received signal at the destination is

$$Y_r^i = \sqrt{P_{r_i} G_{r_i,d}} X_{r_i} + n_d', \tag{13.5}$$

where

$$X_{r_i} = \frac{Y_{r_i}}{|Y_{r_i}|} \tag{13.6}$$

is the energy-normalized transmitted signal from the source to the destination in Phase 1, $G_{r_i,d}$ is the channel gain from relay i to the destination, and n_d' is the received noise in Phase 2. Substituting Equation 13.2 into 13.6, we can rewrite Equation 13.5 as

$$Y_r^i = \frac{\sqrt{P_{r_i} G_{r_i,d}} \left(\sqrt{P_s G_{s,r_i}} X_s + n_{r_i}\right)}{\sqrt{P_s G_{s,r_i} + \sigma^2}} + n_d'. \tag{13.7}$$

Using Equation 13.7, the relayed SNR at the destination for the source, which is assisted by relay node i, is given by

$$\Gamma_{s,r_i,d}^{AF} = \frac{P_{r_i} P_s G_{r_i,d} G_{s,r_i}}{\sigma^2 (P_{r_i} G_{r_i,d} + P_s G_{s,r_i} + \sigma^2)}. \tag{13.8}$$

Therefore, by Equations 13.4 and 13.8, the channel capacity assuming maximal ratio combining at the destination is given by

$$R_{s,r_i,d}^{AF} = \frac{1}{2} W \log_2 \left(1 + \Gamma_{s,d}^{DT} + \Gamma_{s,r_i,d}^{AF}\right). \tag{13.9}$$

If multiple relay nodes (say, $i \in L$, which $|L| = N$) are available to help the source, then we have

$$R_{s,L,d}^{AF} = \frac{1}{N+1} W \log_2 \left(1 + \Gamma_{s,d}^{DT} + \sum_{r_i \in L} \Gamma_{s,r_i,d}^{AF}\right). \tag{13.10}$$

Here the $\frac{1}{2}$ in Equation 13.9 and the $\frac{1}{N+1}$ in Equation 13.10 are due to the fact that the relays need extra orthogonal channels for transmission.

3. Decode-and-forward (DF) CT: Here, the relay decodes the source information in Phase 1 and relays it to the destination in Phase 2. The destination combines the direct transmission information with the relayed information. The achievable rate in this case can be calculated by the following maximization:

$$R_{s,r_i,d}^{DF} = \max_{0 \le \rho \le 1} \min\{R_1, R_2\}, \tag{13.11}$$

where

$$R_1 = \log\left[1 + (1 - \rho^2)\frac{P_{s,d}G_{s,r_i}}{\sigma^2}\right], \tag{13.12}$$

and

$$R_2 = \log\left(1 + \frac{P_s G_{s,d}}{\sigma^2} + \frac{P_{r_i}G_{r_i,d}}{\sigma^2} + \frac{2\rho\sqrt{P_s G_{s,d}P_{r_i}G_{r_i,d}}}{\sigma^2}\right). \tag{13.13}$$

4. Estimate-and-forward (EF) CT: Here, in Phase 2, the relay sends an estimate of the received signal of Phase 1. The destination uses the relay's information as side information to decode the direct transmission of Phase 1. From Refs. [5,6], the channel capacity resulting from this approach can be written as

$$R_{s,r_i,d}^{EF} = W\log_2\left(1 + \Gamma_{s,d}^{DT} + \Gamma_{s,r_i,d}^{EF}\right), \tag{13.14}$$

where

$$\Gamma_{s,r_i,d}^{EF} = \frac{P_s P_{r_i} G_{s,r_i} G_{r_i,d}}{\sigma^2[P_{r_i}G_{r_i,d} + P_s(G_{s,d} + G_{s,r_i}) + \sigma^2]}. \tag{13.15}$$

5. Coded cooperation: This type of CT protocols integrate relay cooperation with channel coding [7]. Instead of exactly repeating the received information, the relay decodes the partner's transmission and transmits additional parity symbols (e.g., incremental redundancy) according to a certain overall coding scheme. The destination receiver conducts channel decoding by concatenating the data from the direct transmission and relay transmission, so that the channel gain can be obtained.

6. Distributed space-time coded cooperation: Space-time coding has been shown to significantly improve the link performance in multiple-input/multiple-output (MIMO) systems. Distributed space-time cooperative diversity protocols [8] exploit the spatial diversity available among a collection of distributed nodes that relay messages for one another, in such a manner that the destination terminal can combat the fading. Those relays can fully decode the transmission from the source and then code the data by using a space-time code to cooperatively relay to the destination. At the destination, the space-time code implemented by the source and relays can achieve full diversity.

7. Incremental relaying [4]: Here, the destination will broadcast ACK or NACK information after the first stage. The relay retransmits only after receiving a NACK. By doing this, the bandwidth efficiency can be greatly improved, because the only bandwidth increase occurs when the direct transmission link fails. However, the gain comes with additional implementation costs for the feedback mechanism.

8. Cognitive relaying [9,10]: Cognitive radio is a revolutionary paradigm with high spectral efficiency involving wireless communication in occupied spectrum without interfering with existing band occupants. Cooperative communication protocols can help cognitive users to reduce the detection time for clear spectral band and thus to increase their agility. On the other hand, relays can monitor the spectrum cognitively so as to improve the source-to-destination link.

13.2.2 CURRENT STATE OF THE ART AND IMPACT ON DIFFERENT LAYERS

CT began as a physical layer protocol and most work in this area focuses on its merits in the physical layer. However, the impact of CT on the design of the higher layers is obviously also of importance, although it is not yet well-understood. We now briefly discuss the impact of cooperative communications on different layers. Specifically, we divide the discussion as follows:

- Capacity analysis and new cooperative communication protocols: The major concerns here are to analyze how much gain CT can bring to a link and to the overall network [11–13] or how to implement CT under practical constraints [6,7,14].
- Relay selection and power control: When there are several relays, a question arises as to which one to select for a given retransmission. After such relay selection, the next issue is how limited power resources should be distributed over sources and relays [15,16]. These questions have also been addressed in systems using multiuser detection [17] and orthogonal frequency division multiplexing (OFDM) [18].
- Routing protocols: CT can provide extra routes for network protocols so that the network performance can be significantly improved. These routes can be found through traditional routes [19,20] or from cooperative routes [21]. It has been shown that network lifetime can be significantly improved via such consideration [22]. Multi-hop CT can be considered as a special case of routing in [23–26].
- Distributed resource allocation: Game theoretic approaches are natural for distributed cooperative resource allocation problems, as the individual nodes can use only local information to optimize cooperative communications [27]. Moreover, as shown in [28], cooperative game theory and CT can be used to improve packet-forwarding networks with selfish nodes.
- Others: CT has been considered jointly with other problems such as source coding [29] and energy-efficient broadcast [30].

In the following three sections, we concentrate on the impact of CT on the network layer for ad hoc networks, sensor networks, and general networks, respectively.

13.3 COOPERATIVE ROUTING IN AD HOC NETWORKS

In wireless packet-forwarding networks with selfish nodes, applications of a repeated game can induce the nodes to forward each others' packets, so that the network performance can be improved. However, the nodes on the boundary of such networks cannot benefit from this strategy, as the other nodes do not depend on them. This problem is sometimes known as the curse of the boundary nodes. In this section, following Ref. [28], an approach to this problem based on coalition games is discussed, in which the boundary nodes can use CT to help the backbone nodes in the middle of the network. In return, the backbone nodes forward the boundary nodes' packets. We discuss the stability of such coalitions using the concept of a core. Then two types of fairness, namely, the min–max fairness using nucleolus and the average fairness using the Shapley function are described. Finally, we discuss a protocol that uses both repeated games and coalition games. Simulation results show how boundary nodes and backbone nodes form coalitions according to different fairness criteria. The proposed protocol can improve the network connectivity by about 50 percent, compared with pure repeated game schemes.

13.3.1 MOTIVATION AND BASIC PROBLEM

In wireless networks with selfish nodes such as ad hoc networks, the nodes may not be willing to fully cooperate to accomplish the overall network goals. Specifically for the packet-forwarding problem, forwarding of other nodes' packets consumes a node's limited battery energy. Therefore, it may not be in a node's best interest to forward other's arriving packets. However, refusal to forward other's packets noncooperatively will severely affect the network functionality and thereby impair a node's own performance. Hence, it is crucial to design a mechanism to enforce cooperation for packet forwarding among greedy and distributed nodes.

The packet-forwarding problem in ad hoc networks has been extensively studied in the literature. The fact that nodes act selfishly to optimize their own performance has motivated many researchers to apply game theory [31,32] in solving this problem. Broadly speaking, the approaches used to encourage packet forwarding can be categorized into two general types. The first type makes use of virtual payments. Pricing [33] and credit-based method [34] fall into this first type. The second type of approach is related to personal and community enforcement to maintain the long-term relationship among nodes. Cooperation is sustained because defection against one node causes personal retaliation or sanction by others. Watchdog and pathrater are proposed in Ref. [35] to identify misbehaving nodes and deflect traffic around them. Reputation-based protocols are proposed in Refs. [36,37]. In Ref. [38], a model is considered to show cooperation among participating nodes. The packet-forwarding schemes using "TIT for TAT" schemes are proposed in Ref. [39]. In Ref. [40], a cartel maintenance framework is constructed for distributed rate control for wireless networks. In Ref. [41], self-learning repeated game approaches are constructed to enforce cooperation and to study better cooperation. Some recent works for game theory to enhance energy-efficient behavior in infrastructure networks can be found in Refs. [42–45].

A wireless packet-forwarding network can be modeled as a directed graph $G(L, A)$, where L is the set of all nodes and A is the set of all directed links $(i, l), i, l \in L$. Each node i has several transmission destinations, which are included in set D_i. To reach the destination j in D_i, the available routes form a depending graph G_i^j whose nodes represent the potential packet-forwarding nodes. The transmission from node i to node j depends on a subset of the nodes in G_i^j for packet forwarding. Note that this dependency can be mutual. One node depends on the other node, while the other node can depend on this node as well. In general, this mutual dependency is common, especially for backbone nodes at the center of the network. Next, we discuss how to make use of this mutual dependency for packet forwarding using a repeated game, and then we explain the curse of boundary nodes.

13.3.1.1 Repeated Games for Mutually Dependent Nodes

A repeated game is a special type of dynamic game (a game that is played multiple times). When the nodes interact by playing a similar static game (which is played only once) numerous times, the game is called a repeated game. Unlike a static game, a repeated game allows a strategy to be contingent on the past moves, thus allowing reputation effects and retribution, which give possibilities for cooperation. The game is defined as follows:

DEFINITION 1

A T-period repeated game is a dynamic game in which, at each period t, the moves during periods $1, \ldots, t - 1$ are known to every node. In such a game, the total discounted payoff for each node is computed by $\sum_{t=1}^{T} \beta^{t-1} u_i(t)$, where $u_i(t)$ denotes the payoff to node i at period t and where β is a discount factor. Note that β represents the node's patience or on the other hand how important the past affects the current payoff. If $T = \infty$, the game is referred as an infinitely repeated game. The average payoff to node i is then given by

$$u_i = (1 - \beta) \sum_{t=1}^{\infty} \beta^{t-1} u_i(t). \tag{13.16}$$

It is known that repeated games can be used to induce greedy nodes in communication networks to show cooperation. In packet-forwarding networks, if a greedy node does not forward the packets of other nodes, it can enjoy benefits such as power saving. However, this node gets punishment from the other nodes in the future if it depends on the other nodes to forward its own packets. The benefit of greediness in the short term will be offset by the loss associated with punishment in the future. So the nodes will rather act cooperatively if the nodes are sufficiently patient. From the Folk theorem below, we infer that in an infinitely repeated game, any feasible outcome that gives each node a better payoff than the Nash equilibrium [31,32] can be obtained.

THEOREM 1   (Folk Theorem [31,32])

Let $(\hat{u}_1, \ldots, \hat{u}_L)$ be the set of payoffs from a Nash equilibrium and let $(u_1, \ldots, u_L)$ be any feasible set of payoffs. There exists an equilibrium of the infinitely repeated game that attains any feasible solution $(u_1, \ldots, u_L)$ with $u_i \geq \hat{u}_i, \forall i$ as the average payoff, provided that β is sufficiently close to 1.

In the literature on packet-forwarding wireless networks, the conclusion of the above Folk theorem is achieved by several approaches. Tit-for-tat [38,39] is proposed so that all mutually dependent nodes have the same set of actions. A cartel maintenance scheme [40] has closed-form optimal solutions for both cooperation and noncooperation. A self-learning repeated game approach is proposed in Ref. [41] for individual distributed nodes to study the cooperation points and to develop protocols for maintaining them.

13.3.1.2   The Curse of Boundary Nodes

However, for packet-forwarding networks, there exists the so-called curse of boundary nodes. The nodes at the boundary of the network must depend on the backbone nodes in the middle of the networks to forward their packets. On the other hand, the backbone nodes will not depend on the boundary nodes. As a result, the backbone nodes do not worry about retaliation or lost reputation for not forwarding the packets of the boundary nodes. This fact causes the curse of the boundary nodes, an example of which is shown in Figure 13.2. Suppose node 1 needs to send data to node 3 and node 2 needs to send data to node 0. Because node 1 and node 2 depend on each other for packet forwarding, they are obliged to do so because of the possible threat of retaliation from the other node. However, if node 0 wants to transmit to node 2 and node 3 or node 3 tries to communicate with node 0 and node 1, the nodes in the middle have no incentive to forward the packets due to their greediness. Moreover, this greediness cannot be punished in the future because the dependency is not mutual. This problem is especially severe for the nodes on the boundary of the networks, so it is called the curse of boundary nodes.

On the other hand, if node 0 can form a coalition with node 1 and help node 1's transmission (e.g., to reduce the transmitted power of node 1), then node 1 has the incentive to help node 0 transmit as a reward. A similar situation arises for node 3 to form a coalition with node 2. We call nodes like 1 and 2 backbone nodes, while nodes like 0 and 3 are boundary nodes. In Section 13.3.2, we study how coalitions can be formed to address the above issue using CT.

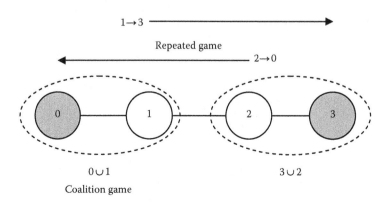

FIGURE 13.2   Example of the curse of boundary nodes.

13.3.2   COALITION GAME WITH CT

In this subsection, we first study a CT technique that allows nodes to participate in coalitions. Then, we formulate a coalition game with CT. Furthermore, we investigate the fairness issue and propose two types of fairness definitions. Finally, a protocol for packet forwarding using repeated games and coalition games is constructed.

13.3.2.1   Coalition Game Formation for Boundary Nodes

Next, we study possible coalitions between the boundary nodes and the backbone nodes, for situations in which the boundary nodes can help relay the information of the backbone nodes using CT. In the following, we first define some basic concepts that are needed in our analysis.

DEFINITION 2

A coalition S is defined to be a subset of the total set of nodes $\mathbb{N} = \{0, \ldots, N\}$. The nodes in a coalition want to cooperate with each other. The coalition form of a game is given by the pair $(\mathbb{N}, v)$, where v is a real-valued function, called the characteristic function. $v(S)$ is the value of the cooperation for coalition S with the following properties:

1. *$v(\emptyset) = 0$.*
2. *Super-additivity: if S and Z are disjoint coalitions ($S \cap Z = \emptyset$), then $v(S) + v(Z) \leq v(S \bigcup Z)$.*

The coalition states the benefit obtained from cooperation agreements. However, we still need to examine whether or not the nodes are willing to participate in the coalition. A coalition is called stable if no other coalition will have the incentive and power to upset the cooperative agreement. Such division of v is called a point in the core, which is defined by the following definitions.

DEFINITION 3

A payoff vector $\boldsymbol{U} = (U_0, \ldots, U_N)$ is said to be group rational or efficient if $\sum_{i=0}^{N} U_i = v(\mathbb{N})$. A payoff vector $\boldsymbol{U}$ is said to be individually rational if the node can obtain the benefit no less than acting alone, i.e., $U_i \geq v(\{i\})$, $\forall i$. An imputation is a payoff vector satisfying the above two conditions.

DEFINITION 4

An imputation $\boldsymbol{U}$ is said to be unstable through a coalition S if $v(S) > \sum_{i \in S} U_i$, i.e., the nodes have incentive for coalition S and upset the proposed $\boldsymbol{U}$. The set C of a stable imputation is called the core, i.e.,

$$C = \{\boldsymbol{U} : \sum_{i \in \mathbb{N}} U_i = v(\mathbb{N}) \quad \text{and} \quad \sum_{i \in S} U_i \geq v(S), \forall S \in \mathbb{N}\}. \qquad (13.17)$$

In the economics literature, the core gives a reasonable set of possible shares. A combination of shares is in the core if there is no sub-coalition in which its members may gain a higher total outcome than the combination of shares of concern. If a share is not in the core, some members may be frustrated and may think of leaving the whole group with some other members and form a smaller group.

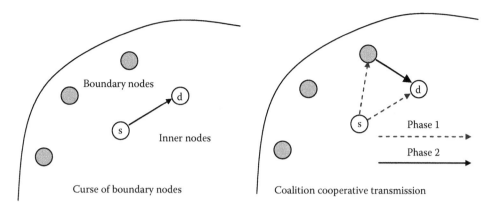

FIGURE 13.3 Cure for the curse of boundary nodes.

In the packet-forwarding network as shown in Figure 13.3, we first assume one backbone node to be the source node (node 0) and the nearby boundary nodes (node 1 to node N) to be the relay nodes. We discuss the case of multiple source nodes later. If no CT is employed, the utilities for the source node and the relay nodes are

$$v(\{0\}) = -P_\mathrm{d} \quad \text{and} \quad v(\{i\}) = -\infty, \forall i = 1, \dots, N. \tag{13.18}$$

With CT and a grand coalition that includes all nodes, the utilities for the source node and the relay nodes are

$$U_0 = -P_0 - \sum_{i=1}^{N} \alpha_i P_\mathrm{d}, \tag{13.19}$$

and

$$U_i = -\frac{P_i}{\alpha_i}, \tag{13.20}$$

where α_i is the ratio of the number of packets that the backbone node is willing to forward for boundary node i, to the number of packets that the boundary node i relays for the backbone node using CT. Here we use negative power as the utility so as to be consistent with the conventions used in the game theory literature. Smaller α_i means the boundary nodes have to relay more packets before realizing the rewards of packet forwarding. The other interpretation of the utility is as the average power per transmission for the boundary nodes.* The following theorem gives conditions under which the core is not empty, i.e., in which the grand coalition is stable.

THEOREM 2

The core is not empty if $\alpha_i \geq 0$, $i = 1, \dots, N$, and α_i are such that $U_0 \geq v(\{0\})$, i.e.,

$$\sum_{i=1}^{N} \alpha_i \leq \frac{P_\mathrm{d} - P_0}{P_\mathrm{d}}. \tag{13.21}$$

---

* Note that we omit the transmitted power needed to send the boundary node's own packet to the backbone node, because it is irrelevant to the coalition.

Proof: First, any relay node will get $-\infty$ utility if it leaves the coalition with the source node, so no node has incentive to leave coalition with node 0. Then, the inclusion of relay nodes will increase the received SNR monotonically. So P_0 will decrease monotonically with the addition of any relay node. As a result, the source node has the incentive to include all the relay nodes, as long as the source power can be reduced, i.e., $U_0 \geq v(\{0\})$. A grand coalition is formed and the core is not empty if Equation 13.21 holds.

The concept of the core defines the stability of a utility allocation. However, it does not define how to allocate the utility. For the proposed game, each relay node can obtain different utilities by using different values of α_i. Next, we study how to achieve min–max fairness and average fairness.

13.3.2.2 Min–Max Fairness of a Game Coalition Using Nucleolus

We introduce the concepts of excess, kernel, and nucleolus [31,32]. For a fixed characteristic function v, an imputation **U** is found such that, for each coalition S and its associated dissatisfaction, an optimal imputation is calculated to minimize the maximum dissatisfaction. The dissatisfaction is quantified as follows:

DEFINITION 5

*The measure of dissatisfaction of an imputation **U** for a coalition S is defined as the excess:*

$$e(U, S) = v(S) - \sum_{j \in S} U_j. \tag{13.22}$$

*Obviously, any imputation **U** is in the core, if and only if all its excesses are negative or zero.*

DEFINITION 6

*A kernel of v is the set of all allocations **U** such that*

$$\max_{S \subseteq N-j, i \in S} e(U, S) = \max_{T \subseteq N-i, j \in T} e(U, T). \tag{13.23}$$

If nodes i and j are in the same coalition, then the highest excess that i can make in a coalition without j is equal to the highest excess that j can make in a coalition without i.

DEFINITION 7

*The nucleolus of a game is the allocation **U** that minimizes the maximum excess:*

$$U = \arg \min_U (\max \, e(U, S), \, \forall S). \tag{13.24}$$

The nucleolus of a game has the following property: The nucleolus of a game in coalitional form exists and is unique. The nucleolus is group rational and individually rational. If the core is not empty, the nucleolus is in the core and kernel. In other words, the nucleolus is the best allocation under the min–max criterion.

Using the above concepts, we prove the following theorem to show the optimal α_i in Equation 13.19 to have min–max fairness.

THEOREM 3

The maximal α_i to yield the nucleolus of the proposed coalition game is given by

$$\alpha_i = \frac{P_d - P_0(\mathbb{N})}{NP_d}, \tag{13.25}$$

where $P_0(\mathbb{N})$ is the required transmitted power of the source when all relays transmit with transmitted power P_{max}.

Proof: Because for any coalition other than the grand coalition, the excess will be $-\infty$, we need only consider the grand coalition. Suppose the min–max utility is μ for all nodes, i.e.,

$$\mu = -\frac{P_i}{\alpha_i}. \tag{13.26}$$

From Equation 13.21 and because U_i is monotonically increasing with α_i in Equation 13.20, we have

$$\alpha_i = \frac{P_i}{\sum_{i=1}^{N} P_i} \cdot \frac{(P_d - P_0)}{P_d}. \tag{13.27}$$

Because P_0 for SNR is a monotonically increasing function of P_i, to achieve the maximal α_i and μ, each relay transmits with the largest possible power P_{max}. Notice here that we assume the backbone node can accept an arbitrarily small power gain to join the coalition.

13.3.2.3 Average Fairness of Game Coalition Using the Shapley Function

The core concept defines the stability of an allocation of payoff and the nucleolus concept quantifies the min–max fairness of a game coalition. Next, we study another average measure of fairness for each individual using the concept of a Shapley function [31,32].

DEFINITION 8

A Shapley function ϕ is a function that assigns to each possible characteristic function v a vector of real numbers, i.e.,

$$\phi(v) = (\phi_0(v), \phi_1(v), \phi_2(v), \dots, \phi_N(v)), \tag{13.28}$$

where $\phi_i(v)$ represents the worth or value of node i in the game. There are four Shapley axioms that $\phi(v)$ must satisfy

1. *Efficiency axiom:* $\sum_{i \in \mathbb{N}} \phi_i(v) = v(\mathbb{N})$.
2. *Symmetry axiom: If node i and node j are such that $v(S \bigcup \{i\}) = v(S \bigcup \{j\})$ for every coalition S not containing node i and node j, then $\phi_i(v) = \phi_j(v)$.*

3. *Dummy axiom: If node i is such that $v(S) = v(S \bigcup \{i\})$ for every coalition S not containing i, then $\phi_i(v) = 0$.*
4. *Additivity axiom: If u and v are characteristic functions, then $\phi(u + v) = \phi(v + u) = \phi(u) + \phi(v)$.*

It can be proved that there exists a unique function ϕ satisfying the Shapley axioms. Moreover, the Shapley function can be calculated as

$$\phi_i(v) = \sum_{S \subset \mathbb{N}-i} \frac{(|S|)!(N - |S|)!}{(N + 1)!} [v(S \cup \{i\}) - v(S)]. \tag{13.29}$$

Here $|S|$ denotes the size of set S and $\mathbb{N} = \{0, 1, \ldots, N\}$.

The physical meaning of the Shapley function can be interpreted as follows. Suppose one backbone node plus N boundary nodes form a coalition. Each node joins the coalition in random order. So there are $(N + 1)!$ different ways that the nodes might be ordered in joining the coalition. For any set S that does not contain node i, there are $|S|!(N - |S|)!$ different ways to order the nodes so that S is the set of nodes that enter the coalition before node i. Thus, if the various orderings are equally likely, $|S|!(N - |S|)!/(N + 1)!$ is the probability that, when node i enters the coalition, the coalition of S is already formed. When node i finds S ahead of it as it joins the coalition, then its marginal contribution to the worth of the coalition is $v(S \cup \{i\}) - v(S)$. Thus, under the assumption of randomly ordered joining, the Shapley function of each node is its expected marginal contribution when it joins the coalition.

We consider the case in which the backbone node is always in the coalition, and the boundary nodes randomly join the coalition. We have $v(\{0\}) = -P_d$ and

$$v(\mathbb{N}) = P_d - P_0(\mathbb{N}) - \sum_{i \in \mathbb{N}} \alpha_i P_d, \tag{13.30}$$

which is the overall power saving. The problem here is how to find a given node's α_i that satisfies the average fairness, which is addressed by the following theorem.

THEOREM 4

The maximal α_i that satisfies the average fairness with the physical meaning of the Shapley function is given by

$$\alpha_i = \frac{P_i^s}{P_d}, \tag{13.31}$$

where P_i^s is the average power saving with random entering orders, which is defined as

$$P_i^s = \frac{1}{N}[P_d - P_0(\{i\})] + \frac{\sum_{j=1, j \neq i}^{N}[P_0(\{j\}) - P_0(\{i,j\})]}{N(N - 1)} + \cdots. \tag{13.32}$$

Proof: The maximal α_i is solved by the following equations:

$$\begin{cases} \dfrac{\alpha_i}{\alpha_j} = \dfrac{\phi_i}{\phi_j}, \\ v(\mathbb{N}) \geq 0. \end{cases} \tag{13.33}$$

The first equation in Equation 13.33 is the average fairness according to the Shapley function, and the second equation in Equation 13.33 is the condition for a nonempty core. Similar to min–max fairness, we assume that the backbone node can accept an arbitrarily small power gain to join the coalition.

If boundary node i is the first to join the coalition, the marginal contribution for power saving is $\frac{1}{N}[P_d - P_0(\{i\}) - \alpha_i P_d]$, where $\frac{1}{N}$ is the probability. If boundary node i is the second to join the coalition, the marginal contribution is $\frac{\sum_{j=1,j\neq i}^{N}[P_0(\{j\})+\alpha_j P_d - P_0(\{i,j\}) - (\alpha_i+\alpha_j)P_d]}{N(N-1)}$. By means of some simple derivations, we can obtain the Shapley function ϕ_i as

$$\phi_i = -\alpha_i P_d + \frac{1}{N}[P_d - P_0(\{i\})] + \frac{\sum_{j=1,j\neq i}^{N}[P_0(\{j\}) - P_0(\{i,j\})]}{N(N-1)} + \cdots, \qquad (13.34)$$

and then we can obtain

$$\alpha_i = \frac{[P_d - P_0(\mathbb{N})]P_i^s}{P_d \sum_{j=1}^{N} P_j^s}. \qquad (13.35)$$

Because

$$P_d - P_0(\mathbb{N}) = \sum_{j=1}^{N} P_j^s, \qquad (13.36)$$

we prove Equation 13.31.

Notice that different nodes have different values of P_i^s, due to their channel conditions and abilities to reduce the backbone node's power. Compared with the min–max fairness in the previous subsection, the average fairness using the Shapley function gives different nodes different values of α_i according to their locations.

Using the above analysis, we now develop a packet-forwarding protocol based on repeated games and coalition games based on the following steps.

13.3.2.4 Packet-Forwarding Protocol Using Repeated Games and Coalition Games

1. Route discovery for all nodes
2. Packet-forwarding enforcement for the backbone nodes, using threat of future punishment in the repeated games
3. Neighbor discovery for the boundary nodes
4. Coalition game formation
5. Packet relay for the backbone nodes with CT
6. Transmission of the boundary nodes' own packets to the backbone nodes for forwarding

First, all nodes in the network undergo route discovery. Then each node knows who depends on it and on whom it depends for transmission. Using this route information, the repeated games can be formulated for the backbone nodes. The backbone nodes forward the other nodes' information because of the threat of future punishment if these packets are not forwarded. Due to the network topology, some nodes' transmissions depend on the others while the others do not depend on these nodes. These nodes are most often located at the boundary of the network. In the next step, these boundary nodes try to find their neighboring backbone nodes. Then, the boundary nodes try to form coalitions with the backbone nodes, so that the boundary nodes can be rewarded for transmitting their own packets. CT gives an opportunity for the boundary nodes to pay some "credits" first to the backbone nodes for the rewards of packet forwarding in return. On the other hand, competition among the backbone nodes prevents the boundary nodes from being forced to accept the minimal payoffs.

13.3.3 NUMERICAL STUDY

We model all channels as AWGN channels with a propagation factor of 3; that is, power falls off spatially according to an inverse-cubic law. The maximal transmitted power is 10 dBm and the thermal noise level is −60 dBm. The minimal SNR γ is 10 dB. In the first setup, we assume that the backbone node is located at (0 m, 0 m), and the destination is located at either (100 m, 0 m) or (50 m, 0 m). The boundary nodes are located on an arc, with angles randomly distributed from 0.5π to 1.5π and with distances varying from 5 to 100 m.

In Figure 13.4a, we study the min–max fairness and show the average α_i over 1000 iterations as a function of distance from the relays to the source node. Due to the min–max nature, all boundary nodes have the same α_i. When the distance is small, i.e., when the relays are located close to the source, α_i approaches $\frac{1}{N}$. This is because the relays can serve as a virtual antenna for the source, and the source needs very low power for transmission to the relays. When the distance is large, the relays are less effective and α_i decreases, which means that the relays must transmit more packets for the source to earn the rewards of packet-forwarding. When the destination is located at 50 m, the source–destination channel is better than that at 100 m. When $N = 1$ and the source–destination distance is 50 m, the relays close to the source have larger α_i and the relays farther away have lower α_i than that in the 100 m case. In Figure 13.4b, we show the corresponding P_0 for the backbone node. We can see that P_0 increases when the distances between the boundary nodes to the backbone node increase.

If we consider the multiple backbone (multiple core) case with min–max fairness, Figure 13.4a and b provide the boundary nodes a guideline for selecting a backbone node with which to form a coalition. First, a less crowded coalition is preferred. Second, the nearest backbone node is preferred. Third, for $N = 1$, if the source–destination channel is good, the closer backbone node is preferred; otherwise, the farther one can provide larger α_i.

Next, we investigate the average fairness using the Shapley function. The simulation setup is as follows. The backbone node is located at (0 m, 0 m) and the destination is located at (−50 m, 0 m). Boundary node 1 is located at (20 m, 0 m) and (50 m, 0 m), respectively. Boundary node 2 moves from (5 m, 0 m) to (100 m, 0 m). The remaining simulation parameters are the same. In Figure 13.5a, we show maximal α_i for two boundary nodes. We can see that when boundary node 2 is closer to the backbone node than boundary node 1, $\alpha_2 > \alpha_1$, i.e., boundary node 2 can help relay fewer packets for backbone node 1 before being rewarded. The two curves for α_1 and α_2 for the same boundary node 1

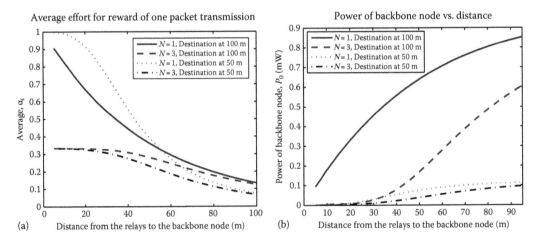

FIGURE 13.4   (a) α for different channels and no. of nodes, min–max fairness. (b) P_0 for different channels and no. of nodes, min–max fairness.

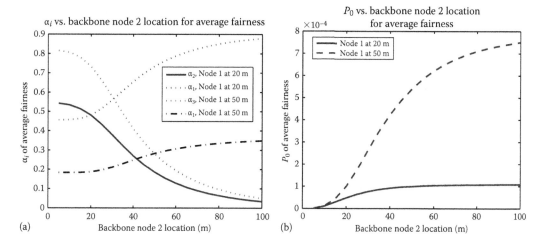

FIGURE 13.5 (a) α_i of average fairness for different users' locations. (b) P_0 of average fairness for different users' locations.

location cross at the boundary node 1 location. The figure shows that the average fairness using the Shapley function gives greater rewards to the boundary node whose channel is better and which can help the backbone node more. When boundary node 2 moves from (20 m, 0 m) to (50 m, 0 m), α_1 becomes smaller, but α_2 becomes larger. This is because the backbone node must depend on boundary node 2 more for relaying. However, the backbone node will pay less for the boundary nodes. Notice that α_i at the crossover point is lower. This is because the overall power for the backbone node is high when boundary node 2 is far away, as shown in Figure 13.5b.

Finally, we examine the degree to which the coalition game can improve the network connectivity. Here we define the network connectivity as the probability that a randomly located node can connect to the other nodes. All nodes are randomly located within a square of size $B \times B$. In Figure 13.6, we show the network unconnectivity as a function of B for the numbers of nodes equal to 100 and 500. With increasing network size, the node density becomes lower and more and more nodes are located at the boundary and must depend on the others for packet forwarding. If no coalition game is formed, these boundary nodes cannot transmit their packets due to the selfishness of the other nodes. With the coalition game, the network connectivity can be improved by about 50 percent. The only chance that a node cannot connect to the other nodes is when this node is located too far away from any other node. We can see that the game coalition cures the curse of the boundary nodes in wireless packet-forwarding networks with selfish nodes.

On the whole, we have proposed a coalition game approach to provide benefits to selfish nodes in wireless packet-forwarding networks using CT, so that the boundary nodes can transmit their packets effectively. We have used the concepts of coalition games to maintain stable and fair game coalitions. Specifically, we have studied two fairness concepts: min–max fairness and average fairness. A protocol has been constructed using repeated games and coalition games. From simulation results, we have seen how boundary nodes and backbone nodes form coalitions according to different fairness criteria. We can also see that network connectivity can be improved by about 50 percent, compared with the pure repeated game approach.

13.4 COOPERATIVE ROUTING IN SENSOR NETWORKS

Extending network lifetime of battery-operated devices is a key design issue that allows uninterrupted information exchange among distributive nodes in wireless sensor networks [46]. In Ref. [47], a data routing algorithm has been proposed with an aim to maximize the minimum lifetime over all nodes

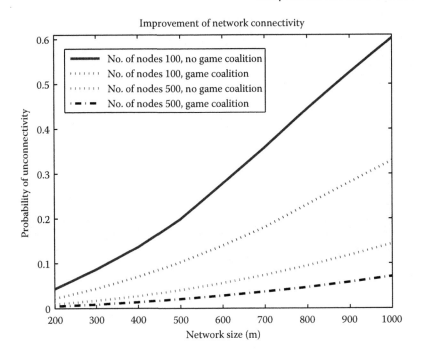

FIGURE 13.6   Network connectivity vs. network size.

in wireless sensor networks. A survey of energy constraints for sensor networks has been studied in Ref. [48]. In Ref. [49], the network lifetime has been maximized by employing the accumulative broadcast strategy. The work in Ref. [50] has considered provisioning additional energy in the existing nodes and deploying relays to extend the lifetime.

Like CT, collaborative beamforming (CB) has recently emerged to enable and leverage effective resource sharing among collaborative/cooperative nodes. In this section, following Ref. [22], we consider how to maximize the lifetime of sensor networks by using the idea that closely located nodes can use CB/CT to reduce the load or even avoid packet-forwarding requests to nodes that have critical battery life. First, we discuss the effectiveness of CB/CT to improve the signal strength at a faraway destination using energy in nearby nodes. Then, a 2D disk case is investigated to assess the resulting performance improvement. For general networks, if information-generation rates are fixed, the new routing problem is formulated as a linear programming problem; otherwise, the cost for routing is dynamically adjusted according to the amount of energy remaining and the effectiveness of CB/CT. From the analysis and simulation results, it is seen that the proposed schemes can improve the lifetime by about 90 percent in the 2D disk network and by about 10 percent in the general networks, compared with existing schemes.

13.4.1   MOTIVATION AND BASIC PROBLEM

To illustrate the problem, we assume that a group of sensors is uniformly distributed with a density of ρ. Each node is equipped with a single ideal isotropic antenna. There is no power control for each node, i.e., the node transmits with power either P or 0. There is no reflection or scattering of the signal. Thus, there is no multipath fading or shadowing. The nodes are sufficiently separated and any mutual coupling effects among the antennas of different nodes are negligible.

For traditional direct transmission, a node tries to reach another node at a distance of A. The SNR is given by

$$\gamma = \frac{PC_0 A^{-\alpha}|h|^2}{\sigma^2},$$ (13.37)

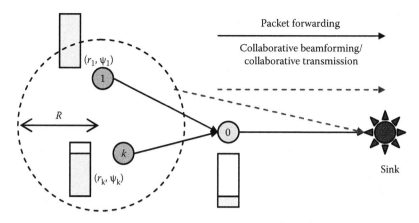

FIGURE 13.7 System model.

where

C_0 is a constant that incorporates effects such as antenna gains
α is the propagation loss factor
h is the channel gain
σ^2 is the thermal noise level

We define the energy cost of such a transmission for each packet to be one unit.

In Figure 13.7, we show the system model with CB/CT. In traditional sensor networks, the only choice a node has is to forward its information toward the sink. This will deplete the energy of the nodes near the sink, because they have to transmit many other nodes' packets. To overcome this problem, we propose another choice for a node consisting of forming CB/CT with the nearby nodes so as to transmit further toward the sink. By doing this, we can balance the energy usage of the nodes with different locations and different remaining energy.

13.4.2 NEW COOPERATIVE ROUTE DISCOVERY

In this subsection, we study how effectively CB and CT can improve the link quality; then in the next subsection we formulate the lifetime optimization problem and proposed algorithms to obtain the solutions.

13.4.2.1 Effectiveness of Collaborative Beamforming

Suppose there is a total of N users for CB within a disk of radius R. We have

$$N = \lfloor \rho \pi R^2 \rfloor. \tag{13.38}$$

Each node has polar coordinate (r_k, ψ_k) to the disk center. The distance from the center to the beamforming destination is A. The Euclidean distance between the kth node and the beamforming destination can be written as

$$d_k = \sqrt{A^2 + r_k^2 - 2r_k A \cos(\phi - \psi_k)}, \tag{13.39}$$

where ϕ is the Azimuth direction and is assumed to be a constant. By using loop control or the global positioning system (GPS), the initial phase of node k can be set to

$$\psi_k = -\frac{2\pi}{\lambda} d_k(\phi), \tag{13.40}$$

where λ is the wavelength of the radio frequency carrier.

Define $\mathbf{z} = [z_1, \ldots, z_N]^{\mathrm{T}}$ with

$$z_k = \frac{r_k}{R}\sin(\psi_k - \phi/2). \tag{13.41}$$

The array factor of CB can be written as

$$F(\phi|\mathbf{z}) = \frac{1}{N}\sum_{k=1}^{N}e^{-j4\pi R\sin(\frac{\phi}{2})z_k/\lambda}. \tag{13.42}$$

The far-field beam pattern can be defined as

$$P(\phi|\mathbf{z}) = |F(\phi|\mathbf{z})|^2$$
$$= \frac{1}{N} + \frac{1}{N^2}\sum_{k=1}^{N}e^{-ja(\phi)z_k}\sum_{l\neq k}e^{ja(\phi)z_l}, \tag{13.43}$$

where

$$a(\phi) = \frac{4\pi R\sin\frac{\phi}{2}}{\lambda}. \tag{13.44}$$

Define the directional gain $D_{\mathrm{av}}^{\mathrm{CB}}$ as the ratio of radiated concentrated energy in the desired direction over that of a single isotropic antenna. From Theorem 1 in Ref. [51], for large $\frac{R}{\lambda}$ and N, the following lower bound for far-field beamforming is tightly held:

$$\frac{D_{\mathrm{av}}^{\mathrm{CB}}}{N} \geq \frac{1}{1 + \mu\frac{N\lambda}{R}}, \tag{13.45}$$

where $\mu \approx 0.09332$.

Considering this directional gain, we can improve the direct transmission by a factor of $D_{\mathrm{av}}^{\mathrm{CB}}$. Notice that this transmission distance gain for one transmission is obtained at the expense of consuming a total power of N units from the nearby nodes.

13.4.2.2 Effectiveness of Cooperative Transmission

Similar to the CB case, we assume that N users are uniformly distributed over a radius of R. The probability density function of the users' radial coordinate r is given by

$$q(r) = \frac{2r}{NR^2}, \quad 0 \leq r \leq R, \tag{13.46}$$

and the users' angular coordinate ψ is uniformly distributed between $[0, 2\pi)$.

Suppose that at the first stage, node 1 transmits to the next hop or sink. Then in subsequent stages, node 2 to node N relay the node 1's information if they decode it correctly. The received signals at node 2 to node k at stage 1 can be expressed as

$$z_k = \sqrt{Pr_k^{-\alpha}}h_k^r x + n_k^r, \quad k = 2, \ldots, N, \tag{13.47}$$

and the received signals at the destination in subsequent stages are

$$y_k = \sqrt{Pd_k^{-\alpha}}h_k x + n_k. \tag{13.48}$$

Here P is the transmitted power, h_k^r and h_k are the channel gains of source–relay and relay–destination, which are modeled as independent zero mean circularly symmetric complex Gaussian random variables with unit variance, x is the transmitted data with unit power, and n_k and n_k^r are independent thermal noises with noise variance σ^2.

THEOREM 5

Define D_{av}^{CT} to be the energy enhancement at the destination node due to CB. Under the far-field condition and the assumption that channel links between source and relays are sufficiently good, we have the following approximation:

$$\frac{D_{av}^{CT}}{N} \approx \frac{1 + (N-1) \, _2F_1\left(\frac{2}{\alpha}, -L; \frac{\alpha+2}{\alpha}; \frac{\sigma^2 R^\alpha}{4P}\right)}{N},$$ (13.49)

where L is the frame length and $_2F_1$ is the Hypergeometric function

$$_2F_1(a, b; c; z) = \sum_{n=0}^{\infty} \frac{(a)_n (b)_n}{(c)_n} \frac{z^n}{n!},$$ (13.50)

where $(a)_n = a(a+1)\cdots(a+n-1)$ is the Pochhammer symbol.

Proof: The SNR received by the kth user at stage one can be written as

$$\gamma_k = \frac{PC_0 r_k^{-\alpha} |h_k^r|^2}{\sigma^2},$$ (13.51)

where $|h_k^r|^2$ is the magnitude square of the channel fade and follows an exponential distribution with unit mean.

Without loss of generality, we suppose that binary phase shift keying modulation is used and $C_0 = 1$. The probability of successful transmission of the packet with length L is given by

$$P_r^k(r) = \left(\frac{1}{2} + \frac{1}{2}\sqrt{\frac{P}{P + \sigma^2 r_k^\alpha}}\right)^L.$$ (13.52)

For fixed (r_k, ψ_k), the average energy that arrives at the destination can be written as

$$D^{CT} = \sum_{k=1}^{N} P d_k^{-\alpha} P_r^k.$$ (13.53)

Because for node 1, $r_1 = 0$, we can write the average energy gain in the following generalized form:

$$D_{av}^{CT} = \sum_{k=1}^{N} \int_0^R \int_0^{2\pi} \frac{A^\alpha}{2\pi} d_k^{-\alpha} P_r^k(r_k) q(r_k) dr_k \, d\psi_k.$$ (13.54)

Because each user is independent of the others, we omit the notation k and can rewrite Equation 13.54 as

$$D_{av}^{CT} = 1 + (N-1) \int_0^{2\pi} \int_0^R \frac{2r}{A^{-\alpha} R^2} (A^2 + r^2 - 2rA \cos \psi)^{-\frac{\alpha}{2}} P_r(r) dr \, d\psi.$$ (13.55)

With the far field assumption, we have

$$\int_0^{2\pi} (A^2 + r^2 - 2rA\cos\psi)^{-\frac{\alpha}{2}} d\psi \approx A^{-\alpha}. \tag{13.56}$$

The average energy gain is approximated by

$$D_{av}^{CT} \approx 1 + (N-1)\frac{2}{R^2}\int_0^R rP_r(r)dr. \tag{13.57}$$

With the assumption of sufficiently good channels between sources and relays, we have the following approximation of Equation 13.52:

$$P_r(r) \approx \left(1 - \frac{\sigma^2 r^\alpha}{4P}\right)^L. \tag{13.58}$$

Because

$$\int_0^R r\left(1 - \frac{\sigma^2 r^\alpha}{4P}\right)^L dr = \frac{1}{2}R^2 \,_2F_1\left(\frac{2}{\alpha}, -L; \frac{\alpha+2}{\alpha}; \frac{\sigma^2 R^\alpha}{4P}\right), \tag{13.59}$$

we can obtain Equation 13.49.

13.4.3 SENSOR NETWORK LIFETIME MAXIMIZATION

In this subsection, we first define the lifetime of sensor networks and formulate the corresponding optimization problem. Then, by using a 2D disk case, we demonstrate analytically the effectiveness of lifetime saving using CB/CT. Finally, two algorithms are proposed for general network configurations.

13.4.3.1 Problem Formulation

In Figure 13.8, we show the routing model with CB/CT. A wireless sensor network can be modeled as a directed graph $G(M, \mathbb{A})$, where M is the set of all nodes and $\mathbb{A}$ is the set of all links $(i,j), i, j \in M$. Here the link can be either a direct transmission link or a link with CB/CT. Let S_i be the set of nodes

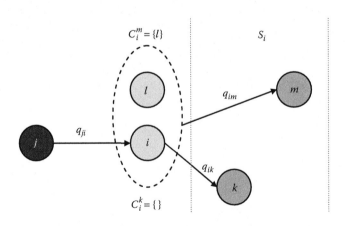

FIGURE 13.8 CB/CT routing model.

that the ith node can reach by direct transmission. Denote by C_i^m the set of nodes that node i needs to apply CB/CT to reach node m. In the example in Figure 13.8, $C_i^m = \{l\}$ and $C_i^k = \{\}$. A set of origin nodes O where information is generated at node i with rate Q_i can be written as

$$O = \{i|Q_i > 0, i \in M\}. \tag{13.60}$$

A set of destination nodes is defined as D where

$$D = \{i|Q_i < 0, i \in M\}. \tag{13.61}$$

Define $\mathbf{q} = \{q_{ij}\}$ to represent the routing and the transmission rate. There are many types of definitions for the lifetime of sensor networks. The most common ones are the first node failure, the average lifetime, and α lifetime. In this section, we use the lifetime until first node failure as an example. Other types of lifetime can be examined in a similar way. Suppose node i has remaining energy of E_i. The lifetime for each node can be written as

$$T_i(\mathbf{q}) = \frac{E_i}{\sum_{j \in S_i} q_{ij} + \sum_{i \in C_j^m, \forall j,m} q_{jm}}, \tag{13.62}$$

where the first term in the denominator is for direct transmission and the second term in the denominator is for CB/CT. Notice that C_j^m is not a function of $\mathbf{q}$. We formulate the problem as

$$\max_{\mathbf{q}} \min T_i \tag{13.63}$$

$$\text{s.t.} \begin{cases} q_{ij} \geq 0, \forall i,j \\ \sum_{j,i \in S_j} q_{ij} + Q_i = \sum_{k \in S_i} q_{ik}, \end{cases}$$

where the second constraint is for flow conservation.

13.4.3.2 2D Disk Case Analysis

Next, we study a 2D disk case network. Users with the same remaining energy are uniformly located within a circle of radius B_0. One sink is located at the center location $(0,0)$. Each node has a unit amount of information to transmit. Here we assume that the user density is large enough, so that each node can find enough nearby nodes to form CB/CT to reach the faraway node.

For traditional packet forwarding without CB/CT, the number of packets needing transmission for each node at the distance B to the sink is given by

$$N_{\text{pf}}(B) = \sum_{n=0}^{\left\lfloor \frac{B_0 - B}{A_0} \right\rfloor} \left(1 + \frac{nA_0}{B}\right). \tag{13.64}$$

where A_0 is the maximal distance over which a minimal link quality γ_0 can be maintained, i.e., $\gamma(A_0) = \gamma_0$.

If all nodes use their neighbor nodes to communicate with the sink directly, we call this scheme pure CB/CT. To achieve the range of B, we need $N_{\text{CB/CT}}(B)$ for CB/CT, i.e.,

$$D_{\text{av}}\left(\sqrt{\frac{N_{\text{CB/CT}}(B)}{\rho\pi}}\right) = \left(\max\left(\frac{B}{A_0}, 1\right)\right)^\alpha. \tag{13.65}$$

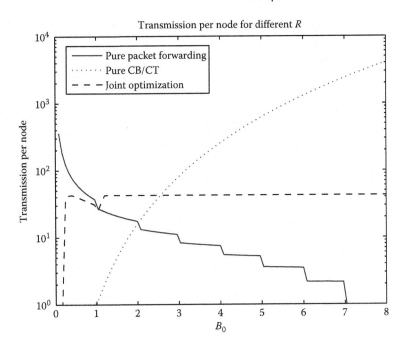

FIGURE 13.9 Analytical results for the 2D disk case.

For CB, we can calculate

$$N_{CB}(B) \geq \frac{1}{2}\left(c_0(2 + c_0 c_1^2) + c_0^{1.5} c_1 \sqrt{4 + c_0 c_1^2}\right), \tag{13.66}$$

where $c_0 = (\max(\frac{B}{A_0}, 1))^\alpha$ and $c_1 = \mu\lambda\sqrt{\rho\pi}$. For CT, numerical results need to be used to obtain the inverse of $_2F_1$ in Theorem 1. Notice that if the node density is large enough, then $D_{av}/N \longrightarrow 1$.

In Figure 13.9, we show the average transmission per node versus the disk size B_0. We can see that for traditional packet forwarding, the node closest to the sink has the most transmissions per node, i.e., it has the lowest lifetime if the initial energy is the same for all nodes. On the other hand, for the pure CB/CT scheme, more nodes need to transmit to reach the sink directly when B_0 is larger. The transmission is less efficient than packet forwarding, because the propagation loss factor α is larger than 1. The above facts motivate the joint optimization case where nodes transmit packets with different probabilities over traditional packet forwarding and CB/CT.

For traditional packet forwarding, nodes near the sink have lower lifetimes. If the faraway nodes can form CB/CT to transmit directly to the sink and bypass these life depleting nodes, the overall network lifetime can be improved. Notice that in this special case, if the faraway nodes form CB/CT to transmit to nodes other than the sink, the lifetime will not be improved. For each node with distance B to the sink, and supposing the probability of using CB/CT is $P_r(B)$, we have

$$N_{joint}(B) = (1 - P_r(B) + N_{CB/CT}(B)P_r(B))$$

$$\sum_{n=0}^{\lfloor\frac{B_0 - B}{A_0}\rfloor} \left(1 + \frac{nA_0}{B}\right) \pi_{j=1}^n(1 - P_r(B + jA_0)), \tag{13.67}$$

where the first term on the right-hand side (RHS) is the necessary energy for transmitting one packet, and the second term is the number of packets for transmission. The goal is to adjust $P_r(B)$ such that the lifetime is maximized, i.e.,

$$\min_{1 \geq P_r(B) \geq 0} \max N_{\text{joint}}(B). \tag{13.68}$$

Notice that $N_{\text{CB/CT}} \geq 1$, and in Equation 13.67 the second term on the RHS depends on the probabilities of CB/CT being larger than B. Therefore, we can develop an efficient bisection search method to calculate Equation 13.68. We define a temperature κ that is assumed to be equal or greater than $N_{\text{joint}}(B)$, $\forall B$. We can first calculate $N_{\text{joint}}(B)$ from the boundary of the network where the second term on the RHS of Equation 13.67 is one. Then we can derive all $N_{\text{joint}}(B)$ by reducing B. If κ is too large, most information is transmitted by CB/CT, and the nodes faraway from the sink waste too much power for CB/CT; on the other hand, if κ is too small, the nodes close to the sink must forward too many packets. A bisection search method can find the optimal values of κ and $N_{\text{joint}}(B)$.

In Figure 13.9, we show the joint optimization case where the node density is sufficiently large. We can see that to reduce the packet-forwarding burdens of the nodes near the sink, the faraway nodes form CB/CT to transmit to the sink directly. This will increase the number of transmissions per node for them, but reduce the transmissions per node for the nodes near the sink. In Table 13.1, we show the maximal $N_{\text{joint}}(B)$ and the lifetime saving over the traditional packet forwarding. We can see that the power saving is around 90 percent.

13.4.3.3 General Case Algorithms

Next, we first consider the case in which the information generation rates are fixed for all sensors and develop a linear programming method to calculate the routing table. Here to simplify the calculation of set C_j^m, we assume its size equals one. Obviously, this is suboptimal for Equation 13.63. Then we select the nearest neighbor for CB/CT. $C_j^m = 1$, if node j's nearest neighbor can help node j to reach node m. Define $\hat{q}_{ij} = Tq_{ij}$. The problem can be written as a linear programming problem:

$$\max T \tag{13.69}$$

$$\text{s.t.} \begin{cases} \hat{q}_{ij} \geq 0, \forall i, j; \\ \left(\sum_{j \in S_i} \hat{q}_{ij} + \sum_{i \in C_j^m, \forall j, m} \hat{q}_{jm} \right) \leq E_i, \forall i; \\ \sum_{j, i \in S_j} \hat{q}_{ji} + TQ_i = \sum_{k \in S_i} \hat{q}_{ik}, \forall i \in M - D, \end{cases}$$

where the second constraint is the energy constraint and the third constraint is for flow conservation.

If the information rate is random, each sensor dynamically updates its cost according to its remaining energy and with consideration of CB/CT. Some heuristic algorithms can be proposed to update the link cost dynamically. Here the initial energy is E_i. Define the current remaining energy as $\underline{E}_i$. We define the cost for node i to communicate with node j as

$$\text{cost}_{ij} = \left(\frac{E_i}{\underline{E}_i} \right)^{\beta_1} + \sum_{l \in C_i^j} \left(\frac{E_l}{\underline{E}_l} \right)^{\beta_2}, \tag{13.70}$$

TABLE 13.1

Lifetime Saving vs. Disk Size

R_0	2	4	6	8	10
max $N_{\text{joint}}(B)$	2.82	10.25	23.4	42.5	64.5
Saving %	94.56	93.33	90.86	88.13	85.98

---

where β_1 and β_2 are positive constants. Their values determine how the packets are allocated between the energy sufficient and energy depleting nodes and between the direct transmission and CB/CT. Notice that Equation 13.70 can be viewed as an inverse barrier function for $\underline{E}_i \geq 0$.

13.4.4 NUMERICAL STUDY

We assume that nodes and one sink are randomly located within a square of size $\mathbb{L} \times \mathbb{L}$. Each node has a power of 10 dBm and the noise level is -70 dBm. The propagation loss factor is 4. The minimal link SNR is 10 dB. The initial energy of all users is assumed to be unit and information rates for all users are 1.

In Figure 13.10a, we show a snapshot of a network of five sensor nodes and a sink with $\mathbb{L} = 50$ m. Here node 1 is the sink. The solid lines are the links for the direct transmission, and the dotted line from node 6 to the sink is the CB/CT link with the help of node 5. For a traditional direct packet-forwarding scheme, the best flow is

$$
\hat{q}_{ij} =
\begin{bmatrix}
0 & 0 & 0 & 0 & 0 & 0 \\
1 & 0 & 0 & 0 & 0 & 0 \\
0 & 0.2 & 0 & 0 & 0 & 0 \\
0 & 0 & 0 & 0 & 0.1 & 0.1 \\
0 & 0.3 & 0 & 0 & 0 & 0 \\
0 & 0.3 & 0 & 0 & 0 & 0
\end{bmatrix}
\tag{13.71}
$$

with the resulting energy consumed for all nodes given by $[0, 1.0, 0.2, 0.2, 0.3, 0.3]$. Because node 2 is the only node that can communicate with the sink, the best lifetime of this routing is 0.2 before node 2 runs out of energy.

With CB/CT, the best flow is

$$
\hat{q}_{ij} =
\begin{bmatrix}
0 & 0 & 0 & 0 & 0 & 0 \\
1 & 0 & 0 & 0 & 0 & 0 \\
0 & 0.321 & 0 & 0 & 0 & 0.012 \\
0 & 0 & 0 & 0 & 0 & 0.333 \\
0 & 0.23 & 0 & 0 & 0 & 0.103 \\
0.667 & 0.115 & 0 & 0 & 0 & 0
\end{bmatrix}
\tag{13.72}
$$

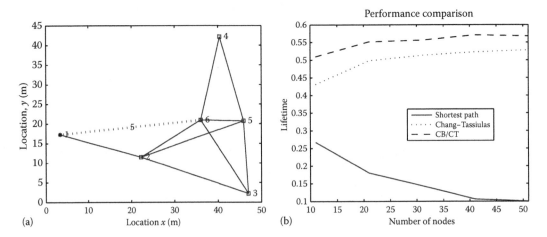

FIGURE 13.10 (a) Snapshot of CB/CT routing; (b) lifetime improvement comparison.

with the energy consumed for all nodes given by [0, 1.000, 0.333, 0.333, 1.0000, 0.782]. Here some flow can be sent to the sink through node 6. Because of CB/CT, node 5 has to consume its power. The lifetime becomes 0.333, which is 67 percent improvement over direct packet forwarding.

In Figure 13.10b, we compare the performance of three algorithms, the shortest path, the algorithm in Ref. [47], and the proposed CB/CT algorithm. Here $\mathbb{L} = 100\,\text{m}$. As the number of users increases, the performance of the shortest path algorithm decreases. This is because more users will need packet forwarding by the nodes near the sink. Consequently, they die more quickly. Compared with the algorithm in Ref. [47], the proposed schemes have about 10 percent performance improvement. This is because of the alternative routes to the sink that can be found by CB/CT.

On the whole, we have studied the impact of CB/CT on the design of higher-level routing protocols. Specifically, using CB/CT, we have proposed a new idea based on bypassing energy depleting nodes that might otherwise forward packets to the sink, to improve the lifetime of wireless sensor networks. From the analytical and simulation results, we have seen that the proposed protocols can increase lifetime by about 90 percent in a 2D disk case and about 10 percent in general network situations, compared with existing techniques.

13.5   COOPERATIVE ROUTING IN GENERAL NETWORKS

In this section, we study the impact of cooperative communications on the network layer for general wireless networks. Then we briefly discuss the current standard development for IEEE 802.16j WiMAX relay networks.

13.5.1   COOPERATIVE ROUTING ALGORITHMS

In the network layer, routing algorithms select multi-hop links between a source and destination with minimal cost in terms such as overall power or with maximal gain in terms such as throughput. For each hop, metrics can be defined to cover such information as bandwidth, delay, hop count, path cost, load, reliability, and communication cost. A routing metric is a value used by a routing algorithm to determine whether one route should perform better than another. Then a routing algorithm, such as the shortest path algorithm (e.g., Dijkstra's algorithm), can find the optimal route among all possible connections from the source to the destination.

With CT, the cooperative routing problem has been recently considered in the literature [19, 20,24,52,53]. Most of the current cooperative-based routing algorithms, such as the cooperation along the minimum energy noncooperative path (CAN) [20], progressive cooperative (PC) [20], and cooperative routing along truncated noncooperative route (CTNCR) [19], are implemented in two consecutive steps. First, a noncooperative route is constructed using any shortest-path routing algorithm. Second, cooperative-communication protocols are applied on some or all of the nodes along the established route. Indeed, these routing algorithms do not fully exploit the merits of CT, because the optimal cooperative routes might not be along the noncooperative routes.

One simple example is shown in Figure 13.11, in which a regular grid topology is studied. To illustrate the routes selected by different routing schemes, we assume that the source is node 0 and the destination is node 7. The shortest path routing algorithm chooses one of the possible shortest routes. For instance, the chosen shortest route is {(0, 1), (1, 5), (5, 6), (6, 7)}, where (i, j) denotes the direct transmission mode from node i to node j. Figure 13.11a shows the route chosen by the shortest path routing algorithm, where the solid line between each two nodes indicates the direct transmission mode. The cooperative route based on the shortest path algorithm applies cooperation among each of the three consecutive nodes on the shortest route, and the resulting route is {(0, 1, 5), (5, 6, 7)}, where (x, y, z) denotes the CT mode between sender x, relay y, and destination z. Figure 13.11b shows the route chosen by this routing algorithm. The solid lines indicate the sender–destination transmissions and the dashed lines indicate the sender–relay and relay–receiver transmissions. Finally, we can find that the optimal cooperative route is given by {(0, 5, 1), (1, 2, 6), (6, 11, 7)} as shown in Figure 13.11c.

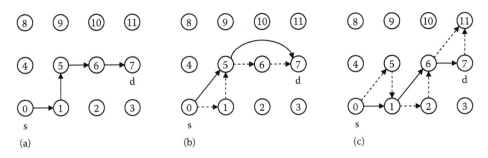

FIGURE 13.11   (a) Shortest path route; (b) cooperative route based on shortest path; (c) optimal cooperative route.

In this example, we can visually notice the difference between the routes chosen by the optimal cooperative routing and the cooperative routing algorithms based on the shortest path.

This example provides motivations to propose a one-step cooperative routing algorithm, where the routing decision is based on CT directly. The design process can be summarized as follows:

1. The one-hop cost function is to be designed while considering CT. The cost function can be the cost for traditional direct transmission or the cost for the source and relays together for CT. A certain quality of service (QoS) needs to be guaranteed. The major challenge here for the cooperative cost function is the relay selection, which is a complicated integer optimization problem. Some heuristics should be designed to reduce the computation complexity and still achieve near optimal performance.
2. Then the optimum route is defined as the route, which requires the minimum overall cost. Any routing algorithms can be used to calculate the optimal route.

In Ref. [21], the one-hop cost is defined as the power while a certain bit error rate (BER) is ensured. The heuristic is to select only one relay. It is shown that this one-step algorithm outperforms the two-step algorithms [19,20] by 10 percent.

13.5.2   WiMAX IEEE 802.16J

Finally, we discuss the impact of CT on wireless network standards. WiMAX, based on the IEEE 802.16 standard for wireless metropolitan area networks (WMANs), is expected to enable broadband speeds over wireless networks at a cost that enables mass market adoption and thereby make the vision of pervasive connectivity a reality. WMANs are designed for relatively large-scale networks, such as a large corporate or university campus or an entire city. The IEEE 802.16 standard has helped to pave the way for WMAN technology globally and, because its first inception, has been expanded considerably. Next, we discuss one of the expansions related to CT.

Current deployments of IEEE 802.16 standards suffer from problems such as limited spectrum, low signal-to-interference-plus-noise ratio at cell edges, coverage holes due to shadowing, and nonuniformly distributed traffic loads. To address these issues, the IEEE instituted the work on the standard 802.16j "mobile multi-hop relay" (MMR) in 2006 [54,55]. The basic idea behind MMR is to allow WiMAX base stations to impose a demanding performance requirement on relay stations (RSs). These relays will functionally serve as an aggregating point on behalf of the base station (BS) for traffic collection from, and distribution to, multiple mobile stations (MSs) associated with the relays and thus naturally incorporate a notion of "traffic aggregation." On the one hand, this approach will of course reduce the bandwidth available to users in the cells involved in relaying packets. On the other hand it is an elegant way to reduce costs and extend network coverage into areas where connecting a base station directly to the network through a fixed line connection is economically or

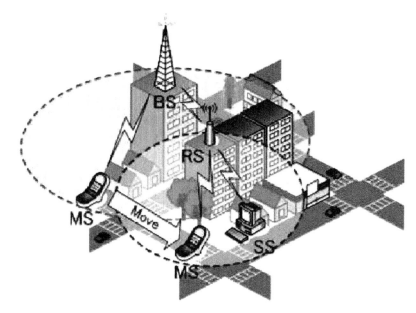

FIGURE 13.12   IEEE 802.16j overview.

technically not feasible. The goal is to enhance coverage, throughput, and system capacity of exist-ing IEEE 802.16 networks by specifying multi-hop relay capability and functionality of the relay stations and base stations. One illustrated example is shown in Figure 13.12. Some of the design requirements of IEEE 802.16j are listed as follows:

- Backward compatibility with the existing structure
- Definition of relay frequency and channel bandwidth
- Support from relays for network entry of mobile stations
- Support of QoS and hybrid automatic repeat request (ARQ)
- Support of handover and mobility
- Deployment of multiple antennas for the relay link
- Support of multiple hops between the BS and MS
- Enhancement of link reliability

According to the different network scenarios, there are three types of WiMAX relay stations as shown in Figure 13.13.

- Fixed relay stations (FRSs): Permanently installed at fixed locations.
- Nomadic relay stations (NRSs): Location fixed for periods of time; but can be moved around. NRSs are used for situations such as special events.
- Mobile relay stations (MRSs): For use in mobile environments.

It is anticipated that there will be no change to WiMAX subscriber devices used with the above WiMAX relay stations. But 802.16 BS is being updated to support MMR functions and to be backward compatible with the current version of WiMAX subscriber services. For the frames to transmit between the BS and the destination through the relay stations, in Figure 13.14, we illustrate the frame structure for IEEE 802.16j. In this example, the message from the BS can be forwarded to the destination through three relay stations. However, this brings the following challenges for the network design.

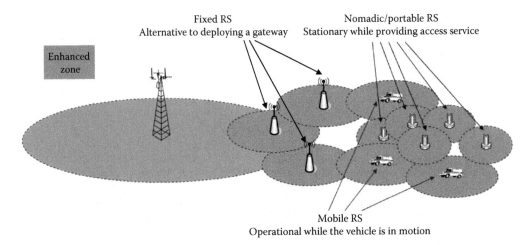

FIGURE 13.13   IEEE 802.16j three types of relay stations.

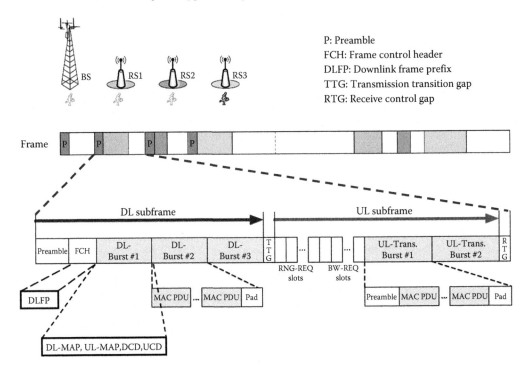

FIGURE 13.14   IEEE 802.16j frame structure with relay stations.

- System configuration/management. To optimally deploy the relay stations, the network topology needs to be known. Moreover, for dynamic scenarios with mobile relay stations, neighbor detection is necessary. A relay path management process (such as a path selection algorithm and path recovery) can eliminate coverage holes. For areas with high subscriber densities, congestion control at the network level needs to be implemented. In addition, connection management determines how the BS can be connected to the destination nodes. Finally, QoS provisioning needs to be considered especially for the multimedia payloads.
- Network entry, bandwidth management, and scheduling. When a node enters the network, admission control synchronization, ranging, and authorization need to be implemented.

TABLE 13.2
Comparison of Repeater, MMR, and BS

	Conventional Repeater	MMR	Base Station
Concept	Dummy repeater	Smart repeater	Radio access station
Function	AF	Decode-and-forward	Encode-and-decode
Cost	Low	Reasonable	High
Coverage	Narrow-wide	Narrow-wide	Wide
Performance	Severe degradation	Mild degradation	No degradation
Resource Management	Controlled by BS	Collaborative with BS	Self-controlled
Interference	Unmanaged	Managed by BS	Self-managed

When nodes request bandwidth, the BS and RS need to perform bandwidth allocation. The scheduling of packets from the BS, RSs, and mobile users can be difficult to implement centrally if the number of relays is large. In this case, distributed scheduling is preferred.

- Data delivery. For MAC protocol data unit (PDU) processing, information is delivered as a unit among peer entities of a network and may contain control information, address information, or data. The transmitted data can be classified as unicast/multicast/broadcast data. To ensure successful delivery of data, ARQ or Hybrid ARQ processes are needed. Finally, if the receiver can combine the data in different frames, cooperative communication techniques can also be used to improve the communication links.
- Mobility management. Algorithms need to be constructed for MS handover and MRS handover to support mobility. The handoff can be intra-MMR cell handover in which only one MMR-BS participates or inter-MMR cell handover, which involves multiple MMR BS.

A comparison of MMS with conventional repeaters and BS is shown in Table 13.2.

In summary, 802.16j is under development for coverage extension and throughput enhancement for existing WiMAX. There are many open issues relating to MMR systems such as system configuration and management, network admission, bandwidth management, scheduling, data delivery, and mobility management.

13.6 CONCLUSION

Overall, cooperative communications have attracted considerable attention as a transmit technology for future wireless networks, by efficiently taking advantage of the broadcast nature of wireless networks as well as by exploiting the inherent spatial and multiuser diversities by treating cooperative relays as virtual antenna arrays. Most work in this area focuses on how to improve link quality in the physical layer, while the networking issues have been less well studied.

The cross-layer impact of cooperative communications lies in the fact that it offers a new degree of freedom for traditional communication problems. For example, for power control, the relay's power can determine the performance at the destination. With different power control strategies, rate adaptation can be adjusted to fully use spectrum. With limited spectrum and multiple users, relay selection and channel allocation address the problems of multiple access and spectrum access. Moreover, in the network layer, the routing metrics can be significantly different from traditional ones under CT, because the routes with cooperative users can greatly improve the route performance. The impact of cooperative communications can also be seen in the application layer, with applications such as cooperative video transmission [29,56].

In this chapter, we have concentrated on the impact of CT on the network layer for wireless networks. Specifically, we have examined the three cases of ad hoc networks, sensor networks, and general networks. The key point we would like to deliver is that, by using CT, new methods can emerge to improve network performance. We have given several examples of how CT can affect different layers to improve wireless network design.

REFERENCES

[1] http://ieee802.org/16/relay/

[2] A. Sendonaris, E. Erkip, and B. Aazhang, User cooperation diversity, Part I: System description, *IEEE Transactions on Communications*, 51(11), 1927–1938, November 2003.

[3] A. Sendonaris, E. Erkip, and B. Aazhang, User cooperation diversity, Part II: Implementation aspects and performance analysis, *IEEE Transactions on Communications*, 51(11), 1939–1948, November 2003.

[4] J. N. Laneman, D. N. C. Tse, and G. W. Wornell, Cooperative diversity in wireless networks: Efficient protocols and outage behavior, *IEEE Transactions on Information Theory*, 50(12), 3062–3080, December 2004.

[5] T. M. Cover and A. El Gamal, Capacity theorems for the relay channel, *IEEE Information Theory*, 25(5), 572–584, September 1979.

[6] M. A. Khojastepour, A. Sabharwal, and B. Aazhang, On the capacity of 'cheap' relay networks, in *Proceedings of the 37th Annual Conference on Information Sciences and Systems*, Baltimore, MD, March 2003.

[7] T. E. Hunter and A. Nosratinia, Performance analysis of coded cooperation diversity, in *Proceedings of 2003 International Conference on Communications ICC'03*, Vol. 4, pp. 2688–2692, Seattle, WA, May 2003.

[8] J. N. Laneman and G. W. Wornell, Distributed space-time coded protocols for exploiting cooperative diversity in wireless networks, *IEEE Transactions on Information Theory*, 49(10), 2415–2525, October 2003.

[9] A. K. Sadek, K. J. R. Liu, and A. Ephremides, Cognitive multiple access via cooperation: Protocol design and performance analysis, *IEEE Transactions on Information Theory*, 53(10), 3677–3696, October 2007.

[10] K. Lee and A. Yener, Outage performance of cognitive wireless relay networks, in *Proceedings of the IEEE Global Telecommunications Conference*, San Francisco, CA, November 2006.

[11] W. Su, A. K. Sadek, and K. J. R. Liu, SER performance analysis and optimum power allocation for decode-and-forward cooperation protocol in wireless networks, in *Proceedings of IEEE Wireless Communications and Networking Conference (WCNC)*, New Orleans, LA, March 13–17, 2005.

[12] A. Host-Madsen, Upper and lower bounds for channel capacity of asynchronous cooperative diversity networks, *IEEE Transactions on Information Theory*, 50(4), 3062–3080, December 2004.

[13] A. Host-Madsen, A new achievable rate for cooperative diversity based on generalized writing on dirty paper, in *Proceedings of IEEE International Symposium Information Theory*, p. 317, Yokohama, Japan, June 2003.

[14] T. E. Hunter, S. Sanayei, and A. Nosratinia, Outage analysis of coded cooperation, *IEEE Transactions on Information Theory*, 52(2), 375–391, February 2006.

[15] J. Luo, R. S. Blum, L. J. Greenstein, L. J. Cimini, and A. M. Haimovich, New approaches for cooperative use of multiple antennas in ad hoc wireless networks, in *Proceedings of the IEEE Vehicular Technology Conference*, Vol. 4, pp. 2769–2773, Los Angeles, CA, September 2004.

[16] Y. Zhao, R. S. Adve, and T. J. Lim, Improving amplify-and-forward relay networks: Optimal power allocation versus selection, in *Proceedings of the IEEE International Symposium on Information Theory*, Seattle, WA, July 2006.

[17] Z. Han, X. Zhang, and H. V. Poor, Cooperative transmission protocols with high spectral efficiency and high diversity order using multiuser detection and network coding, in *Proceedings of the IEEE International Conference on Communications*, Glasgow, Scotland, June 2007.

[18] Z. Han, T. Himsoon, W. Siriwongpairat, and K. J. Ray Liu, Energy efficient cooperative transmission over multiuser OFDM networks: Who helps whom and how to cooperate, in *Proceedings of the IEEE Wireless Communications and Networking Conference*, Vol. 2, pp. 1030–1035, New Orleans, LA, March 2005.

[19] Z. Yang, J. Liu, and A. Host-Madsen, Cooperative routing and power allocation in ad-hoc networks, in *Proceedings of the IEEE Global Telecommunications Conference*, Dallas, TX, November 2005.

[20] A. E. Khandani, E. Modiano, L. Zheng, and J. Abounadi, Cooperative routing in wireless networks, in *Advances in Pervasive Computing and Networking*, Eds. B. K. Szymanski and B. Yener, Kluwer Academic Publishers, Norwell, MA, 2004.

[21] A. S. Ibrahim, Z. Han, and K. J. R. Liu, Distributed power-efficient cooperative routing in wireless ad hoc networks, in *Proceedings of the IEEE Global Telecommunications Conference*, Washington, DC, November 2007.

[22] Z. Han and H. V. Poor, Lifetime improvement in wireless sensor networks via collaborative beamforming and cooperative transmission, to appear, *IEE Proceedings of Microwaves, Antennas and Propagation, Special Issue on Antenna Systems and Propagation for Future Wireless Communications*, 1(6), 1103–1110, December 2007.

[23] J. Boyer, D. D. Falconer, and H. Yanikomeroglu, Cooperative connectivity models for wireless relay networks, *IEEE Transactions on Wireless Communications*, 6(6), 1992–2000, June 2007.

[24] F. Li, K. Wu, and A. Lippman, Energy-efficient cooperative routing in multi-hop wireless ad hoc networks, in *Proceedings of the IEEE International Performance, Computing, and Communications Conference*, pp. 215–222, Phoenix, AZ, April 2006.

[25] A. K. Sadek, W. Su, and K. J. R. Liu, A class of cooperative communication protocols for multi-node wireless networks, in *Proceedings of the IEEE International Workshop on Signal Processing Advances in Wireless Communications (SPAWC)*, New York, June 2005.

[26] A. Bletsas, A. Lippman, and D. P. Reed, A simple distributed method for relay selection in cooperative diversity wireless networks, based on reciprocity and channel measurements, in *Proceedings of the IEEE Vehicular Technology Conference*, Vol. 3, pp. 1484–1488, Stockholm, Sweden, May 2005.

[27] B. Wang, Z. Han, and K. J. Ray Liu, Distributed relay selection and power control for multiuser cooperative communication networks using buyer/seller game, in *Proceedings of the Annual IEEE Conference on Computer Communications, INFOCOM'07*, Anchorage, AK, May 2007.

[28] Z. Han and H. V. Poor, Coalition game with cooperative transmission: a cure for the curse of boundary nodes in selfish packet-forwarding wireless networks, in *Proceedings of the 5th International Symposium on Modeling and Optimization in Mobile, Ad Hoc, and Wireless Networks, (WiOpt07)*, Limassol, Cyprus, April 2007.

[29] D. Gunduz and E. Erkip, Joint source-channel cooperation: Diversity versus spectral efficiency, in *Proceedings of 2004 IEEE International Symposium Information Theory*, p. 392, Chicago, IL, June–July 2004.

[30] I. Maric and R. D. Yates, Cooperative multihop broadcast for wireless networks, *IEEE Journal on Selected Areas in Communications*, 22(6), 1080–1088, August 2004.

[31] G. Owen, *Game Theory*, 3rd ed., Academic Press, St. Louis, MO, 2001.

[32] R. B. Myerson, *Game Theory, Analysis of Conflict*, Harvard University Press, Cambridge, MA, 1991.

[33] J. Crowcroft, R. Gibbens, F. Kelly, and S. Ostring, Modelling incentives for collaboration in mobile ad hoc networks, *Performance Evaluation*, 57(4), 427–439, August 2004.

[34] S. Zhong, J. Chen, and Y. R. Yang, Sprite: A simple, cheat-proof, credit-based system for mobile ad-hoc networks, in *Proceedings of the Annual IEEE Conference on Computer Communications, INFOCOM*, pp. 1987–1997, San Francisco, CA, March 2003.

[35] S. Marti, T. J. Giuli, K. Lai, and M. Baker, Mitigating routing misbehaviour in mobile ad hoc networks, in *Proceedings of the ACM/IEEE Annual International Conference on Mobile Computing and Networking (Mobicom)*, pp. 255–265, Boston, MA, August 2000.

[36] S. Buchegger and J.-Y. Le Boudec, Performance analysis of the CONFIDANT protocol (cooperation of nodes-fairness in dynamic ad-hoc networks), in *Proceedings of the ACM International Symposium on Mobile Ad Hoc Networking and Computing (MobiHoc)*, pp. 80–91, Lausannae, Switzerland, June 2002.

[37] P. Michiardi and R. Molva, A game theoretical approach to evaluate cooperation enforcement mechanisms in mobile ad hoc networks, in *Proceedings of IEEE/ACM International Symposium on Modeling and Optimization in Mobile, Ad Hoc, and Wireless Networks (WiOpt)*, Sophi Antipolis, France, March 2003.

[38] E. Altman, A. A. Kherani, P. Michiardi, and R. Molva, *Non-cooperative Forwarding in Ad Hoc Networks*, Springer Berlin/Heidelberg, Germany, May 2005.

[39] V. Srinivasan, P. Nuggehalli, C. F. Chiasserini, and R. R. Rao, Cooperation in wireless ad hoc networks, in *Proceedings of the Annual IEEE Conference on Computer Communications (INFOCOM)*, San Francisco, CA, March 2003.

[40] Z. Han, Z. Ji, and K. J. R. Liu, Dynamic distributed rate control for wireless networks by optimal cartel maintenance strategy, in *Proceedings of the IEEE Global Telecommunications Conference*, pp. 3742–3747, Dallas, TX, November 2004.

[41] Z. Han, C. Pandana, and K. J. R. Liu, A self-learning repeated game framework for optimizing packet forwarding networks, in *Proceedings of the IEEE Wireless Communications and Networking Conference*, pp. 2131–2136, New Orleans, LA, March 2005.

[42] F. Meshkati, H. V. Poor, S. C. Schwartz, and N. B. Mandayam, An energy-efficient approach to power control and receiver design in wireless data networks, *IEEE Transactions on Communications*, 53(11), 1885–1894, November 2005.

[43] F. Meshkati, M. Chiang, H. V. Poor, and Stuart C. Schwartz, A game-theoretic approach to energy-efficient power control in multi-carrier CDMA systems, *IEEE Journal on Selected Areas in Communications - Special Issue on Advances in Multicarrier CDMA*, 24(6), 1115–1129, June 2006.

[44] F. Meshkati, D. Guo, H. V. Poor, and S. C. Schwartz, A unified approach to energy-efficient power control in large CDMA systems, *IEEE Transactions on Wireless Communications* (to appear).

[45] F. Meshkati, H. V. Poor, S. C. Schwartz, and R. Balan, Energy-efficient resource allocation in wireless networks with quality-of-service constraints, *IEEE Signal Processing Magazine: Special Issue on Resource-Constrained Signal Processing, Communications and Networking*, May 2007.

[46] I. F. Akyildiz, W. Su, Y. Sankarasubramaniam, and E. Cayirci, A survey on sensor networks, *IEEE Communications Magazine*, 40(8), 102–114, August 2002.

[47] J. H. Chang and L. Tassiulas, Energy conserving routing in wireless ad-hoc networks, in *Proceedings of the Annual IEEE Conference on Computer Communications, INFOCOM 2000*, pp. 22–31, Tel-Aviv, Israel, March 2000.

[48] A. Ephremides, Energy concerns in wireless networks, *IEEE Wireless Communications*, 9(4), 48–59, August 2002.

[49] I. Maric and R. D. Yates, Cooperative broadcast for maximum network lifetime, in *Proceedings of the Conference on Information Sciences and Systems*, Vol. 1, pp. 591–596, Princeton, NJ, March 2004.

[50] Y. T. Hou, Y. Shi, H. D. Sherali, and S. F. Midkiff, On energy provisioning and relay node placement for wireless sensor networks, *IEEE Transactions on Communications*, 4(5), 2579–2590, September 2005.

[51] H. Ochiai, P. Mitran, H. V. Poor, and V. Tarokh, Collaborative beamforming for distributed wireless ad hoc sensor networks, *IEEE Transactions on Signal Processing*, 53(11), 4110–4124, November 2005.

[52] M. Sikora, J. N. Laneman, M. Haenggi, D. J. Costello, and T. E. Fuja, Bandwidth and power-efficient routing in linear wireless networks, *IEEE Transactions on Information Theory*, 52(6), 2624–2633, June 2006.

[53] J. Luo, R. S. Blum, L. J. Greenstein, L. J. Cimini, and A. M. Haimovich, New approaches for cooperative use of multiple antennas in ad hoc wireless networks, in *Proceedings of the IEEE 60th Vehicular Technology Conference*, Vol. 4, pp. 2769–2773, September 2004.

[54] IEEE 802.16j Mobile Multihop Relay Project Authorization Request (PAR), Official IEEE 802.16j Website: http://standards.ieee. org/board/nes/projects/802-16j.pdf, March 2006.

[55] http://www.ieee802.org/16/relay/index.html

[56] A. Kwasinski, Z. Han, and K. J. R. Liu, Cooperative Multimedia Communications: Joint Source Coding and Collaboration, in *Proceedings of the IEEE Global Telecommunications Conference*, St. Louis, MO, November 2005.

14 Cooperative Relaying in Multihop Cellular Networks

Zaher Dawy

CONTENTS

In traditional cellular networks, mobile stations (MSs) communicate via direct connections with base stations (BSs). With the technology advances and the proposed solutions for distributed communications in ad hoc networks, significant performance gains can be achieved by enabling cooperation among MSs in cellular networks. One form of cooperation is to allow MSs to communicate with BSs using other stations as relays. In this chapter, a detailed overview is presented on various aspects related to the design of relaying in multihop cellular networks with emphasis on the type of relays whether fixed or mobile. Moreover, various enhancement techniques at the relay level are proposed to improve the overall network performance. Selected design aspects and enhancement techniques are then analyzed based on Monte-Carlo simulations for typical scenarios. Finally, useful insights and open research issues are identified and discussed.

14.1 INTRODUCTION

In multihop cellular networks (MCNs), a mobile station (MS) that meets a predefined set of conditions can cooperate with other mobile or fixed stations to communicate with its base station (BS). As a result, the path between an MS and its BS can be composed of multiple hops as demonstrated in Figure 14.1. Multihop transmission can lead to significant transmit power reduction compared with direct single-hop transmission even if only one relay node is used. Moreover, due to peak transmit power constraints, multihop transmission can also lead to significant coverage extensions by dividing a path into a set of shorter hops. Another advantage of relaying is its ability to avoid obstacles and, thus, combat the dead spots problem, which leads to severe signal degradation and occurs in places with high shadowing and lack of line of sight (Figure 14.2). In addition to coverage enhancement, relaying can also be used to enhance user capacity and quality of service (QoS) by reducing interference, increasing frequency reuse, enabling cooperative diversity and path redundancy, reducing call blocking probability by routing traffic between cells, etc. It adds flexibility to the design of cellular networks, and it is expected to play a central role in the development of 4G wireless standards [1–6].

On the other hand, there are several challenges that need to be addressed when designing MCNs. These include capabilities and complexity of relay stations (RSs), availability of additional channels for relaying, distributed and centralized intelligence for functionalities such as relay selection and handoffs, data security and integrity, collaboration incentives, and billing models.

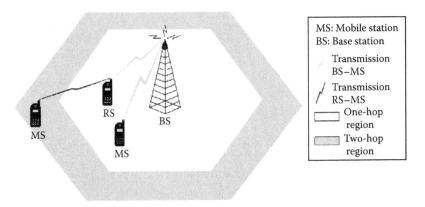

FIGURE 14.1   Node outside the cell coverage area relayed by another MS.

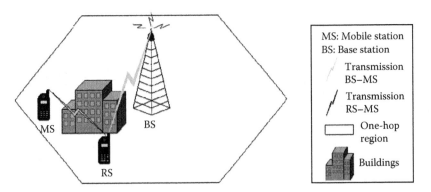

FIGURE 14.2   Node in a dead spot relayed by another MS.

This chapter presents an overview of various design aspects related to the integration of relaying in conventional cellular networks (CCNs). Various enhancement techniques for fixed RSs are then proposed to further improve the network performance. To demonstrate the gains of relaying with/without enhancement techniques, a typical system model is considered, and Monte-Carlo simulation results for practical network scenarios are presented and analyzed. Finally, open research issues are identified and discussed.

14.2 MULTIHOP CELLULAR NETWORKS: DESIGN ASPECTS

The performance of MCNs highly depends on the RSs' capabilities and the selection of multihop paths between MSs and their BSs. In this section, these aspects are addressed and design alternatives are presented.

14.2.1 RELAY STATIONS

14.2.1.1 Analog vs. Digital Relays

Analog relays, also called repeaters, have been traditionally used to extend coverage to dead spot areas not covered by the cellular network. An analog relay amplifies the received signal before forwarding it without performing any signal detection or advanced processing. As a result, noise is also amplified, which in many practical scenarios is a performance limitation. On the other hand, digital or regenerative relays detect and decode the received signal and then reencode the information before forwarding it. For each hop along the multihop path, this allows for cooperative error correction and detection [7], retransmission and diversity combining [8,9], and adaptive modulation and coding [10]. As a result, digital relaying has been widely adopted in the existing literature on MCNs due to the added design flexibility and the possible gains of intermediate data detection and processing.

14.2.1.2 Mobile vs. Fixed Relays

A mobile relay station (MRS) is a normal nonactive MS (i.e., not transmitting own traffic) operating as a relay for other MSs. The main advantage of mobile relaying is the use of existing MSs without the need for any extra infrastructure. It is especially useful in scenarios with relatively high density of MSs to guarantee the availability of candidate RSs throughout the network. The higher the density of MSs, the higher the number of possible paths between an MS and its BS, which in turn results in higher cooperative diversity gains. Other advantages of MRSs include high robustness to traffic variation and scalable coverage extension. On the other hand, the RSs in this case are simple devices that are required to have low complexity and, thus, cannot support advanced enhancement techniques. Furthermore, mobile relaying results in energy consumption at the MSs acting as RSs due to foreign traffic from the relayed MSs; this necessitates some sort of compensation or incentives from the network operators. In addition, mobile relaying might result in instable performance and increased control overhead because MSs dynamically enter and leave the network.

A fixed relay station (FRS) is similar to a mini BS in terms of structure but with much more limited functionality and connectivity, much less complexity and cost, and much smaller size. FRSs are placed by network operators at specific locations inside the network, for example, near to cell boundaries or in specific locations throughout the cells (for example, on street lights). As a result, energy consumption is not a limitation due to the existence of power supplies, the overhead of control traffic for functionalities such as RS tracking and RS selection is relatively low, and QoS can be better guaranteed. Moreover, FRSs can support more complexity, which allows for advanced enhancement techniques at the relay level. The main drawbacks of FRSs are placement constraints in the network (number and locations of FRSs) in addition to extra deployment and maintenance costs [11].

14.2.2 MULTIHOP PATHS

14.2.2.1 Number of Hops

Existing research shows that allowing up to two or three hops between an MS and its BS achieves most of the relaying gains with an acceptable complexity [2,5,12,13]. As the number of hops increases, the number of required channels or time slots should also increase in case of frequency division multiple access (FDMA)- or time division multiple access (TDMA)-based networks and the number of simultaneous transmissions would also increase in the case of code division multiple access (CDMA)-based networks. In both cases, each new hop would require its own additional transmission power, which limits the overall performance gains [14]. Moreover, as the number of hops increases, the intelligence in the network becomes more complicated, the control overhead increases, and additional channel resources are required. The results presented in this chapter assume CDMA-based MCN with a maximum of two hops between a given MS and its BS. Limiting the number of hops to two achieves notable gains and results in minimal complexity for the relay selection and channel allocation schemes.

14.2.2.2 Duplexing Modes

An RS in an MCN should transmit and receive on two radio interfaces. Assuming one RS between an MS and a BS (two hops), the first interface is for the MS–RS link (or hop) and the other interface is for the RS–BS link. To avoid signal masking problems, proper design is required to guarantee orthogonality between the different transmitted and received signals at the RS. For each radio interface alone (MS–RS or RS–BS), the uplink and downlink can be separated using conventional techniques such as time division duplexing (TDD) or frequency division duplexing (FDD). TDD allows reciprocal channel estimation and offers more flexibility for dynamic asymmetric resource allocation between the uplink and the downlink. Between the two radio interfaces at a given RS (MS–RS and RS–BS), orthogonality can be achieved either by using different time slots or different frequency bands. The former option requires strict synchronization, which is practically complex, whereas the latter option requires extra spectrum, which is normally expensive. Another solution that is suitable only for FRSs is to use antennas that are spatially separated using shields to overcome signal masking.

14.2.2.3 Relaying Channels

The introduction of relaying in cellular networks requires new channels for the extra hops. In FDMA-, TDMA-, or orthogonal frequency division multiple access (OFDMA)-based networks, a channel selection scheme has to be defined to reuse frequency channels, time slots, or subcarriers among cells. Either idle channels from neighboring cells can be reused or specific channels can be reserved for relaying purposes. In CDMA-based networks, the new hops can be directly allocated own spreading codes and, as a result, their transmissions will contribute to the total interference level in the network.

Another option would be to use unlicensed frequency bands for the relaying links, for example, via a WLAN interface [1,15,16]. The drawbacks of using an 802.11-based protocol for relaying are the following: QoS is more difficult to be guaranteed due to the random accessing, throughput is lower due to collisions and backoff timers [17], and complexity is higher due to the coupling of different technologies at the relaying nodes.

14.2.2.4 Path Selection

A path-selection scheme should be designed to select an appropriate multihop path (set of RSs) between an MS and its BS, for example, see Ref. [18]. The authors in Ref. [19] present different possible suboptimal path-selection schemes: path selection based on geographical distance, arbitrary path selection, and path selection based on minimal pathloss and shadowing. Important related aspects include possibilities for simultaneously communicating over multiple multihop paths and for handing over between different multihop paths depending on specific handoff mechanisms.

14.3   MULTIHOP CELLULAR NETWORKS: ENHANCEMENT TECHNIQUES

In this section, various enhancement techniques that can be implemented at RSs in MCNs are presented and discussed [20,21]. These enhancement techniques add complexity and cost to the RSs and, thus, are more suitable for FRSs.

14.3.1   COOPERATIVE DIVERSITY

Cooperative multihop diversity is based on cooperation between the nodes forming the multihop paths between an MS and its BS. Several publications have proposed practical schemes for cooperative diversity with varying performance and complexity properties, for example, see Refs. [8,9,22–26]. The main concept is to combine multiple concurrent receptions due to the transmission of the same signal by various nodes along the selected multihop path(s). The authors in Ref. [5] studied multihop diversity in MCNs at the system level. They showed in their proposed system model that cooperative relaying with a diversity gain of 1 dB can realize gains of up to 40 percent with respect to conventional cellular networks compared to a maximum gain of 20 percent with respect to conventional MCNs.

The results presented and analyzed in this chapter assume a simple cooperation scheme where the RS decodes and forwards the signal of the relayed MS. As a result, the BS receives over different channels multiple uncorrelated copies of the relayed MS signal and combines them using maximum ratio combining to achieve distributed space diversity gain. The concept of the multihop diversity scheme is illustrated in Figure 14.3. Because a wireless device cannot simultaneously transmit and receive at the same frequency channel, transmissions (a) and (b) should be orthogonal in time, frequency, or code. The latter option is not practical because it requires tight synchronization. The first option consists of dividing the transmission time into two slots. The nodes would share the information to be transmitted in the first time slot and then synchronously transmit the same signal to the BS in the second time slot. This option requires a well-distributed synchronization, which makes its implementation difficult in practice. Therefore, the third option is selected where the two hops use two orthogonal frequency channels and the BS receives at both channels and benefits from the broadcast nature of the transmissions from the relayed MS and from the RS.

14.3.2   MULTIUSER DETECTION

In CDMA-based cellular networks, the signal of a given MS is subject to interference from other MSs, which causes performance degradation. Assuming the spreading sequences of the interfering MSs are known at the BS, part of the intracell interference can be detected and canceled using multiuser detection (MUD). Optimal multiuser detectors result in the cancellation of a major part of the interference. However, they exhibit exponential complexity in the number of users. Suboptimal MUD techniques like interference cancellation receivers offer a practical trade-off between complexity and performance, which makes them appealing for practical applications, for example, see Refs. [27–29].

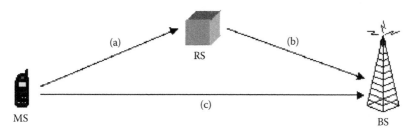

FIGURE 14.3   Multihop diversity in a two-hop relaying scenario. In a time division case, (a) occurs in the first time slot whereas (b) and (c) in the second time slot. In a frequency division case, (a) and (c) occur on one frequency channel whereas (b) on a second orthogonal frequency channel.

In MCNs, multiuser detection can be additionally applied at the relay level on condition that RSs posses the necessary information. For example, to apply intracell interference cancellation at a given RS, it should know the spreading sequences of all relayed MSs in the same cell. This information can be either provided by the BS over a control channel or can be estimated at the RS by performing correlations with possible spreading sequences to acquire the used codes.

14.3.3 MULTIPLE RELAYING

Most publications that study MCNs assume a maximum of one relayed MS per RS. Routing or relay selection schemes assign relays based on some metric, which might result in a specific RS being best suited for multiple MSs. Therefore in the case of simple RSs (for example, mobile RSs), only one of the MSs would be assigned this RS. The allocation outcome would not be optimal because the other MSs are not assigned their best option. Consequently, allowing multiple MSs per RS should yield better performance from a system-level point of view.

14.3.4 ADVANCED ANTENNAS

An important design aspect in MCNs is the type of antennas used at the BSs and RSs. The existing work on MCNs normally assumes that RSs are equipped with omnidirectional antennas. Using omnidirectional antennas would be suitable for highly mobile scenarios due to difficulties in tracking MSs. Moreover, omnidirectional antennas require relatively low complexity at the MSs. On the other hand, a main drawback of omnidirectional antennas is the high interference because the power of the signal is uniformly spread and not beamed toward specific directions.

Directional antennas and adaptive antenna arrays have proven to achieve great benefits in ad hoc networks and in conventional cellular networks, for example, see Refs. [30–33]. These advanced antenna configurations result in efficient spatial reuse, interference reduction, and range extension. There are several design issues related to using advanced antennas at FRSs in MCNs, which include antenna pattern and orientation strategy for directional antennas, number and placement of antenna elements for adaptive antennas, and maximum antenna gain.

14.3.4.1 Directional Antennas

In the case of directional antennas, the antenna pattern is directional but with fixed orientation. A design parameter is the orientation of the directional antenna at the RSs. Assuming that all RSs use the same orientation, the following three orientation strategies are proposed (Figure 14.4):

- Street adjacent: Antenna patterns are adjacent to the streets with alternating directions going from one street side to another
- BS centripetal: Antenna patterns are directed in opposite directions to the BS
- Integrated: RS adopts the BS centripetal orientation if it is closer to the BS than the cell boundary and the street adjacent orientation otherwise

14.3.4.2 Adaptive Antenna Arrays

The primary goal of adaptive antenna arrays is the generation of antenna beams (beamforming) that track a desired signal through linear combining of the signals captured by the different antenna elements [34,35]. In this chapter, uniform linear arrays with M isotropic elements are assumed. The pattern of the adaptive beams depends on the number and separation of the antenna elements. The higher the number of antenna elements, the beam normally becomes more directive and narrower (Figure 14.5).

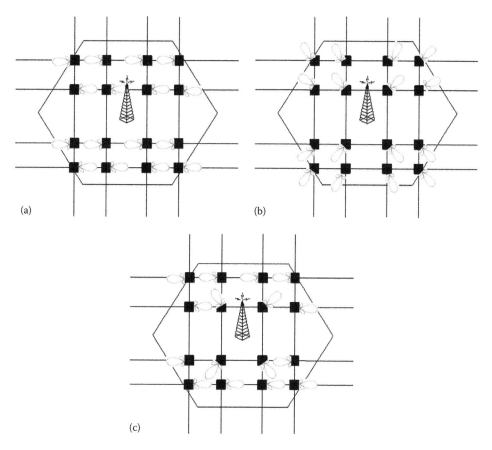

FIGURE 14.4 Directional antenna orientation strategies: (a) street adjacent strategy, (b) BS centripetal strategy, and (c) integrated strategy.

14.3.5 TRANSMISSION ADAPTATION

Power control and rate control are two of the main forms of transmission adaptation suitable for MCNs. Power control ensures that each node receives and transmits just enough power to communicate at its target quality while causing minimal interference. On the other hand, in rate control or adaptive modulation and coding, the modulation level and coding rate are dynamically adapted according to channel-state information. The possible gains of transmitter adaptation techniques are lower power consumption, higher network throughput, and better QoS guarantees, for example, see Refs. [10,36].

14.4 RESULTS, ANALYSIS, AND INSIGHTS

To study the implications of the various design aspects and the gains of the proposed enhancement techniques at the relay level, Monte-Carlo simulation results for various MCN scenarios are presented and discussed in this section.

14.4.1 MCN TYPICAL SYSTEM MODEL

The uplink of a CDMA-based cellular network is considered. All MSs in the network are assumed to belong to the same service class with spreading factor SF. In addition, a maximum of two hops per multihop path is allowed with the hops orthogonal in the frequency domain. Limiting the number

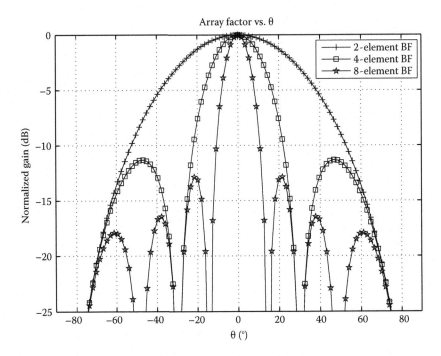

FIGURE 14.5   Adaptive antenna array patterns as a function of the number of antenna elements (interelement separation of $\lambda/2$ where λ is wavelength).

of hops to two results in minimal complexity for the relay selection and channel allocation schemes and has been shown theoretically to achieve notable gains.

Two modes of relaying are considered: mobile relaying where the RSs are normal MSs that are in an idle state and fixed relaying where the RSs are fixed nodes installed by the network operator at fixed locations.

Mobile relaying: The following are some related system model parameters: K users are uniformly distributed per cell. Assuming that a user transmits with probability p_{active}, K_{active} users are active MSs with $K_{active} = p_{active} \cdot K$. Furthermore, p_{relay} of the remaining users act as potential mobile RSs. Therefore, the number of candidate RSs is given by $K_{relay} = p_{relay} \cdot (K - K_{active})$ and, thus, increases with the value of K.

Fixed relaying: In this relaying mode, the candidate RSs are fixed stations placed on a Manhattan grid of streets with width w and bloc side length of l because it is assumed that they are installed on street lights. The parameters for the fixed RS deployment are n_h, n_v, and s. n_h and n_v are the number of equidistant fixed RSs between two corners on a horizontal and vertical street border, respectively. s can be either 1 to denote that the actual positions of the RSs alternate on the street or 2 to indicate that all possible positions on the street are occupied by RSs. An example allocation is shown in Figure 14.6. Similarly, Figure 14.7 is a 2D representation of the network, showing the different types of nodes, the Manhattan grid of streets, and the cellular structure.

A relay is chosen based on a minimum pathloss criterion with prioritized selection. The BS measures the total pathloss of all possible paths, direct and two-hop, from any given MS. Moreover, it keeps track of all the two-hop paths that have a smaller loss than that with the direct path. Each of these two-hop paths is a good option because it reduces the pathloss compared with the case where no relaying is used. Let Γ be the maximum number of MSs that an RS can support (in case multiple relaying is supported). The priority for using an RS is given to the Γ MSs that have the minimum number of good options.

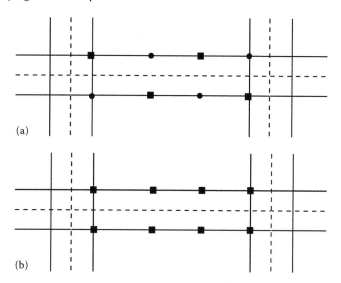

FIGURE 14.6 Fixed RSs allocation example: (a) $n_h = 4$, $n_v = 2$, and $s = 1$; (b) $n_h = 4$, $n_v = 2$, and $s = 2$. The squares represent installed fixed RSs and the circles represent street lights (possible positions) without fixed relays.

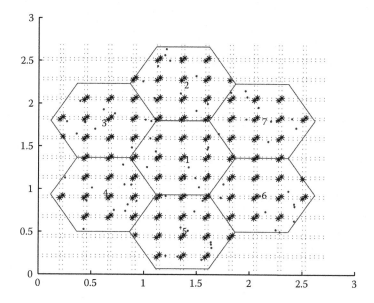

FIGURE 14.7 A 2D view of the network: $K = 35$, $p_{active} = 0.3$, $R_{cell} = 0.5$ km, $w = 30$ m, $l = 200$ m, $n_h = 2$, $n_v = 2$, and $s = 1$. The stars represent the fixed RSs, the dots represent the network MSs, the numbers represent BSs, and the dashed lines represent the streets.

As for the allocation of resources, it is assumed that the RSs cannot simultaneously transmit and receive on the same frequency band. Consequently, the direct hops (MS–BS and RS–BS) and the relay hops (MS–RS) are assigned orthogonal frequency channels. To keep the resource allocation simple and practical, all direct links in the network are assigned one frequency channel and all relay links in the network are assigned a second frequency channel. Therefore, the proposed MCN resource allocation scheme requires the use of two channels. This is not a limitation because most cellular operators get multiple frequency licences to operate a network in any given area. It is assumed that the enhancement techniques can only be employed in the case of fixed relaying because they require complex processing at the relay level.

14.4.2 ANALYTICAL MODEL

The distance-based pathloss between a transmitter t and a receiver r is given by

$$\rho_{t,r} = c \cdot d_{t,r}^e, \tag{14.1}$$

where

  c is the pathloss constant
  e is the pathloss exponent
  $d_{t,r}$ is the separation distance

The parameters c and e are calculated from the COST 231 extension of the HATA model using the proper parameters [37]. The total equivalent pathloss can be expressed as

$$\alpha_{t,r} = \frac{G_t \cdot G_r}{\prod_x L_x \cdot \prod_y M_y \cdot \rho_{t,r}} \cdot 10^{\psi_{t,r}/10}, \tag{14.2}$$

where

  G_t and G_r are the transmitter and receiver antenna gains, respectively
  L represents link budget losses (that is, body loss, cable loss, car loss, etc.)
  M represents link budget power margins (that is, fading margin, handover margin, etc.)
  $\psi_{t,r}$ is the shadowing random variable in dB, with zero mean and standard deviation equal to σ_s

  It is given by

$$\psi_{t,r} = a \cdot \psi_t + b \cdot \psi_r, \tag{14.3}$$

where

  a and b are correlation weighing factors such that $a^2 + b^2 = 1$
  ψ_t and ψ_r are the shadowing random variables in dB, with zero mean and standard deviation σ_s, at the transmitter and receiver, respectively

  It is normally assumed that $a = b = 1/\sqrt{2}$.
  The antenna gains are given by

$$G = \begin{cases} G_0 & \text{for omnidirectional antenna;} \\ G_0 \cdot g(\theta) & \text{for directional or adaptive antennas,} \end{cases}$$

where

  G_0 is the maximum gain of the antenna pattern
  $g(\theta)$ is the normalized pattern for an angle θ (that is, $0 < g(\theta) < 1$ for $-\pi < \theta < \pi$)

In all considered cases, an omnidirectional antenna is assumed at the BSs and at the MSs. For the RSs, there are three cases:

1. Omnidirectional antennas: The antenna gain at the RSs is also constant equal to G_0.
2. Directional antennas: The antenna has a fixed directional pattern with a fixed orientation. The antenna is used for both MS-RS and RS-BS links.
3. Adaptive antennas: The adaptive antenna is used only on the MS–RS link with maximum gain always directed toward the relayed MS of the given RS. This is based on the assumption

that the angle between the MS and the RS is known at the RS. On the RS–BS link, an omnidirectional antenna is assumed at the RS to keep the complexity low. Using an antenna array also on the RS–BS link would result in additional gains.

Taking into account the proposed resource allocation scheme, MUD can be applied at a given RS to combat the interference from all the relayed MSs in the same cell and at a given BS to combat the intracell interference from direct MSs and RSs in its cell. Note that stations with direct link to the BS (MS–BS and RS–BS links) and MS–RS links do not interfere with each other because they use different frequency channels. To apply an MUD technique at the relay level, the RSs should possess the necessary information. For example, to apply serial interference cancellation at a given RS, it should know the spreading sequences of all relayed MSs in the same cell.

The number of active users per cell, K_{active}, is divided into K_d direct MSs and K_r relayed MSs. K_d and K_r are determined by the relay selection scheme. The number of RSs is $K_u \leq K_r$ with equality in case of single relaying (one relayed MS per RS). Let N_k, $k \in [1, K_u]$ be the number of relayed MSs per RS k. Finally, let $K_t = K_d + K_u$ represent the number of stations directly transmitting to a given BS with $K_t = K_{\text{active}}$ in the case of single relaying. Without loss of generality, stations active in the cell of the central BS, BS_1, are considered as reference. The average signal-to-interference ratio (SIR) of direct station k (either direct MS or RS) at BS_1 is given by

$$\text{SIR}_{k,1}^{\text{BS}} = \frac{\text{SF} \cdot P_k/\alpha_{k,1}}{\sigma^2 + \lambda_{\text{BS}} \sum_{l=1,l\neq k}^{K_t} P_l/\alpha_{l,1} + \sum_{l=K_t}^{\delta_d} P_l/\alpha_{l,1}}, \tag{14.4}$$

where
P_k is the transmit power of station k
σ^2 is the thermal noise variance at the receiver
$\alpha_{k,1}$ is the pathloss between station k and BS_1
δ_d is the total number of direct stations in the network
λ_{BS} is the interference cancellation factor modeling the use of MUD at the BS level

For the conventional RAKE receiver $\lambda_{\text{BS}} = 1$ (no intracell interference is canceled), whereas for typical interference cancellation receivers $0.1 \leq \lambda_{\text{RB}} \leq 0.7$ [38]. Note that the third term in the denominator models the intercell interference from direct stations connected to BSs other than BS_1.

Similarly, the average SIR of relayed MS i at its RS j is given by

$$\text{SIR}_{i,j}^{\text{RS}} = \frac{\text{SF} \cdot P_i/\alpha_{i,j}}{\sigma^2 + \lambda_{\text{RS}} \sum_{l=1,l\neq i}^{K_r} P_l/\alpha_{l,j} + \sum_{l=K_r}^{\delta_r} P_l/\alpha_{l,j}}, \tag{14.5}$$

where
δ_r is the total number of relayed MSs in the network
λ_{RS} is the interference cancellation factor modeling the type of receiver at the RS level

The third term in the denominator models the intercell interference coming from all relayed MSs in other cells. Similarly, $0.1 \leq \lambda_{\text{RS}} \leq 0.7$ when MUD is applied at the RS and $\lambda_{\text{RS}} = 1$ otherwise.

Let SIR_{BS} and SIR_{RS} represent the SIR targets at the BSs and RSs, respectively. Performing ideal power control and setting the SIRs at all receiving stations equal to their target values, we can obtain the required transmit power of each station in the network by solving a set of equations. The total transmit power in the reference cell can then be calculated as

$$P_T = \sum_{l=1}^{K_t+K_r} P_l. \tag{14.6}$$

14.4.2.1 Multihop Diversity Gain Model

The adopted multihop diversity scheme results at the BS in diversity gain against fast fading in addition to antenna gain due to an increase in the effective average SIR. These gains are achieved because the BS receives and combines two copies of the relayed MS signal over two independent frequency channels. Therefore, the main benefit is for the relayed MSs. However, reducing the power consumption on the two-hop paths would also reduce the total interference in the network and, thus, also benefit the direct MSs.

The diversity gain is modeled by reducing the SIR target at the BS for relayed MSs by the term $G_{\text{diversity}}$. The transmit power of relayed MS i is still determined by the target SIR, SIR_{RS}, at its RS j. Let the SIR of the received signal from relayed MS i at its BS, BS_1, be denoted as $\text{SIR}_{i,1}^{\text{BS}}$. Note that without multihop diversity, the signals transmitted by the relayed MSs are ignored at the BSs and just detected at their RSs. The new SIR target of RS j at BS_1 is then given by

$$\text{SIR}_{\text{BS,div}} = \text{SIR}_{\text{BS}} - G_{\text{diversity}} - \text{SIR}_{i,1}^{\text{BS}}. \tag{14.7}$$

With multihop diversity, the SIR targets of the stations at their destinations can be summarized as

$$\text{SIR}_{\text{target}} = \begin{cases} \text{SIR}_{\text{BS}} & \text{for direct MSs;} \\ \text{SIR}_{\text{RS}} & \text{for relayed MSs;} \\ \text{SIR}_{\text{BS,div}} & \text{for RSs.} \end{cases}$$

These SIR targets are then used with ideal power control to calculate the total power consumption.

14.4.3 MCN Simulation Tool

A simulation tool has been developed to evaluate the performance gains of the proposed enhancement techniques in realistic MCN scenarios. Monte-Carlo-driven simulations are performed whereby K users per cell are uniformly distributed at each iteration with p_{active} of them set as active MSs. In case of fixed relaying, the fixed relays are then deployed across the streets. The relay selection algorithm is then applied to connect the active MSs to the corresponding BSs (direct connection or two-hop connection). Then, ideal power control is performed to find the total power transmitted in the reference cell. The total transmit power metric is used because it reflects the level of interference in the network, the user capacity and area coverage performance gains, and the level of energy consumption, which is a scarce resource in handheld devices. A large number of iterations is performed and the results are averaged out to get reliable statistics.

The allocation of fixed RSs in the network is an issue to be considered. Some publications study the optimal locations of the fixed relays whereas others place them in strategic locations around the cell. For example, the authors in Ref. [1] present an MCN model in which fixed RSs are placed near cell boundaries to relay traffic between cells and, hence, achieve load balancing. However, the positions of the fixed RSs are subject to environment constraints. In other words, the permissable positions of the RSs in real scenarios may be different from the theoretical positions around the cell. Consequently, it is assumed in the simulator that the area to be covered is modeled by a Manhattan grid because many cities have this topology of street distribution. Furthermore, fixed relays are placed on street lights as a practical and realistic positioning. At any point in time, only a subset of the available relays is used depending on the active MSs in the network.

The simulation tool is divided into three modules: network module, computational module, and graphical module. The network module is responsible for setting the network (cells and streets), deploying the nodes (BSs, MSs, and RSs), and applying the mobility model. The network module passes all this information to the computational module to perform the necessary computations, simulate the communication between the nodes, apply the required algorithms/schemes, plot a simple 2D representation of the network, and plot the calculated performance statistics. The computational

FIGURE 14.8 Snapshot of the demonstration mode fancy 3D display of an MBCN in urban environment (distant view).

module is the most important part of the simulator because it implements all the possible design options and simulates the relay selection and power control schemes. The 2D network characteristics from the network module and the transmit powers from the computational module are passed to the graphical module to generate the fancy 2D and 3D views of the cellular network. These fancy views allow the user to get a better grasp of what happens in a real MCN. It is important to emphasize that all three network visualizations (basic 2D, fancy 2D, and fancy 3D) display the same information (nodes, streets, mobility, connectivity, etc.) and show the same mobility steps because they take their inputs from the same modules.

The fancy 3D view visualizes the network components in a pseudo real 3D environment with the characteristics of a city modeled as accurately as possible without overloading the processing capabilities of the simulator (for example, see Figures 14.8 and 14.9). Nodes (MSs and mobile RSs) allocated on streets are modeled by mobile cars whereas nodes allocated off streets are modeled by pedestrians either standing or inside buildings/houses. Fixed RSs are installed on street lights located on the borders of the streets. If a fixed RS is used then an emission sphere in green color is added to the corresponding street light. BSs are represented by antenna towers in yellow color with drum shaped top placed at the center of the cells. The user has the option to remove the street lights, cell boundaries, streets, and even the nodes to obtain different views of the network area. Finally, signals between the different nodes are modeled using different colors and intensity widths depending on transmitted power level.

14.4.4 Performance Results

The default set of parameters used for the simulation results presented in this section is summarized in Table 14.1.

Figure 14.10 presents real-time simulation results for the number of MSs with two-hop connections to the BS as a function of the percentage of candidate RSs, assuming mobile relaying. Note

FIGURE 14.9 Snapshot of the demonstration mode fancy 3D display of an MBCN in urban environment (close view).

TABLE 14.1

Simulation Parameters

Parameter	Value
Spreading factor	64
Target SIR at BSs and RSs	3.2 dB
Chip rate	3.84 Mcps
Transmit antenna gain at MSs and RSs	0 dBi
Receive antenna gain at BSs	10 dBi
Link budget losses	4 dB
Link budget power margins	4 dB
Pathloss constant	137 dB
Pathloss exponent	3.52
Shadowing standard deviation	8 dB
Cell radius	500 m
p_{active}	30 percent
p_{relay}	80 percent
Street width w	30 m
Street bloc side length l	200 m

that as p_{relay} increases, the number of candidate RSs increases and, thus, the average number of used RSs increases, which in turn reduces the overall power consumption.

Figure 14.11 presents the results of the total transmit power (by all MSs and RSs) in the central cell for different types of receiver structures at the BS and RS levels. Because the proposed relaying model uses two frequency channels, results are presented for the CCN with two channels (carriers) per cell (assuming same number of users per carrier) to obtain a fair comparison. It can be noticed first

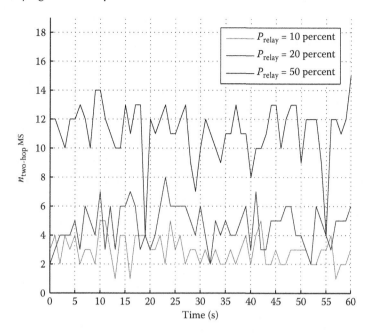

FIGURE 14.10 Number of two-hop nodes statistics with mobile relaying.

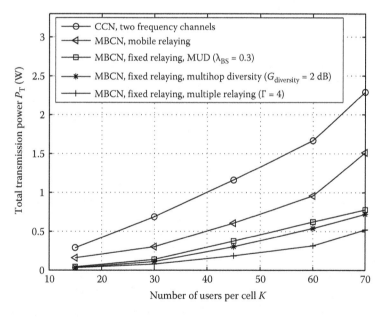

FIGURE 14.11 Total transmission power P_T vs. total number of users per cell K for different types of receiver structures. $\lambda_{BS/RS}$ is the intracell intereference cancellation factor at the BSs/RSs ($\lambda_{BS/RS} = 1$ indicates no multiuser detection). MBCN stands for multihop based cellular network (equivalent to MCN).

that independent of the type of receiver structure, fixed relaying MCNs outperform CCNs. In fact, for MCNs with fixed relaying and no enhancement techniques, a gain up to 60 percent is achieved with $K = 45$ users.

Using MUD at the RS level adds little to the gain of MCNs because the contribution of the relayed MSs to the total transmit power is not that significant. In fact, most of the power is generated

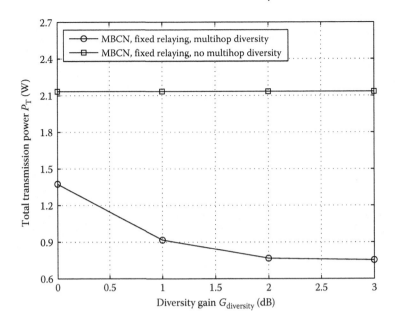

FIGURE 14.12 Total transmission power P_T vs. diversity gain $G_{\text{diversity}}$ for $K = 70$ and without MUD receivers ($\lambda_{\text{BS}} = \lambda_{\text{RS}} = 1$).

by the larger number of MSs and RSs with direct connection to the BSs. On the other hand, it can be seen that using MUD at the BS level results in remarkable gains for both CCNs and MCNs. Without MUD at the BS level, the gain of fixed relaying decreases as K increases. However, using MUD at the BS level linearizes the power increase and the gain of fixed relaying increases as K increases. In fact, the gain of fixed relaying with MUD at the BS is 65 percent for $K = 70$ in comparison with 7 percent without MUD at the BS. In summary, the best configuration is to use MUD receivers at the BSs for MCNs with fixed relaying.

Figure 14.12 presents the influence of multihop diversity gain on the total transmit power in MCNs. It can be noted that multihop diversity achieves a gain over basic fixed relaying that approaches 65 percent for $G_{\text{diversity}} = 3$ dB. The achieved power reduction gain increases with the diversity gain from a minimum of 35 percent to a maximum of 65 percent. However, it is not linear in the diversity gain, which indicates that, for example in this scenario, a 2 dB diversity gain is enough and, thus, there is no need to use more advanced cooperation schemes that can achieve more gain but with higher complexity.

Figure 14.13 presents the effect of the maximum number of MSs that can be relayed by a single RS on the total transmit power. It can be noted that increasing the allowed number of MSs per RS improves the choice of the appropriate multihop path. As a result, the total pathloss is reduced, which translates into a reduction in the transmission power of the nodes and, hence, a decrease in the interference and an improvement in performance. In fact, the gain achieved over single relaying reaches 78 percent for $\Gamma = 8$. Similar to the case of multihop diversity, the performance improvement becomes less sensitive to the allowed number of MSs per RS as Γ increases. In fact, most of the gain is obtained for $\Gamma = 4$.

Figure 14.14 presents a comparison between MRSs and FRSs with various enhancement techniques. The cases with the best trade-off between performance and complexity are chosen for each of the enhancement techniques using the results of Figures 14.11 through 14.13. It can be noticed first that MCN with both FRSs and MRSs has better performance than CCN. Second, FRSs with enhancement techniques outperform MRSs and the gaps increase with K despite the fact that the

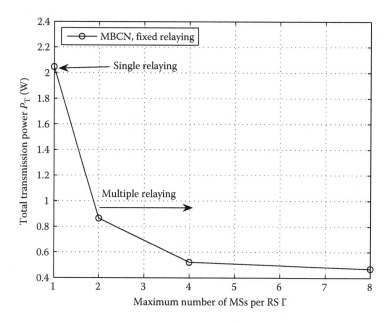

FIGURE 14.13 Total transmission power P_T vs. maximum number of MSs per RS Γ for $K = 70$, and without MUD receivers ($\lambda_{BS} = \lambda_{RS} = 1$).

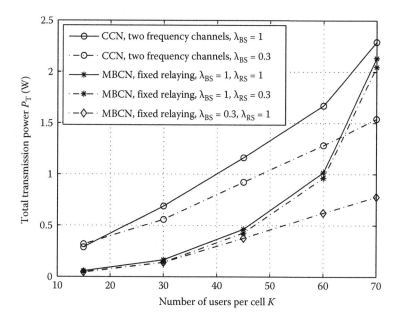

FIGURE 14.14 Total transmission power P_T vs. total number of users per cell K for MRSs (mobile relaying) and advanced FRSs (fixed relaying).

number of candidate relays in the case of MRSs increases with K unlike the case of FRSs (determined by the number of allocated FRSs). In addition, among the chosen parameters for the enhancement techniques, multiple relaying outperforms MUD and multihop diversity in terms of power reduction. However, the three enhancement techniques exhibit somehow close gains with respect to CCNs

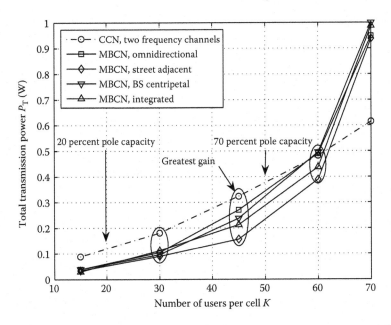

FIGURE 14.15   Total transmission power P_T vs. the total number of users per cell K ($\sigma_s = 4$ dB).

ranging between 66 percent and 78 percent for $K = 70$. Finally, it is worth mentioning that using different enhancement techniques together will result in even more gains.

Figure 14.15 presents a performance comparison between conventional cellular networks and different cases of MCNs employing fixed RSs for different directional antenna configurations. The total transmission power in the network (logarithmic scale) is plotted vs. the number of users per cell K. For fairness, the comparison is done with the CCN operating on two carriers because the adopted multihop network assumes the use of two frequency channels. It should be noted that all antenna orientation strategies add gains to the cellular network, with a maximum gain of 50 percent, except near the pole capacity ($K = 70$). In fact, the transmission power of a user is determined by the pathloss to its destination and the interference from other nodes. This interference depends on the pathloss and on the network load modeled by the number of users. With low shadowing, the main role of relaying is to reduce the distance-based pathloss. In addition, at high network load, the total transmission power is load determined. Consequently, dividing the load over two carriers reduces more P_T. However, as shadowing increases, the role of relaying becomes more important because shadowing cannot be combated in conventional networks and deeply affects the performance of the network. As a result, the relative gain of two carrier CCNs diminishes over a large range of node densities.

In addition, it can be seen that the street adjacent orientation strategy performs the best among the three directional antenna models. On the other hand, the worst performance is for the BS centripetal antenna orientation. In the latter case, the RSs antennas in a cell are directed toward neighboring cells, which increases the intercell interference, especially when the RSs are near the cell boundary. In addition, the gain is minimum toward the BS, which results in an increase in the required transmitter power and hence an increase in the effective transmitted power amplified by the greater gains in the interfering directions. On the other hand, the street adjacent performs better because the intracell interference on the RSs is reduced because the maximum gain of an RS is directed toward the minimum gain of the neighboring RS, which suppresses the interference coming from the former one. Similarly, the intercell interference is reduced because the antennas are not radially directed toward the neighboring cells as in the case of the BS centripetal. Finally, the integrated orientation exhibits a somehow average performance with respect to the other two strategies, and this is expected because it is a combination of both models.

14.5 CONCLUSIONS AND OPEN ISSUES

This chapter presented an overview of various design aspects and enhancement techniques for relaying in MCNs. Moreover, it presented results for several advanced enhancement techniques at FRSs and compared them to CCNs in addition to MCNs with MRSs. The following is a summary of the main insights:

- MCNs with fixed and mobile RSs achieve significant power reduction gains when compared with CCNs. These power reduction gains can be mapped to user capacity gains, area coverage gains, or electrosmog reduction gains. This strongly supports incorporating multihop transmission capabilities as part of 4G wireless standards.
- Using multiuser detection at fixed RSs does not yield notable gains unlike using multiuser detection at BSs, which is of higher benefit to MCNs than to CCNs.
- Multihop diversity realizes notable gains even with simple and practical cooperation schemes. The system gain does not increase linearly with the multihop diversity gain and saturates after a certain limit. This motivates incorporating simple cooperative diversity schemes in MCNs.
- Multiple relaying achieves significant gains, which are a little better than the gains achieved by multiuser detection and multihop diversity. Allowing up to four MSs to be relayed by a single RS achieves most of the possible gains, which provides an interesting trade-off between performance and complexity.
- Fixed RSs are in general more powerful than mobile RSs because they allow the use of more advanced enhancement techniques. The gains are expected to increase by increasing the number of fixed RSs in the network, using other enhancement techniques, such as multiple antennas or adaptive modulation and coding, and jointly using multiple enhancement techniques. On the other hand, using FRSs results in an increase in the infrastructure cost for network operators.
- Using directional antennas at fixed RSs achieves notable gains at medium loads, especially with proper antenna orientation. The gains are expected to increase by using adaptive antenna arrays with beamforming techniques.

In practice, the final decision on which relaying mode to use and which enhancements to deploy will highly depend on the network scenario, affordable cost, and allowed complexity.

There are several open research directions related to the design and deployment of MCNs, which include the following:

- Development of distributed and adaptive schemes to combat interference and improve receiver performance at RSs. For example, distributed collaborative multiuser detection can be used in which clusters of neighboring MSs collaborate to exchange helpful information to cancel part of the interference. The challenge is to be able to exchange a relatively small amount of information that is carefully selected to achieve overall performance gains. Another direction would be to develop distributed collaborative precoding algorithms to combat interference in MCNs. Most distributed cooperative schemes require some kind of synchronization among multiple nodes in the network to exploit the possible gains. Two approaches are possible to tackle this requirement: either to develop algorithms/protocols to synchronize the cooperating nodes or to develop asynchronous cooperative algorithms, for example, see Ref. [39]. Each approach has its own advantages and disadvantages in terms of complexity requirements and achievable performance gains.

- Performing crosslayer design and performance evaluation, taking into account application layer QoS characteristics in terms of source coding, data framing and multiplexing, end-to-end delay constraints, and so on.
- Radio network planning and optimization algorithms are an essential step in the deployment of conventional wireless cellular networks [40]. The problem of cellular radio network planning can be formulated as a multi objective constrained optimization problem with input capacity and coverage constraints. The typical outputs of a radio network planning process are normally the number of required BSs, the locations of the BSs, the configuration of the BSs (sectorization, antenna tilting, etc.), and frequency/code plans. For the successful deployment of multihop functionality in next generation cellular networks, there is a need to develop a new framework for radio network planning algorithms that takes into account the additional constraints and gains of RSs for both mobile and fixed relaying models.
- Most of the presented analyses and insights in the literature are based on either analytical derivations or simulation studies. The reported gains are important in identifying the capabilities of different proposed concepts to develop optimized designs. However, there is a need to accompany this theoretical work with real prototype implementation studies. This includes on one hand the development of customized channel models for mobile to mobile communications (for example, see Ref. [41]) and on the other the development of hardware testbeds to identify some of the practical challenges of implementing cooperative relaying at FRSs or MRSs, for example, see Refs. [42,43].

ACKNOWLEDGMENTS

Special thanks to Sami Arayssi, Ibrahim Abdel Nabi, and Ahmad Husseini for their contributions to this work. This research was supported in part by the University Research Board of the American University of Beirut.

REFERENCES

[1] H. Wu, C. Qiao, S. De, and O. Tonguz, Integrated cellular and ad hoc relaying systems: iCAR, *IEEE Journal on Selected Areas in Communications*, 19, 2105–2115, October 2001.
[2] Z. Dawy, S. Davidovic, and I. Oikonomidis, Coverage and capacity enhancement of CDMA cellular systems via multihop transmission, in *IEEE Globecom 2003*, San Francisco, CA, December 2003.
[3] H. Yanikomeroglu, Cellular multihop communications: Infrastructure-based relay network architecture for 4G wireless systems, in *Affordable Wireless Services and Infrastructure, 1st Annual Workshop*, Kingston, Ontario, Canada, June 2004.
[4] R. Pabst, B. H. Walke, D. C. Schultz, P. Herhold, H. Yanikomeroglu, S. Mukherjee, H. Viswanathan, M. Lott, W. Zirwas, M. Dohler, H. Aghvami, D. D. Falconer, and G. P. Fettweis, Relay-based deployment concepts for wireless and mobile broadband cellular radio, *IEEE Communications Magazine*, 42, 80–89, September 2004.
[5] P. Herhold, E. Zimmermann, and G. Fettweis, Relaying and cooperation—A system perspective, in *IST Mobile and Wireless Communications Summit*, Lyon, France, June 2004.
[6] L. Le and E. Houssein, Multihop cellular networks: Potential gains, research challenges, and a resource allocation framework, *IEEE Communicaitons Magazine*, 45, 66–73, September 2007.
[7] A. Stefanov and E. Erkip, Cooperative coding for wireless networks, *IEEE Transactions on Communications*, 52, 1470–1476, September 2004.
[8] E. Zimmermann, P. Herhold, and G. Fettweis, On the performance of cooperative relaying protocols in wireless networks, *European Transactions on Telecommunications*, 16(1), 5–16, January 2005.
[9] J. Boyer, D. D. Falconer, and H. Yanikomeroglu, Multihop diversity in wireless relaying channels, *IEEE Transactions on Communications*, 52, 1820–1830, October 2004.
[10] S. Hares, H. Yanikomeroglu, and B. Hashem, Diversity- and AMC (adaptive modulation and coding)-aware routing in TDMA peer-to-peer multihop networks, in *IEEE GLOBECOM 2003*, San Francisco, CA, December 2003.

[11] B. Timus, Cost analysis issues in a wireless multihop architecture with fixed relays, in *IEEE 61st Vehicular Technology Conference (VTC Spring 2005)*, Stockholm, Sweden, May–June 2005.

[12] E. Kudoh and F. Adachi, Transmit power efficiency of a multihop virtual cellular system, in *IEEE VTC 2003 Fall*, Orlando, FL, October 2003.

[13] Z. Dawy and H. Kamoun, The general Gaussian relay channel: Analysis and insights, in *5th International ITG Conference on Source and Channel Coding (SCC 2004)*, Erlangen, Germany, January 2004.

[14] Z. Dawy, Power allocation in wireless multihop networks with application to virtual antenna arrays, in *15th IEEE International Symposium on Personal, Indoor, and Mobile Radio Communications (PIMRC)*, Barcelona, Spain, September 2004.

[15] I. Gruber and H. Li, A novel ad hoc routing algorithm for cellular coverage extension, in *Networks 2004*, Vienna, Austria, June 2004.

[16] D. Zhao and T. D. Todd, Cellular cdma capacity with out-of-band multihop relaying, *IEEE Transactions on Mobile Computing*, 5, 170–178, February 2006.

[17] S. Xu and T. Saadawi, Does the IEEE 802.11 MAC protocol work well in multihop wireless ad hoc networks? *IEEE Communications Magazine*, 39(6), 130–137, June 2001.

[18] Y. Liu, R. Hoshyar, X. Yang, and R. Tafazolli, Integrated radio resource allocation for multihop cellular networks with fixed relay stations, *IEEE Journal on Selected Areas in Communications*, 24, 2137–2146, November 2006.

[19] V. Sreng, H. Yanikomeroglu, and D. Falconer, Effect of relay node selection on coverage improvement in a two-hop cellular relaying radio network, in *Affordable Wireless Services and Infrastructure, 1st Annual Workshop*, Tammsvik, Bro, Stockholm, Sweden, June 2003.

[20] Z. Dawy and S. Arayssi, Advanced fixed relaying in multihop based cellular networks, in *IEEE ICC 2006*, Istanbul, Turkey, June 2006.

[21] Z. Dawy, S. Arayssi, I. A. Nabi, and A. Husseini, Fixed relaying with advanced antennas for CDMA cellular networks, in *IEEE GLOBECOM 2006*, San Francisco, CA, December 2006.

[22] S. Barbarossa and G. Scutari, Cooperative diversity through virtual arrays in multihop networks, in *IEEE ICC 2003*, Hong-Kong, April 2003.

[23] I. Daou, E. Kudoh, and F. Adachi, Transmit power efficiency of multi-hop MRC diversity for a virtual cellular network, *IEICE Transactions on Communications*, E88-B, 3643–3648, October 2005.

[24] P. Liu, Z. Tao, Z. Lin, E. Erkip, and S. Panwar, Cooperative wireless communications: A cross-layer approach, *IEEE Wireless Communicaitons Magazine*, 13, 84–92, August 2006.

[25] O. Oyman, J. N. Laneman, and S. Sandhu, Multihop relaying for broadband wireless mesh networks: From theory to practice, *IEEE Communicaitons Magazine*, 45(11), 116–122, 2007.

[26] A. Adinoyi and H. Yanikomeroglu, Cooperative relaying in multi-antenna fixed relay networks, *IEEE Transactions on Wireless Communications*, 6, 533–544, February 2007.

[27] Z. Dawy and A. Seeger, Coverage and capacity enhancement of multiservice WCDMA cellular systems via serial interference cancellation, in *IEEE ICC 2004*, Paris, France, June 2004.

[28] J. Andrews, Interference cancellation for cellular systems: A contemporary overview, *IEEE Wireless Communications Magazine*, 12, 19–29, April 2005.

[29] Z. Dawy, S. Davidovic, and A. Seeger, The coverage-capacity tradeoff in multiservice WCDMA cellular systems with serial interference cancellation, *IEEE Transactions on Wireless Communications*, 5, 818–828, April 2006.

[30] H. Singh and S. Singh, A MAC protocol based on adaptive beamforming for ad hoc networks, in *14th IEEE Proceedings on Personal, Indoor and Mobile Radio Communications (PIMRC 2003)*, Beijing, China, September 2003.

[31] S. Yi, Y. Pei, and S. Kalyanaraman, On the capacity improvement of ad hoc wireless networks using directional antennas, in *ACM MobiHoc'03*, Annapolis, MD, June 2003.

[32] A. K. Saha and D. B. Johnson, Routing improvement using directional antennas in mobile ad hoc networks, in *IEEE Globecom 2004*, Dallas, TX, December 2004.

[33] R. Vilzmann, C. Bettstetter, and C. Hartmann, On the impact of beamforming on interference in wireless mesh networks, in *IEEE Workshop on Wireless Mesh Networks*, Santa Clara, CA, September 2005.

[34] A. Alexiou and M. Haardt, Smart antenna technologies for future wireless systems: Trends and challenges, *IEEE Communications Magazine*, 42(9), 90–97, September 2004.

[35] B. Alen and M. Ghavami, *Adaptive Array Systems: Fundamentals and Applications*. West Sussex, U.K.: Wiley, 2005.

[36] Z. Lin, E. Erkip, and M. Ghosh, Adaptive modulation for coded cooperative systems, in *IEEE 6th Workshop on Signal Processing Advances in Wireless Communication*, June 2005.

[37] H. Holma and A. Toskala, *WCDMA for UMTS*. West Sussex, U.K.: Wiley, 2000.

[38] Z. Dawy, S. Davidovic, and A. Seeger, A performance measure for WCDMA serial interference cancellation receivers, *Arabian Journal for Science and Engineering*, 28, 81–97, December 2003.

[39] S. Wei, D. L. Goeckel, and M. C. Valenti, Asynchronous cooperative diversity, *IEEE Transactions on Wireless Communications*, 5, 1547–1557, June 2006.

[40] J. Laiho, A. Wacker, and T. Novosad, *Radio Network Planning and Optimisation for UMTS*. West Sussex, U.K.: Wiley, 2002.

[41] M. Patzold, B. O. Hogstad, and N. Youssef, Modeling, analysis, and simulation of MIMO mobile-to-mobile fading channels, *IEEE Transactions on Wireless Communications*, 7(2), 510–520, February 2008.

[42] A. Bletsas and A. Lippman, Implementing cooperative diversity antenna arrays with commodity hardware, *IEEE Communications Magazine*, 44, 33–40, December 2006.

[43] B. Can, M. Portalski, H. Simon, D. Lebreton, S. Frattasi, and H. Suraweera, Implementation issues for OFDM-based multihop cellular networks, *IEEE Communicaitons Magazine*, 45, 74–81, September 2007.

15 Cooperative Radio Resource Management for Heterogeneous Networks

*Albena Mihovska, Elias Tragos, Jijun Luo,
and Emilio Mino*

CONTENTS

This chapter discusses cooperation and coexistence mechanisms for radio resource management (RRM) of heterogeneous systems. The concepts are based on research work carried out within the IST project WINNER (phase 1 and phase 2) [1]. The reference WINNER system is a candidate next generation wireless communication system. Section 15.1.1 gives an overview of the current heterogeneous networks and their requirements. A brief overview of the platforms for heterogeneous RRM management is also given. Section 15.1.2 describes the state of the art in cooperative RRM. In particular, we describe combined RRM, common RRM, layered RRM and, as an advance of the state of the art, hybrid RRM. Section 15.2 describes an architecture that is proposed for the support of cooperative RRM including the functionalities of the physical and logical entities. Three different approaches to the cooperation architecture are described: the distributed approach, the centralized approach, and the hybrid approach. Further, Section 15.2 describes the decision-making mechanisms required for the execution of the cooperative RRM mechanisms. Section 15.3 describes the cooperative system functions, for support of active/idle mobility management, location service and updates, congestion avoidance control, quality of service (QoS), and security. Section 15.4 describes the various algorithms proposed for cooperative RRM. These are algorithms for mobility management, congestion avoidance control, QoS, and security. Both, evolutionary from traditional schemes and novel methods (based on computational intelligence) for RRM are described. The algorithms are assessed in terms of advantages and disadvantages for different scenarios. Section 15.5 describes how the proposed algorithms can be implemented onto a suitable platform. Here, we describe the required low-level and high-level entities and their functionalities. Section 15.6 concludes the chapter and proposes directions for future challenges and research. Suitable cooperation mechanisms are currently being investigated with respect to vertical handover, authorization, authentication, and accounting (AAA), and common RRM.

15.1 INTRODUCTION

Current types of network infrastructure of telecommunication are unable to support ubiquity of communications because of problems such as heterogeneity in protocols, air interfaces, inter-RAT interference, network congestion, spectrum limitations, and so on. Another difficulty is the heterogeneity of radio access technologies (RATs) in ownership because it leads to complex systems and interworking problems, which can be seen from the challenges to establish and maintain connections with required quality, in fault detection and location, in resource allocation, and in charging of the usage of the network's resources. Systems that are adaptive and flexible and provide coverage in various deployment modes such as wide area (WA), metropolitan area (MA), and local area (LA) can help overcome some of these challenges. Cooperation and coexistence then require interworking between different PHY-layer modes of the same radio system [1] as well as cooperation and coexistence at higher layers, including intersystem cooperation. To ensure seamless user mobility, one of the main challenges is to consider how and which information to extract from a multitude of parameters, some not directly mapped to each other and belonging to different domains, for ensuring the availability of the required radio resources. Another aspect is to ensure that an interworking model is flexible and generic enough to include any new type of network that might be designed in the future and that will require inclusion in the heterogeneous domain. Finally, ubiquitous communication systems and the challenge to deploy them successfully also bring the problem of how to manage the spectrum to ensure successful interworking of the various communication links.

15.1.1 HETEROGENEOUS NETWORKS

The new converged and access-independent network model, dubbed next generation networks (NGN) [2], is based on the extensive use of IP and is designed to accommodate the diversity of applications inherent in emerging broadband technologies. Effort is already heavily committed toward the development of the necessary standards to bridge disparate networks and domains and enable them to interoperate [3]. Heterogeneity of networks means that connections span several networks that deploy different transport technologies or that the networks are owned and operated by separate organizations. When, for example, a customer subscribes to an Internet connection or a virtual private network (VPN) service with specified quality, it is almost impossible for the customer to make sure that the promised quality is fulfilled. On the other hand, it is also difficult for the Internet service provider (ISP) to guarantee the promised quality in a multitechnology and multioperator environment. In case the service does not meet the promised quality, it is equally difficult to identify the cause for that. The rising demand for fast, scalable, efficient, and robust data transfer over the air has produced a variety of RATs. Future wireless systems are supposed to provide high-bit-rate services in IP-based, real-time, person-to-person as well as machine-to-machine multimedia communications. Such systems will include a number of coexisting subnetworks with different RATs [2]. The interworking between radio subnetworks and, especially, the tight cooperation between them are very interesting for providing system capacity optimization. The RATs are often divided into different "generations" of mobile radio networks and they also refer to different transport technologies. Around the year 2001, there was a heated debate whether wireless local area networks (WLAN) [4] systems like IEEE 802.11x [5] and HIPERLAN/2 [6] or the Universal Mobile Telecommunications System (UMTS) [7] would be more efficient in terms of costs for planning, building, and operation. It was mentioned that these systems could be an extension to UMTS [8]. Hence, two or more of these systems may peacefully coexist in such a way that the user would always be served with the best possible QoS at the lowest possible price. UMTS is currently gaining a momentum with increased deployment worldwide. At areas with a high volume of traffic, the so-called hot spots or hot areas, systems with high data rates are deployed. In addition, new RATs covering a wide range of services and operating in different deployment concepts are the focus of research and development [1]. The trend toward services over heterogeneous networks requires support not just of simple mobile voice and data services but also of access to mobile Internet-based services with varying bandwidth and QoS requirements. The operative word in such a scenario is transparency of the delivered services, ubiquitous coverage, and uninterrupted user mobility. The increasing demand for wireless access to packet-based services has, in turn, driven the need for call admission and intersystem handover control strategies to make network roaming more seamless.

15.1.1.1 Requirements

The heterogeneous systems must be capable of forwarding data streams and session context among each other, and a (vertical) handover should happen seamlessly. The IST project WINNER [1,9] proposed an air interface based on the orthogonal frequency division multiplexing (OFDM) technology whose system comprises various functions that are intended to avoid data loss and minimize delay during handover as well as ensure coverage as recommended by the International Telecommunications Union (ITU) [10] deployment scenarios. This is possible by the WINNER simplified radio access network (RAN) architecture and supporting cooperation mechanisms. One requirement is that a ubiquitous radio system has to be self-contained, allowing it to target the chosen requirements without the need for interworking with other systems. Another requirement is that cooperation, whenever required, must be successfully ensured between any new or legacy systems. Existing RANs can at this moment be modified or updated for cooperation only at higher layers of the mobile network, which translates into routing at the radio network controllers (RNC) or an equivalent network element. Radio resource management (RRM) has been extensively studied over the years, with the aim of providing maximum resource efficiency for particular air interface technologies; therefore, the developed architectures are optimized for a network that operates using a single-layer technology [11,12]. In future heterogeneous wireless networks, RRM must be coordinated across a number of access technologies coexisting within the same network. Inter-RRM signaling is also required to transfer the information between RRM entities on which resource allocation and admission control decisions can be based.

15.1.1.2 Platforms for Heterogeneous Network Management

For efficient management of the resources, the location of the RRM functions within the network architecture is an essential issue, because it can affect the performance of the network. An RRM platform and its implementation for the management of heterogeneous networks were proposed in Refs. [13–16]. The platform is based on real-time monitoring of RANs and support of service requests and user-/system-initiated intermode and intersystem handover, as well as congestion management and QoS guarantees. The main monitoring module in this platform provides all the information to a central entity. Each RAN is provided with a specific RRM module that has some monitoring functionalities. The functionalities that characterize the main monitoring module are as follows:

- Receive real-time traffic measurements (RTTM)
- Calculate key performance indicators (KPIs)
- Compare KPIs

The monitoring module not only communicates with the central RRM unit to send the information toward the unit but also receives status information requests. It also communicates with the specific RRM module and based on the received information, alarms are sent that execute the RRM techniques specific to the system. A successfully designed RRM platform should be able to handle the following tasks:

- Monitor each RAN separately.
- Detect congestion situations that may occur in those RANs.
- Apply techniques locally to each network to tune resource management for alleviating/eliminating the effects of overloads and centrally when this is necessary.
- Ensure stable transition from the congested state to the normal state.
- Apply techniques for inter network resource management when the congestion cannot be effectively handled locally inside a network. This also includes roaming between different RANs, owned by the same network operator or by different network operators.

15.1.2 RADIO RESOURCE MANAGEMENT TECHNIQUES

15.1.2.1 Mobility Management

Mobility management techniques support the user mobility, including the traffic balancing that is essential for a network to use the resources of the system efficiently. Intersystem (that is, vertical) handovers are key mechanisms to implement traffic balancing strategies. In legacy systems, these algorithms are mainly based on coverage criteria. In future heterogeneous systems, the cooperation at RRM level between different RANs will be an integrated feature while the RRM algorithms will be able to use more metrics and information as inputs (triggers), because information will be exchanged between networks and new metrics becoming available, allowing sophisticated traffic balancing strategies. Examples of such metrics are load- and service-based criteria, location, velocity, user's environment (indoor, outdoor, etc.), terminal capabilities, and handover statistics. Fast handovers are another challenging task. Mobility management also comprises roaming, paging, and routing. Therefore, the localization of users is closely related to mobility management [17]. For example, a handover to another RAT can be initiated based on the position of a user; in this case the decision will be based on the time-variant densities of users, and the temporary allocation of resources can be managed to offer the limited resources where they are needed [18]. To maintain the intially obtained service and radio bearers (that is, QoS), the handover management should achieve the following:

- Minimize packet loss and delay during a handover (seamless handover).
- Make use of any "triggers" available (for example, information from the mobile host or from the network that a handover is imminent), in order that action can be taken in advance of the actual handover (planned handover).
- Allow the possibility of transferring context (QoS, security, header compression state, link layer states) and also any buffered packet (tunneling) from the old to the new access router.
- Ensure that a planned handover can fall back gracefully to an unplanned one (in case it fails) and that the same actions can happen (transferring buffered packets and context).
- Allow intertechnology handover if the mobile host supports different technologies (vertical handover).

15.1.2.2   Congestion Avoidance Control

Congestion situations can be avoided by applying a set of techniques that set a mechanism defined as "congestion avoidance control." The task of this mechanism is to control the load of the network by restricting the admission of new user sessions and resolving unwanted overload situations. The congestion avoidance control mechanism consists of admission control (AC) and load control (LC). AC is responsible for controlling the load of the system so that the available capacity can be exploited without compromising system stability. AC algorithms can be power-based, which means that they rely on periodic measurements of the transmitted power to compute the interference at the receiver and make the decision of admitting or rejecting the user [19,20]. Other algorithms monitor the load of the system against predefined thresholds, above which the network will be considered as overloaded [21]. In throughput-based algorithms, the throughput that can be delivered by the system is determined according to some dimensioning calculations, assuming the existence of certain conditions in the system [22]. There are also algorithms that use the equivalent capacity of aggregated traffic, which is an estimation of the arrival rate of a class of traffic [23]. Finally, there exist algorithms that focus on the bandwidth and delay constraints of each flow [22]. LC acts in an overload situation to bring the load back to the targeted levels. Some possible actions include interfrequency or intersystem handover for some users, quality decrease for some connections, throughput decrease, and controlled dropping of low-priority users. If the packet scheduling and the AC mechanisms have been designed properly, the system is supposed to operate within the desired load range. However, the AC and the packet scheduling are based on estimations and predictions of the load in the system and therefore errors are common.

15.1.2.3   Cooperative Radio Resource Management

Cooperative RRM would include the mechanisms described here but for optimized performance there should be a tight cooperation between them. Coordination among the different RATs is mandatory. Cooperative RRM can be ensured by a centralized approach where coordination is done by a single functional entity. Two examples of a centralized cooperative RRM scheme can be given here: the common RRM (CRRM) defined within 3GPP to allow for better interworking between UMTS and GSM/GPRS networks, further defined by Release 99 [24] and the joint RRM (JRRM) as defined in the IST project SCOUT [25] for interworking between HIPERLAN2 and UMTS. Different architectures are capable of enabling CRRM, and several solutions for the mapping of functional entities into physical entities have been proposed in Ref. [26]. An example of a cooperative RRM mechanism to support the coexistence of various RANs including the WINNER RAN and based on the CRRM approach is shown in Figure 15.1 [14]. The RRM architecture has a hierarchical structure with the following elements: an interface management unit (IMU), a resource management unit (RMU), and a common management unit (CMU). The centralized CMU is a core component providing global resource management to the RANs and further help to the RMU tasks. It collects information about connectivity and availability of each RAN and distributes the traffic accordingly. CRRM

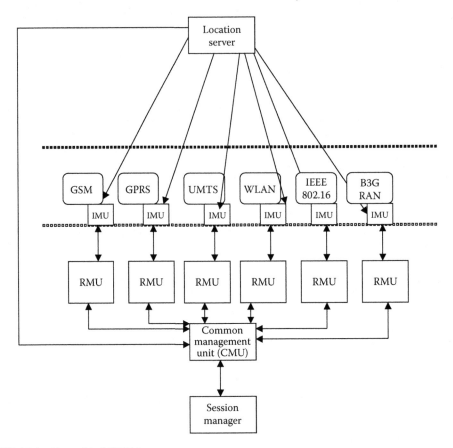

FIGURE 15.1   Example of CRRM.

can be exploited for interworking between GSM and UMTS RANs, for example, this is the so-called integrated CRRM because it integrates the functionality into the existing UTRAN/GERAN nodes [24]. However, 3GPP [26] considers that the RRM algorithms cannot be completely moved to an external CRRM. Some part of the functionality will always reside in the other entities. This means that all three systems need to be tuned to achieve optimal performance, making the system tuning more cumbersome.

A wireless network might have four specific cases of interaction, namely, single technology–single domain; single technology–multidomain; multitechnology–single domain; multitechnology–multidomain. Current RRM solutions consider the first case, where radio resources are managed solely at the link layer (L2). With a single technology but multiple domains it is also possible to have an L2 solution. On the other hand in a native-IP environment, this could cause conflicts with the network layer (L3) interactions. Therefore, communication with L3 entities is very important. When multiple technologies are introduced, different link layers (L2) interact with each other and there should be a layer that will be the bridge between the technologies. This layer could be the IP layer (L3) through the IP2W interface (Figure 15.2). At L3 a decision can be made on the best resource management across the multiple technologies. In the multitechnology–multidomain case (the case shown in Figure 15.2), L3 decisions are needed not only to allow for cross-technology RRM but also to remove any inter domain management conflicts at L3. Figure 15.2 shows a framework for the multi-layered approach based on the use of JRRM [25]. Entities related to RRM are found at both L2 and L3 whereas IP2W provides the generic interface between L2 and L3. The multilayered approach has a manager function that manages interactions between access technique (A-T) specific RRM entities,

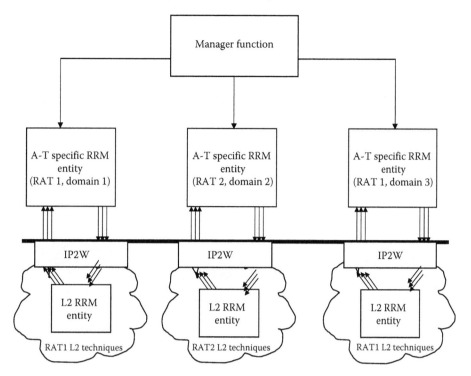

FIGURE 15.2 Multilayered approach.

such as coordination of handover. The JRRM architecture is based on the assumption of coexisting of different RATs with different profiles [25]. Each RAT needs an efficient interworking between traffic volume, measurement function, traffic scheduler, load control, and admission control functions. The traffic estimation (TREST) module informs the administrative entity session/call admission control (SAC) in every subnetwork on the predicted traffic and planned traffic information to update the priority information of each connection and the admission decision within the network.

15.2 COOPERATION ARCHITECTURE

15.2.1 GENERAL OVERVIEW

Many current cellular systems use a centralized RRM approach to provide advanced system functionalities controlled by the operator. Such an architecture, however, has been inherited from the early usage scenarios when the only traffic component was voice, an application that does not need high data rates. Today, the wireless network has to support an increasing demand of IP traffic and user data have to go through different nodes to reach the Internet (for example, in UMTS through the RNC, the serving GPRS support node [SGSN], and the gateway GPRS support node [GGSN]). There is an increasing tendency, therefore, in new cellular systems, such as 3GPP long-term evolution (LTE) [8], WiMAx Mobile [27], IEEE 802.20 [28], and WINNER [10], to increase the system performance for higher throughput and lower latency. This can be achieved by a reduction in the number of system nodes and going for more decentralized architecture. As well as an all-IP architecture to avoid transcoding delays (for example, in UMTS IP to asynchronous transfer mode (ATM) interfaces in the transport network), 3GPP works on the air interface evolution LTE system [8] and the system architecture evolution (SAE) [29]. The network architecture is an all-IP architecture. At the radio access level, only two elements are proposed: the base station, named eNodeB, and the gateway termed as a GW interfacing the core network. The eNodeB terminates most of the radio resource control

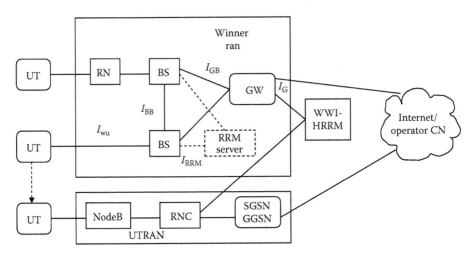

FIGURE 15.3 Functional architecture for WINNER and UMTS.

(RRC)/RRM protocols, the management functionality, and the outer automatic repeat request (outer ARQ). There is a logical connection between eNodeBs, transmitting packets, and context transfer for mobility management with minimal latency. LTE and SAE use radio handover for efficient and minimum packet losses. Here, we describe an architecture for a beyond 3G system designed for the radio interface of the WINNER system [1,30] and harmonized in some aspects with the LTE architecture. However, two additional nodes are incorporated in this architecture, namely, the relay node (RN) for support of multihop communications and an optional RRM server. The architecture implements functionalities for centralized and distributed RRM and is composed of several logical nodes that, in some cases, could be merged to implement physical nodes. Figure 15.3 presents the WINNER functional architecture cooperating with a legacy RAN, here, represented by UMTS.

15.2.1.1 Logical Entities

The following logical nodes can be described for the architecture in Figure 15.3. The gateway (GW) provides access to external data networks (for example, Internet, other wireless networks) and operator core network and services (for example, MMS, MBMS) via the I_G interface. The GW terminates the data flows on the network side, serves as anchor point for external routing, and terminates user plane protocols on the user terminal (UT) side. It handles the idle UTs. The base station (BS) controls all radio-related functions for UTs and RNs, granting the needed radio resources to these nodes. The BS terminates the transport network layer protocols on the network side and the control plane radio protocols on the UT and RN side (MAC and PHY layers). It manages the logical relay nodes connected to it and handles UTs in active mode. The RNs request radio resources to the parent BS (or another RN), and they organize their granted radio resources to serve the associated UTs. The motivation of the RN is to extend the BS coverage or to increase the signal-to-noise ratio (SNR) of the radio signal, increasing the available throughput. The UT comprises all functionality necessary to communicate directly with the network, that is, a BS or an RN or GW. It contains functions to handle UT mobility in active and idle modes as well as functionality to perform an initial access to the network and to initiate the flow establishment. The World Wireless Initiative-Hybrid Radio Resource Management (WWI-HRRM) is the entity enabling inter-WINNER-legacy system interworking. It permits the intersystem resource allocation at the GW level. When a system is coupled through a WWI-HRRM entity, the policy for service provisioning to the end users is mapped onto this entity to allow for efficient resource allocation after intersystem handover. The WWI-HRRM entity also hosts a function enabling interoperator interworking including roaming or spectrum sharing.

The RRM server is a control plane node, which contains centralized system functionality that can be invoked on demand. This entity controls the inter-BS control plane signaling but without interfering with the user plane data (conversely to the UMTS RNC). Therefore, at RRM level two differentiated specific RRM entities have been defined, the RRM server for intra-WINNER RRM coordination, without connection with external networks, and the WWI-HRRM server, a centralized entity that arbitrates network cooperation. In the case of intra-WINNER RRM, it has been studied that some functionalities and network circumstances using a hybrid mixture of centralized and distributed RRM could be advantageous [31]. A hybrid approach supports scalability from low-traffic networks that could not justify the use of this node to high-traffic networks when the use of this node is necessary to effectively exploit the network in congestion situations, such as in the case of LA scenarios with reduced RRM functionality. It has been assumed that this node will be activated by the BS in cases of medium to high congestion. In the cases of low traffic it would be inactive, because in this case it will not increase the system performance significantly.

15.2.1.2   Interfaces between Logical Nodes

The definition of logical interfaces, and logical entities, is important for communication networks, because these interfaces are the connection points of equipments of different providers, which allow the operators to buy interoperable equipment from multiple vendors. The architecture should be modular enough to allow the different physical node implementation, obtained by combining the logical entities, but on the other hand the number of logical nodes and interfaces should be reduced as much as possible. The I_G interface in Figure 15.3 connects the RAN with the public IP network and with the operator services to ensure interworking with other systems and a connection to the AAA server, roaming service, advanced services such as IMS, and other operator services. The I_{GB} is the interface between the GW and the BS, and the I_{BB} is the interface of BSs belonging to the same pool. These are a multi-to-multi interface, that is, there is a pool of GWs connected to a pool of BSs. Figure 15.4 shows a possible implementation, where a central routing entity implements virtual connections between the BS and the GW. The virtual connection can be established using IP

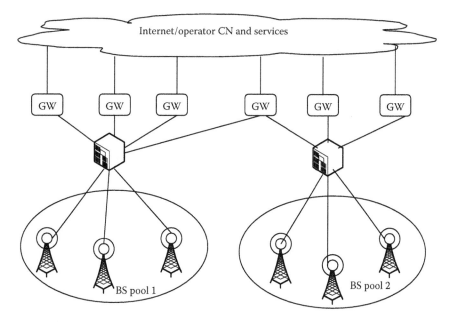

FIGURE 15.4   Interfaces between pools of gateways and pools of BSs.

tunneling, allowing for a handover between BSs without the change of UT IP address. All user data are transported via this interface. On the control plane the interface provides means to establish a flow context in a BS by a GW. Furthermore, it supports UT mobility functions. The BS informs the GW about the user mobility (for example, handover and paging updates). Similarly, the BS provides means to forward the paging messages to a certain UT.

The I_{BB} interface supports distributed RRM functions, such as active mode mobility, interference management schemes, and other distributed inter-BS control and negotiation functions (for example, load balancing between BS). Several spectrum functions are located in the BS that would potentially need this interface. It is foreseen that the interface between the RRM server and the BSs hosts only RRC layer messages. Detailed radio link control and MAC layer context at the user plane are not controlled by this entity. In addition, the interface between the RRM server and the GW is absent, because the RRM server is responsible only for the active UTs, for which all the RRM functions supported by the RRC messages are terminated in the BS.

15.2.1.3 Physical Entities

The WINNER system has been designed to operate over a wide range of scenarios, covering current rural cellular, metropolitan cellular, and WLANs, referred to in Section 15.1 as WA, MA, and LA, each one using certain scenario-specific designs and parameter settings. The logical nodes allow a wide variety of physical implementations and bring in flexibility to the system. A physical implementation (a physical entity) may be composed of one or more logical nodes, depending on the application scenario (for example, requirements of WA scenarios are quite different from those that emerge from an LA scenario). In the case of home networks, a physical node should be provided with low cost and a limited functionality (for example, handover does not have to be supported). In this case the foreseen implementation includes the integration of the GW and the BS in only one element with direct connection to the Internet and with a connection to the operator core (the RRM server and the RN are not relevant for this case). This is shown in Figure 15.5.

In the case of WA and MA deployments, it is foreseen that all logical nodes could be implemented in individual physical entities (that is, BSs), while the RNs would be implemented for support in rural population sites. For low traffic it could be advantageous to have a spectrum server allowing spectrum sharing between operators and systems. A more elegant solution, however, would be to allow that every operator has its own GWs, sharing a common pool of BSs. Figure 15.6 shows

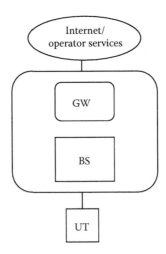

FIGURE 15.5   Implementation of logical entities in a home node physical entity.

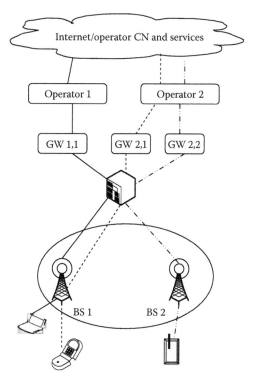

FIGURE 15.6 Implementation of logical entities with sharing of BS between operators.

a scenario in which BS1 is shared by Operator 1 and Operator 2, using respectively GW1,1 and GW2,1. The logical association between a UT and the GW is kept during a handover between BSs, thus avoiding roaming. Normally, the association between a UT and the GW is preserved to avoid a change of IP address in the UT, but if, because of some reasons (for example, load balancing between GWs) it should be changed, then an IP handover using mobile IP will take place.

15.2.1.4 Centralized, Distributed, and Hybrid RRM

Simplified and distributed control architectures are meant to minimize the signaling exchange, increasing the user plane performance in terms of delay and throughput. Centralized architectures, however, offer some gains for specific circumstances; therefore, an optimal solution must be obtained between these two opposite in essence architectures, to satisfy a wide range of scenarios. The definition/understanding of centralized and decentralized RRM is described as follows. Decentralized RRM is one where all RRM decisions are made at cell level, basically by the BS, and independently from other cells (but considering information available from other cells, for example, measurements). Centralized RRM means that some RRM decisions are at least influenced (not necessarily made) by higher hierarchical RRM entities (for example, a RRM server) considering information reported from other cells. In the hybrid RRM some decisions are centralized and other decisions could be decentralized. This will depend on the existence or not of a central RRM server. Even in the case of existence of this node, in some cases, (for example, in the case of low network load and adequate service—mode mapping), it will not be necessary to consult the central RRM server. Conversely in the case of congestion, with possible inter-RAN HO, the RRM server advice will be very beneficial. This concept is called RRM scalability.

15.2.2 DISTRIBUTED COOPERATION ARCHITECTURE

A distributed cooperation architecture will exclude the RRM server entity. The BSs communicate directly without the control of a centralized node. A distributed cooperation architecture has the following advantages:

- Faster decisions in some cases (for example, low network load, decisions based on fewer variables). Because decisions are made locally (that is, between two BSs), no delay limit is required for communications with the RRM server.
- Lower signalization overhead as no communications with the RRM server is needed.
- Higher robustness. If the central RRM server fails (all) cells might be affected. When the RRM server is optional, the network will not be affected by the absence of this node, and the impact could be that some decisions cannot take into account some intercell, intermode, or inter-RAN information that the RRM server has.
- Higher flexibility. Adaptations and changes could be made locally for each cell without evoking signaling exchange with other cells.
- Better suitable for work in heterogeneous environment. For example, if two neighboring cells belong to different providers, they can adapt their decisions to each other using decentralized RRM without relying on a (nonexisting) joint RRM server.
- In cases of low network load and intramode (and in some cases of intermode) handover, the BSs do not have to communicate with the RRM server.

15.2.3 CENTRALIZED COOPERATION ARCHITECTURE

An RRM scheme is centralized by one or a set of central units. Referring to the function allocation architecture in Figure 15.3, the system includes the RRM server, which controls the resources of the entities under its supervision. The advantages of centralized RRM can be summarized as follows:

- Possibility to achieve better decisions, considering intercell, intermode, and intersystem information and also more parameters to have a correct decision (UT mobility, type of service, positioning, etc.) because information for all cells, networks (for example, network load in different BSs and GWs), and UT information could be available at one place.
- In cases of higher network load and to support advanced RRM functions, decisions can be taken in an efficient way by the central RRM entity, avoiding multiple queries and their associated delay to neighboring BSs. To operate efficiently with high network load, a congestion avoidance functionality is convenient; this functionality could use other system functionalities as handover (forcing some UTs to handover) and AC (not admitting new UT/flows). These functionalities need a central RRM entity able to orchestrate the overall network.
- Efficient coordination of radio resources between multiple modes (and RATs) working together (for example, frequency reuse in WINNER is 1) with effective interference mitigation.
- Spectrum refarming (WINNER-legacy) and assignment (inter-WINNER modes).
- Load balancing (resource partitioning between modes and RATs).
- Congestion avoidance gain (using HO and AC).
- Multiplexing gain given by multimode UT (simultaneous reception in several modes or RATs).
- Additional multiuser diversity gain with more potential resources available from different deployment modes.
- Higher trunking gain thanks to more instantaneous available radio resources.

- QoS enhancement gain for scalable services, for example, scalable video coding (SVC).
- Bridge of WINNER RAN with the OM subsystem.
- Service scalability (services provided using the most adequate mode).

15.2.4 Hybrid Cooperation Architecture

The hybrid cooperation architecture provides the best features of the distributed or centralized RRM architectures [31]. The system performance is optimized as much as possible and by default by use of a decentralized RRM but whenever required, RRM algorithms can be triggered by the centralized RRM server. The centralized RRM server provides centralized RRM gains to the baseline distributed architecture, and because this node only includes the control plane functionality its complexity will be much lower than the current centralized node (for example, RNC in UMTS). The assumption is that in the case when such a server does not exist, some system functionalities will not be supported and some decisions would not be optimal due to the absence of a central intelligence. The centralized RRM will be activated only in the case of a medium to high congestion. The RRM server can provide network scalability from initial deployments without lower load and without advanced system functions to networks with medium to high load and a complete set of network services. To provide scalability in the BSs, the interface with the RRM server (I_{RRM}) should be mandatory. Figure 15.7 shows the general RRM architecture, including centralized cooperation with other networks through the WWI-HRRM and hybrid cooperation for the interworking within the WINNER RAN.

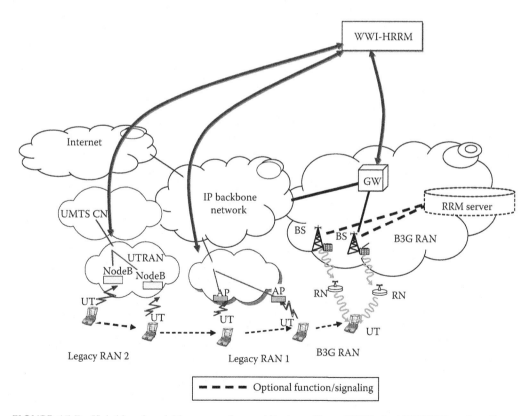

FIGURE 15.7 Hybrid and scalable cooperation architecture. (From IST Project WINNER, at http://www.ist-winner.org; IEEE 802.20 Mobile Broadband Wireless Access (MBWA), at http://www.ieee802.org/20/. With permission.)

15.2.5 DECISION-MAKING MECHANISMS FOR COOPERATION

The architecture in Figure 15.7 provides interoperator, intersystem, interoperation mode, interentity, and interlayer cooperation. To reduce the unnecessary signaling, decision-making mechanisms including triggers, monitoring, and autonomous functions are used. The cooperation mechanisms rely on interentity interworking, which typically guarantees the seamless connectivity for the UTs associated with different network nodes. The various parts of the decision-making mechanisms are described in the following sections.

15.2.6 TRIGGERS FOR COOPERATION

In general, triggers include internal signaling between layers. In the cooperation architecture in Figure 15.7, the cooperative RRM algorithms are initiated by a given trigger. Typical RAN triggers are allocated at L1/L2. An L2 trigger is understood as a signal of an L2 event. An L2 event regarding the L2 handover may be the early notice of an upcoming change in the L2 point of attachment of the UT to the access network. Another possible event is the completion of relocation of the UT's L2 point of attachment to a new L2 access point. Although specific L2 information has to be exchanged, an L3 protocol should be kept independent of any specific L2 feature. This means that an L2 trigger that is exchanged with L3 must be generic and technology independent. Apart from that, triggers should only enable performance enhancements and not be used for correct protocol operation. In most of the cases, an L2 trigger is a kind of notification from L2 (potentially including parameter information) to an adjacent layer that a certain event has happened or is about to happen. In this way, the process of triggering is related to information exchange and communication between layers. However, a trigger might also be a physical probe, the outcome of an algorithm or the transformation of a requirement, all of which are L2 self-autonomous transactions with no interlayer communication [30].

A trigger may be implemented and applied in a variety of ways. An L2 trigger used for the activation of an RRM mechanism is not associated with any specific link layer technique but is rather based on the kind of L2 information that is or could be available from a wide variety of radio link protocols. The basic property of a trigger, however, is that its appearance is closely connected to the handover question. A "traditional" handover is located in L2, because L2 is responsible for the physical and the logical connectivity of higher layers. In fact, every L3 handover is usually preceded by a L2 handover [30]. The only problem is that the L3 handover is not aware of the L2 handover due to the independency of the layers. The reference B3G system of Figure 15.3 combines both the link layer triggers and the triggers of load information at each GW to trigger a joint L2 and L3 handover. For example, to detect whether a UT must change its association point, the BSs periodically (for example, every 2 s) broadcast a unique message, a so-called advertisement. To be able to detect a new advertisement, the MAC will establish first the new radio link, after which an algorithm to reassociate to the new access router (AR) is started. With a constant inter arrival time of 2 s of the advertisements, the UT will take a small portion of the period on average to detect the movement to a new AR. This information is available at the link layer immediately after the link is established. The link layer may now send a trigger and inform the network layer where a load scan may take place. This reduces on average the IP handover execution by 1 s. This is only one example where link layer triggers may be used. Some possible triggers are the PHY triggers (for example, signal-to-interference-noise ratio [SINR] and real signal strength interference [RSSI]), location information triggers, self-organized information provisioning (scanning triggers), triggers based on information exchange within the same or with other systems; unbalanced load between GWs, and so forth. A detailed analysis of a selection of triggers is available in Ref. [30].

15.2.7 INTELLIGENT MONITORING

The most common way to gather the required information is to perform monitoring and scanning of the air interface, that is, gaining information from the PHY to detect possible systems to switch over to [32]. The advantage of this method is that it can be realized self-autonomously by the UT, which is important for newly switched-on devices or terminals that suddenly have lost their old connection. Furthermore, especially for the hybrid handover, the UT does not need to possess additional hardware components, because the same device can be used in different cells of the same type (that is, roaming). The disadvantage of this approach is that the scanning procedure needs special coordination and resources, such as additional battery. Also, the monitoring/scanning takes time, which usually cannot be used for communication. Another way is to exchange information within the same or with other systems. Information gathering within the same system is required for horizontal handover (HHO), semi-vertical handover (semi-VHO), and VHO. Semi-VHO is a handover between different modes of the system. VHO is the intersystem handover and it requires information gathering between different types of systems. This approach offers a great economic potential because the scanning procedures can be minimized. Each system collects data about the current state within the covered cell and provides this information on request to UTs that are willing to change their connection within the same system (HHO and semi-VHO) or different systems (VHO). For the HHO case this is partly already realized by means of broadcast channels, on which a BS periodically transmits. The broadcast channels, however, do not include information about neighboring or different systems. If this information is provided as well, then, especially, for the VHO-case this means a remarkable gain [30].

15.2.8   AUTONOMOUS COOPERATION

Autonomous adaptive BS for the selection of the most suitable RAN for the user was proposed in Ref. [33]. In a more advanced manner, the system can be designed to introduce an autonomous cooperation (AuCo) mechanism that allows for automatic handshake and cooperation between entities that are not by default interworking with each other. An AuCo function communicates with a distributor proxy and maps the communication on the application level to the generic transport. Incoming measurements are forwarded to three different blocks that do basic preparation of measurements for future or immediate use, namely

- Link status aggregation: This entity is responsible for data mining. Current position estimations or predictions are matched against database entries to generate a measure of expected link quality in complementary systems.
- Traffic estimation and prediction: Information from the OMC subsystem and cognitive methods could be used to estimate traffic distribution within the system. This information can be used to aid joint call admission control in their decisions.
- Position tracking and estimation: Provides prediction of the UT's movement. This can be used by the VHO algorithms to prepare handovers in a make-before-break style.

Publications on the association topic are forwarded to the node registry. This component keeps track of each UT to provide the addressing information. The main purpose of AuCo is to provide attached RANs with advice on intersystem RRM. Entities that are able to receive the publication triggers are able to join the AuCo circle. The VHO recommendation component can generate handover recommendations by incorporating position information of UTs, link state maps, and link quality estimations as well as information from call admission control within an algorithm. One example for this is the use of the center of gravity (CoG) [34,35] algorithm.

15.3   DESCRIPTION OF COOPERATIVE SYSTEM FUNCTIONS

15.3.1   MOBILITY MANAGEMENT

Mobility management is an essential cooperation mechanism that ensures the successful deployment of any new RAN. It is assumed that a UT attached to a next generation RAN could hand over to a legacy RAN when it loses the coverage of its current system. For example, in the WINNER system (see, Figure 15.7), the decision about the handover execution or the triggers for a handover could come from either the network or the UT. This means that in WINNER the handover is terminal- or network-initiated. If the network decides about the handover execution, the UT can still influence the decision and this is a terminal-assisted handover. The intersystem handover is expected to take place either when there is a loss of coverage of the current system or in the case of overlapping coverage due to user/operator preferences or traffic congestion. Examples for triggers assisting a UT decision in the case of complementary coverage are signal strength; interference level; carrier-to-interference ratio (C/I); and bit error rate (BER)/packet error rate (PER). Example triggers in the case of overlapping coverage, where the link quality is not expected to be a frequent trigger for a handover, are the required service available in the other system (network decision); the cheaper service cost (UT and network decision); the UT location (network decision); the UT velocity (terminal and network decision); and the congestion in the current cell and surrounding cells of current system (network decision).

15.3.2 LOCATION SERVICE SUPPORT

Location context includes the concurrent location, the future location, and the mobility parameter of the corresponding UT [36]. A location-service-based system comprises a communication subsystem, a location information subsystem, and the service policy subsystem [37,38]. For simplicity, we integrate the latter two subsystems into one, namely, the enhanced hybrid information system (eHIS). The communication subsystem can include a number of radio systems, following different standards, as depicted in Figure 15.8.

The basic property of the eHIS is to map incoming measurement reports to specific locations. Therefore, it is of existential importance to support the location of the feeding clients. In principle,

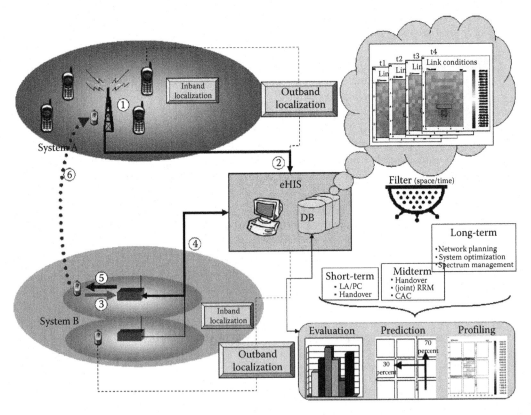

FIGURE 15.8 Location service enabled communication system.

two localization alternatives are possible, namely, either feeding clients perform inband localization, indicated by the two inband localization boxes on the left side of Figure 15.8 or the eHIS system acquires the feeding clients' position by outband localization, indicated by the respective box in the eHIS. The inband localization is performed by the signal processing technique inside the communication subsystem, whereas the outband localization is performed by an auxiliary function not included in the communication system [39]. Besides basic location entries stored in the eHIS database, more parameters could be added. Extended entries consist of terminal-related properties such as velocity, moving direction, current service consumption, and others. As such, the extended entries support personalized service provision. By evaluation of that data it is possible to set up user profiles and predict user requirements. A cooperative RRM mechanism could exploit the eHIS information to prepare a planned handover to another serving BS/AP. Due to the interference maps, the current link condition in the target system is well known, so it is possible to determine whether extra bandwidth needs to be reserved for UT handover. Moreover, if the geographic target is covered by several vertical systems, the eHIS may even trigger intersystem support for the terminal and support joint RRM in this way. An interesting point with respect to usability of the information stored by the eHIS is the use case, for which respective updated data shall be applied to.

The concept of eHIS distinguishes between short-term, midterm, and long-term data, (see the right part of Figure 15.8). Short-term data are meant to support real-time requests from the information clients. As soon as a feeding client provides new measurement reports, the essence is extracted and stored in the eHIS database (DB). Short-term data, therefore, reflect the latest entries in the data base. The respective information will serve as decision basis for the short-dated handover triggers. Midterm data are less time critical. They are based on the short-term input but due to respective filtering and averaging time, selective fading effects are equalized. Nonetheless, midterm data are of interest for ongoing communication because they serve as a set value, especially, for predictable actions. Especially, in combination with prediction and profiling, midterm data are useful for a planned handover triggering, (joint) RRM, or connection admission control (AC). The period for the validity of long-term data is supposed to be longer than one day. Together with the measurement report the location of the reporting UT is stored in the eHIS. A UT that intends to perform a handover sends a request to its BS. The BS in turn acquires the corresponding measurement report from the eHIS, depending on the current location of the UT and signals the handover (HO) decision (respectively related information that allows the UT to take the decision) to the UT. The UT can then perform the HO to the proper BS or the mode. In the above example, measurements that are inherently available for each system are made available to heterogeneous systems as well to support the interworking between heterogeneous systems. Depending on the new target system and the current location of the UT, the mobile is supplied with state reports of the same system type (for horizontal handover, HHO) or a vertical system (VHO) and subsequently may perform the (V)HO, which is referred to as location-based VHO because the location of the mobile is exploited in the HO process. The eHIS is an intelligent concept facilitating intersystem cooperation and a means to allow for context transfer between different systems.

The purpose of the eHIS is to perform accurate detection of complementary systems and to initiate optimal VHO execution by respective triggering. In all cases, the eHIS approach offers a great economic potential because participating devices can minimize or even avoid self-driven scanning. The eHIS entails a decision unit that takes into account trigger origins as input and produces handover recommendations (triggers) as output. The advantage is that the eHIS is not restricted to local and system-specific trigger origins. Besides incorporation of a multiple number of systems, eHIS supports load balancing and joint RRM. Further, due to its backbone connection specific user preferences may be requested from, for example, the home network provider and incorporated in any decision process. Thus, eHIS supports intelligent intersystem-control by combined evaluation of various trigger origins. As the location is a key parameter in the location-based HO, the less accurate and precise the location information is, the larger the difference between the anticipated, that is, retrieved measurement reports and the real link condition in the target system after the HHO, followed by VHO. The estimation of the precision and accuracy of typical location techniques evaluated for the WINNER system has been presented in Ref. [36].

15.3.3 CONGESTION AVOIDANCE CONTROL

The required interworking between the various entities implementing a CAC mechanism is shown in Figure 15.9. The CAC mechanism requires close interworking of the AC and LC modules. Most algorithms reported in the literature give priority to handover calls in contrast to new calls. The terms "new session" and "handover session" in future heterogeneous wireless networks are slightly different from those in legacy networks and could be described as follows:

- New session is a session request in a specific network that cannot be admitted in that network and it is checked if another network (and which one) can meet the session requirements.
- Handover session is an ongoing session, (which means that it has already a vital connection with a network) and for some reasons it needs to handover to another cell, deployment mode, or network.

The basic assumptions for CAC in future heterogeneous networking environments are

- An accepted call that has not been completed in the current deployment mode/RAT may have to be handed off to another deployment mode/RAT.
- New calls and handover calls normally have to be treated differently in terms of resource allocation.
- Handover calls are normally assigned higher priority over the new calls.
- Traffic will be routed through the cooperating deployment modes and systems according to the restrictions and advantages of each mode and system.
- Different levels of service calibration can be identified for the users.
- Mobile users can alternatively access different deployment modes (intermode HO) or RATs during a call (intersystem handovers).
- Coexisting modes and RATs jointly cooperate together.

CAC can be implemented in a centralized, distributed, or hybrid manner. Given that most of the communications systems candidates for IMT-advanced [2,38] are characterized by a flat architecture, it can be assumed that a hybrid approach will be a common trend.

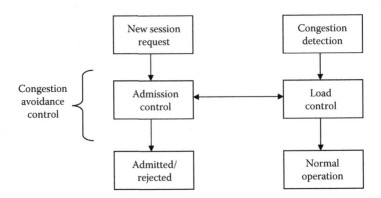

FIGURE 15.9 Functional blocks of the CAC mechanism.

15.3.4   QoS AND SECURITY

QoS is a term with many interpretations, but it should reflect how end users perceive the quality of their communication. The requirements are different from application to application. Many parameters influence QoS, for instance: BER/PER, delay, bandwidth, packet retransmission rate. In this context, focus is on how to exploit the information coming from lower layers and residing at the BS/AP. The quality of the radio channel is the traditional parameter that impacts QoS. Here, approaches adopted to improve the quality of transmission [40,41] should be taken into account along with whether these might create additional problems for certain applications. Imperfect mobility and handover procedures will also affect QoS. Information about these will be aggregated at a special entity where user profiles related to mobility, capacity, etc. will be created [42]. In this context common indicators could be identified to allow grouping, for example, these can be indicators for real-time traffic and indicators for non-real-time traffic. Also, the proposed QoS scheme should be evaluated to determine compatibility with the proposed mobility management and CAC schemes. Security procedures (authentication, access control, encryption) may also cut transmission and make communication quality unacceptable. This should be taken into account when selecting a RAN/RAT for a requested service. In a heterogeneous scenario, delivery of QoS for a requested service will depend on the user requirements and service class. Additionally, each session is characterized by specific requirements for the quality of transmission link. Delivery of QoS then means that the network performance profiles will be evaluated versus the specific user profile to take a satisfactory decision for all parties. User profiles can be described by an analytical model that corresponds to the real traffic observed on the network [43]. The QoS requirements are strongly linked to the performance and the load of the RANs. Therefore, it is important to monitor and manage the resources of each RAN both separately and jointly.

In a heterogeneous environment networks should be ready to convey interdomain QoS traffic before customers can initiate end-to-end service-level specification negotiations (for example, interdomain routing). A possible QoS management scheme can be based on the one described in Section 15.1 CRRM approach. The scheme should implement a central QoS management functionality to obtain information on the status of users and the availability/status of networks and a distribution functionality that can be triggered as a response to act upon requests from cooperating networks and neighboring cells to handle the arisen situation. The scheme will comprise two schedulers, one located at the BS/AP and the other scheduler will perform scheduling of inter-RAN packet flows (located at the GW). That means that when a service is requested, the scheduling mechanism will prioritize the flow that corresponds to the provision of the service according to the implemented RAN selection/prioritizing algorithm [44] also described in Section 15.4.

15.4   ALGORITHMS FOR COOPERATIVE RRM

To achieve cooperative RRM, interaction is required between functional entities of the RAN architecture [45]. The required interactions are shown in Figure 15.10. We can deduct the following requirements for the cooperative RRM: context and location awareness planning, and coping with multiple operators, and deployments. Mobility management, for example, should include aspects related to handling the session/application, load balancing, multihoming, security, and all sides of mobility, that is, end points, sessions, flows, interfaces, network/groups, flexible spectrum utilization, and group and signaling managements.

The following subsections describe algorithms proposed for cooperative RRM and how each algorithm interacts with the rest of the cooperative RRM mechanisms as outlined in Section 15.3.

15.4.1   ALGORITHMS FOR MOBILITY MANAGEMENT

In this section, we define an intersystem handover mechanism that enables the cooperation between a B3G system and legacy networks. The proposed algorithm takes only load criteria into account,

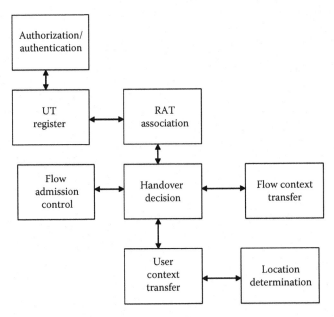

FIGURE 15.10 Interactions between functional entities for cooperative RRM.

but it can also include service criteria. Figure 15.11 shows the principle of intersystem handover and its interworking with the AC algorithm. The intersystem handover algorithm is implemented in the RAN and has two phases: first the decision to trigger the handover and prepare it, then the selection of the most suitable target network to execute the handover. Both criteria can be based on various inputs such as measurements coming from UTs or heterogeneous networks, or service attributes. These criteria and the intersystem handover algorithm are detailed below. Once the target network for handover has been selected by the handover algorithm, the admission control on that target network will check if the call or session can actually be accepted. If not, another system must be chosen, but the handover algorithm should be defined so as to minimize the rejection of calls/sessions by the admission control.

We consequently assume that load information is obtained, when required, through a special RRM RAN-related entity. The only limiting point may be signal strength or quality measurements performed by the UT on the candidate target system or mode. Use of neighboring cell lists and a measurement optimization process can restrict the measurements related to handover to the cells [44]. The optimization process for the handover algorithms can be implemented as follows.

First of all, the handover triggering decision can be made by the UT. Let us assume that both the current and target systems or mode can provide the UT with its requested service. The algorithm deciding on the target system can be based on the following:

- Service requirements of the UT (for example, speech users cannot get connected to WLAN and users requesting high bit rate cannot be connected to GPRS)
- User priority (some users can be rejected because their subscription profiles do not allow such a connection)
- User velocity (for example, fast-moving UTs will preferably stay in a macrocell)

In the following, we assume that the target system or mode choice is not based on the UT signal strength or quality measurements on the target system or mode. These measurements need to be performed only when it is decided that an intersystem or intermode handover algorithm is necessary. Furthermore, such handover is triggered by measurements on the current system only. It is, therefore,

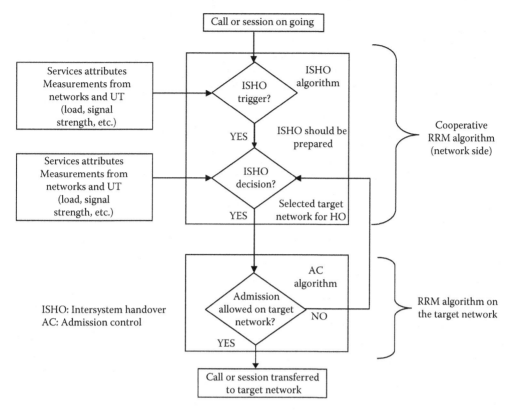

FIGURE 15.11 Intersystem handover diagram.

not necessary to perform measurements on the candidate target systems or modes before the handover decision has been taken. For example, the following rules can be applied:

If $M_{Best,Current} < Th_{Current,1}$ then measurements are triggered on the neighboring cells identified for the target system (1).

If $M_{Best,Current} < Th_{Current,2}$ and $M_{Best,Target} > Th_{Target}$, then handover is performed to the best measured neighboring cell (2).

With this algorithm, as only coverage information is used, it is not possible to restrict the neighboring cell lists because of the required load or QoS information. Therefore, it is important that the neighboring cell lists are defined as accurately as possible.

Another set of rules can define a different algorithm, one that supports decisions during handover from the current system/mode to the target system/mode, and is based on load and coverage criteria. It is composed of the following rules:

AlgoLoad1 (1): This algorithm does not use load as discriminating information, but as ordering information. Consequently, it is not possible to use it to discard neighboring cells.

If $M_{Best,Current} < Th_{Current,1}$ then ask for the list of neighboring cells on the target system and perform measurements on these cells.

If $M_{Best,Current} < Th_{Current,2}$ and $M_{celli,Target} > Th_{Target}$ then handover will be performed to Cell*i* of the target system with the lowest load among the cells that have sufficient level (defined by Th_{Target}).

AlgoLoad2 (2): This algorithm uses load to keep cells with low enough load only. It is applied to decrease the number of the neighboring cells to measure.

If $M_{Best,Current} < Th_{Current,1}$ then ask for the list of neighboring cells on the target system.

If $M_{Best,Current} < Th_{Current,2}$

For each of these cells Cell$i$, if $Load_{Celli,Target} > Th_{load}$ then discard Cell$i$ from the list of neighboring cells. Perform measurements on the remaining neighboring cells. Then keep Cell$j$, if $M_{Celli,Target} > Th_{Target}$ then perform handover to Cell$j$ with the best level.

The evaluation of the performance results that can be achieved with these algorithms has been given elsewhere [46]. The criteria to choose the users (and their number) that perform the handover must be determined; these could be service-based criteria, resource-based criteria, or users-based criteria.

15.4.1.1   Intelligent Monitoring for Handover Decisions

In an interference-constrained environment, several Layer 1 and Layer 2 performance measures can be used to trigger a handover between two different access networks or between BSs serving different deployment areas (for example, MA and LA). Because the signal level measure alone may not necessarily reflect the level of interference between the devices operating in the same band and the device that needs to be handed over, a number of packets retransmitted at the MAC layer can be used as a measure of the number of packets that have been discarded due to packet collisions. As the number of packets exceeds the threshold value, the mobile device may trigger a handover. This measure captures the percentage of packet loss at the receiver. In addition, this measure can indirectly provide information about packet collisions at the receiver at the other end. For example, if a UT correctly receives data packets but observes that the BS is trying to send the same data packets several times, it means that the acknowledgments are lost at the BS. In that case, the cognitive radio enabled device can trigger a handover.

This decision process can be assisted further by capturing the specific application requirements in terms of bandwidth, delay, and packet loss that also relate to the overall QoS the user expects [47]. Different threshold requirements can be devised and quickly evaluated by means of inference and learning techniques implemented at the convergence layer, and these results can be made available at the lower layers (for example, link layer) [47]. For example, to support an RT video streaming application, the delay between each packet received can be monitored, and a handover can be triggered if the delay variance (jitter) goes beyond a predefined threshold. Also, the monitoring process for the collection of measurements can be provided with a self-learning capability that uses outcomes of decisions and observations and learns from these to characterize and predict the system or user behavior and thus decrease the number of required handovers, that is, user-context transfers. Intelligent monitoring and self-learning are possible by introducing the following additional functionalities:

- Cognitive sensing in the radio segment
- Collecting and mining information and status from the network and full detection functions
- Cognition learning at decision level (located in a central RRM entity)

For the implementation of intelligent monitoring, the BS and GW nodes should be provided with autonomous management capabilities based on the use of cognitive radio technology and cognitive routing, elevating the process of network management to a cognitive state, for improved network performance (for example, improved handover performance) [47]. An example of the implementation of a learning block at the central RRM entity is shown in Figure 15.12. The decision-making process is assisted by the use of fuzzy logic rules. Fuzzy logic is a simple and fast solution to provide a conclusion from imprecise, noisy, or incomplete inputs. Fuzzy logic is based on simple "IF X AND Y THEN Z" rules rather than complicated mathematical models. System behavior can be tuned, simply, by modifying the appropriate rules, and it is possible to compare quantities from

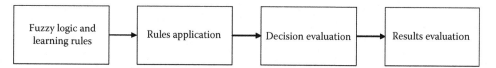

FIGURE 15.12 Example of a learning block implementation at the central RRM entity by use of fuzzy logic rules.

heterogeneous RANs. In complex systems, fuzzy models, based on simple IF-THEN rules, give more easily assimilated information than that in precise models. Nevertheless, rule definition requires a good knowledge of the systems and prior field experience. Therefore, learning techniques are needed to provide full knowledge about the system parameters.

15.4.1.2 Algorithms for Congestion Avoidance Control

15.4.1.2.1 Mechanisms for Admission Control
The AC algorithm maximizes the number of admitted or in-session traffic sessions supported over the cooperating RANs, while guaranteeing their QoS requirements and ensuring that a new session does not violate the QoS of the ongoing sessions. In the case of joint wireless systems that include multiple legacy networks, with different characteristics and a future network that provides many different services to the users (for example, B3G RANs can have defined at least 18 different service classes for the users [48]), the admission decision is a complex procedure, involving many important factors that have to be taken into account. The decisions to accept or reject a new session are based on the following different criteria:

- Network load: The predicted load of the network after the admission of the new session is computed, and if it remains under a certain threshold, then the new session can be accepted; otherwise, it will be rejected. The load is not computed in the same way in all networks and the algorithm takes into account the distinct characteristics of each network before performing the measurements.
- User's QoS requirements: QoS parameters such as mean throughput, bandwidth demands, BER requirements, service class, and priority of each session are taken into account to decide whether to admit the session or not.
- User's context: The algorithm normally gives priority to handover sessions that require lower blocking probability in relation to new sessions [46]. There is an exception for emergency calls. Also, there is an option of grouping the users according to their subscription. For example, we may have classes of gold, silver, and bronze users who are assigned priorities in a descending order (that is, gold users get the highest and bronze the lowest). The overall priority of the session will be then be a cross evaluation of the service class priority and the user's priority.
- Link quality: If the admission of the new session results in a decrement of the link's quality under a desired value, then the session is rejected. Link quality refers to the quality of the radio link between the base station and the mobile terminal. Link quality is measured based on the received signal strength at the mobile terminal and by the interference caused to this link by other mobile terminals in the same area.
- User's preference: The preference of the users should also be taken into account. A user may prefer to be connected always to the WINNER WA deployment mode, to WINNER MA deployment mode, or to a legacy network, and this preference should be acknowledged and taken into account by the admission control algorithm. Therefore, all the other options should either be discarded or left for the worst-case scenario that the user cannot be admitted to his mode/network of preference.

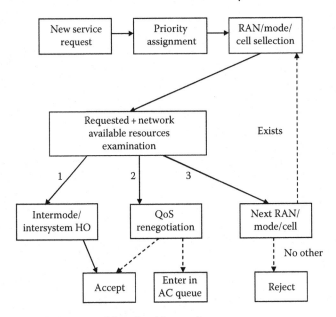

FIGURE 15.13 Main admission control functional interactions.

A basic efficient AC algorithm procedure for future wireless networks includes the following steps and interactions (Figure 15.13).

15.4.1.2.2 Mechanisms for Load and Congestion Control

When already admitted users cannot satisfy their guaranteed QoS to their services for a specific percentage of the time, then the network is considered to be in an overload/congestion situation. If an overload situation occurs, the load can be decreased by performing several actions that exploit the cooperation between the networks. When there is a congestion situation detected in the current network, the load control mechanism tries to solve the problem locally, initially within the affected network, and if it fails, then it sends a request to the central RRM for global resolution, taking into account the other available cooperating networks. The LC mechanism is a composite mechanism that is divided into three phases. The load control functions and interactions are shown in Figure 15.14. and are explained as follows:

Load monitoring phase: The congestion control algorithm continuously monitors the network and periodically checks the load to detect if a congestion situation has occurred in any of the cells. It is considered that a cell of a deployment mode or a network is overloaded if the load factor is over a certain predefined threshold during a certain amount of time:

if $n \geq n_{thC}$, for an amount of time ΔT_C , then the cell is overloaded and the congestion control algorithm is triggered.

For a B3G RAN of the type described in Ref. [1], the ways to detect a congestion situation can be summarized as follows:

- In the MAC user plane (associated to the gateway or BS/RN)
  a. Number of used chunks divided by the total number of chunks available for data transmission exceeds a threshold
  b. High retransmission rate, which is the number of retransmitted MAC packets divided by the total number of transmitted packets
  c. Increased interference level in chunks
  d. High block error rate (BLER)

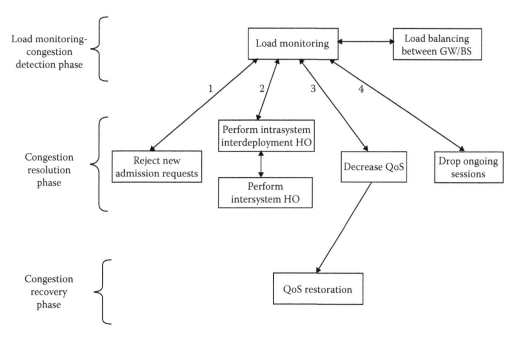

FIGURE 15.14 Load control functions and interactions.

- In the service level controller (SLC)
  a. Increased buffer occupancy, inferred by the total, average, and variance of buffer occupancy (of the different buffers)
  b. High delay, inferred by the total, average, and variance of packet delay,
  c. Low available bandwidth (throughput)
- In the radio resource control plane (RRC)
  a. High number of users in the admission control queue
  b. Requests for short-term resource assignment
  c. Requests for proactive routing (more than two hops)

Congestion resolution phase: After congestion detection, the LC mechanism enters the congestion resolution phase where it tries to resolve the problem that causes the overload situation. Initially, the LC must interact with the AC to decode what action is required in accordance with the AC performed. There is only one exception, namely, for emergency sessions, which should be handled in any case and at any cost based on available resources. The next step is to check if multimode UTs can handover to another deployment mode/cell. For example, if the congestion is detected in a BS in the WA deployment, then the users that can be served by an LA BS are redirected to that BS. The same action can be performed if a neighbor WA BS can provide wider coverage than the LA BS. The following step would perform the intersystem handover. From the ongoing users, those users who can be served by a legacy network are selected and are handed over to a neighbor BS of that network. The users that will perform this type of handover should of course be subscribed also to that network or there should exist agreements between the network operators. Also, these users should have UTs with the required capabilities. If the forced handovers cannot resolve the congestion situation, then an attempt would be made to decrease the load by changing the QoS parameters of some users. This can be done in many ways, but mainly by decreasing the data rate of selected users. The ultimate action for resolving the congestion situation is to start dropping ongoing sessions that have low priority. After all of these actions are performed, the load of the network is recomputed to determine if the congestion has been resolved. A network is considered to have overcome the congestion situation if the load remains underneath a certain threshold for a specific amount of time or if $n < n_{thN}$, for an amount of time ΔT_N .In this case the network enters the congestion recovery phase. It must be noted that the mechanism takes into account the group that the users belong to according to their subscription profiles (that is, gold, silver, and bronze) and can degrade their QoS accordingly.

Congestion recovery phase: In the congestion recovery phase, the CAC tries to restore the degraded QoS of the sessions to avoid the violation of the service agreements of the users. The critical point is to decide how the bit rate of the sessions should be restored because the network might return to the congestion state. This is done by computing the amount of load needed to restore the transmission rate of the affected sessions.

The information that is exchanged between the entities during the procedure of the AC and LC can be summarized as follows:

- BS load / GW load
- User location information
- Service class/user information
- Neighboring cells information from RRM server (optional)
- AAA information
- RSSI/SINR

For the load or congestion control algorithm, the inputs are summarized as follows:

- Congestion status and congestion control command
- Load information of the BS or the GW or the relay nodes
- Load measurements

The output of the AC and load/congestion control algorithm is a selection of the needed RRM function that can resolve the admission/congestion (that is, intersystem handover, spectrum exchange with legacy RANs, resource repartitioning, and so forth) and the load balancing between the networks.

15.4.1.3 Algorithms for QoS and Security

QoS and security mechanisms can be user-oriented and network-oriented algorithms. The user-oriented algorithms specifically guarantee the user's QoS, whereas network-oriented algorithms are used to avoid or resolve congestion situations and to improve the network performance. From the user's point of view, each session is initiated with the connection process and terminated by the disconnection process, in case the connection was successful. Between connection and disconnection processes, the user can request as many applications as the UT allows it; each application request is handled independently by the AC process. In both types of algorithms, a scheduling process should be introduced to avoid problems resulting from a massive users arrival and to prioritize users. Independently from the user-oriented processes, the network-oriented algorithms take place as background processes. Figure 15.15 shows the interactions between the different QoS algorithms. Figure 15.15 considers the case of a generic user called User 1. User 1 gets connected to the B3G RAN through the connection process and requests an application, application 1. This action triggers the AC process. A request for a new application, application 2, goes through the AC process again. Let us assume that application 2 is a messaging type of application and that application 1 is a video streaming type of application. Application 2 ends before application 1. If during the connection time of User 1, the network enters congestion state, the congestion control mechanism will redirect User 1 to another RAN or another mode of the same RAN (that is, an intersystem handover is performed).

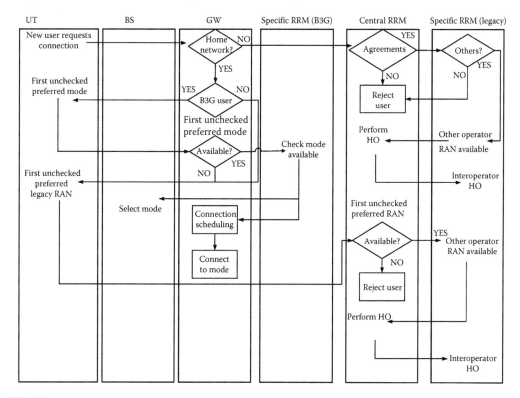

FIGURE 15.15 Interactions between user-oriented and network-oriented algorithms (From Villemont, A. and Ingrain, T., Implementation of a QoS algorithm for the efficient cooperation of heterogeneous RANs, MSc thesis, Aalborg, Denmark, June 2006. With permission.)

There, User 1 will run an application n. Then, the user will leave the area and the disconnection process will terminate the session. Each time a new user enters the network an alarm is triggered corresponding to the release of a place in the scheduling queue, which means that one waiting user can come in. This scheduling process does not take into account the user priorities, because it assumes that the connection process does not require many resources. Therefore, here we can apply a simple first-come-first-served algorithm [49].

15.5 PLATFORM IMPLEMENTATION OF COOPERATIVE RRM

This section gives an example of a practical implementation of cooperative RRM. Figure 15.16 shows the demonstration setup. It shows cooperation between two RANs, RAN 1 (WLAN) and RAN 2, which emulate a B3G RAN [1].

The AP uses the 802.11g wireless protocol that connects it to the UT. The setup in Figure 15.15 supports handover from RAN 1 to RAN 2 and from RAN 2 to RAN 1 for a one-user scenario. The scenario can be enhanced to include multiple users and mobility features. A real-time video streaming is the chosen requested service for the user. The central RRM and the application server are independent of the RANs and normally should be located outside the RAN architecture. An HDTV camera captures video and sends it to the application server that streams it to the UT. The UT has two network interfaces, an Ethernet card and a wireless card. When the UT is connected to the RAN 2, the Ethernet card is enabled and the wireless is disabled and the opposite happens when the terminal is connected to the legacy network through the access point. The Ethernet card can be replaced by a wireless card. In general, another implementation can associate different service sets

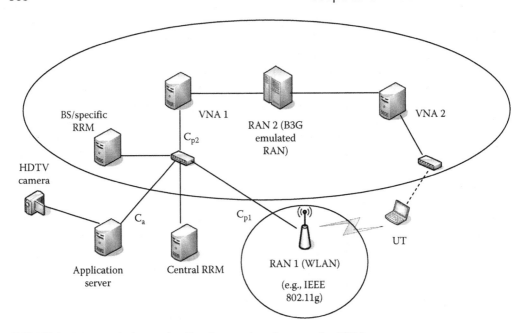

FIGURE 15.16 Practical example of implementation of cooperative RRM.

with the scenario of Figure 15.16. The service sets are a grouping of applications requiring different data rates, delays, and ranges. The service sets can be stored in a database located or associated with the network management system (not depicted) but which can be made a part of a larger functional block including also a traffic generator and a reporting module (RMU) for the transmission of real-time traffic measurements (RTTMs). The logical functionality of the central RRM unit handles the common functionalities of the systems for each RAN. This unit provides a common interface toward upper layer functions/protocols. The location of the central RRM unit is flexible. It is possible to associate it with the network management system module. It is also possible to add a control functionality that can control the access to the radio interface resources. The RAN-specific RRM functions are located at the BS-specific RRM module in Figure 15.16. This action is realized at lower layers; in consequence the specific RRM is associated to the corresponding entity depending on the type of RAN (for example, the RNC, BTS, and interworking unit (IWU) of UMTS, GSM, and IEEE 802.11 networks, respectively). The central RRM requires interfaces to the specific RRM entities of all the RANs (for monitoring and for actuation purposes). The following interfaces are shown in Figure 15.16:

- C_a interface: This is the interface between the specific RRM units of the RANs. Over this interface, RTTMs are transferred for monitoring purposes. In addition, when the local RRM techniques are applied by the specific RRM module, the commands are forwarded to the control module over this interface. It is also used to transfer the user service requests for intrasystem management.
- C_{p2} interface: This is the interface between the central RRM and the specific RRM. This interface is used to send alarms to the central RRM, to provide the network status to the central RRM, on demand, and allows the messages for the specific RRM techniques (including handover command) from the central RRM. It is also used to forward new service requests (that cannot be handled inside the RAN) to the central RRM entity.
- C_{p1} interface: This interface is meant for the exchange of alarm and the status demand messages from the specific RRM associated with RAN 1.

A brief description of the involved entities is given below. The BS extracts information from the RAN2 on predefined intervals configured by the system. This information is delivered as RTTMs, which show network parameters and actual operating modes of the RAN2. In addition, user information, such as the amount of users under a specific mode and the type of service, which is used (per user), is also extracted. This is the monitoring part of the BS. This monitoring part also forwards the statistical information extracted by the RTTMs via structured text (for example, in XML format) to the central RRM entity, which has the functionality of the global monitoring of the system. The BS will be controlled by an application with a graphical user interface (GUI), where it connects to the RAN2 and the central RRM entity and shows the statistical output of the RTTMs. The other entity is the RRM server, which is split into two parts. The first is the monitoring unit and the second the RRM-s unit. The RRM server is the basic entity that executes the intrasystem RRM algorithms. It accepts information from the coupled BS entity in the structured XML format. From this input it calculates the KPIs, which are then used for the monitoring of the RAN2 and for the application of the intrasystem RRM algorithms. Its role is more passive than active; on one side it constantly reads the flow of information from the BS, in the mean time based on formulas, it calculates the KPIs and outputs the results to the intrasystem RRM unit or the central RRM entity. The results lead to two different sections of the monitoring unit. The first is driven by the monitoring unit, where in every cycle of calculation, a comparison of the results is performed between each KPI and the predefined threshold values. When the value of the calculation is higher or lower than the threshold value (the latter is based on whether the KPI is increased or decreased to the maximum value), then alarm signals are created and a list of the current information is sent to the RRM-s for local management inside RAN2 or to the central RRM for global management. The list contains a summary of the time of calculation, which happens in the monitoring unit by providing information on the KPIs, the number of users who are currently connected, the type of service per user, and the current mode of the BS. The second section of the monitoring unit is related to continuously sending the results of the XML messages from the BS and the calculated KPIs to the central RRM entity for the global monitoring of the networks and the application of the RRM-g algorithms. The information exchange is the same as before but the only difference is the type of message that has arrived at the central RRM. The RRM-s unit performs the specific RRM techniques for handling user requests inside RAN2 and for performing intramode handovers. It also gets alarms that trigger the RRM-s techniques for managing the performance of the network. If the RRM-s techniques are not successful then the alarms are forwarded to the central RRM for triggering generic RRM (RRM-g) techniques.

The central RRM is the central element of the RRM platform. It carries out three different tasks. The first task is to monitor and control the network status and alarm information, the second one is the handling of the RRM, and the third one is to handle on-demand service and handover requests from the UT. The alarm and status information provide the driving force for handling the cooperation between RAN2 and RAN1. This information is extracted directly from the RAN1 AP as structured format-based information and from the RRM server RAN2. The central RRM handles only the RRM-g techniques. This case is mainly adopted for the handover requirements between RAN2 and RAN1. This type of RRM is used to handle the users to provide the desired bandwidth and QoS demanded for a service. Another task of the central RRM is to perform and request authentication between RAN2 and RAN1. The central RRM communicates with the specific RRM associated to RAN1 and requests an approved authentication for a specific user before the HO sequence. To realize this process, initially, the monitoring information is transferred to the central RRM entity either from the RRM server (associated to RAN2) or directly from the AP (RAN1). If congestion is detected, the central RRM entity triggers an alarm indicating the status and user information. This information is filtered and based on the result, the central RRM executes RRM-g techniques to initiate user-triggered intersystem handover.

The UT is platform independent; the only limitations are the hardware capabilities and the number of simultaneously active connections. The capabilities of the UT are divided into two sections. The UT will be able to request a handover into another RAN2 mode or to RAN1 (legacy network), so it needs to be able to communicate with both the specific and central RRMs. This communication will be done by sending special messages in a structured XML format. At the same time the UT is able to receive ACK messages from the specific and central RRM entities, which again are XML structured information providing an approval of the request and a summary of information of the handover command. A very important functionality of the UT is the new service request.

15.6 CONCLUSIONS

This chapter discussed the main challenges related to cooperation of heterogeneous RANs and technologies. Cooperation among RANs requires optimized handling of radio resources. Heterogeneous network management can be achieved by implementing cooperative RRM algorithms implemented in a distributed or a centralized cooperation architecture. The RRM algorithms include mechanisms related to mobility management, admission control, congestion and load control, and QoS. An approach combining both types of architectures can help overcome problems related to network scalability. In a heterogeneous mobile environment, it is difficult to rely on exact mathematical descriptions or on prior knowledge for all of the processes and interactions in each system. In fact, for some of these, it is impossible from the practical point of view to describe mathematically or predict a relationship. Further, a large portion of the information related to the system performance and radio resources allocation is to be found in the users and applications of that system. In other words, data from the edge of the system must be combined with data from other parts of the system to understand the complete sequence of events. Cooperation, therefore, can further be optimized by introducing computational intelligence at various levels of the communication process. Such an approach can increase the service quality through user and application context awareness, while allowing for more efficient use of network resources. A new intelligent monitoring scheme was proposed for network and radio resource management in heterogeneous environments. A self-learning capability from monitoring and from outcomes of the autonomic systems own decisions can elevate network management into a "cognitive" state. The heterogeneous environment and each RAN can be optimized according to an automated continuous improvement paradigm, whereas the current practice is to deploy a quite static resource allocation and to tolerate traffic variations through a certain level of overdimensioning.

A cooperative approach toward heterogeneous RAN management has a number of benefits. For example, by exploiting resources of multiple access networks, traffic requests from congested areas can be accommodated and the network will increase its resilience to sudden peaks of overload traffic. Services that cannot be satisfactorily provided (for example, because of resource unavailability) through the primary wireless network of given subscribers can be satisfied by a different network, thus leading to increased subscriber satisfaction. Finally, resources that would be otherwise left unused may be allocated to service provision requests made in other networks, thus improving resource management.

Open research issues are related to exploring further the advantages of introducing computational intelligence and cognitive radio for cooperative RRM and the optimal architecture implementing it. The expected output is an improved network performance in terms of maximized throughput and secure accessibility to services, minimized transport delay, improved network coverage, simplified security management, and satisfied users. Computational intelligence can be applied for improved handover, admission and congestion control, and QoS management.

REFERENCES

[1] IST Project WINNER, at http://www.ist-winner.org/
[2] International Telecommunications Union (ITU), ITU-T Recommendation Y.1540, at www.itu.int.
[3] ETSI White Paper No. 1, Security for ICT—the Work of ETSI, at http://etsi.org/
[4] WLAN, at http://en.wikipedia.org/wiki/Wireless_LAN

[5] IEEE 802.11, The working group setting the standards for wireless LANs, at http://www.ieee802.org/11

[6] HIPERLAN standards, at http:// www.etsi.org/

[7] UMTS, at http://www.umts-forum.org

[8] 3GPP, at http://www.3gpp.org

[9] IST project WINNER, Concept Group WA, D6.13.1, Intermediate concept proposal and evaluation, at http://www.ist-winner.org

[10] International Telecommunications Union, ITU, at www.itu.int

[11] IST-2000-25133 Project ARROWS, Advanced radio resource management for wireless services, at www.cordis.lu/ist/ka4

[12] IST Project CAUTION++, Capacity and network management platform for increased utilization of wireless systems of next generation++, at www.cordis.lu/ist/ka4

[13] S. Giordano, et al., Advanced QoS provisioning in IP networks: The European premium IP projects, *IEEE Communications Magazine*, 41(1), January 2003.

[14] A. Mihovska, S. Kyriazakos, E. Gkroutsiotis, and J. M. Pereira, QoS management in a heterogeneous environment, *Proceedings of WPMC 05*, Aalborg, Denmark, September 2005.

[15] S. Kyriazakos, A. Mihovska, E. Tragos, G. T. Karetsos, E. Gkroustiotis, and E. Mino, WINNER emulator for cooperative RRM, IST-2003-507581 WINNER, at www.ist-winner.org

[16] A. Villemont and T. Ingrain, Implementation of a QoS algorithm for the efficient cooperation of heterogeneous RANs, MSc thesis, Aalborg University, Denmark, June 2006.

[17] A. Mihovska, H. Laitinen, and P. Eggers, Location and time aware multi-system mobile network, *Proceedings of Mobile Location Workshop'03*, Aalborg, Denmark, May 2003.

[18] E. Mino, et al., D4.4, Impact of cooperation schemes between RANs-A final study, Deliverable 4.4 IST Project WINNER, November 2005.

[19] K. I. Pedersen and P. Mogensen, Directional power-based admission control for WCDMA systems using beamforming antenna array systems, *IEEE Transactions on Vehicular Technology*, 51(6), 1294–1303, November 2002.

[20] J. Korhonen, *Intoduction to 3G Mobile Communications*, Norwood, MA: Artech House, 2001.

[21] S. L. Ramirez, et al., Performance evaluation of policy-based admission control algorithms for a joint radio resource management environment, Proceedings of IEEE MELECON, Malaga, Spain, May 16–19, 2006.

[22] S. Kyriazakos and G. Karetsos, *Practical Radio Resource Management in Wireless Systems*, Norwood, MA: Artech House, 2004.

[23] G. Razzano, L. Guipponi, and R. Cusani, Wireless local area network: An effective call admission control algorithm to support quality of service, IST Project WIND-FLEX, 2003.

[24] Release 99, at http://www.3gpp.org/Releases/3GPP_R99-contents.doc.

[25] IST Project SCOUT, at http://www.cordis.lu/ist/ct/neweve/pubs-reports.htm

[26] 3GPP Specifications—Release contents and functionality, at http:// www.3gpp.org/specs/releases-contents.htm

[27] WiMAx Forum, at http://www.wimaxforum.org/technology/documents/

[28] IEEE 802.20 Mobile Broadband Wireless Access (MBWA), at http://www.ieee802.org/20/

[29] 3GPP TR 23.882, 3GPP System architecture evolution (SAE): Report on technical options and conclusions, at http://www.3gpp.org/ftp/Specs/html-info/23882.htm

[30] IST-2003-507581 WINNER D4.3, Identification, definition and assessment of cooperation schemes between RANs, June 2005, at http://www.ist-winner.org

[31] E. Mino, et al., Scalable and hybrid RRM for future wireless networks, in *Proceedings of IST Mobile Submit 2007*, Budapest, Hungary, July 2007.

[32] S. Kyriazakos and G. Karetsos, *Practical Radio Resource Management in Wireless Systems,* Norwood, MA: Artech House, 2004.

[33] K. Akabane, et al., An autonomous adaptive base station that supports multiple wireless network systems, *Proceedings of the Second International IEEE Symposium on DySPAN*, April 2007, pp. 85–88.

[34] K. Dimou, Enhancements of packet access for WCDMA, PhD thesis, AAU, January 2004.

[35] K. Chang and Y. Han, QoS adaptive scheduling for a mixed service in HDR system, *Proceedings of PIMRC'02*, Portugal, September 2002.

[36] Ch. Mensing, et al., Location determination using in-band signaling for mobility management in future networks, *IEEE PIMRC'07*, Athens, Greece, September 2007.

[37] M. Lott, et al., Cooperation of 4G radio networks with legacy systems, *Proceedings of IST Mobile Summit 2005*, Dresden, Germany, June 2005.

[38] International Telecommunications Union (ITU), RECOMMENDATION ITU-R M.1645, Framework and overall objectives of the future development of IMT 2000 and systems beyond IMT 2000, at www.itu.int

[39] S. Goebbels, M. Siebert, M. Schinnenburg, and M. Lott, Simulative evaluation of location aided handover in wireless heterogeneous systems, *15th International Symposium on Personal, Indoor and Mobile Radio Communications (PIMRC2004)*, Barcelona, Spain, September 5–8, 2004.

[40] K. Navaie and H. Yanikomeruglu, Optimal downlink resource allocation for non-real time traffic in cellular CDMA/TDMA networks, *IEEE Communications Letter*, 10(4), 278–280, April 2006.

[41] IST Project WINNER, Deliverable 3.4.1, The WINNER II air interface: refined spatial-temporal processing solutions, December 2006, at http://www.ist-winner.org

[42] A. Mihovska, et al., A novel flexible technology for intelligent base station architecture support for 4G systems, *Proceedings of WPMC'02*, Honolulu, Hawaii, October 2002.

[43] A. Mihovska, S. Kyriazakos, and E. Gkroustiotis, QoS management in heterogeneous environments, *WPMC'05*, Aalborg, Denmark, September 19–22, 2005.

[44] IST Project WINNER, Deliverable 4.2 D4.2: Impact of cooperation schemes, IST-2003-507581 WINNER, February 2005.

[45] A. Mihovska, et al., Policy-based mobility management for heterogeneous networks, IST Mobile Summit 2007, Budapest, Hungary, July 2007.

[46] A. Mihovska, et al., Requirements and algorithms for cooperation of heterogeneous access networks, in *Springer International Journal of Wireless Personal Communications*, online publication, August 2008.

[47] A. Mihovska, et al., Cognitive ubiquitous mobile communications, 2nd CTIF Workshop, Aalborg, Denmark, May 2007.

[48] P. Karamolegkos, et al., A methodology for user requirements definition in the wireless world, *Proceedings of IST Mobile Summit 2006*, Mykonos, Greece, June 2006.

[49] IST Project WINNER II Deliverable 6.12.4, Limited proof-of-concept of WINNER physical layer with live multimedia application, November 2007.

16 Cooperative Caching in Wireless Multimedia Sensor Networks

Nikos Dimokas, Dimitrios Katsaros, and Yannis Manolopoulos

CONTENTS

The advancement in wireless communication and electronics has enabled the development of low-cost wireless sensor networks. Especially, the production of cheap complementary metal–oxide–semiconductor (CMOS) cameras and microphones, which are able to capture rich multimedia content, have fueled a new research and development area, that of wireless multimedia sensor networks (WMSNs). WMSNs will boost the capabilities of current wireless sensor networks (WSNs), and several novel applications, like multimedia surveillance sensor networks, storage of relevant activities, and so on, will be developed. WMSNs introduce several new research challenges, mainly related to mechanisms to deliver application-level quality-of-service (QoS) (e.g., latency minimization). Such issues have almost completely been ignored in traditional WSNs, where the research focused on energy consumption minimization.

To address this goal in an environment with extreme resource constraints, with variable channel capacity and with requirements for multimedia in-network processing, the efficient and effective caching of multimedia data, exploiting the cooperation among sensor nodes is vital. This chapter describes a cooperative caching scheme particularly suitable for WMSNs. The presented scheme has been evaluated extensively in an advanced simulation environment, and it has been compared to the state-of-the-art cooperative caching algorithm for mobile ad hoc networks.

16.1 INTRODUCTION

WSNs [2,18] have emerged during the recent years due to the advances in low-power hardware design and the development of appropriate software which enabled the creation of tiny devices that are able to compute, control, and communicate with each other. A WSN consists of wirelessly interconnected devices that can interact with their environment by controlling and sensing "physical" parameters. WSNs attracted a huge interest from both the research community and the industry, which continues to grow. This growing interest can be attributed to the many new exciting applications that were born as a result of the deployment of large-scale WSNs. Such applications range from disaster relief to environment control and biodiversity mapping, to machine surveillance, intelligent building, precision agriculture, pervasive health applications, and to telematics.

The support of such a huge range of applications will be (rather) impossible for any single realization of a WSN. Nonetheless, certain common features appear, with regard to the characteristics and the required mechanisms of such systems, and the realization of these characteristics is the major challenge faced by these networks. The most significant characteristics shared by the aforementioned applications concern [18]

- Lifetime: Usually, sensor nodes rely on a battery with limited lifetime, and their replacement is not possible due to physical constraints (they lie in oceans or in hostile environments) or it is not interesting for the owner of the sensor network.
- Scalability: The architecture and protocols of sensor networks must be able to scale up (or to exploit) any number of sensors.
- Wide range of densities: The deployment of sensor nodes might not be regular and may vary significantly, depending on the application, on the time and space dimension, and so on.
- Data-centric networking: The target of a conventional communication network is to move bits from one machine to another, but the actual purpose of a sensor network is to provide information and answers, not numbers [17].

The production of cheap CMOS cameras and microphones, which can acquire rich media content from the environment, created a new wave in the evolution of WSNs. For instance, the Cyclops imaging module [30] is a light-weight imaging module that can be adapted to MICA2 (http://www.xbow.com) or MICAz sensor nodes. Thus, a new class of WSNs came to the scene, the WMSNs [1]. These sensor networks, apart from boosting the existing application of WSNs, will create new applications: (a) multimedia surveillance sensor networks, which will be composed by miniature video cameras [22], will be able to communicate, process, and store data relevant to crimes and terrorist attacks; (b) traffic avoidance and control systems will monitor car traffic and offer routing advices to prevent congestion; (c) industrial process control will be realized by WMSNs that will offer time-critical information related to imaging, temperature, pressure, and so on.

The novel applications of WMSNs challenged the scientific community because, as emphasized in Ref. [1], these applications force the researchers to rethink the computation–communication paradigm of traditional WSNs. This paradigm has mainly focused on reducing the energy consumption, targeting to prolong the longevity of the sensor network. However, the applications implemented by

WMSNs have a second goal, as important as the energy consumption, to be pursued; this goal is the delivery of application-level QoS and the mapping of this requirement to network layer metrics, such as latency. This goal has (almost) been ignored in mainstream research efforts on traditional WSNs.

The goal of Internet QoS in multimedia content delivery has been pursued in architectures such as Diffserv and Intserv, but these protocols and techniques do not face the severe constraints and hostile environment of WSNs. In particular, WMSNs are mainly characterized by

- Resource constraints: Sensor nodes are battery-, memory-, and processing-starving devices.
- Variable channel capacity: The multi-hop nature of WMSNs, which operate in a store-and-forward fashion because of the absence of base stations, implies that the capacity of each wireless link depends on the interference level among nodes, which is aggravated by the broadcasting operations.
- Multimedia in-network processing: In many applications of WMSNs, a single sensor node is not able to answer a posed question, but several sensors must collaborate to answer it. For instance, a sensor node with a camera monitoring a moving group of people cannot count their exact number and determine their direction, but it needs the collaboration of nearby sensors to cover the whole extent of the group of people. Therefore, sensor nodes are required to store rich media, for example, image and video, needed for their running applications, and also to retrieve such media from remote sensor nodes with short latency.

Under these restrictions/requirements, the goal of achieving application-level QoS in WMSNs becomes a challenging task. There could be several ways to attack parts of this problem, for example, channel-adaptive streaming [14] and joint source-channel coding [11]. However, none of them can provide solutions to all of the three aforementioned issues. In this chapter, the solution of cooperative caching multimedia content in sensor nodes is being inevstigated to address all three characteristics. In cooperative caching, multiple sensor nodes share and coordinate cache data to cut communication cost and exploit the aggregate cache space of cooperating sensors.

Because the battery lifetime can be extended if someone managed to reduce the "amount" of communication, caching the useful data for each sensor either in its local store or in the near neighborhood can prolong the network lifetime. Additionally, caching can be very effective in reducing the need for network-wide transmissions, thus reducing interference and overcoming the variable channel conditions. Finally, it can speed-up the multimedia in-network processing, because, as emphasized in Ref. [1], the processing and delivery of multimedia content are not independent and their interaction has a major impact on the levels of QoS that can be delivered.

This chapter investigates the technique of caching in the context of WMSNs. The need for effective and intelligent caching policies in sensor networks has been pointed out several times [10,25] in the literature, but no appropriate sophisticated policies have been proposed, although there are many caching protocols in other fields (see relevant work in Section 16.2). This chapter also presents a novel and high-performance cooperative caching protocol, the NICoCa protocol which is the abbreviation of Node Importance-based Cooperative Caching, and the comparison with the state-of-the-art cooperative caching policy for mobile ad hoc networks (MANETs), which is the "closer" competitor. An experimental evaluation of the two methods is presented at the end of the chapter.

The rest of this chapter is organized as follows: in Section 16.2 the solutions proposed so far in the area of cooperative caching in WSNs and MANETs are described and the benefits achieved in terms of communication cost, query latency, energy dissipation, and network lifetime prolongation are investigated. The details of the NICoCa protocol are discussed in Section 16.3, and the results of the performance evaluation of the methods are presented in Section 16.4; finally, Section 16.5 concludes the chapter.

16.2 RELEVANT WORK

16.2.1 CACHING ON THE WEB

The technique of caching has been widely investigated in the context of operating systems and databases and is still an attractive research area [24]. Similarly, caching on the Web has been thoroughly investigated for cooperative [12] and for noncooperative [19,26] architectures. Wessels and Claffy [39] introduced the Internet cache protocol (ICP) to support communication between caching proxies by using message exchange. Cache digests [31] and summary cache [12] enable proxies to exchange information about cached content. In Ref. [6], a cooperative hierarchical Web caching architecture was studied. However, the problem addressed in wireless networks is different from that in wired networks. The above architectures and protocols usually assume a fixed network topology and require powerful computation and communication capabilities. Due to the constrained resources (that is, bandwidth, battery power, and computing capacity) of sensor nodes, theses techniques do not adapt well to the wireless sensor network.

16.2.2 CACHING IN WIRELESS CELLULAR NETWORKS

In the context of wireless broadcast cellular networks, a number of caching approaches have been proposed [20,38]. These policies assume more powerful nodes than the sensor nodes and one-hop communication with resource-rich base stations, which serve the needed data.

16.2.3 CACHING IN MANETS AND WSNS

A number of data replication schemes [15,16] and caching schemes [32,36,40] have been proposed to facilitate data access in MANETs. Data replication studies the issue of allocating replicas of data items to meet access demands. These techniques normally require a priori knowledge of the network topology.

Caching schemes, however, do not facilitate data access based on the knowledge of distributed data items. In SimpleCache [40] the requested data item has always been cached by the requester node. The node uses the cached copy to serve subsequent requests when they arrive. The requester node has to get the data from the data center in case of cache miss. However, increasing the hop distance between the requester node and caching node increases the response time for the request.

In the research area of MANETs a number of caching protocols have been developed. The proposed caching protocols exploit the cooperation between mobile caches to decrease query latency and energy dissipation. The main motive for the development of these protocols is the mobility of the nodes, and thus they all strive to model it or exploit it. A cooperative caching scheme, called CoCa, was proposed in Ref. [7,8]. The CoCa framework facilitates mobile nodes in sharing their cached contents with each other to reduce the number of server requests and the number of access misses in a single-hop wireless mobile network. The authors extended CoCa with a group-based cooperative caching scheme, called GroCoCa, in Ref. [9]. According to GroCoCa, the decision of whether a data item should be cached depends on two factors of the access affinity on the data items and the mobility of each node. The mobile support station performs an incremental clustering algorithm to cluster the mobile nodes into tightly coupled groups based on their mobility patterns. In GroCoCa also the similarity of access patterns is captured by frequency-based similarity measurement. GroCoCa improves system performance at the cost of extra power consumption. Papadopouli and Schulzrinne [27] suggested the 7DS architecture. The authors deployed a couple of protocols to facilitate sharing and dissemination of information among users. It operates on two modes. The first one is a prefetch mode, based on the information and user's future needs and the second one is an on-demand mode, which searches for data items on a one-hop multicast basis. Depending on the collaborative behavior, a peer-to-peer (P2P) and server-to-client mode are used. This strategy focuses also on single-hop wireless environment and on data dissemination. Sailhan and Issarny [33] proposed a collaborative cache management strategy among mobile terminals interacting via an ad hoc network. The issue that the authors addressed was setting an ad hoc network of mobile terminals that cooperate to exchange Web pages. The proposed solution aims at improving the Web latency on mobile terminals while optimizing associated energy consumption. It is implemented on top of zone routing protocol (ZRP). The authors proposed a fixed broadcast range based on the underlying routing protocol.

The zone cooperative (ZC) [5], the cluster cooperative (CC) [4], and the energy efficient cooperative caching with optimal radius (ECOR) [34] protocols attempt to form clusters of nodes based either on geographical proximity or using widely known node clustering algorithms for MANETs [3]. In ZC, mobile nodes belonging to the neighborhood (zone) of a given node form a cooperative cache system for this node because the cost for communication with them is low both in terms of energy consumption and message exchanges. Each node has a cache to store the frequently accessed data items. The data items in the cache satisfy not only the node's own requests, but also the data requests passing through it from other nodes. For a data miss in the local cache, the node first searches the data in its zone before forwarding the request to the next node that lies on a path toward the data center. As a part of cache management, a value-based replacement policy based on popularity, distance, size, and time-to-live was developed to improve the data accessibility and reduce the local cache miss ratio. Simulation experiments revealed improvements in cache hit ratio and average query latency in comparison with other caching strategies. In CC, the authors present a scheme for caching in MANETs. The goal of CC is to reduce the cache discovery overhead and provide better cooperative caching performance. The authors partition the whole MANET into equal size clusters based on the geographical network proximity. In each cluster, CC dynamically chooses a "super" node as cache state node to maintain the cluster cache state information of different nodes within its cluster domain. The cache state node is defined as the first node that enters the cluster. The cluster cache state for a node is the list of cached data items along with their time-to-live field. The cache replacement policy is similar to that in ZC. In ECOR, each mobile node forms a cooperation zone (CZ) with mobile nodes in proximity by exchanging messages to share their cached data items to minimize bandwidth and energy cost for each data retrieval. When a data request arrives, the node first searches the data in its CZ before forwarding the request to the data center. The authors developed an analytical model to determine the optimal radius of the CZ based on the mobile node's location, data popularity, and node density. According to ECOR each node broadcasts every modification of the cached data items to nodes that belong to CZ. Each node maintains a cache hint table for the cache information of all nodes in its proximity.

The only protocols that tried to exploit both data and node locality in a homogeneous manner are described in Ref. [40] and are the following: CachePath, CacheData, and HybridCache. In CacheData, intermediate nodes may cache data to serve future requests instead of fetching data from the "Data Center." An intermediate node caches passing by data item locally when it finds that data item is popular and does not cache the data item if all requests for it are from the same node. This rule is designed to reduce the cache space requirement because the mobile nodes have limited cache spaces. In CachePath, a mobile node may cache the information of a path to a nearby data requester while forwarding the data and use the path information to redirect future requests to the nearby caching site. By caching the data path for each data item, the bandwidth and the query delay can be reduced because the requested data can be obtained through fewer number of hops. The authors also proposed some optimization techniques. An intermediate node can save only the destination node information because the path from the current router to the destination node can obtained by the underlying routing protocol. Also, the intermediate node needs to record the data path when it is closer to the caching node than the data center. One major drawback of CachePath according to the authors is that the cached path may not be reliable and using it may increase the overhead. A cached path may not be reliable because either the data item has become obsolete or the caching node cannot be reached. The hybrid protocol HybridCache combines CacheData and CachePath while avoiding their weaknesses. In HybridCache, when a mobile node forwards a data item, it caches the data or the path based on some criteria. These criteria include the data item size, the time-to-leave of the data item, and the number of hops that a cached path can save, denoted as H_{save}. H_{save} value is the difference between the distance to the data center and the distance to the caching node. H_{save} must be greater than a system tuning threshold. However, according to these methods the caching information of a node cannot be shared if the node does not lie on the path between the data requester and the data source. Moreover, the threshold values used in these heuristics must be set carefully to achieve good performance.

The only studies on caching in wireless sensor networks concern the placement of caches [29,37] and thus they are not examined in this chapter. Remotely related to the topic of this chapter are the caching policies for MANET routing protocols [41].

16.2.4   Why a New Caching Protocol?

The protocols proposed so far for cooperative caching in MANETs present various limitations. Those protocols, which first perform a clustering of the network and then exploit this clustering (the cluster-heads, [CH]), to coordinate the caching decisions, inherit the shortcomings of any CH selection. For instance, in Refs. [4,5], the nodes that form the cluster are assumed to reside within the same communication range, that is, they are with on-hop distance from the other nodes of the cluster. Additionally, the nodes do not cache the data originating from a one-hop neighbor. Thus, CHs that do not reside in a significant part of data routes cannot serve their cluster members efficiently, because they do not have fast access (short latency) to requested data. The CZ, which is formed in Ref. [34] by selecting an optimal radius, implies a large communication overhead, because every node within that radius must send/receive any changes to the caches of the other nodes within the radius. Finally, the HybridCache policy is tightly coupled to the underlying routing protocol, and thus if a node does not reside in the route selected by the routing protocol, it cannot cache the data/path, or conversely, cannot serve the request even if it holds the requested data.

This chapter presents a cooperative caching policy that takes into account the unique requirements of the WMSNs, which are mainly static and not mobile, and tries to avoid the shortcomings of the current cooperative protocols. The cooperative caching policy is based on the idea of exploiting the sensor network topology, so as to discover which nodes are more important than the others, in terms of their position in the network or in terms of residual energy. Incorporating both factors into the design of the caching policy, the authors ensure both network longevity and short latency in multimedia data retrieval. In summary, the core features that are presented in this chapter are the following:

- Definition of a metric for estimating the importance of a sensor node in the network topology, which will imply short latency in retrieval.
- Description of a cooperative caching protocol, which takes into account the residual energy of the sensor nodes.
- Development of algorithms for discovering the requested multimedia data and maintaining the caches (cache replacement policy).
- Performance evaluation of the protocol and comparison with the state-of-the-art cooperative caching protocol for MANETs, using an established simulation package (J-Sim).

16.3   NICoCa PROTOCOL FOR WMSNs

One of the main parts of the proposed protocol is the estimation of the importance of sensors relative to the network topology and the cooperation among network nodes achieved through them. The intuition is that if someone discovers those nodes that reside in a significant part of the (short) paths connecting other nodes, then these are the "important" nodes; then they may be selected as coordinators for the caching decisions, that is, as "mediators" to provide information about accessing the requested data or even as caching points. The mediator nodes constitute the main point of the cooperative caching protocol. The cooperation among neighboring nodes and their energy balancing are managed by mediator nodes. The mediator nodes include indices about the cached data in the neighborhood, the remaining energy, and the free cache space of its neighboring node. Thus, whenever a data request reaches a mediator node, it decides which neighboring node with the highest remaining energy could answer it. Mediator nodes also take place during the cache replacement phase that happens in a neighboring node. The cooperation between the node that replaces a cached data and the mediators takes place not to permanently evict the data item from the neighborhood. According to the free cache space and the cached data items that neighboring nodes have, mediator nodes decide where to place the replaced cache data. If the replaced data item exists somewhere else in neighborhood, then it is simply deleted.

16.3.1 MEASURING SENSOR NODE IMPORTANCE

A WMSN is abstracted as a graph $G(V, E)$, where V is the set of its nodes and E is the set of radio connections between the nodes. An edge $e = (u, v)$, $u, v \in E$ exists if and only if u is in the transmission range of v and vice versa. All links in the graph are bidirectional, that is, if u is in the transmission range of v, v is also in the transmission range of u. The network is assumed to be in a connected state. The set of neighbors of a node v is represented by $N_1(v)$, that is, $N_1(v) = \{u : (v, u) \in E\}$. The set of two-hop nodes of node v, that is, the nodes that are the neighbors of node v's neighbors except for the nodes that are the neighbors of node v, is represented by $N_2(v)$, that is, $N_2(v) = \{w : (u, w) \in E$, where $w \neq v, w \notin N_1$, and $(v, u) \in E\}$. The combined set of one-hop and two-hop neighbors of v is denoted as $N_{12}(v)$. Definition [Local network view of node v]. The local network view, denoted as LN_v, of a graph $G(V, E)$ w.r.t. a node $v \in V$ is the induced subgraph of G associated with the set of vertices in $N_{12}(v)$.

A path from $u \in V$ to $w \in V$ has the common meaning of an alternating sequence of vertices and edges, beginning with u and ending with w. The length of a path is the number of intervening edges. The distance between u and w is denoted as $d_G(u, w)$, that is, the minimum length of any path connecting u and w in G, where by definition $d_G(v, v) = 0$, $\forall v \in V$, and $d_G(u, w) = d_G(w, u)$, $\forall u, w \in V$. The distance is not related to network link costs (for example, latency), but it is a purely abstract metric measuring the number of hops.

Let $\sigma_{uw} = \sigma_{wu}$ denote the number of shortest paths from $u \in V$ to $w \in V$ (by definition, $\sigma_{uu} = 0$). Let $\sigma_{uw}(v)$ denote the number of shortest paths from u to w that some vertex $v \in V$ lies on. Then, the node importance index $\mathcal{NI}(v)$ of a vertex v is defined as

$$\mathcal{NI}(v) = \sum_{u \neq v \neq w \in V} \frac{\sigma_{uw}(v)}{\sigma_{uw}}. \tag{16.1}$$

Large values for the $\mathcal{NI}$ index of a node v indicate that this node v can reach others on relatively short paths or that the node v lies on considerable fractions of shortest paths connecting others. Illustration of this metric is presented in Figure 16.1 (with well-formed and vague node clusters).

A very informative picture of which nodes reside in a large number of shortest paths between other nodes is obtained when estimating the $\mathcal{NI}$ index for each sensor node using the whole network topology. The picture about the relative importance of the nodes also remains very accurate, even when the $\mathcal{NI}$ indexes of the nodes is calculated taking into account only their k-hop ($k = 2$ or 3) neighborhood. For $k = 1$ the $\mathcal{NI}$ index of a sensor node is equivalent to its degree. For more information concerning the calculation and the use of this metric in broadcasting protocols, the interested user can consult the work reported in Ref. [21].

16.3.2 HOUSEKEEPING INFORMATION IN THE NICOCA PROTOCOL

Without lost of generality and adopting the model presented in [40], the cooperative caching protocol assumes that the ultimate source of multimedia data is a Data Center. This is not restrictive at all and simply guarantees that every request, if it is not served by other sensor nodes and if it does not expire, will finally be served by the Data Center.

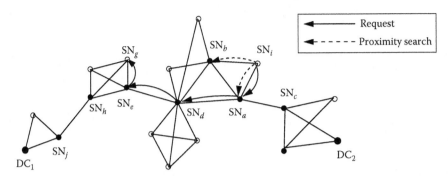

FIGURE 16.1 Calculation of $\mathcal{NI}$ for two sample graphs. The numbers in parentheses denote the $\mathcal{NI}$ index of the respective node considering the whole WMSN topology.

First, it is assumed that each node is aware of its 2-hop neighborhood. This information is obtained through periodic exchange of "beacon" messages. It is also considered an assignment of time slots to the sensor nodes such that no interference occurs, that is, no two nodes transmit in the same time slot. Such a scheme can be found using the D2-coloring algorithm from Ref. [13]. Then, every node calculates the $\mathcal{NI}$ index of its 1-hop neighbors. The node uses this information to characterize some of its neighbors as mediator nodes; the minimum set of neighbors with the larger $\mathcal{NI}$ that "cover" its 2-hop neighborhood are the mediator nodes for that node; The node is responsible for notifying its neighbors about which of them are its mediators. Thus, a node can be either a mediator or an ordinary node.

The sending of requests for data is carried out by an ordinary sensor (or ad hoc) routing protocol, for example, ad hoc on-demand distance vector routing (AODV). A node always caches a datum which has been requested for. A node is aware of its remaining energy and of the free space in its cache. Each sensor node stores the following metadata related to a cached multimedia item:

- The dataID, and the actual multimedia data item.
- The data size (s_i).
- A TTL interval (time-to-live).
- For each cached item, the timestamps of the K most recent accesses to that item. Usually, $K = 2$ or 3.
- Each cached item is characterized either as O (that is, own) or H (that is, hosted). If an H-item is requested by the caching node, then its state switches to O.

When a node acquires the multimedia datum it has requested for, it caches it and broadcasts a small index packet containing the dataID and the associated TTL, its remaining energy, and its free cache space. The mediator nodes, which are also one-hop neighbors of this node, store this broadcasted information. This set of mediator nodes includes the mediators that the broadcasting node has selected and also any other mediators that have been selected by nearby nodes. In summary, every mediator node stores the remaining energy and the free cache space for each one of its one-hop neighbors, and for each dataID that has been heard through the broadcasting operation, the TTL of this datum and the nodes that have cached this datum.

16.3.3 COOPERATIVE CACHE DISCOVERY COMPONENT PROTOCOL

When a sensor node issues a request for a multimedia item, it searches its local cache. If the item is found there (a local cache hit), then the K most recent access timestamps are updated. Otherwise (a local cache miss), the request is broadcasted and received by the mediators. If none of them responds (a "proximity" cache miss), then the request is directed to the Data Center.

When a non-one-hop mediator node receives a request, it searches its local cache. If it deduces that the request can be satisfied by a neighboring node (a remote cache hit), then it stops the request's route toward the Data Center and forwards the request to this neighboring node. If more than one nodes can satisfy the request, then the node with the largest residual energy is selected. This happens to achieve energy balancing among network nodes. If the request cannot be satisfied by this mediator node, then it does not forward it recursively to its own mediators. This is due to the fact that these mediators will most probably be selected by the routing protocol as well (AODV) and thus a great deal of savings in messages is achieved. Therefore, during the procedure of forwarding a request toward the Data Center, no searching of other nodes is performed apart from the nodes that reside on the path toward the Data Center.

Based on the above idea, we describe the cache discovery protocol and the cooperation among nodes, that is taking place, to determine the path to the sensor node with the requested item or to the data center. For example, suppose that sensor node SN_i in Figure 16.2 issues a request for data item x that is placed in data center DC_1 and has been cached by sensor nodes SN_g and SN_h. The shaded nodes are considered to be the mediator nodes of the sensor network. In the beginning sensor node SN_i searches its own cache. If it deduces that data item is not available in local cache, it sends a proximity search request in neighboring mediator nodes SN_a and SN_b. Upon receiving the search request, each mediator searches in the proximity cache table. If data item is found, each mediator replies with an index packet that contains the dataID and the remaining energy of the sensor node that has the uppermost battery power and has cached the data item. SN_i, upon the receipt of index packets, selects the sensor node that has the smallest energy consumption and sends the request packet. The caching node responds with a reply packet containing the requested data item.

If no neighboring sensor node caches the data item, SN_i sends a request packet to the data center DC_1, as shown in Figure 16.2. When SN_x ($x \in \{d, e\}$) receives a request packet, it searches in local cache and in the proximity cache table. If the data item is not found, SN_x forwards the request through the path to the DC_1. When sensor node SN_e finds that the requested data item has been cached by some neighboring nodes, it chooses the node that has the smallest energy dissipation and redirects the request packet to the caching node. The caching node sends a reply packet containing the data item x along the routing path until it reaches the original requester. Once the requester node receives the data item, it notifies its one-hop mediators about the new caching item by sending an index packet containing the item's dataID. In case of not enough cache capacity, it triggers the cache replacement protocol to determine the data items that should be evicted from the cache.

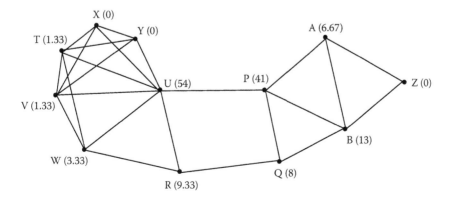

FIGURE 16.2 A request packet from sensor node SN_i is forwarded to the caching node SN_g.

For every issued request one of the following four cases may take place:

1. Local hit (LH): The requested datum is cached by the node that issued the request. If this datum is valid (the TTL has not expired) then the NICoCa is not executed.
2. "Proximity" hit (PH): The requested datum is cached by a node in the two-hop neighborhood of the node that issued the request. In this case, the mediator(s) return to the requesting node the "location" of the node that stores the datum.
3. Remote hit (RH): The requested datum is cached by a node and this node has at least one mediator residing along the path from the requesting node to the Data Center.
4. Global hit (GH): The requested datum is acquired from the Data Center.

16.3.4 CACHE REPLACEMENT COMPONENT PROTOCOL

Even though the cache capacity of individual sensors may be in the order of gigabytes (for example, NAND flash) the development of an effective and intelligent replacement policy is mandatory to cope with the overwhelming size of multimedia data generated in WMSNs. The NICoCa protocol employs the following four-step policy:

STEP 1. In case of necessity, before purging any other data from cache, each sensor node first purges the data that it has cached on behalf of some other node. Each cached item is characterized either as O (that is, own) or H (that is, hosted). In case of a local hit, then its state switches to O. If the available cache space is still smaller than the required, execute Step 2.

STEP 2. Calculate the following function for each cached datum i: $\text{cost}(i) = \frac{s_i}{\text{TTL}_i} * \frac{\text{now} - t_{K-\text{th access}}}{K}$. The candidate cache victim is the item that incurs the largest cost.

STEP 3. Inform the mediators about the candidate victim. If it is cached by some mediator, then this information returns to the node and purges the datum. If the datum is not cached by some mediator(s), then it is forwarded to the node with the largest residual energy and the datum is purged from the cache of the original node. In any case, the mediators update their cached metadata about the new state.

STEP 4. The node that caches this purged datum informs the mediators with the usual broadcasting procedure and the cached item is characterized as H (that is, hosted).

The pseudocode for the complete algorithm NICoCa is presented in the Appendix.

16.4 PERFORMANCE EVALUATION

The performance of the NICoCa protocol has been evaluated through simulation experiments. A large number of experiments with various parameters, and a comparison between the performance of NICoCa with the state-of-the-art cooperative caching policy for MANETs, namely HybridCache [40], are presented in this chapter.

16.4.1 SIMULATION MODEL

A simulation model based on the J-Sim simulator [35] has been developed. In the simulations, the AODV [28] routing protocol is deployed to route the data traffic in the wireless sensor network. Also, IEEE 802.11 has been chosen as the MAC protocol and the free space model as the radio propagation model. The wireless bandwidth was 2 Mbps.

The protocols have been tested for a variety of sensor network topologies, to simulate sensor networks with varying levels of node degree, from 4 to 10. Experiments have also been conducted by choosing the number of nodes between 100 and 1000. In addition, the experiments evaluate the protocols' efficiency under two different set of data item sizes. Each data item has size that is uniformly distributed from 1 to 10 KB for the first set, and from 1 to 5 MB for the second.

The network topology consists of many square grid units where one or more nodes are placed. The number of square grid units depends on the number of nodes and the node degree. The topologies are generated as follows: the location of each of the n sensor nodes is uniformly distributed between the point ($x = 0$, $y = 0$) and the point ($x = 500$, $y = 500$). The average degree d is computed by sorting all $n * (n - 1)/2$ edges in the network by their length, in increasing order. The grid unit size corresponding to the value of d is equal to $\sqrt{2}$ times the length of the edge at position $n * d/2$ in the sorted sequence. Two sensor nodes are neighbors if they are placed in the same grid or in adjacent grids. The simulation area is assumed to be of size 500×500 m and is divided into equal-sized square grid units. Beginning with the lower grid unit, the units are named as $1, 2, \ldots$ in a column-wise fashion.

The client query model is similar to those used in previous studies [40]. Each sensor node generates read-only queries. After a query is sent out, if the sensor node does not receive the data item, it waits for an interval (t_w) before sending a new query. The access pattern of sensor nodes is (a) location independent, that is, sensor nodes decide independently the data of interest; each sensor node generates accesses to the data following the uniform distribution, and (b) Zipfian with $\theta = 0.8$, where groups of nodes residing in neighboring grids (25 grids with size 100×100 m) have the same access pattern. The protocols have been tested both for Zipfian access pattern and for uniform access pattern. For the case of the Zipfian access, the experiments were conducted with varying θ values between 0.0 and 1.0.

Similar to [40], two data centers are placed at opposite corners of the simulation area. Data Center 1 is placed at point ($x = 0$, $y = 0$) and Data Center 2 is placed at point ($x = 500$, $y = 500$). There are $N/2$ data items in each data center. Data items with even ids are stored at Data Center 1 and data items with odd ids are stored at Data Center 2. The size of each data item is uniformly distributed between S_{min} and S_{max}. The data items are considered static, that is, not updated. The data centers serve the queries on an FCFS (first-come-first-served) basis. The system parameters are listed in Table 16.1.

TABLE 16.1
Simulation Parameters

Parameter	Default Value	Range
# items (N)	1000	
S_{min} (KB)	1	
S_{max} (KB)	10	
S_{min} (MB)	1	
S_{max} (MB)	5	
# requests per node	250	200–300
# nodes (n)	500	100–1000
Bandwidth (Mbps)	2	
Waiting interval (t_w)	10 s for items with KB size	
	100 s for items with MB size	
Client cache size (KB)	800	200–1200
Client cache size (MB)	125	25–250
Zipfian skewness (θ)	0.8	0.0–1.0

16.4.2 PERFORMANCE METRICS

The measured quantities include the number of hits (local, remote, and global), the average latency for getting the requested data, and the message overhead. It is evident that a small number of global hits implies less network congestion and thus fewer collisions and packet drops. Moreover, a large number of remote hits proves the effectiveness of cooperation in reducing the number of global hits. A large number of local hits does not imply an effective cooperative caching policy, unless it is accompanied by a small number of global hits, because the cost of global hits takes away the benefits of local hits.

16.4.3 EVALUATION

A large number of experiments were performed by varying the size of the sensornet (in terms of the number of its sensor nodes), varying the access profile of the sensor nodes, and the cache size relative to the aggregate size of all data items. In particular, experiments were performed for 100, 500, and 1000 sensors, for cache size equal to 1, 5, and 10 percent of the aggregated size of all distinct multimedia data, for access pattern with θ equal to 0.0 (uniform access pattern) to 1.0 (highly skewed access pattern), for average sensor node degree equal to 4, 7 (very sparse and spare sensornet), and 10 (dense sensornet), and for data item size equal to a few kilobytes (KB) and also equal to a few megabytes (MB). For each different setting, the performance measure includes the number of hits (local, remote, and global), the latency, and the message overhead. (The latency is measured in seconds, which does not correspond to the usual time metric, but to an internal simulator clock.) The remote hits comprise of proximity hits and remote hits. For a HybridCache scheme, proximity hits are always zero. Proximity hits determine the number of hits that are generated when a sensor node inside the requester node's two-hop neighborhood responds to the request. In the sequel of the chapter we present only a representative set of the results, because there are many independent parameters and three dependent performance metrics. The graphs were partitioned into two large groups w.r.t. whether they deal with small KB-sized files or large MB-sized multimedia files.

16.4.3.1 Experiments with MB-Sized Data Items

The purpose of a first set of experiments is to investigate the performance of the caching algorithms when they have to deal with large multimedia files, for example, video files, queried by the sensornet.

All figures show that both schemes exhibit better average query latency, hit ratio (local, remote, and global), and message overhead when varying the cache size from 25 to 250 MB. This is because more required data items can be found in the local cache as the cache gets larger. Figures 16.3 and 16.4 show the number of hits achieved by the two protocols for a small (sparse and dense, respectively)

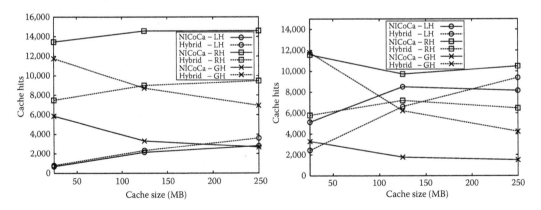

FIGURE 16.3 Impact of sensor cache size on hits (MB-sized files, $\theta = 0.0$ and $\theta = 0.8$) in a sparse WMSN ($d = 7$) with 100 sensors.

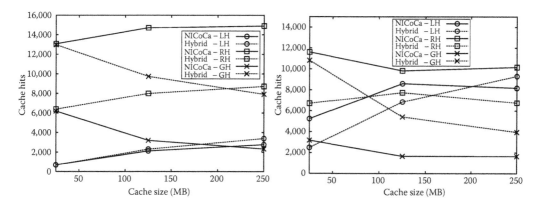

FIGURE 16.4 Impact of sensor cache size on hits (MB-sized files, $\theta = 0.0$ and $\theta = 0.8$) in a dense WMSN ($d = 10$) with 100 sensors.

sensornet for both uniform and skewed access pattern. The NICoCa scheme achieves always higher number of remote hits than HybridCache scheme because the chances of some neighboring nodes tuning in the required data item is higher due to cooperative caching. When the cache size is small, more required data can be found in local and proximity cache for NICoCa scheme compared with HybridCache scheme, which uses only the local cache, thus alleviating the need for remote and global cache access. It is worth noting that NICoCa always reaches the near optimum performance when the cache size is equal to 125 MB. This demonstrates the low cache-space requirement.

A significant observation is that, as expected, the number of local hits increases for both protocols as the access pattern becomes more skewed. The interesting point is that, although for uniform access patterns HybridCache is slightly better than NICoCa w.r.t. the local hits, the situation is reversed when the requests are concentrated to a smaller number of files, which can be attributed to the more efficient replacement and admission policy of the NICoCa. With respect to the number of global hits, NICoCa achieves half that of hybrid and the performance gap widens as we move to dense sensor deployments; actually NICoCa maintains the number of global hits almost constant. The reason behind this is the relative performance of the algorithms w.r.t. the remote cache hits. For sparser deployments NICoCa is two times better than HybridCache, and this difference becomes more evident for denser networks. Thus, it proves to be a more effective cooperation scheme, due to the fact that it strives to exploit the network topology. These relative performance results are straightforwardly reflected in the access latency incurred by the algorithms (Figures 16.5 and 16.6).

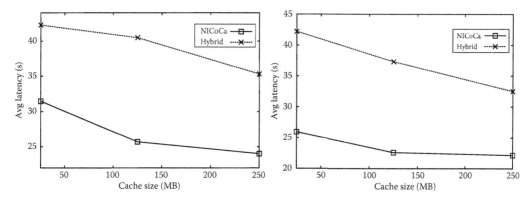

FIGURE 16.5 Impact of sensor cache size on latency (MB-sized files, $\theta = 0.0$ and $\theta = 0.8$) in a sparse WMSN ($d = 7$) with 100 sensors.

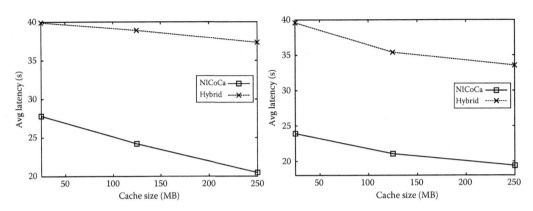

FIGURE 16.6 Impact of sensor cache size on latency (MB-sized files, $\theta = 0.0$ and $\theta = 0.8$) in a dense WMSN ($d = 10$) with 100 sensors.

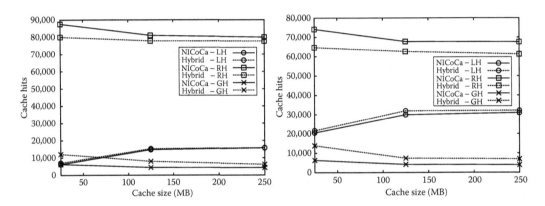

FIGURE 16.7 Impact of sensor cache size on hits (MB-sized files, $\theta = 0.0$ and $\theta = 0.8$) in a sparse WMSN ($d = 7$) with 500 sensors.

NICoCa achieves low average query latency compared with Hyrbid Cache. Nodes can access most of the required data items from local and *proximity* cache, therefore reducing the query latency. When the cache size increases, the average query latency decreases for both schemes. This happens because more data are cached near to requester nodes. In this case, NICoCa also outperforms over HybridCache. The performance of NICoCa improves as the node density increases, while the performance of HybridCache maintains the same or is slightly better in some cases. With an increase in node density, the size of each proximity region increases and thus the requested data item could be retrieved from the neighboring nodes than from data center or remote caches. This results in decreasing the average query latency.

When we move to larger sensornets with 500 (Figure 16.7) and 1000 nodes (Figure 16.10), the performance gains of the NICoCa caching algorithm in terms of hits are still evident, but the results are not so impressive, because more sensor nodes are dispersed in the same geographical region, thus creating replicas of the same data. This performance gain is reflected in the access latency as well (Figure 16.8). Moving to larger sensornets, the latency gradually increases, because the denser deployment (more nodes in the same region) has a negative effect on the efficiency of communication, aggravating the collisions and packet drops.

At this point it is interesting to note the total number of messages that are communicated between the sensor nodes, which is also the metric that models the total network energy dissipated

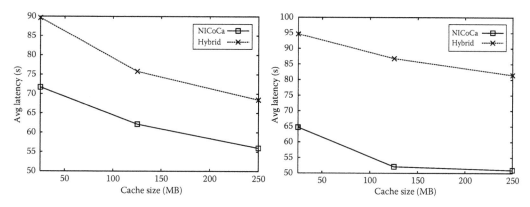

FIGURE 16.8 Impact of sensor cache size on latency (MB-sized files, $\theta = 0.0$ and $\theta = 0.8$) in a sparse WMSN ($d = 7$) with 500 sensors.

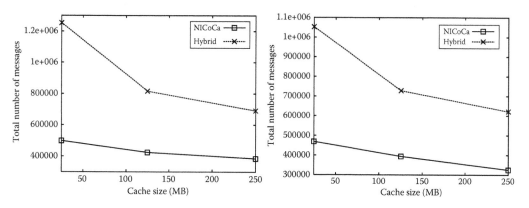

FIGURE 16.9 Impact of sensor cache size on number of messages (MB-sized files, $\theta = 0.0$ and $\theta = 0.8$) in a dense WMSN ($d = 10$) with 500 sensors.

(Figures 16.9). For a dense sensornet with uniform and skewed access pattern, NICoCa sends at most half of the messages sent out by HybridCache, and the situation becomes better for NICoCa as the access pattern becomes more skewed, which is expected. These results are confirmed for larger sensornets with 1000 nodes (Figures 16.10). The reason is that due to cache cooperation within a proximity region NICoCa gets data from nearby nodes instead of far away data center or remote caches. Therefore, the data requests and replies need to pass by a significantly smaller number of hops, and sensor nodes have to process lower number of messages.

16.4.3.2   Experiments with KB-Sized Data Items

A significant question arises whether these relative results still hold when the sensornet has to deal with smaller multimedia files, with size equal to a few. Although it is expected that WMSNs will deal with MB-sized images of video files, it might be the case that the sensor nodes will exchange smaller images as well. To investigate the performance of the cooperative caching protocols for this case, the same set of experiments was performed but for KB-sized files and here we present the results.

The general observations that were recorded for the case of large MB-size files still hold for this case; NICoCa achieves a significantly smaller number of global hits and larger number of remote

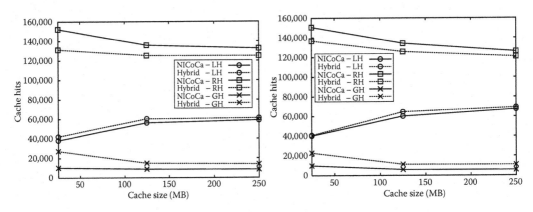

FIGURE 16.10 Impact of sensor cache size on hits (MB-sized files, $\theta = 0.8$) in a dense WMSN ($d = 10$) with 1000 sensors.

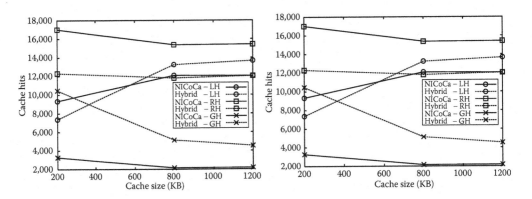

FIGURE 16.11 Impact of sensor cache size on hits (KB-sized files, $\theta = 0.8$) in a sparse and dense WMSN ($d = 7$ and $d = 10$) with 100 sensors.

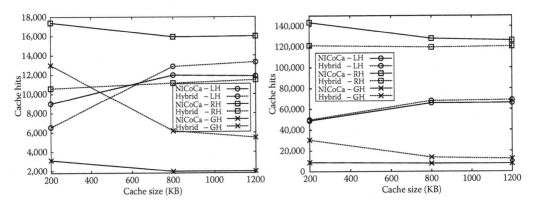

FIGURE 16.12 Impact of sensor cache size on hits (KB-sized files, $\theta = 0.8$) in a sparse and dense WMSN ($d = 7$ and $d = 10$) with 1000 sensors.

hits than HybridCache does. It is not worth commenting on each individual performance graph (Figures 16.11 and 16.12), because in all cases NICoCa has a better performance; it achieves again 25 percent more remote hits and 50 percent less global hits than HybridCache, which is only marginally better than NICoCa in terms of local hits.

16.5   CONCLUSION

The recent advances in miniaturization, the creation of low-power circuits, and the development of cheap CMOS cameras and microphones, which are able to capture rich multimedia content, gave birth to what is called WMSNs. WMSNs are expected to fuel many new applications and boost the existing ones. The unique features of WMSNs call for protocol designs that will provide application-level QoS, an issue that has largely been ignored in traditional wireless sensor networks. Taking a first step toward this goal, this chapter presented a cooperative caching protocol, the NICoCa protocol, suitable for deployment in WMSNs. The protocol "detects" which sensor nodes are most "central" in the network neighborhoods and gives to them the role of mediator to coordinate the caching decisions. The NICoCa protocol is evaluated with J-Sim and its performance is compared with that of a state-of-the-art cooperative caching protocol for MANETs. The results attest the performance gains of the NICoCa protocol, which is able to reduce the global hits at an average of 50 percent and increase the remote hits due to the effective sensor cooperation at an average of 40 percent. The performance of the protocol is particularly high for the delivery of large multimedia data.

ACKNOWLEDGMENTS

Research supported by a $\Gamma\Gamma ET$ grant in the context of the project "Data Management in Mobile Ad Hoc Networks" funded by $\Pi Y\Theta A\Gamma OPA\Sigma\ II$ national research program.

APPENDIX

The NICoCa cooperative caching protocol

// d_i: data item i, $i \in [1 \dots 1000]$
// request(d_i): Request for data item i
// N_i: Node i
// FS: Free cache space
// RE: Remaining energy
// PCT: Proximity Cache Table
// ipacket: An index packet that contains d_i's id, FS and RE

(A) Cache Discovery Algorithm
if(d_i is in local cache of requester node) **then**
    send ipacket to CHs;
    return;
if(requester node is CH and d_i's id in PCT) **then**
    select caching node with largest RE;
    send request(d_i) to caching node;
else
    requester node sends request(d_i) to CHs;
when CHs answers or time elapsed
if(caching nodes found) **then**
    select caching node with largest RE;
    send request(d_i) to caching node;
else
    send request(d_i) to data center;
when N_i receives request(d_i)
if(N_i has a valid copy) **then**
    send d_i to requester node;
else if(N_i is CH **and** d_i's id in PCT) **then**
    select caching node with largest RE;
    redirect request(d_i) to caching node;

else
    forward request(d_i) to caching node;

(B) Replacement Policy
while(current node has not enough FS)
    Select a valid d_i with largest value and store it temporary;
    Send to CHs d_i's id;
    Remove the valid d_i;
when a CH gets d_i's id
if(CH gets d_i's id and d_i's id not in PCT) **then**
    select caching node with largest RE and FS;
send answer to requester node;
when current node get answers from CHs
foreach(temporary stored d_i)
    if(there is no other caching node) **then**
        Select caching node with least RE and largest FS;
        Send d_i to new caching node;
    Remove temporary stored d_i;

(C) Cache Admission Policy
when the packet with d_i obtained from current node
if(current node is packet's destination) **then**
    if(there is enough FS) **then**
        cache d_i;
        send ipacket to CHs;
    else
        call Replacement Policy ;
when CH gets an ipacket
if(CH get ipacket) **then**
    store d_i's id, RE and FS in PCT;

REFERENCES

[1] I. Akyildiz, T. Melodia, and K. R. Chowdhury. A survey of wireless multimedia sensor networks. *Computer Networks*, 51(4):921–960, 2007.

[2] I. Akyildiz, W. Su, Y. Sankarasubramaniam, and E. Cayirci. A survey of wireless sensor networks. *IEEE Communications Magazine*, 40(8):102–116, 2002.

[3] S. Basagni, M. Mastrogiovanni, A. Panconesi, and C. Petrioli. Localized protocols for ad hoc clustering and backbone formation: A performance comparison. *IEEE Transactions on Parallel and Distributed Systems*, 17(4):292–306, 2006.

[4] N. Chand, R. C. R. C. Joshi, and M. Misra. A zone co-operation approach for efficient caching in mobile ad hoc networks. *International Journal of Communication Systems*, 19:1009–1028, 2006.

[5] N. Chand, R. C. R. C. Joshi, and M. Misra. Cooperative caching strategy in mobile ad hoc networks based on clusters. *Wireless Personal Communications*, 43(1):41–63, 2007.

[6] H. Che, Y. Tung, and Z. Wang. Hierarchical Web caching systems: Modeling, desing and experimental results. *IEEE Journal on Selected Areas in Communications*, 20(7):1305–1314, 2002.

[7] C.-Y. Chow, H. V. Leong, and A. T. S. Chan. Cache signatures for peer-to-peer cooperative caching in mobile environments. In *Proceedings of the IEEE International Conference on Advanced Information Networking and Applications (AINA)*, Vol. 1, pp. 96–101, 2004.

[8] C.-Y. Chow, H. V. Leong, and A. T. S. Chan. Peer-to-peer cooperative caching in mobile environments. In *Proceedings of the IEEE International Conference on Distributed Computing Systems Workshops (ICDCSW)*, pp. 528–533, 2004.

[9] C.-Y. Chow, H. V. Leong, and A. T. S. Chan. GroCoca: Group-based peer-to-peer cooperative caching in mobile environment. *IEEE Journal on Selected Areas in Communications*, 25(1):179–191, 2007.

[10] Y. Diao, D. Ganesan, G. Mathur, and P. Shenoy. Rethinking data management for storage-centric sensor networks. In *Proceedings of the Conference on Innovative Data Systems Research (CIDR)*, pp. 22–31, 2007.

[11] Y. Eisenberg, C. E. Luna, T. N. Pappas, R. Berry, and A. K. Katsaggelos. Joint source coding and transmission power management for energy efficient wireless video communications. *IEEE Transactions on Circuits and Systems for Video Technology*, 12(6):411–424, 2002.

[12] L. Fan, P. Cao, and A. Z. Almeida, J. M. Broder. Summary cache: A scalable wide-area web cache sharing protocol. *IEEE/ACM Transactions on Networking*, 8(3):281–293, 2000.

[13] R. Gandhi and S. Parthasarathy. Fast distributed well connected dominating sets for ad hoc networks. Technical Report CS-TR-4559, Computer Science Department, University of Maryland at College Park, 2004.

[14] B. Girod, M. Kalman, Y. J. Liang, and R. Zhang. Advances in channel-adaptive video streaming. *Wireless Communications and Mobile Computing*, 2(6):573–584, 2002.

[15] T. Hara. Replica allocation methods in ad hoc networks with data update. *ACM/Kluwer Mobile Networks and Applications*, 8(4):343–354, 2003.

[16] T. Hara and S. K. Madria. Data replication for improving data accessibility in ad hoc networks. *IEEE Transactions on Mobile Computing*, 5(11):1515–1532, 2006.

[17] G. T. Huang. Casting the wireless sensor net. *Technology Review*, pp. 51–56, July 2003.

[18] H. Karl and A. Willig. *Protocols and Architectures for Wireless Sensor Networks*. John Wiley & Sons, New York, 2006.

[19] D. Katsaros and Y. Manolopoulos. Caching in Web memory hierarchies. *Proceedings of the ACM Symposium on Applied Computing*, pp. 1109–1113, 2004.

[20] D. Katsaros and Y. Manolopoulos. Web caching in broadcast mobile wireless environments. *IEEE Internet Computing*, 8(3):37–45, 2004.

[21] D. Katsaros and Y. Manolopoulos. The geodesic broadcast scheme for wireless ad hoc networks. In *Proceedings of the IEEE International Symposium on World of Wireless, Mobile Multimedia (WoWMoM)*, pp. 571–575, 2006.

[22] P. Kulkarni, D. Ganesan, P. Shenoy, and Q. Lu. SensEye: A multi-tier camera sensor network. In *Proceedings of the ACM International Conference on Multimedia (MM)*, pp. 229–238, 2005.

[23] G. Mathur, P. Desnoyers, D. Ganesan, and P. Shenoy. Ultra-low power data storage for sensor networks. In *Proceedings of the ACM International Conference on Information Processing in Sensor Networks (IPSN)*, pp. 374–381, 2006.

[24] N. Megiddo and D. S. Modha. ARC: A self-tuning, low overhead replacement cache. In *Proceedings of the USENIX Conference on File and Storage Technologies (FAST)*, pp. 115–130, 2003.

[25] S. Nath and A. Kansal. FlashDB: Dynamic self-tuning database for NAND flash. In *Proceedings of the ACM International Conference on Information Processing in Sensor Networks (IPSN)*, pp. 410–419, 2007.

[26] G. Pallis, A. Vakali, and J. Pokorny. A clustering-based prefetching scheme on a Web cache environment. *Computers & Electrical Engineering*, 34(4):309–323, 2008.

[27] M. Papadopouli and H. Schulzrinne. Effects of power conservation, wireless coverage and cooperation on data environments. In *Proceedings of ACM Symposium on Mobile Ad Hoc Networking and Computing (MOBIHOC)*, pp. 117–127, 2001.

[28] C. E. Perkins and E. M. Royer. Ad hoc on-demand distance vector routing. In *Proceedings of the IEEE Workshop on Mobile Computing Systems and Applications*, pp. 90–100, 1999.

[29] K. S. Prabh and T. F. Abdelzaher. Energy-conserving data cache placement in sensor networks. *ACM Transactions On Sensor Networks*, 1(2):178–203, 2005.

[30] M. Rahimi, R. Baer, O. I. Iroezi, J. C. Garcia, J. Warrior, D. Estrin, and M. Srivastava. Cyclops: In situ image sensing and interpretation in wireless sensor networks. In *Proceedings of the ACM International Conference on Embedded Networked Sensor Systems (SenSys)*, pp. 192–204, 2005.

[31] A. Rousskov and D. Wessels. Cache digests. *Computer Networks and ISDN Systems*, 30(22–23):2155–2168, 1998.

[32] F. Sailhan and V. Issarny. Energy-aware Web caching for mobile terminals. In *Proceedings of the IEEE International Conference on Distributed Computing Systems Workshops (ICDCSW)*, pp. 820–825, 2002.

[33] F. Sailhan and V. Issarny. Cooperative caching in ad hoc networks. In *Proceedings of the IEEE International Conference on Mobile Data Management (MDM)*, pp. 13–28, 2003.

[34] H. Shen, S. K. Das, M. Kumar, and Z. Wang. Cooperative caching with optimal radius in hybrid wireless networks. In *Proceedings of the International IFIP-TC6 Networking Conference (NETWORKING)*, volume 3042 of *Lecture Notes on Computer Science*, pp. 841–853, 2004.

[35] A. Sobeih, J. C. Hou, L.-C. Kung, N. Li, H. Zhang, W.-P. Chen, H.-Y. Tyan, and H. Lim. J-Sim: A simulation and emulation environment for wireless sensor networks. *IEEE Wireless Communications Magazine*, 13(4):104–119, 2006.

[36] M. Takaaki and H. Aida. Cache data access system in ad hoc networks. In *Proceedings of the IEEE Spring Semiannual Vehicular Technology Conference (VTC)*, Vol. 2, pp. 1228–1232, 2003.

[37] B. Tang, S. Das, and H. Gupta. Cache placement in sensor networks under update cost constraint. In *Proceedings of the (ADHOC-NOW)*, volume 3738 of *Lecture Notes on Computer Science*, pp. 334–348, 2005.

[38] L. Tassiulas and C. Su. Optimal memory management strategies for a mobile user in a broadcast data delivery system. *IEEE Journal on Selected Areas in Communications*, 15(7):1226–1238, 1997.

[39] D. Wessels and K. Claffy. ICP and the Squid Web cache. *IEEE Journal on Selected Areas in Communications*, 16(3):345–357, 1998.

[40] L. Yin and G. Cao. Supporting cooperative caching in ad hoc networks. *IEEE Transactions on Mobile Computing*, 5(1):77–89, 2006.

[41] X. Yu. Distributed cache updating for the dynamic source routing protocol. *IEEE Transactions on Mobile Computing*, 5(6):609–626, 2006.

17 Cooperative Security in Peer-to-Peer and Mobile Ad Hoc Networks

Esther Palomar, Juan M. E. Tapiador,
Julio C. Hernández-Castro, and Arturo Ribagorda

CONTENTS

The extreme decentralization, dynamism, and self-organization of a number of emerging environments, including teamwork, pure peer-to-peer (P2P), and mobile ad hoc networks (MANETs), where nodes are involved in the process of sharing and collaboration without relying on central authorities, enforce the overall cooperation to play an essential role. Particularly, ad hoc networks rely upon the cooperation among individual nodes to carry out essential tasks such as packet forwarding. P2P file sharing systems face a similar situation. This collaboration-based nature of the communication layers imposes a number of challenges in the provision of other services, especially concerning security. In fact, most of the difficulties found to apply classic security solutions are just related to the inherent lack of central authorities, such as public key infrastructures (PKIs), and new proposals have to deal with avoiding such a centralization by exploring alternative paradigms which, in turn, require cooperation among peers. In this chapter, we elaborate on the concept of cooperation-based security services, such as authentication and access control, and is concluded by pointing out some open research issues and emerging trends in this research area.

17.1 INTRODUCTION

Extremely decentralized and self-organized environments, such as pure P2P and ad hoc networks, are receiving special attention at present due to the emerging range of new applications. Providing efficient security services in such networks is an active research area, which prompts many challenges. It is particulary challenging because of their unique characteristics, especially mobility and heterogeneity, dynamic node joins and leaves, and routing issues. Moreover, generally, a high transient topology of the physical network, formed and maintained independently by the peers, can cause significant problems.

Most of the difficulties found to apply classic security requirements (for example, authentication and authorization services) are just related to the inherent lack of fixed infrastructures. In particular, the absence of a centralized authority (typically a certification authority or CA) disables the application of common cryptographic primitives such as digital signature. In fact, researchers have to adapt common cryptographic techniques, for example, threshold and public key cryptography, to highly dynamic environments to ensure that even when some nodes are unavailable, others can still perform the task through the coalition of cooperating parties.

Nearly all the solutions proposed so far are one way or another based on the idea of replacing the whole PKI by a collaboration scheme. Put simply, a subgroup of peers must cooperate to perform tasks such as generating (or verifying) a digital signature or negotiating a group key for secure communications. Nevertheless, collaboration-based applications have to deal with the social and rational dilemma, which motivates a possible tendency of entities toward self-interested behavior. For example, a node may decide not to cooperate to save its resources while still using the network. Most research efforts have modeled and quantified the incentives and disincentives for cooperation in P2P networks [1]. On the other hand, anonymity and a highly transient population incur additional complexity for peers to determine others' identities, making difficult (or very inefficient) some tasks such as authentication and accounting.

In this chapter, a brief overview of P2P and MANETs is presented. First, we discuss the need for basic security solutions for P2P and ad hoc networks, where cooperation among nodes is required. We also present the main security solutions based on cooperation strategies. Finally, some conclusions and future research directions are highlighted.

17.1.1 P2P AND MANETs

A fundamental feature of P2P networks is the honest collaboration among a heterogeneous community of participants. After Napster's success, the first P2P file sharing application massively used, advances in this area have been intense, with the proposal of many new architectures and applications for content and computing sharing and collaborative working environments. However, the inherent differences between the P2P model and the classic client–server paradigm have rendered the many security solutions developed for the latter not applicable or, in the best case, they have to be carefully adapted. In the following subsections, we elaborate on the influence of ad hoc environments nowadays, their benefits, and why collaboration-based security services are in the spotlight.

17.1.1.1 Important Concepts and Terminology

P2P is often described as a type of distributed computing paradigm where nodes communicate directly with each other to exchange information. P2P applications allow users to communicate synchronously, supporting tasks such as instant messaging, working on shared documents or sharing files, among many others. As a result, the P2P paradigm provides users with the capability of integrating their platforms within a distributed environment with a broad range of possibilities.

A P2P network has neither clients nor servers; each individual node could act simultaneously as a client and as a server for the rest of the nodes in the network. Within this paradigm, any node can initiate or complete a transaction, and it can also play an active role in the routing operations. In general, nodes will be users' personal computers, instead of typical elements of the network infrastructure, but they can present heterogeneous characteristics regarding the local configuration, processing power, connection bandwidth, storage capacity, and so forth.

P2P systems are characterized by being extremely decentralized and self-organized. These properties are essential in collaborative and ad hoc environments, in which dynamic and transient population prevails. Perhaps the most popular application of P2P systems, which has been rather successful and has attracted considerable attention in the last years, is file sharing. This kind of systems are typically made up of millions of dynamic peers involved in the process of sharing and collaboration without relying on central authorities. One of the main advantages of file sharing systems is their capability to offer replicas of the same content at various locations. Replication has important properties, such as the possibility of accessing contents even when some nodes are disconnected.

P2P networks are useful not only for relatively simple file sharing systems, in which the main goal is directly exchanging contents with others. However, large P2P distribution networks will be more robust against attacks and range to more sophisticated structures, which self-organize into ad hoc network topologies, with the purpose of sharing resources such as content and CPU cycles (GRID), of maintaining secure and efficient storage, indexing, searching, updating, and retrieving data. In fact, a number of factors, such as the increasing popularity of wireless networks, the opportunities offered by 3G services, and the rapid proliferation of mobile devices, have stimulated a general trend toward extending P2P characteristics to wireless environments. As a result, the P2P paradigm has begun to migrate to pervasive computing scenarios. So far, the most straightforward approach consists in mounting a mobile P2P (M-P2P) system over a MANET. Nevertheless, nodes in a MANET are constrained by a limited amount of energy, storage, bandwidth, and computational power. These factors limit, among others, the type of security measures that can be deployed.

17.1.2 System Model

The study of security issues in P2P networks becomes more difficult due to the diversity and heterogeneity of existing P2P architectures. With the aim of providing a general specification, we have identified three elements common to every P2P system:

1. The user community (nodes). The user community is characterized by a high node transience, the total ignorance of the node's intentions, and the lack of a centralized authority. These issues have been tackled by different models, protocols, and systems, which mostly stress cooperation playing an essential role in the network performance.
   The evolution of cooperation is the creation of social profiles of P2P virtual communities [2], which are addressed by recent investigations focused on the establishment of incentives that motivate users to behave well. Typically, at the heart of these proposals operate traditional reputation schemes as polling-based algorithms. A solution to encourage resource sharing is to force each peer to contribute before being served. This collaboration is evaluated for computing a user participation level, rewarding the most collaborative peers with, for example, high priorities for their queries or by decreasing the transmission delays of their desired services.

2. The overlay architecture that defines the logical structure of the network over the underlying communication layers. Essentially, the overlay network manages the aspects related to node location and message routing.

Overlay networks can be classified in terms of their degree of centralization and structure. There are three categories concerning the former:

- Purely decentralized architectures: There is no central coordination point of the distribution activities. Nodes are referred as "servents" due to their dual nature (SERVers+cliENTS).
- Partially centralized architectures: Special roles are assumed for some nodes called "supernodes," which carry out special tasks mainly aimed at improving the performance of network routing.
- Hybrid decentralized architectures: A central server provides the interaction among the nodes, because indexes that support data searches and node identification are centralized, but the data are distributed.

On the other hand, P2P networks are categorized in terms of their structure as unstructured or structured. The first category of overlay models was popularized by `Napster`, which showed some scalability limits, but reduced the network dependence to a small number of highly connected, easy-to-attack peers. Peers join the network without any prior knowledge of the topology. Searching mechanisms include brute force methods, such as flooding the network with propagating queries to locate highly replicated contents. Other well-known unstructured systems are `Gnutella` and `FastTrack/KaZaA`. On the contrary, structured P2P networks provide a mapping between content and location in the form of a distributed routing table. Queries can be efficiently routed to the node having the desired content, and data items can be discovered using the given keys. The overlay network assigns keys to data items and organizes its peers into a graph that maps each data key to a peer. Maintenance of this graph is not easy, especially due to the high transience of nodes.

3. The information (content) stored at nodes and accessible through the services offered by the network. The most commonly used file replication strategy in P2P systems simply makes replicas of objects on the requesting peer, upon a successful query/reply.

17.1.2.1 Cooperation-Based Interactions in Mobile Scenarios

P2P systems for mobile and ad hoc networks introduce a number of new issues related to naming, discovery, communication, and security. In particular, as many mobile devices cannot still store many large files and the network infrastructure have remarkably low capabilities, M-P2P services will differ notably, at least by now, from current P2P applications. Smaller contents, such as games, video clips, photographs, and news, to name a few, are more likely to be distributed. Indeed, there are some future collaboration-based applications in which peers could share data with each other using mobile devices in a P2P fashion. Some works present middleware implementations in Java and Simple Object Access Protocol (SOAP) to provide M-P2P services in several wireless/mobile technologies, including Wireless Local Area Network (WLAN), Universal Mobile Telecommunications System (UMTS), General Packet Radio Service (GPRS), and Global System for Mobile Communications (GSM). However, the highly dynamic, decentralized, and self-organizing nature of MANETs does not fit well with many approaches developed in the P2P world.

Research on MANET technology has addressed the efficiency of data search and routing, a major problems because the topology continuously and unpredictably changes due to frequent joins and leaves and because the network performs physical broadcast. Basically, the efficiency of routes often depends on the honest collaboration among nodes, who may serve also as routers, and generally the trustworthiness of routes is kept to exchange them among different peers. Obviously, this raises a scalability problem. Some recent works and surveys on MANETs study these fundamental problems (for example, Refs. [3,4]). On the other hand, it is also necessary to identify vulnerabilities and threats, establish the essential security requirements, and define appropriate mechanisms for M-P2P applications. Moreover, most interactions and main operational tasks carried out in a MANET (for example, forwarding, recommending/accusing, and, according to the service provided, participating) are mainly built around trust and have to necessarily rely on the cooperation of all the participants. Nevertheless, this fact also incurs a problem; malicious nodes and whitewashers may impersonate others and use the spoofed identities to launch false accusations and purify the bad reputation accumulated under their previous identity.

Regarding adversary models, collusion is one of the most dangerous threats, because dishonest peers can act together in efforts to impact honest decisions. For instance, Sybil attack may involve a stream of colluding recommenders boosting the trust of one badly behaved principal. This should be limited by using identity certification. Thus, peers need to be sure that the other party (especially the source or the recommender) is really who he or she claims to be. On the contrary, anonymity has been an important consideration in the earlier P2P designs and especially in mobile environments where anonymity incurs an unacceptable overhead of flooding and uncertainty about the credibility of the potential participant. Recent solutions consider the existence of a self-organized PKI to ensure correct identification, such as that proposed in Ref. [5].

17.1.3 SECURITY GOALS: MOBILE VS. WIRED

Basically, P2P security schemes show two main characteristics: they are inefficient and, in the absence of central authorities, require nodes to cooperate. In fact, their inefficient performance is currently an obstacle to the acceptance and usage of several cryptographic solutions. Particulary, it does not seem easy to deploy security solutions for mobile architectures due, among other reasons, to the inherent limitations of the peers' devices and their potentially sporadic interaction with other peers. Under those assumptions, a significant challenge was how to establish a decentralized trust management system.

Despite the advances in P2P technology, security-related issues have remained systematically unaddressed or, at best, handled without a global perspective. Classic approaches have concentrated on specific points, such as mitigating attacks against three main system properties: availability, integrity, and anonymity. Availability is measured by how often object requests are successfully served, and, in particular, mapping two factors: the number of peers (average node availability) and the number of object replicas (replica storage size). An important problem is how to deal with an overestimated number of copies that could cause serious security conflicts, like denial-of-service (DoS), self-replication (false information distribution), man-in-the-middle, and pseudospoofing attacks. It is necessary to balance the total network anonymity (see Ref. [6,7]) and the need of preventing network abuse, to assure the content's high quality. On the other hand, the integrity of information in a P2P system may be attacked through the introduction of degraded-quality content or by misrepresenting the identity of the content (for example, falsely labeling).

Moreover, research efforts have also focused on the study of DoS attacks and the abuse of multiple identities (Sybil attack) [8]. Other problems have been recently identified, as those associated to the transience of peers (churn) or how to combat the selfish behavior exhibited by nodes that do not share their resources (free-riding) [9].

17.1.3.1 Security Requirements

Authentication and access control are fundamental to secure the system from unauthorized actions. In this regard, several works have already shown how to provide authentication, confidentiality, integrity, and non-repudiation services in ad hoc domains, based on identity, reputation and trust, proximity, and also public key cryptography [10]. Traditional cryptographic primitives can be actually used, perhaps with some restrictions, in mobile architectures. However, the low capabilities of wireless nodes have reinforced the use of trust-based solutions rather than the inclusion of regular cryptographic schemes. The main concern with these approaches is that, in most of them, nodes are trusted by default, and therefore they are susceptible to attacks. Concretely, a correct node identification is critical, and the lack of control on it could yield to vulnerabilities in the main processes. Node identification (and its relationship with anonymity) is an intense research area due to the potential risk of performing traffic analysis attacks and the traceability of communications among nodes. An example is the problem during churn, which involves a large number of potentially malicious peers in the P2P system to certify the peers identities. The simplest (but unrealistic) scheme for assigning an ID to each node is to have a centralized authority providing cryptographic certificates, which is only consulted when new nodes join.

The existence of an underlying mechanism for providing keys is also a problematic issue. Approaches based on the creation and distribution of a common key or on the inference of a strong key from a weak shared password have some problems with scalability and mobility [3]. Thus, an encrypted channel can be created if nodes share a key. If the common key can be stored and shared by devices, the problem would have an easy solution. Otherwise, a pair-wise shared key has to be established on the fly, without requiring the use of any online key distribution center. Current solutions envisage probabilistic key sharing and threshold secret sharing. Other solutions suggest the combination of centralization and key agreement techniques [11]. In fact, the key agreement protocol is only executed between a subset of nodes, which play a connecting role within the community. Then, the main idea is to cluster nodes in service-oriented communities, generally according to their physical position [12]. Apart from this, problems such as Sybil and pseudospoofing attacks, extensively studied in the context of P2P networks, appear whenever authentication protocols use opaque identifiers in favor of anonymity [13]. For this reason, most mobile solutions require the existence of some kind of control, such as a CA or a key sharing mechanism.

The establishment of a key management service using a hierarchy of CAs seems unsuitable for the moment, because a naive delegation and replication of the CA's responsibilities makes the service more vulnerable. To solve this problem, some works suggest the use of trust as a principal building block to address public key management. Particularly, schemes similar to Pretty Good Privacy (PGP [14]) are fully distributed and self-organized [5]. Nodes can generate their keys, and their distribution can be done without relying on external directories, such as in friend-to-friend protocols (F2F). Nevertheless, this approach presents some inconveniences, for example, those derived from highly transient communities with a high number of sporadic participants. On the other hand, proposals based on threshold cryptography present a completely different approach, distributing the CA's functionality over selected nodes [15]. Briefly, this scheme is secure if the adversary cannot compromise more than k out of the n members in any period of time. These schemes have also a number of drawbacks, mainly related to their communication overhead (any client need to contact at least k different nodes to get a certificate). Furthermore, the work presented in Ref. [16] relies on threshold signature to protect both routing and data forwarding.

On the contrary, trust negotiation solves the problems associated with classical authentication and authorization schemes by allowing individuals to safely agree on participation admissions for resources and services. The key idea is that each node has a token issued by its local neighbors. Other approaches use a similar idea but using mobile agents [17].

17.1.4 CHALLENGES IN COOPERATION

Besides security contributions and constraints mentioned above, mostly provided by ad hoc computing, cooperation-based services have been successfully addressed to create opportunities for new forms of approaches that involve interaction among nodes and to support different constraints, such as those listed below:

1. Support for decentralization, fault-tolerance, and scalability
   Because centralized control on ad hoc networks has poor performance for their single point of failure, research advances have had either to adapt classic schemes or to explore

novel mechanisms, most in the way of using informal collaborations to solve complex problems.

2. Support for trust-related tasks

   Almost all reputation systems have been designed for motivating peers to positively contribute to the network, just as in punishing adversaries who try to disrupt the system. In fact, users of current P2P systems actually appreciate the notion of reputation as an incentive in view of future interactions [18]. There has also been a trend in the research community toward the use of trust models to address some security concerns, particularly for ad hoc and self-organized environments. However, trust-based solutions generally rely on the exchange of reputation feedback (submit ratings on performance of their mutual transaction) among community nodes to globally/locally evaluate a certain node's trust value. These interactions with many other participants necessarily require collaboration.

3. Support for incentive and fairness models

   Both concepts begin to be taken into account to control aggressive behavior (antisocial) between peer interactions, and also both look to enforce fair resource sharing in which, at best, users are willing to cooperate even without being explicitly rewarded by the system for their contributions. Some studies have pointed out the benefits of employing incentive-based mechanisms as a means to foster cooperation among peers. Even though these incentives can rarely substitute the strength provided by a cryptographic primitive, they can still be very useful as a trade-off for noncritical applications. In fact, node participation in a network has been recently addressed in Ref. [19] by adopting a game theoretical approach. From the results presented, the authors conclude that even if nodes perceive a cost in sharing their resources, this may induce node participation.

17.2   PUBLIC KEY PRIMITIVES FOR P2P AND MANETS

For readability and completeness, we first discuss some of the most representative cryptographic primitives to provide security through cooperation.

The provision of security services in a fully distributed system (for example, a pure P2P network) becomes quite difficult if public key cryptography cannot be used —at least, in the way it is employed in classic distributed systems. Nearly all the solutions proposed so far are one way or another based on the idea of replacing the whole PKI by a collaboration scheme. Put simply, a subgroup of peers must cooperate to perform tasks such as generating or verifying a digital signature. Threshold cryptography has been repeatedly pointed out as an appropriate technique to achieve these purposes, even though its computational cost can disable its application for restricted devices (for example, mobile scenarios) [20]. We provide an overview of basic operational characteristics of this technique among others.

In a public key (or asymmetric) cryptosystem, users have a pair of cryptographic keys, a public key and a private key, related mathematically, but the private key cannot be practically derived from the public key. The private key is kept secret, while the public key may be widely distributed. Encryption with public key cryptography has some interesting properties. A a message encrypted with a recipient's public key cannot be decrypted by anyone except the recipient possessing the corresponding private key. This is used to ensure confidentiality. Conversely, in a secret key or symmetric cryptosystem both entities must agree to use a single secret key for both encryption and decryption. On the other hand, to ensure authenticity and integrity, a message can be digitally signed with a sender's private key. The digital signature can be verified by anyone who has access to the signer's public key, thereby proving that the sender signed it and that the content has not been modified [21].

The critical problem in public key cryptography is how to become convinced that a certain public key is authentic and has not been tampered with or replaced by a malicious third party. To face this problem a PKI is commonly applied, in which one or more (configured in hierarchy) third parties (CAs) certify the ownership of key pairs by means of binding public keys with respective user identities, and several other attributes, in public key certificates.

However, although Trusted Third Parties (TTPs) are acceptable in the client–server computing paradigm, they are not suitable to P2P models for a number of reasons. The viability of a PKI between P2P communicating parties is limited by practical problems such as uncertain certificate revocation, CA conditions for certificate issuance and reliance, variability of regulations and evidentiary laws by jurisdiction, and trust. In fact, another approach is the use of trust metrics, also known as "web of trust," used by PGP. We further elaborate on this primitive below.

DEFINITION 1   Public-key cryptosystem [22]

Generally, a public-key encryption scheme, (G, E, D), consists of three main phases:

1. *G is a key generator that on input n, the security parameter, outputs a pair (e, d), where e is the public key written in a public file and d is the private key.*
2. *E is an encryption mechanism that given a message m, the public key e produces a ciphertext c.*
3. *Finally, D is the decryption mechanism that on input e the public key, d the private key, and a ciphertext e produces a plaintext message m.*

As an example:

Let B be a receiver who publishes his public encryption key e_B (say, in some public file), while keeping secret the private decryption key d_B. Then, whoever wants to send a plaintext message m secretly to B, will pick e_B from the public file, encrypt the plaintext, and send him the resulting ciphertext $E_{e_B}(m)$. This way, only B can decrypt the message by applying the decryption key $D_{d_B}(E_{e_B}(m)) = M$.

Specifically, if a public key can be securely associated to a party, a typical scenario is to ensure the integrity of a content generated by her, as follows. First, the source A

1. Computes a hash value from the content, $h(m)$.
2. Encrypts $h(m)$ with her private key, obtaining a digital signature, $s_A(m)$.
3. The m, $h(m)$, and $s_A(m)$ are enclosed together and sent to the CA.
4. CA should validate the sender's digital signature. CA will sign $h(m)$ and send this signature, $s_{CA}(m)$, to A.

Then, the receiver

1. Retrieves m and the CA's signature
2. Decrypts the digital signature enclosed by using the CA's public key certificate, thus generating a first hash value
3. Computes a second hash value from the received content
4. Compares both hash values to confirm the nonalteration of the content

As shown, distributed online CA-based systems count on TTPs to generate public keys and issue digital certificates. Research on ad hoc computing has had to adapt such schemes to such scenarios in the absence of CAs [16]. For further details on introduction to cryptography, the reader is referred to Ref. [23].

17.2.1 Threshold Cryptography

The goal of secret sharing schemes and threshold cryptography is to distribute the provision of basic security services/tasks (for example, the authority to sign a file), in fully distributed environments without the existence of any fixed infrastructure and in the presence of malicious nodes [24]. Such techniques offer better fault tolerance and resilience with respect to crashes of some of the system components because even if some nodes are unavailable, others can still perform the task. Indeed, threshold cryptography exploits cooperation in a way that a coalition of cooperating parties can jointly perform a critical action.

DEFINITION 2 (t,n)**–Threshold signature scheme [20]**

At any time, a node i may request a signature of message m from other parties in the community, by sending her identity, parameters t, n, and the message m to be signed. Let $(\mathcal{P}_{kg}, \mathcal{P}_{sgn})$ be an execution of threshold signature scheme, where

1. *A randomized distributed threshold key generation protocol, denoted as $\mathcal{P}_{kg}$, is run by the players $P_1 \ldots P_n$. The output of this phase is a string sk_i, pk_i for each party, who use as input the common parameters and a different random string. Each party will output its signature share.*
2. *The execution of t-out-of-n signature protocol, denoted as $\mathcal{P}_{sgn}$, produces a pair (sig, out) for the node i.*

In a MANET, the scheme should satisfy the following properties: correctness, unforgeability, and robustness.

Concretely, n parties share the ability of performing a cryptographic operation (for example, creating a digital signature), and any subset of at least t out of n parties can jointly perform the operation. On the other hand, any $t - 1$ (or less) parties cannot perform the operation. Moreover, any $t < \frac{n}{2}$ parties cannot prevent the remaining honest parties from performing the task.

We can classify threshold schemes into two types: those that need a trusted centralized authority or dealer and those that do not require this kind of TTP. Verifiable secret sharing and common Rivest, Shamir and Adelman Encryption Standard (RSA) or Digital Signature Standard (DSS) threshold signature schemes belong to the former, whereas others provide joint secret sharing without a dealer. With the aim of not violating the nature of the P2P and ad hoc network, recent approaches let peers themselves to locally handle main procedures without any dealer [25].

The establishment of the threshold t is usually a main matter. With a fixed threshold policy, an adversary may try to manage and permanently compromise t group members to expose the secret. In such cases, the reduction of value t involves associated problems such as when a large number of members leave the group, resulting in a new group size, probably less than t. Thus, threshold schemes need to securely and reliably determine the number of current participants on each task.

Several approaches have addressed the provision of fault-tolerant authentication, most in the way of protocols relying on threshold cryptography that use key sharing and agreement techniques. Practically they all share a common idea: to use a quorum of participants to create, for example, a digitally signed public key certificate. Particularly, the work presented in Ref. [26] provides an ubiquitous authentication service by following a certificate-based approach. In this case, no single node has the power of providing full certification services (for example, certificate issuing, renewal, and revocation), because none of them holds the complete certificate signing key. Instead, multiple nodes collaborate in a network locality, by means of establishing a temporary trust relationship via a localized trust model. In general, a node is trusted if any fraction k (a system-wide parameter) of trusted nodes (typically the neighboring nodes of the entity) claims so. The required k sets the global acceptance criteria because a locally trusted entity is globally accepted anywhere. The latter incurs a problematic requirement concerning node identification reliability.

17.2.2 BYZANTINE AGREEMENT

The Byzantine Generals Problem basically consists in "deciding" a common battle plan; a group of distributed Byzantine generals, camped around an enemy city, must agree upon "attack" or "retreat" the enemy. Each base communicates with others sending conflict information, with the vulnerability of traitors and enemies who try to prevent the loyal generals from reaching a plan. As shown in Ref. [27], using oral messages, this problem is solvable if and only if more than $\frac{2}{3}$ of the generals are loyal, or it is the same; no solution with fewer than $3m + 1$ generals can cope with m traitors; but with signed messages, the problem is solvable for any number of generals and possible traitors. Consequently, if all the participants are honest, they will reach consensus on validity.

Regarding computation, we can model two types of processes/participants: honest and faulty. Obviously, a participant is faulty if it does not follow, accidentally or maliciously, the specified algorithm. There is no solution if the upper bound on the number of faults exceeds one-third of the nodes. Note that if the actual number of faulty processes is larger than the upper bound, then the algorithm may fail to reach Byzantine Agreement without alerting any correct process to that fact. Therefore, the immediate sender of any message must be identified, and also there must be a low threshold that prevents potential collusion of faulty nodes from introducing fake information.

Several works have applied this idea to provide Byzantine Fault Tolerance [28], while others apply the scheme to support high-level security services based on consensus. Pathak and Iftode [29] apply the same idea to provide public key authentication in ad hoc networks. This work postulates that a correct authentication depends on an honest majority of a particular subgroup of the peers' community, labeled "trusted group." However, in P2P systems an authenticated peer could create multiple fake identities and act maliciously in the future (Sybil attack [8]). For this reason, the classification of the rest of the community maintained by each node (Figure 17.1) has to be proactive and should be periodically flushed. A periodic pruning of the trusted group will ensure honest majority. Thus, honest members from trusted groups are used to provide a functionality similar to that of the PKI through a consensus procedure.

The authentication protocol consists in the four phases that are briefly discussed below (Figure 17.2). Interested readers can find further details in Ref. [29].

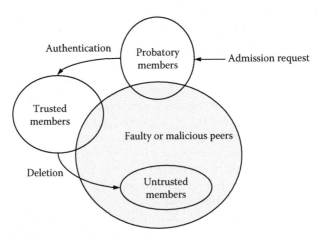

FIGURE 17.1 Community structure according to the authentication state of each node. (From Pathak, V. and Iftode, L., *Comput. Netw.*, 50, 579, March 2006. With permission.)

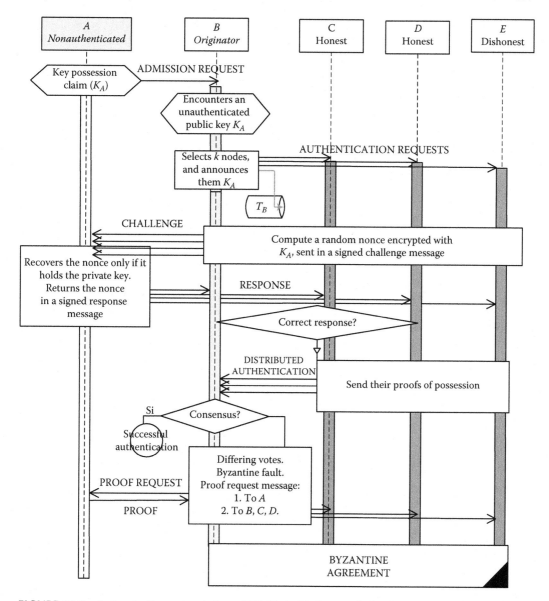

FIGURE 17.2 Authentication protocol phases [29]. Node *A* belongs to "others" group. Node *B* authenticates *A* using its trusted peers, and one of them turns malicious (black node) that tries to prevent authentication of *A*, *C*.

1. Admission request. The protocol begins when *B* (Bob) runs into a newly discovered peer *A* (Alice), which claims to be the owner of an unauthenticated public key K_A. Then, *B* asks a subgroup of his trusted group for helping him in verifying the authenticity of K_A. Finally, *B* sends K_A to those trusted peers that agree.

2. Challenge response. Each notified peer challenges Alice by sending a random nonce encrypted with Alice's supposed public key. Alice is able to return each received nonce if and only if she holds the corresponding private key, K_A^{-1}. Each challenger checks if the received response is correct, thus obtaining a proof of possession of K_A.

3. Distributed authentication. Each peer helping Alice sends her proof of possession to Bob. If all peers are honest, then there will be a consensus, so Bob gets the authentication result: K_A

belongs to Alice or not. However, some of the peers summoned by Bob could be malicious or faulty, which may result in Alice receiving different opinions about the authenticity of K_A. In this case, Bob must initiate the Byzantine agreement phase.

4. Byzantine agreement. First, Bob verifies if Alice is malicious by sending to her a proof request message. Alice must respond with all challenge messages received and the respective responses sent by her. If A is honest, she can provide a correct response and also demonstrate her good behavior by sending to B the challenges she received and the corresponding responses. If A cannot be proved to be malicious, then some of the peers must be. At this stage, B announces a Byzantine fault to the group. Each group member sends an agreement message to others. At the end of this phase, the honest peers will be able to recognize malicious peers causing the split in authentication votes.

Successful authentication moves a peer to B's trusted group, while encountered malicious peers are moved to the untrusted group. Peers can also be deleted from trusted groups due to inactivity and periodic pruning of the group.

The fundamental limit of this scheme is the following. Assume that N is the number of peers in the community, t the number of malicious or faulty peers, and ϕ a fraction of N, denoting that ϕN peers may not be reached during the protocol execution and another ϕN peers exhibit faulty behavior because the path between the source and them suffers from a man-in-the-middle attack. It can be shown that the community has honest majority if $t < \frac{1-6\phi}{3}N$. As the value of ϕ does not have influence on a random selection, we can consider that a group has honest majority with $3t + 1$ peers [29].

Summarizing, the previous protocol provides us with public key cryptography without relying on CAs in MANETs. This can be viewed as an essential building block upon which more complex security schemes can be developed. As an example, authors mention its application within an e-mail authentication system named SAM (self-authenticating mail).

17.2.3 WEB OF TRUST AND TRUST METRICS

The idea of creating trustworthy communities has attracted considerable attention in the last years [30,31]. Because it is not realistic to assume such a reliance exclusively on a hierarchy of CAs in a P2P network, and especially if it is the case of a MANET, some works use web-of-trust models (like PGP) to provide, among others, authentication mechanisms [32].

This technique has focused largely on public key authentication, digital signatures, and certificates and creates a decentralized fault-tolerant web of confidence for all public keys. It does not require the existence of a single point of trust, but allows the use of email digital signatures for self-publication of public key information. The key idea is simple; trust decisions are in the hands of individual users. This way, it is relatively easy to establish one's own trust community. Nevertheless, how does one authenticate others' public key? The response is a digest value (hash result or fingerprint). A key is validated by verifying the key's fingerprint with the key's owner. By signing the key one certifies it as a valid key [33].

A few security protocols proposed for P2P environments so far are based on cryptographic schemes as diverse as threshold cryptography, distributed consensus, or Byzantine agreement. Nevertheless, most proposals require an underlying trust model [34]. In fact, trust and reputation models have been recently suggested as an effective security mechanism for open environments, showing that rating nodes is an effective approach in distributed environments to improve security, to support decision-making, and to promote node collaboration [35].

On the other hand, the trust metric implemented in PGP is simple and can lead to counter intuitive decisions being made [36]. Concerning trust metrics, different trust models have been developed and their properties have been extensively studied. Because of the dynamic nature of MANETs, trust computation/evaluation may be uncertain and incomplete. Some works give intuitive requirements

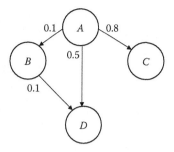

FIGURE 17.3 Example of asymmetry of trust evaluation.

that any trust algorithm should have under that framework as well. Basically, users need not have direct experience with every other user in the mobile network to compute an opinion about them [37]. Instead, they can base their opinion on the indirect recommendations provided by intermediate nodes (and probably neighbors). Obviously, trust schemes in MANETs rely on the cooperation among nodes (or on secure automated agents) and also be robust in the presence of dishonest actions. As many services in MANETs, such a problem is addressed as a generalized shortest path problem on a weighted directed graph. As an example, let the graph plotted in Figure 17.3 be a simple trust network (temporary or not) in which directed edges explicitly contain trust values specified by the evaluator. Node A trusts node B as 0.1 ($\in [0, 1]$), node C as 0.8, and node D as 0.5, while B trusts D as 0.1. In turn, this does not mean that, for example, B trusts A as the same value; arrows represent the direction. C knows no one except for A, so a trust metric can be used to infer the trust values between two nodes that are not directly connected, for example, how much C could trust D. Consequently, there are two different interaction paths from C to D: crossing B or not. A simple metric may be the sum of the trust values obtained directly from own experiences (D has none) together with the recommendation of others, with the aim of choosing the maximum. In the example, the shortest path is also the most trustworthy, by chance, and usually the recommended node will store the recommended trust value (and the recommender) to compute its own trust strategy. These topics are out of the scope of the present chapter, and therefore we do not provide more details either. Note that we have already mentioned some drawbacks, for example, the lack of robustness, of adopting security models which rely, to an extent, on trust in mobile environments.

17.3 COOPERATION-BASED SECURITY SERVICES

As the bandwidth available per user increases and Internet becomes accessible anywhere, the lack of scalability and fault tolerance given by the classic client–server paradigm motivates the exploration of fully distributed schemes, which often require cooperation among the network participants. As single nodes become essential for system operation, their responsibility, costs, and involvement rise, and therefore their vulnerability to malicious attacks and unreliability also increase. Particularly in P2P file-sharing applications, distributing control means relying on the cooperation of participants, and it is a significant challenge to design mechanisms that affect the overall behavior in a population of rational, and therefore selfish, nodes and also to provide incentives contributing to the public good by sharing of files. The aim of this section is to provide an overview of the most representative proposals that use cooperation as a main building block to provide security services.

17.3.1 SECURE ROUTING IN SELF-ORGANIZING OVERLAY NETWORKS

In P2P computing, the general function of an overlay network is to promote and support adaptive self-organization and maintenance of a (un)structured network formed of logical relationships among components. The basic principles to provide self-organization are to dynamically discover potential communication nodes and available services (generally starting from a few basic mechanisms such as broadcasting) and efficiently navigate with independence of the physical network [38].

Most approaches tend to require a priori assumptions on the network configuration that would limit its self-organizing nature and therefore its degree of adaptivity and robustness. In an ad hoc wireless network, no predeployed infrastructure is available for routing packets, instead routing relies on intermediary peers. In particular, a collection of mobile nodes join together and create a network by agreeing to route messages for each other. All networking functions must be then performed by the nodes themselves in a self-organizing way. This operating principle involves cooperation among nodes as an essential requirement [39]. In consequence, instead of considering all peers trusted by default or caring the trustworthiness of the immediate neighbors, ad hoc routing proposals should attach more importance to security requirements [40–42].

17.3.1.1 Stimulating Forwarding

The problem of stimulating cooperation in MANETs has been extensively addressed, because networking nodes are not naturally motivated to cooperate toward a common goal, but for saving own resources (that is, memory, CPU cycles, battery power, etc.). This is especially critical in wireless communications because any node can be used as a relay to forward packets. The work presented by the authors in Ref. [43] is centered on stimulating the packet forwarding function, assuming that every mobile node has a tamper-resistant hardware module. Basically, this security hardware has a counter that is protected from illegitimate manipulation and also establishes cryptographic associations (that is, link-by-link encryption of the packets using a session key) with neighbors. The counter is decreased when the node wants to send its own packets and increased when the node is an intermediary and forwards packets for the benefit of other nodes (Figure 17.4). However, the value of the counter must remain positive whenever the node wants to send own packets, just as the energy of the node. Furthermore, to send own packets, the estimated number of intermediate nodes that are needed to reach a destination will reduce the counter. The key idea is to equilibrate the number of packets (own as well as forwarding) that the node can send using its remaining energy and also maximizing its own benefit, that is, the number of own packets sent.

On the other hand, a critical situation occurs when the arrival rate of forwarding packets is too low, and then the node cannot earn enough to send all of its own packets, even if it forwards all packets received for forwarding [43]. Experiments show that there is usually a small fluctuation in the ratio between the number of forwarding packets and the number of own packets due to the random manner in which the packets arrive. A collection of forwarding rules are proposed to assist decisions about when to forward or drop packets received for forwarding, while own packets that cannot be sent immediately (due to the low value of the counter) are dropped. For example, when the node reaches optimal counter to drain its battery out by sending only its own packets, it is not necessary to forward more packets. Results from simulations show that the more cooperative the rule is, the best the performance it achieves. In summary, it can be seen experimentally that the network

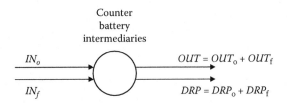

FIGURE 17.4 Model of a node [43]. Two incoming and two outgoing/droping flows of (own and forwarding) packets.

can tolerate less cooperative nodes quite well, even though the throughput of the network decreases as the fraction of less cooperative nodes increases.

17.3.1.2   Authenticated vs. Trusted Routing

Authentication in routing presents several challenges because each user brings to the network his or her own mobile unit, without relying on a centralized policy or control, such as those present in a traditional network. In fact, many different types of attacks have been identified against the most important routing protocols for MANETs, for example, those that are under consideration by the IETF for standardization: the Ad hoc On-demand Distance Vector (AODV) routing protocol and the Dynamic Source Routing (DSR) protocol. Concretely, some analyses focus on DoS attacks based on interception and noncooperation, derived from redirection of network traffic, and by altering control message fields or by forwarding routing messages with falsified values. Modification attacks combined with impersonation and spoofing are also studied. Possible solutions of these attacks range from securing routing through predetermined cryptographic certificates that guarantee end-to-end authentication [44], to applying mechanisms based on watchdogs and distributed reputation systems [42].

An example of such mechanisms is the CONFIDANT protocol [45], which aims at detecting and isolating misbehaving nodes in a DSR-based network, thus making it unattractive not to cooperate. The watchdog is a neighborhood monitoring function that observes the behavior of neighbors and identifies misbehaving nodes by promiscuously listening to communications of nodes in the same transmission range. A reputation system maintains global values for each node according to the information collected.

Nevertheless, these reputation-based solutions present some limitations. First, the ephemeral nature of wireless connections and the presence of collisions may cause the monitoring system to fail. Second, there are severe drawbacks in terms of traffic overhead and the potential spreading of wrong accusations. Finally, they can eventually reach a situation where all the traffic is deviated to well-behaving nodes, with the result of overloading them and the links between them. On the other hand, proactive protocols (for example, DSDV, OLSR [46]) are subject to misbehavior caused by selfish nodes that do not cooperate in the propagation of routing information through the network. We explore and identify the problems potentially affected by self-interested behavior in a subsection bellow.

17.3.2   Admission Control in P2P and MANETs

So far, only a few works have dealt with admission control (or any other form of authorization) in P2P and MANETs [4,19,47]. Most of them provide admission to secure peer groups based on threshold signatures, such as the scheme proposed in Ref. [25]. Newly joining members must acquire a share of the group secret for themselves in a distributed manner. A newly joining member receives t partial secret shares together with a membership certificate from each of the t members, who should admit him or her to the group. The possession of a partial share enables him or her to participate in future voting procedures, to admit new members. The authors examined and compared two threshold signature schemes: one based on RSA signatures and the other based on DSA signatures. We refer the interested reader to the original work for further details.

17.3.3   Content Authentication in P2P Networks

An interesting aspect of P2P systems is the possibility of replicating the same content among different nodes. On many occasions, this task is not performed in a proactive way, but is simply the result of the existence of a file sharing application. By using a search mechanism, users can locate a specific content and then download it from its source. Once a user gets the file, it is usual that a local copy will remain in the node in such a way that future queries will identify the node as one of the various locations from which the content can be obtained.

This fact presents some interesting properties. Faced with different locations of the same content, an application can grant priority to that which offers a less expensive path (for example, in terms of bandwidth or number of network hops). To some extent, replication also guarantees some sort of fault tolerance, because information can be available even if some parts of the network are temporarily disconnected.

In a collaborative working environment, the previous features are highly desirable [48]. However, it is unrealistic to assume that every integrating node will exhibit honest behavior, even if they have always behaved correctly in the past. Once a content is replicated through different locations, the originator loses control over it, and a malicious party can then modify the replica according to several purposes (for example, to claim ownership over the content or to insert malicious software into a highly demanded content).

Secure content distribution protocols are highly desired for these environments. The scheme presented in Ref. [49] shows how digital certificates can be first used for content authentication and also be extended to provide authorization capabilities, much in the way a X.509 public-key certificate can be used as an attribute certificate. Briefly, the main objective of this content authentication protocol is to maintain content integrity, ensuring its authenticity, and avoiding nonauthorized content alterations. This proposal is similar to PGP in the sense that both the content and the corresponding authorization certificates are issued by the users.

However, as opposed to PGP, the proposal does not rely on certificate directories for the distribution of certificates. Instead, in this model, certificates are stored and distributed directly by the users. This is achieved through the collaboration of a fraction of peers in the system. Previously to content distribution, the legitimate owner of a given content generates a content certificate. For this, the owner first selects a subgroup of signing nodes from his or her group of previously contacted peers. The certificate contains a number of fields: The identity of the originator (which ultimately establishes who has generated the content and is its legitimate owner); the identification and hash of the content (ensuring its integrity); and an ordered list of signers, who in turn provide a joint signature certifying that the content is authentic and belongs to its owner. After downloading a replica of a content (and its associated certificate), the user is encouraged to verify its authenticity.

This service can be extended to provide authorization capabilities by increasing the complexity of the certificates (and, therefore, their manipulation). Attributes encoding permissions can be easily inserted into authorization certificates, which are discretionally issued by the owner.

17.3.4 COMBATING SELFISH BEHAVIOR IN P2P NETWORKS

A number of works have focused on quantifying the optimal cost/benefit trade-off that would lead nodes to share their resources, especially in collaboration-based systems where peers are assumed to be rational [50–52]. Schemes based on micropayments protocols tend to meet fairness, assuring that providers are guaranteed to be paid, while requesters are discouraged to behave as freeloaders because they are refunded for each upload. System users are therefore given an incentive to work together toward a common goal [53]. The lack of cooperation (free-riding) and its complex dynamics have also been studied by adopting a game-theoretic approach. As an example, the reader is referred to Ref. [54], where the generalized prisoner's dilemma (a well-known game) is studied from the perspective of its possibilities as a model to encourage cooperation.

Mechanisms to discourage selfish behavior have also been proposed in different environments. For instance, in Ref. [55] the idea of using cryptographic protocols based on proof-of-work has been introduced to increase the cost of sending e-mail and make sending spam unprofitable. Ideas similar to this might be extended to P2P networks, particularly to impede DoS attacks as well as to provide a solution for the free-riding problem [56]. Research on this topic could lead to encouraging fair content distribution using cryptographic puzzles, because sharing can be encouraged by imposing a cost on the downloads. However, to deploy such schemes in mobile networks we must deal with additional concerns, for example, those related to the devices' low capabilities. At the same time, such secure mobile frameworks cannot ignore that the automatization in mobile environments, and the security measures as well, must be proportional to the risk involved in requiring the cooperative interaction between devices.

17.4 EMERGENT TRENDS AND FUTURE RESEARCH DIRECTIONS

It is generally expected that new forms of networking infrastructures and applications become widespread soon. Even though research and technical progress on these areas are nowadays intense, some important problems remain to be solved. In this regard, security concerns are among the most important issues, and, particularly, the problems associated with adapting security solutions that were conceived for completely different scenarios.

Whatever the case may be, it seems clear that, in the absence of fixed and centralized authorities, cooperation will play an essential role in the provision of such services. We have identified three main open research lines: the use of self-issued-certificates and group-based and evolutionary techniques.

17.4.1 CERTIFICATE-BASED SOLUTIONS

Alternatives to threshold schemes range from doting the system with preauthentication via a location-limited channel to using a tamper-resistant storage of group key (specially for nodes with very limited resources, for example, sensors). A more interesting approach from the security standpoint is to use self-issued certificates, that is, nodes issue certificates to trusted nodes. The use of certificates has already been addressed in trust establishment models [35], even under the assumption of existing sparse social relationships among nodes. Nevertheless, the main contribution to this topic is the application of zero knowledge proofs. Briefly, an entity Prover knows a secret, meanwhile entity Verifier wants P to prove his knowledge. Thus, P runs a challenge-response protocol offering a hard problem to V. At the next stage, V proposes a random challenge; P provides the solution. V checks and decides to accept (or reiterate) or reject. This way, after the interaction, Verifier is convinced that Prover knows the secret, but Verifier has *zero knowledge* about the secret itself.

Perhaps, the main drawback is the requirement of a bootstrapping phase with a secret dealer. In the cited work [35], the trust establishment for any two nodes turns into a certificate chain detection problem in a certificate graph, because a node obtains its secret short list (k bindings of its identifier and public key) from a secret dealer and then generates a certificate locally. However, to manage trust and reputation locally with minimal overhead, in terms of extra messages and time delay, is still an open issue. In particular, high latency, bandwidth usage, and energy consumption are not acceptable in situations with strict real-time requirements, like MANETs. In this sense, this proposal tries to restrict flooding to a subset of nodes. How to determine the subset, such that it covers the sufficient number of nodes holding the required trust information, becomes a problem. Consequently, novel works should have to lead toward "one-to-one" models, where an arbitrary transaction (interaction between two nodes, for example, a requester and a provider) is triggered by the requester and may be accepted/rejected individually by the provider [57].

In addition, revocation is an important process that must generally accompany the use of certificate-based models. It refers to the procedure for downgrading or eliminating nodes' privileges. Classic methods based on distributing Certificate Revocation Lists (CRL), Certificate Validation Protocols (SCVP), or Online Certificate Status Protocols (OCSP) may not be applicable for dynamic environments in which there are no centralized repositories. A recent work, presented in Ref. [58], addresses the issue of certificate revocation without input from external entities, employing threshold cryptography and profile tables.

17.4.2   SOCIAL AND GROUP-BASED SOLUTIONS

Some earlier cooperation-based approaches focused on discovering community social patterns. In social networks phenomena, node popularity can be stemmed from the position of a node within the network, that is, the degree (the number of neighbors) and the number of hops to every other networking node. However, because nodes are autonomous, selfish, rational, and strategic, and the optimal action for the individual does not produce the best outcome for the population as a whole. In this situation, most popular nodes (from different studies [18], there are no more than 1 percent of popular nodes) act as centralized hosts for the network and therefore lead to the network losing the benefits of a decentralized architecture. Rather than use a preestablished social network, the network should tend to be dynamically created at runtime.

Intragroup communications have become an important issue in wireless networks. Traditional multicast protocols have been developed and extensively evaluated. To prevent attackers from paralyzing the network and services by manipulating multicast communication, typical scenarios demand the design of protocols that cannot base their security on the existence of setup information and that have no preestablished trust relationship either [20]. In particular, this kind of environments, in which multiple groups may coexist interacting through multicast traffic, security should be enforced and should be protected by means of cryptography, for example, counting on secret keys. In this sense, robust cryptographic solutions are applied more and more in MANETs specially to provide key management schemes [12,59].

17.4.3   USING EVOLUTIONARY ALGORITHMS

The idea of self-organizing cooperation is currently being addressed by different research directions, for example, mobile robots connect with each other in a MANET to coordinate their movements. Other trends, such as the Adaptive Multi-Agent System theory [60], focus on the design of cooperative systems in which agents cooperatively interact. Each agent has the ability of self-organization, and it is locally "cooperative," but not altruistic. Agents also have the ability of locally rearrange their interactions with other agents according to their individual task, and this can indeed lead to a change in the global function.

On the other hand, several approaches have already taken inspiration from a number of natural phenomena, which give potential emergence of possibly self-organizing behaviors and tools to be adopted in decentralized deployments [38]. For example, ant foraging has motivated the design of some adaptive P2P agent-based frameworks by relaying on distributed mobile agents (ants), which can directly interact and cooperate, even reorganize, while leaving and retrieving bunches of data in the visited hosts. Moreover, some proposals in the area of sensor networks exploit learning theories and evolutionary approaches to have systems autonomously learn how to self-organize themselves. However, because no possibility of control is assumed, and due to the complexity (often nonlinearity) of the phenomena involved, it is difficult to prove that the system will behave as needed. In such cases, some proposals are reasonably (that is, probabilistically) confident that the global evolution of the system will eventually lead to the desired coordinated behavior.

Additionally, novel approaches are being studied by means of mechanisms inspired by biological evolution such as mutation and recombination, natural selection, and survival of the fittest. In particular, some works address the evaluation of certain protocol's suitability using evolutionary techniques, with the aim of finding the optimum sequence of premises, for example, using random graphs in a game theory approach. Concretely, candidate solutions to the optimization problem using the role of individuals in a population and the cost function determine the environment within which the solution "lives" [2]. In fact, P2P and MANET's general principle, that is, self-organization, paves the way for considering the use of evolutionary techniques in analysis processes as a open research challenge. In addition, evolutionary techniques mostly involve metaheuristic optimization algorithms such as self-organization. A range of scenarios are already under formation assuming a spontaneously networking environment. Those scenarios may exhibit spontaneously emergent behaviors, which need methodologies to predict them, like a self-adaptive system.

17.5 CONCLUSIONS

Infrastructureless networks, on which, in general, one cannot assume the existence of centralized services such as those provided by TTPs, present a challenge in terms of formalizing collaboration-based security protocols. This chapter presented the most representative cooperation-dependant security approaches using P2P scenarios to highlight the benefits of such a dependence. In our opinion, these schemes outlined in this proposal offer major advantages over other existing ones. First, future research work should not restrict the community dynamics, that is, establishing restrictions over the community new joiners, in the sense that interactions are not based on reputation but on nodes' experience. As shown, although we assume rational behavior, we are able to consider non-collaborative players and to measure the effect they might have on the overall system performance. Finally, we have observed that even in a heterogeneous community profile (having noncooperative, cooperative, and discriminators), the system can reach an equilibrium. On the other hand, providing efficiency and optimization for the nodes is quite challenging, given the computational and communication overhead that cooperation-based security schemes generally incur and also considering the nodes' limitations. Nevertheless, numerous members within the research community are currently working on optimal solutions for maintaining reasonably high levels of cooperation. The future of wireless/mobile ad hoc systems relies on the ability to provide efficient security support that can be performed even in the presence of colluders.

REFERENCES

[1] K. Lai, M. Feldman, I. Stoica, and J. Chuang. Incentives for cooperation in peer-to-peer networks. In *Proceedings of Workshop on Economics of Peer-to-Peer Systems*, Berkeley, CA, June 2003.

[2] T. Ellis and X. Yao. Evolving cooperation in the non-iterated prisoner's dilemma: A social network inspired approach. In *Proceedings of the IEEE Congress on Evolutionary Computation*, pp. 736–743, Singapore, September 25–28, 2007.

[3] D. Djenouri, L. Khelladi, and A.N. Badache. A survey of security issues in mobile ad hoc and sensor networks. *Communications Surveys and Tutorials, 7(4)*, 2–28, 2005, IEEE.

[4] J. Kong, P. Zerfos, H. Luo, S. Lu, and L. Zhang. Providing robust and ubiquitous security support for mobile ad-hoc networks. In *Proceedings of the 9th IEEE International Conference on Network Protocols*, pp. 251–260, California, November 2001.

[5] S. Capkun, L. Buttyán, and J.-P. Hubaux. Self-organized public key management for mobile ad hoc networks. *IEEE Transactions on Mobile Computing, 2(1)*, 52–64, Jan–Mar 2003.

[6] S. Marti and H. Garcia-Molina. Identity crisis: Anonymity vs reputation in P2P systems. In *Proceedings of the 3rd IEEE Peer-to-Peer Computing*, Linköping, Sweden, pp. 134–141, September 2003.

[7] A. Singh and L. Liu. TrustMe: Anonymous management of trust relationships in decentralized P2P systems. In *Proceedings of the 3rd IEEE International Conference on Peer-to-Peer Computing*, Linköping, Sweden, pp. 142–149, September 2003.

[8] J. Douceur. The sybil attack. In *Proceedings of the 1st International Workshop on P2P Systems Workshop*, Cambridge, MA, pp. 251–260, March 2002.

[9] M. Feldman and J. Chuang. Overcoming free-riding behavior in peer-to-peer systems. *ACM Sigecom Exchanges, 6(1)*, 41–50, July 2005.

[10] M. Haque and S.I. Ahamed. Security in pervasive computing: Current status and open issues. *International Journal of Network Security, 3(3)*, 203–214, 2006.

[11] S. Yi and R. Kravets. MoCA: Mobile certificate authority for wireless ad hoc networks. In *Proceedings of the 2nd Annual PKI Research Workshop (PKI 03)* NIST Gaithersburg MD, pp. 65–79, April 2003.

[12] N. Wang and S. Fang. A hierarchical key management scheme for secure group communications in mobile ad hoc networks. *Journal of Systems and Software, 80(10)*, 1667–1677, October 2007.

[13] B. Zhu, S. Jajodia, and M.S. Kankanhalli. Building trust in peer-to-peer systems: A review. *International Journal of Security and Networks, 1(1/2)*, 103–112, 2006.

[14] P. Zimmermann. *The Official PGP User's Guide*, MIT Press, Cambridge, MA, 1995.

[15] L. Zhou and Z.J. Haas. Securing ad hoc networks. *IEEE Network, 13(6)*, 24–30, 1999.

[16] H. Yang, X. Meng, and S. Lu. Self-organized network layer security in mobile ad hoc networks. In *Proceedings of the 1st ACM Workshop on Wireless Security*, pp. 11–20, Atlanta, GA, 2002.

[17] C. Lin, V. Varadharajan, Y. Wang, and Y. Mu. On the design of a new trust model for mobile agent security. In *Proceedings of the 1st International Conference on Trust and Privacy in Digital Business*, pp. 60–69, Springer Verlag, Zaragoza, Spain, September 2004.

[18] J. Cheng, Y. Li, W. Jiao, and J. Ma. A utility-based auction cooperation incentive mechanism in peer-to-peer network. In *Proceedings of the 2nd International Symposium on Network-Centric Ubiquitous Systems (EUC Workshop)*, pp. 11–21, Seoul, Korea, August 2006.

[19] L.A. DaSilva and V. Srivastava. Node participation in ad hoc and peer-to-peer networks: A game-theoretic formulation. In *Proceedings of the 1st Workshop on Games and Emergent Behaviors in Distributed Computing Environments*, Birmingham, U.K., September 2004.

[20] G. Di Crescenzo, R. Geb, and G.R. Arce. Threshold cryptography in mobile ad hoc networks under minimal topology and setup assumptionsstar, open. *Ad Hoc Networks, 5(1)*, 63–75, January 2007.

[21] A. Shamir. Identity-based cryptosystems and signature schemes. *Advances in Cryptology*, Vol. 196 of LNCS, pp. 47–53, Springer-Verlag 1984.

[22] M. Naor and M. Yung. Public-key cryptosystems provably secure against chosen ciphertext attacks. In *Proceedings of the 22nd Annual ACM Symposium on Theory of Computing*, pp. 427–437, Baltimore, MD, 1990.

[23] B. Ryabko and A. Fionov. *Basics of Contemporary Cryptography for IT Practitioners*. World Scientific, Singapore, 2005.

[24] Y. Desmedt. Some recent research aspects of threshold cryptography. In *Proceedings of the 1st International Workshop on Information Security (ISW'97)*, LNCS, Vol. 1196, pp. 158–173, Springer-Verlag, Ishikawa, Japan, September 1997.

[25] M. Narasimha, G. Tsudik, and J.H. Yi. On the utility of distributed cryptography in P2P and MANETs: The case of membership control. In *Proceedings of the 11th IEEE International Conference on Network Protocols*, Atlanta, GA, pp. 336–345, November 2003.

[26] H. Luo and S. Lu. Ubiquitous and robust authentication services for ad hoc wireless networks. Technical Report TR-200030, Department of Computer Science, UCLA, 2000.

[27] L. Lamport, R. Shostak, and M. Pease. The byzantine general problem. *ACM Transactions on Programming Languages and Systems, 4(3)*, 382–401, July 1982.

[28] M. Castro, P. Druschel, A. Ganesh, A. Rowstron, and D.S. Wallach. Secure routing for structured peer-to-peer overlay networks. In *Proceedings of the 5th Symposium on Operating Systems Design and Implementation*, Boston, MA, pp. 299–314, December 2002.

[29] V. Pathak and L. Iftode. Byzantine fault tolerant public key authentication in peer-to-peer systems. *Computer Networks, 50(4)*, 579–596, March 2006, Elsevier.

[30] C. Buragohain, D. Agrawal, and S. Suri. A game theoretic framework for incentives in P2P systems. In *Proceedings of the 3rd IEEE International Conference on Peer-to-Peer Computing*, pp. 48–56, Linköping, Sweden, September 2003.

[31] Th.G. Papaioannou and G.D. Stamoulis. Reputation-based policies that provide the right incentives in peer-to-peer environments. *Computer Networks 50(4)*, 563–578, 2006, Elsevier (special issue on management in peer-to-peer systems: trust, reputation and security).

[32] E. Palomar, J.M. Estevez-Tapiador, J.C. Hernandez-Castro, and A. Ribagorda. A P2P content authentication scheme based on Byzantine agreement. In *Proceedings of the International Conference on Emerging Trends in Information and Communication Security*, pp. 60–72, Germany, June 2006.

[33] S. Balfe, A. Lakhani, and K. Paterson. Trusted computing: Providing security for peer-to-peer networks. In *Proceedings of the 5th IEEE International Conference on Peer-to-Peer Computing*, pp. 117–124, Konstanz, Germany, August 2005.

[34] R. Yahalom, B. Klein, and T. Beth. Trust relationships in secure systems: A distributed authentication perspective. In *Proceedings of the IEEE Computer Society Symposium on Security and Privacy*, Oakland, CA, pp. 150–164, May 1993.

[35] A. Boukerch, L. Xu, and K. EL-Khatib. Trust-based security for wireless ad hoc and sensor networks. *Computer Communications, 30(11–12)*, 2413–2427, September 2007.

[36] L. Eschenauer, V.D. Gligor, and J. Baras. On trust establishment in mobile ad-hoc networks. *Security Protocols, LNCS Vol. 2845/2003*, pp. 47–66, Springer, Berlin/Heidelberg, 2004.

[37] G. Theodorakopoulos and J.S. Baras. Trust evaluation in ad hoc networks. In *Proceedings of the ACM Workshop on Wireless Security*, Philadelphia, pp. 1–10, October 2004.

[38] F. Zambonelli, M. Gleizesb, M. Mameia, and R. Tolksdorf. Spray computers: Explorations in self-organization. *Pervasive and Mobile Computing, 1(1)*, 1–20, March 2005.

[39] V. Srinivasan, P. Nuggehalli, C. Chiasserini, and R. Rao. Cooperation in wireless ad hoc networks. In *22nd Annual Joint Conference of the IEEE Computer and Communications Societies*. San Francisco, CA, March 2003.

[40] M.O. Pervaiz, M. Cardei, and J. Wu. Routing security in ad hoc wireless networks. *Network Security*. 2005, Springer.

[41] M. Haenggi and D. Puccinelli. Routing in ad hoc networks: A case for long hops. *IEEE Communications Magazine*, 43(10), 93–101, October 2005.

[42] S. Marti, T.J. Giuli, K. Lai, and M. Baker, Mitigating routing misbehavior in mobile ad hoc networks. In *Proceedings of MobiCom*, Boston, MA, pp. 255–265, August 2000.

[43] L. Buttyán and J.-P. Hubaux. Stimulating cooperation in self-organizing mobile ad hoc networks. *Mobile Networks and Applications, 8*, 579–592, 2003.

[44] K. Sanzgiri, B. Dahilly, B.N. Levine, C. Shieldsz, and E.M. Belding-Royer. A secure routing protocol for ad hoc networks. In *Proceedings of the 10th IEEE International Conference on Network Protocols*, Paris, France, 2002.

[45] S. Buchegger and J.-Y. Le Boudec. Performance analysis of the CONFIDANT protocol (Cooperation Of Nodes Fairness In Distributed Adhoc NeTworks). In *Proceedings of the ACM Symposium on Mobile Ad Hoc Networking and Computing (MobiHOC)*, Lausanne, Switzerland, 226–236, June 2002.

[46] M. Conti, E. Gregori, and G. Maselli. Cooperation issues in mobile ad hoc networks. In *Proceedings of the 24th International Conference on Distributed Computing Systems Workshops*, Hachioji, Tokyo, Japan, pp. 803–808, 2004.

[47] W. Tolone, G. Ahn, T. Pai, and S. Hong. Access control in collaborative systems. *ACM Computing Surveys, 37(1)*, 29–41, March 2005.

[48] X. Zhang, S. Chen, and R. Sandhu. Enhancing data authenticity and integrity in P2P systems. *IEEE Internet Computing*, 9(6), 42–49, November–December 2005.

[49] E. Palomar, J.M. Estevez-Tapiador, J.C. Hernandez-Castro, and A. Ribagorda. Secure content access and replication in pure P2P networks. *Computer Communications*, 31(2), 266–279, 2008.

[50] R. Gupta and A.K. Somani. Game theory as a tool to strategize as well as predict nodes behavior in peer-to-peer networks. In *Proceedings of the 11th IEEE International Conference on Parallel and Distributed Systems*, Lisbon, Portugal, pp. 244–249, July 2005.

[51] J. Shneidman and D.C. Parkes. Rationality and self-interest in peer to peer networks. In *Proceedings of the 2nd International Workshop on P2P Systems*, pp. 139–148, Berkeley, CA, February 2003, Springer-Verlag.

[52] B. Yang, T. Condie, S. Kamvar, and H. Garcia-Molina. Non-cooperation in competitive P2P networks. In *Proceedings of the 25th IEEE International Conference on Distributed Computing Systems*, Columbus, OH, pp. 91–100, June 2005.

[53] Y. Zhang, L. Lin, and J. Huai. Balancing trust and incentive in peer-to-peer collaborative system. *International Journal of Network Security, 5(1)*, 73–81, July 2007.

[54] M. Feldman, K. Lai, I. Stoica, and J. Chuang. Robust incentive techniques for peer-to-peer networks. In *Proceedings of the 5th ACM Conference on Electronic Commerce*, pp. 17–20, New York, May 2004.

[55] C. Dwork and M. Naor. Pricing via processing or combatting junk mail. In *Proceedings of the 12th Annual International Cryptology Conference on Advances in Cryptology*, Santa Barbara, CA, Vol. 740, pp. 139–147, August 1992.

[56] A. Juels and J. Brainard. Client puzzles: A cryptographic defense against connection depletion attacks. In *Proceedings of the Networks and Distributed Security Systems*, pp. 151–165, California, February 1999.

[57] E. Palomar, J.M. Estevez-Tapiador, J.C. Hernandez-Castro, and A. Ribagorda. Certificate-based access control in pure P2P networks. In *Proceedings of the 6th IEEE International Conference on Peer-to-Peer Computing*, pp. 177–184, Cambridge, U.K., September 2006.

[58] G. Arboit, C. Crépeau, C.R. Davis, and M. Maheswaran. A localized certificate revocation scheme for mobile ad hoc networks. *Ad Hoc Networks, 6(1)*, 17–31, January 2008.

[59] W. Wang and T. Stransky. Stateless key distribution for secure intra and inter-group multicast in mobile wireless network. *Computer Networks, 51(15)*, 4303–4321, October 2007.

[60] B. Biskupski, J. Dowling, and J. Sacha. Properties and mechanisms of self-organizing MANET and P2P systems. *ACM Transactions on Autonomous and Adaptive Systems, 2(1)*, 1–34, March 2007.

18 Application Cooperation in Wireless Mesh Networks

Mohamed El-Darieby, Hazim Ahmed,
Baher Abdulhai, and Yasser Morgan

CONTENTS

There are many challenges for enabling collaborations among municipal wireless networks (a.k.a., wireless mesh networks [WMN]) applications. First, mesh clients are highly mobile nodes with intermittent network connectivity. Second, collaborative applications are challenged by heterogeneity among devices in terms of operating system and hardware capabilities. In this chapter, we describe the role of middleware services in facilitating transparent resource sharing among devices in automated manner. We also describe our implementation of these technologies and how they may be deployed on an open and programmable WMN platform. Specifically, we describe a service-oriented wireless grid architecture (SOWGA) that allow wireless devices to autonomously share resources in a collaborative manner. SOWGA matches an application task's need of resources with the most suitable set of available resources to serve this particular task. Transparent discovery of resources is facilitated by hiding possible heterogeneity of devices and resources. In addition, SOWGA transparently and efficiently manage connectivity so that devices remain optimally connected to required services. These distributed functionalities enable devices that have no prior knowledge of each other to independently share resources and collaboratively execute application tasks that cannot be supported individually. In this chapter, we focus on service discovery, brokering, and roaming components of the architecture. The service discovery enable wireless devices to autonomously share their resources. The service brokering component matches the resource needs of application to available devices. The proposed context-aware Service Roaming Protocol uses an XML-based service profile to efficiently manage services within a highly dynamic and ad hoc wireless networks. The Service Roaming Protocol uses mobile user's geographic location, terminal's capabilities, network topology, and grid devices to provide efficient management of mobile users and services. As a proof of concept, the protocols are implemented for the IEEE 802.15 (Bluetooth) standards running on top a wireless mesh network deployment.

18.1 INTRODUCTION

WMN are poised to be the infrastructure enabling the next generation of value-added applications in public safety, public services, business, and entertainment [5,15]. WMNs augment computing devices with spatial and contextual characteristics of the surrounding environments in addition to mobility and economical broadband connectivity. To fully realize the potential of these capabilities, wireless devices are supplemented by the ability to collaborate with other devices in their vicinity while roaming freely. Collaborative capabilities provide large aggregate capacity that facilitates new types of applications such as dynamic 3D visualization of the device's surrounding environment [1] and virtual mobile communities. Such applications provide the end user with the required services while hiding the process of combining resources from different devices on the network. Examples of resources include CPU cycles, network bandwidth, storage systems, software modules and instruments (for example, sensors), spatial and contextual information, and other data files.

WMNs are also known as municipal wireless networks. WMNs are dynamically self-organized and self-configured systems. Figure 18.1 presents a WMN reference model. WMNs consist of a set of self-configuring mesh routers that form and maintain a wireless infrastructure. Autonomous mesh routers simplify WMN installation and maintenance. Routers exchange data packets over, possibly, mobile multi-hop. Mesh routers may provide gateway access to a fixed structure network or to the Internet. In addition, each mesh router supports a communication zone that is typically accessed by roaming mesh clients. Wireless links in Figure 18.1 can employ the same or variant wireless technologies. Roaming mesh clients (a.k.a. devices) can be desktops, laptops, or cell phones. Generally speaking, WMNs are typically deployed in a quasi stationary manner, where some mesh routers are stationary, such as in municipal metro area networks or networks within buildings such as enterprise buildings or convention centers. In other situations, WMNs are created on demand and

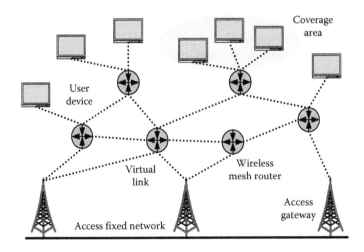

FIGURE 18.1 WMN reference model.

viewed as quasistationary such as in disaster recovery and outdoor situations. It is expected that enabling collaborative applications on top of WMN leverages returns on investments [7].

However, there are many challenges to enabling collaborations among WMN applications. First, the dynamic nature of WMN affects application design. Mesh clients are highly mobile nodes that they may face intermittent network connectivity conditions (for example, changes in network link quality and availability). The current Transmission Control Protocol/Internet Protocol (TCP/IP) architecture does not work well for this nature. Second, collaborative applications are challenged by hetero-geneity among devices in terms of operating system and hardware capabilities. Mesh clients can be handheld devices with limited CPU, memory, and storage capacities. This calls for interface-based loose coupling between collaborating nodes. Other challenges include quality of service (QoS) and experience and security assurance.

To enable collaboration for WMN, we focus on describing the role of middleware services in facilitating transparent resource sharing among devices in automated manner. We describe our implementation of these technologies and how they may be deployed on an open and programmable WMN platform. We also describe two different application examples where these technologies may be used to enable application collaboration.

Specifically, we describe a SOWGA that includes resource discovery, and brokering protocols allow wireless devices to autonomously and independently share resources in a distributed manner. The middleware uses service protocols to match the needs of an application task with the most suitable set of resources to serve this particular task. The proposed model can be extended to describe diverse types of resources ranging from hardware (for example, CPUs) to software services. Such a model facilitates a transparent discovery of resources by hiding possible heterogeneity of devices and resources. It also allows a wireless grid to support a wider range of collaborative applications. In addition, middleware services transparently and efficiently manage connectivity so that devices remain optimally connected to required services. These distributed functionalities enable devices that have no prior knowledge of each other to independently share resources and collaboratively execute application tasks that cannot be supported individually. The protocols enable devices to face challenges imposed by the dynamic nature of WMN such as variable and fading link quality [12]. Traditional software solutions are unsuitable for collaborative applications running in the stringent and dynamic WMN environment.

The chapter is organized as follows. Section 18.2 introduces the importance of supporting collab-orative WMN applications and presents an example WMN application. Section 18.3 describes how service management architecture and components enable such applications. Service discovery and brokering and service roaming management components and algorithms are described. In Section 18.4, we describe how service management takes advantage of the WMN wireless standards and investigate the IEEE 802.15 approach to building WMN. Conclusions are listed in Section 18.6.

This chapter is fundamental in combining the various technologies of middleware and communi-cations standards and defines the use of each technology in addition to defining the interoperability issues. This chapter offers strong bases for researchers investigating solutions in this domain by identifying common terms and interoperability mechanisms

18.2 EXAMPLE APPLICATIONS

One example application is the distribution of a high-definition audio/video file (A/V) within a digital building. The A/V file can be available at the security control room equipped with multi-radio wireless communication interfaces. A security guard can be watching the AV in a central monitoring room and then decide to patrol the building while keeping an eye on all entrances using his/her thin client (for example, mobile phone or watch) as he/she moves to other locations within the building. A similar example would be the distribution of the AV file within a digital home, where the file can be available at the home theater, which is equipped with multi-radio wireless communication interfaces. A user can be watching the AV on his/her TV in the living room and want to continue watching the video and audio streams on his/her thin client (for example, mobile phone or watch) as he or she moves to other locations within the home.

Another dimension for applications collaboration is in the intelligent transportation systems (ITS) [16] where the municipal WMN infrastructure is used to gather and transport real-time traffic data. In this scenario, a car (or a cell phone) can be programmed to discover information about route guidance maps available in other cars (or cell phones) and if desired, the map can be downloaded. Centralized ITS route guidance services find the shortest travel time from origin to destination in real-time. The resulting information can be encoded in an information-rich map that may contain other information such as pictures of traffic and information about shopping areas, schools, and gas stations along the route.

To support such scenarios, the thin client must be able to discover other wireless nodes around and choose to connect and choose between services provided by each. For this to happen autonomously, the services must be described in a standard and expressive manner and the user must be able to define policies/preferences on how to use the service, for example, least amount of data rate on a link to a device. Moreover, the streaming service must be implemented with the possibility of intermittent connectivity in mind. The user may also want to pause the streaming at any moment and continue the streaming with a push of a button. In addition to these requirements, the situation is complicated if the user's device does not support source video format and is limited in computational resources to perform video encoding. Collaboration among other devices, such as a desktop computer in the office room and a Personal digital assistant (PDA) in another room, can help meet user required application.

18.3   RELATED WORK

In this section, we review related communication and middleware technologies. For the communication layers, we discuss wireless technologies used to build WMN and the capabilities they provide to application collaboration. We focus on the IEEE 802.11 and the IEEE 802.15 family of standards and on QoS (for example, packet delay and jitter limits), mobility, and security management functionalities. To review the middleware layers we cover two main paradigms, namely, wireless grids and wireless service-oriented architecture (SOA). The wireless grid paradigm [1,10] is proposed to allow devices with no prior knowledge of each other to share their services to collaborate in executing a given application task. The following are requirements for a wireless grid middleware: resource description, resource discovery, coordination mechanism, trust establishment, and clearing mechanism were identified in Ref. [10]. We believe these requirements are better augmented with resource brokering and roaming to facilitate more effective sharing and collaborations. The SOA considers services as commodities that can be adaptively selected, used, rented, and shared by applications. SOA focuses on defining service description, advertisement, and lookup mechanisms in infrastructure-based and ad-hoc environments. Existing work on these two technologies is surveyed in the following subsections.

18.3.1   WMN COMMUNICATIONS TECHNOLOGIES

Current state-of-the-art WMN research focuses mostly on studying and analyzing communications and networking aspects of WMN [7,12], with emphasis on using WMN as access networks using multi-hop routing techniques. WMN links exhibit variant quality over time. In addition, node mobility affects performance, guarantees, and end-to-end QoS. The socket-based architecture fails to serve the anticipated WMN applications. Research in delay-tolerant networks have suggested using a store-and-forward approach to networking, where information is sent to an intermediate station where it is kept and sent at a later time to the final destination or to another intermediate station.

An ad-hoc network is different from the WMN in the protocols and architectures, although both of them have the common concept of multi-hop traffic routing. Ad-hoc protocols performance is very unsatisfactory when applied to WMNs. The ad-hoc networks are designed for environments of high mobility; however, the WMNs are designed for minimum mobility and most of the nodes are statically deployed or the mobility is not the common behavior of the nodes.

The autonomy of mesh routers simplifies WMN installation and maintenance. Routers exchange data packets over, possibly, mobile multi-hop. WMNs typically deploy two types of routing protocols, namely, proactive and on-demand routing protocols. On the one hand, proactive routing protocols are table-driven (that is, the major part of operations of the protocol depends on building, maintaining, and using the routing table). Whenever a change happens in the topology of the WMN, control messages are propagated through the network to announce this change and update the routing tables maintained in various routers. The flooding of these messages may span the entire network, which may be disadvantageous. Optimized link state routing (OLSR) [19] is a proactive and table-driven routing protocol. It maintains the routing table updates by exchanging messages with neighbor nodes. OLSR was designed with a reduction in the length of the update packet when compared with the design of the original link state routing (LSR) protocol. In addition, OLS uses multipoint relays (MPR) to prevent network flooding. With MPR, each OLSR node selects only some neighbor nodes to forward the updates packets to. This results in a reduction in the total number of control messages, which results in better performance, specifically, in case of deployment in dynamic networks with a large number of nodes [19].

On the other hand, on-demand routing protocols, or reactive protocols, do not require this flooding of update messages. Ad-hoc on-demand distance vector (AODV) [19] routing protocol is an example of the reactive routing protocols. These protocols do not maintain routing tables for the entire network, but only requested routes are maintained. That is, routes are calculated only when there is a request to send data from a source node to a destination node. This process takes more time than looking up a routing table. AODV maintains vectors of destination routes and cost to use. This renders AODV as a better alternative for more static networks. In addition, AODV requires lower memory and processing power.

Current state-of-the-art WMN research is focused mostly on studying and analyzing communications and networking aspects of WMN with emphasis on using WMN as access networks using multi-hop routing techniques. This is important in enhancing the performance of WMN links and QoS [9]. In this chapter, we focus on building applications that exploit the availability and cost-effectiveness of WMN in different municipalities.

The IEEE 802.11 family of standards incorporates essential techniques for supporting collaborative applications. This chapter focuses on the following three standards: First, the inter access point protocol (IAPP) (802.11 F & R), which specifies the information exchanged between 802.11 access points to support mobility and handoff. The handoff is accompanied by context caching to enable faster roaming (around 50 ms) and to improve QoS. Second, the mesh networking (802.11 S), which identifies the self-healing and self-configuring standards including devices discovery and related routing tables updates. Third, the (802.11 U) defines the inter-working with external networks standards that enables 802.11 devices to cross-communicate with non-802 devices.

In parallel to the WMN evolution, a persistent development of IEEE 802.15 (TG5) is fundamental to WMN prospect. The basic idea is to enhance device attachment to the WMN by allowing continuous connectivity and reconfiguration around blocked or broken paths by hopping from node to node. The nodal hopping takes place over a homogeneous wireless technology (homogeneous/horizontal hopping) or over heterogeneous wireless technologies (heterogeneous/vertical hopping). Mesh nodal hopping is essentially a powerful ingredient in sustaining the WMN self-healing property.

18.3.2   WIRELESS GRID PARADIGM

To illustrate the wireless grid concept, McKnight et al. [10] present a distributed ad-hoc resource coordination (DARC*) framework. DARC* is layered on top of a TPC/IP protocol stack to allow devices with no a priori knowledge of each other to collaboratively record and mix audio signals. Resources are described using a simple string and discovered using multicast DNS. Coordination (for example, connecting a sound mixing device with sound recording devices) is performed by exchanging application messages.

A motivating research by Yamin et al., Infraestrutura de Software para as Applications Móveise distribuídas (ISAM) project [1], is aimed at building a pervasive computing environment that supports the development of distributed mobile applications with adaptive behavior. The ISAM architecture considers context to be of paramount importance in managing its applications. ISAM implements a follow-me-semantics based on the support of wired infrastructure and servers. The follow-me semantic is similar to SOWGA service roaming management. The central difference between ISAM project and SOWGA is that SOWGA supports decentralized management and service mobility for mobile devices communicating in a peer-to-peer (P2P) fashion. ISAM project assumes that mobile devices use basic services deployed over a wired fixed network, and consequently its application is limited to fixed wireless networks.

Another approach is the mobile collaboration architecture (MoCA) [15] project. MoCA is a middleware architecture for developing context-processing services and context-sensitive applications for mobile collaboration. The MoCA architecture supports mobile collaboration and implements a flexible and extensible, XML-based service-based environment for developing collaborative applications for fixed wireless networks. The SOWGA approach differs from MoCA in a number of ways. SOWGA context-aware resource management and service roaming protocols support ad-hoc P2P collaboration. SOWGA components are executed on all participating grid devices, such as the decentralized service database, service discovery, context manager, and service brokering and roaming protocol. MoCA relies on fixed infrastructure to manage the mobility for mobile wireless devices, such as the context information service (CIS) executing on one or more nodes of the wired network. SOWGA supports dynamic brokering resources and runtime roaming of services for mobile devices.

The Expeerience project [11] exploits Java-based code mobility to support distribution and execution of tasks in mobile ad-hoc networks. Similar to SOWGA, a common data structure called advertisements is used to describe resources. However, SOWGA diverges from Expeerience by offering explicit middleware support for resource brokering. Furthermore, SOWGA resource management protocols accommodate arbitrary resource access mechanisms. Consequently, it does not require wireless devices to support Java-based technologies such as code mobility, object serialization, and remote method invocation.

SOWGA differs from these architectures in several ways. First, SOWGA employs a more formal and common resource description model. The model can be extended to describe diverse types of resources ranging from hardware devices (for example, Central Processing Unit (CPU)) to software services. Such a scheme facilitates transparent resource discovery by hiding possible heterogeneity of devices and resources. Further, it also allows a wireless grid to support a wider range of collaborative applications. Second, SOWGA supports resource management tasks to help enforcing complex resource usage policies and facilitating effective matchmaking between application tasks and devices. Finally, it also supports a wider range of applications.

In general, existing wireless grid architectures lack sufficient resource management protocols to ensure that mobile users' services are provisioned based on the current context of the operating environment, including terminal capabilities, economical factors, and user's preferences. Context-aware wireless grid architecture, however, can provide such collaboration to efficiently provision mobile services within a wireless grid, and across multiple grids, of various network infrastructures and topologies.

18.3.3 SERVICE-ORIENTED ARCHITECTURE

It is believed that none of the infrastructure-based and ad-hoc-based SOA is designed to take advantage of WMNs.

Infrastructure-based SOA technologies, such as the web services architecture, are designed for fixed enterprise environments. These technologies often require powerful devices with sufficient computation and communication capabilities. Technologies such as Bluetooth service discovery protocol (SDP) [2], and Konark [4] are designed for ad-hoc operations. SDP is designed to handle situations characterized by fairly limited changes to device and service availability. Services other than those described in the Bluetooth specifications cannot be supported by SDP, which is a major challenge in WMN applications. The Konark project proposes a P2P middleware designed for service discovery and consumption in ad-hoc networks. However, Konark is based on the IP protocol and does not provide the full functionality of service provisioning [4]. The reader is referred to Refs. [21,22] for a detailed discussion on SOA in wireless environments.

Notably, Bluetooth-enabled wireless devices have basic resource management capabilities built into them. They can support a set of standard services (for example, printing service, camera service) that provide access to resources. The Bluetooth SDP can be used by devices to discover and consume the services offered by remote devices. SDP is, however, insufficient for the requirements of wireless grids. First, services other than those described in the Bluetooth specifications cannot be supported. SDP is sufficient and successful as a "cable replacement" technology that connects well-known resources (for example, printers, cameras) together. However, in wireless grids devices need the ability to support arbitrary types of services and applications. Second, SDP is designed to handle situations characterized by rapid changes to device and service availability. Finally, Bluetooth SDP is often not expressive enough to facilitate fine-grained control over resource selection [4]. A service is described only by its Universally Unique IDentifier (UUID) or its service class [2]. This limits SDP usage to networks with a smaller number of services. Furthermore, Bluetooth is not focused on resource brokering issues.

Throughout this chapter we use the term Enhanced SDP (or ESDP) [14] to refer to the enhancements to the semantic description of Bluetooth SDP presented in Ref. [14]. The ESDP uses DARPA agent markup language and ontology interface layer (DAML-OIL). The enhanced description is used by a reasoning engine to perform fine-grained matching of service consumers to service providers [14]. Our preliminary experiments show that such reasoning-based service matching incurs significant overhead. This chapter presents a lightweight XML-based service description that affords better expressiveness and extensibility than SDP while limiting the overhead as ESDP. References [21,22] provide a comprehensive review of service description and discovery protocols for mobile ad-hoc systems.

The MIT CarTel project [3] represents one of the most relevant research projects in WMN. CarTel uses traveling vehicles as data collecting sensors. The solution is based on a "carry-and-forward" delay-tolerant network stack that delivers data in intermittently connected networks. The reader is referred to Refs. [7,12] for a detailed survey of these WMN aspects. CarTel uses a simple and centralized programming model that is different from the type of applications this proposal addresses. Also, CarTel ignores issues of collaboration and resource sharing among mesh nodes, which are addressed by wireless grids and SOA.

Finally, resource management mechanisms used in wired grids cannot satisfy all the requirements of wireless grids. Wired grid middleware such as GLOBUS [http://www.globus.org] is designed for enterprise environments with fixed networks and, typically, predefined and static resources. Wireless networks, on the other hand, are characterized by rapid, unpredictable changes in device availability and connectivity. Such an environment requires more dynamic and context-aware resource management architectures. Furthermore, executing wired grid middleware on wireless devices may be computation and power intensive.

18.4 SERVICE MANAGEMENT PROTOCOLS

This section describes how SOWGA allows automatic and transparent resource sharing among WMN devices. The SOWGA service discovery automatically discovers and maintains a list of services available on WMN devices. SOWGA resource broker component matches the requirements of a given application task to the services available on WMN devices. The services are selected to satisfy the collaboration and QoS requirements for the task. The heterogeneities among WMN devices are hidden via an XML-based service description.

SOWGA proposes an autonomic architecture through which a device locates other devices suitable for performing a particular task. We define task as any application activity that consumes device resources. Executing a program on a remote device, collecting measurements from a wireless sensor, and downloading video from a device are all examples of a task. Service is defined as any entity that is required for the completion of a task. Each resource is associated with a service. A service is described in an XML file, which provides detailed information about the service and its related context parameters and management policies. Such context parameters include location, service provider(s), and their power or capabilities, security, and billing information. The service interface allows a task to consume a service by providing a standard way of interacting with the service. A wireless device can support several services. For the rest of the chapter, the terms resource and service are used interchangeably.

18.4.1 SERVICE DESCRIPTION

A common XML-based schema is used to describe services. The schema is used to specify service functionality, capabilities, and usage policies in clear syntax. Using a common schema hides the potential heterogeneity of devices within a wireless grid. The schema also allows flexibility in describing resources. This helps support complex types of resources that may be involved in sophisticated collaborative applications. In addition, service description can include enough information to allow brokers to locate the services best suited for a task. Figure 18.1 shows an example of an XML description of a wireless service. The meanings of XML tags are self-explanatory. The XML schema is designed for ad-hoc networks. That is, although the schema inherits many features of traditional web services, the schema also includes information that suits WMN operations and challenges. For example, the schema may include information pertaining to service mobility and location. The schema may include information on motion plans of service provider, on possible alternate service providers as signal strength of the first provider fades away, and the implied security and trust information.

A service description consists of a set of XML tags that describe service attributes, contextual information, service provider and consumer information, and preference. The service header section of the description contains attributes that are mandatory for all services. These attributes include ServiceType, ServiceID, ServiceName, and ServiceStatus. The ServiceContext section describes the contextual properties of the service including location and mobility status. The ServiceProvider section provides information such as available CPU, memory, and battery power at the providing device. These attributes are used during resource brokering to choose between providers. The User-Preferences section contains attributes that pertain to the individual user preferences and service policies. Examples of such attributes include preferred language destination and personal agenda. The user can include service provisioning policies that will be used to search for the most suitable service available. The ServiceConsumer section provides information about the client devices (service consumer device). Such information includes devices address, current location, and direction. These attributes will be used by the grid and other service providers during service delivery and roaming. In addition, service description must also address security and access control.

18.4.2 SERVICE DISCOVERY

The service discovery (SD) protocol is responsible for maintaining a list of services available in proximity to each wireless device. The list is used to populate a service database (SDB). This component also updates the SDB in response to dynamics of device movements such as devices joining or leaving the grid. We assume that during SOWGA initialization, each SD running on each device creates an SDB for each device and populates it with a list of services available locally on the device. We assume these services are predefined for each device; however, the characteristics of the service may change over time.

The steps for SD are as follows. The SD of a device announces a SD request. The SD on a remote device queries its SDB and compiles a list of available services. The list of services is sent back to the requesting device and is added to the SDB of the requesting device. These steps are shown in Figure 18.2.

Because devices can dynamically join and leave the grid, the SD has to be self-configuring to keep the SDB updated. An SD that wishes to register a new service or deregister an existing service

```
<ServiceProfile>
        <WirelessService>
                <ServiceHeader>
                        <ServiceType> Multi Media -  High Def AV </ServiceType>
                        <ServiceID>0001 </ServiceID>
                        <ServiceName> My Favorite AV </ServiceName>
                        <ServiceVersion> 1.00 </ServiceVersion>
                        <ServiceDescription> For All Ages  </ServiceDescription>
                        <ServiceStatus> Available </ServiceStatus>
                </ServiceHeader>
                <ServiceContext>
                        <ServiceData>
                                <VideoEncoder> MPEG4 Part 10 AVC</VideoEncoder>
                                <CompletionPercentage> 0% </CompletionPercentage>
                        </ServiceData>
                        <ServiceProvider>
                                <ProviderID> Office Computer </ProviderID>
                                <AvailableCPU>900 MHz </AvailableCPU>
                                <AvailableMemory> 300 MB</AvailableMemory>
                                <BTLinkQuality> 150 Kb/s</BTLinkQuality>
                                <Location> Office </Location>
                        </ServiceProvider>
                        <ServiceLocation> Family Room </ServiceLocation>
                        <Security>N/A    </Security>
                        <QoS>
                        </QoS>
                          <Billing>
                                <Cost> Free </Cost>
                        </Billing>
                </ServiceContext>
        </WirelessService>
        <UserPreferences>
                        <Language> English </Language>
                        <Destination> Neighborhood Walk-in Groceries </Destination>
                        <UserProvisioningPolecies>
                                <BTLinkQuality> >= 300 Kb/s </BTLinkQuality>
                                <Provider'sAvailableCPU> >= 500 MHz </Provider'sAvailableCPU>
                                <MonetaryValue> Free </MonetaryValue>
                        </UserProvisioningPolecies>
        </UserPreferences>
        <ServiceConsumer>
                        <ConsumerID>piconet1.BT_ADDR </ConsumerID>
                        <Location> Home </Location>
                        <Direction> Grocery Store </Direction>
        </ServiceConsumer>
</ServiceProfile>
```

FIGURE 18.2 AV service profile.

sends a message to other devices. Other devices update their SDBs accordingly. The SD relies on the communication capabilities of the wireless device protocol stack to perform discovery. For example, in Bluetooth, service announcements may involve L2CAP connections. In a WMN, a device may become unavailable without explicitly deregistering its services. We need to consider this in future work.

18.4.3   RESOURCE BROKER

The resource broker (RB) component matches application task requirements with available WMN services. Services are selected to satisfy the collaboration and QoS requirements for the task. For example, a task that needs to download a video may specify the average download speed for the transfer. Similarly, considering the example of a sensor network monitoring the stresses on a bridge, a device may specify that it needs information from sensors embedded at specific locations. The process is similar to that used by wired grid systems.

A typical service brokerage scenario starts with the RB receiving task requirements from an application. These requirements are expressed in terms of the following:

1. The type of resources required. These are specified as desired service types.
2. For each type of resource, the number of resources required and the duration for which they are required.
3. For each requested resource, the desired properties of the resource. These are expressed in terms of desired values for attributes of the service type corresponding to the resource.

The RB searches the SDB to locate a set of services that manage resources matching these criteria. It then tries to allocate the resource to the task at hand. This involves checking service management policies and current usage of the service resource by other applications. If the resource requirements are satisfied, the RB returns and assigns these to the application.

The RB is also invoked when there is a need to switch service providers in response to a trigger due to changes in contextual service information. The RB follows an algorithm similar to that described here to rebind service consumers and providers in response to, for example, mobility or the availability of a "better" service provider. The criteria for defining a "better" service provider may be defined in the service description.

18.4.4   SERVICE ROAMING PROTOCOL

The service roaming protocol (SRP) enables continuous service consumption in spite of changes that may occur to network links or to a service provider node. This includes issues such as the unavailability of a service provider node due to mobility or shortage in power or to the availability of a "better" service provider (for example, with more CPU power, a higher bandwidth connectivity, or more battery life).

During service execution, the SRP can roam the service between different devices to offer the best available experience to the user as defined in user requirements. The SRP continually searches for the best network resource among available WMN resources. The SRP relies on the service discovery and brokering protocols to efficiently manage available resources.

A context change may trigger roam sequence of a service during consumption. Examples of such context changes or triggers include discovery of a more resourceful provider device (for example, a device with higher bandwidth connectivity or a more stable link). The SRP uses geographic location, terminal's capabilities, network topology, or personal profile to provide efficient management of mobile users and services. Other context changes or triggers include service providers or consumers leaving the grid.

The suitability of a given device to execute a service is based on a set of threshold values of its context information. SRP considers service provisioning policies to efficiently roam services among

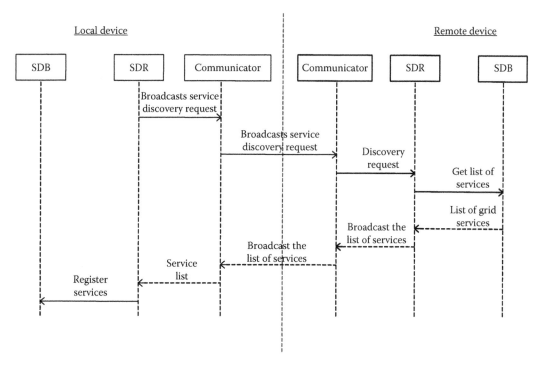

FIGURE 18.3   UML sequence diagram for service discovery and registration.

service providers and across wireless networks. The service provisioning policies answers users' preferences, application requirements, and device capabilities. For example, the user can specify a monetary value for the consumed service or a limited time to provide the service. The protocol can roam services within a grid or across multiple grids. Upon the detection of a local context change, the service discovery and brokerage components are notified to announce the change and to advise other grid devices. The context manager monitors device resources (for example, free CPU, memory, link quality, battery level, location) and issues an event if a threshold value is exceeded. Changes to the status of a service are maintained in a service profile shown in Figure 18.3.

18.5   IMPLEMENTATION AND PERFORMANCE EVALUATION

As a proof-of-concept, we implemented the service discovery, brokering, and roaming protocols with Java [2]. In the current SOWGA implementation, a Bluetooth Piconet corresponds to the wireless grid. The Piconet is configured to enable broadcast capabilities for every device. Typically, a piconet has 10–20 m range. However, high-powered versions have much higher ranges. Application tasks interact with the service interface by establishing object exchange (OBEX) [2] connections. Within each piconet, there is a master node (a.k.a. aceess point) that has Bluetooth and Wi-Fi interfaces. We equipped the wireless devices with ultra-slim Bluetooth V1.2 dongles with USB 2.0 and operation range of 100 m with built-in antenna and maximum data rate of 3 MB.

To implement the SDR and SRP protocols, we used Java 2.0 Micro Edition (J2ME), Java Community Process API for Bluetooth (JSR-082) available http://www.jcp.org/ aboutJava/ communityprocess/final/jsr082/ and described in Ref. [2], kXML [at: http://kxml.sourceforge.net/], and Java NetBeans 4.5 as a development environment. We conducted some preliminary experiments to evaluate the performance of SOWGA service discovery protocol, referred to hereinafter as XSDP.

For our wireless mesh network, we deployed AP of type ASUS WL-500g Premium AP with a 266 MHz and 8M flash CPU, 32M Random Access Memory (RAM), an external dipole antenna, and 2 USB 2.0 ports. Each AP is configured to run the OpenWRT (http://www.openwrt.org) Linux distribution for embedded devices. Each AP is configured to run OLSR. To enable OLSR message exchange, the OpenWRT firewall has to be opened at port 698 and forwarding rules must be added to the firewall configuration files. One of the APs is configured as a gateway to be connected to the Internet. The network model consists of Wi-Fi access points (AP) mounted on posts. The APs can be laid out at arbitrary distances depending on the antenna range and line-of-sign conditions.

OpenWRT is a Linux distribution that is optimized for execution on Wi-Fi access points. It was first released in January 2004 as a GPL source to work on Linksys WRT54G. Updates to the Open-WRT continued since then with two major versions, namely the White Russian and Kamikaze, that were announced. These updates provided a modified OS core and hardware drivers, and many more (more than 1354) Linux packages that are customized to work on OpenWRT. The new OpenWRT version also supports (more than 100) access points. OpenWRT provides an easy to use package management interface that is used to install precompiled packages. In addition, it provides a software development kit (SDK) that is used to compile custom code into a package to be installed on different access points.

We attach Bluetooth dongles to AP through USB ports. We integrated BlueZ, which is an open-source Linux-based implementation of the Bluetooth stack, in each AP. BlueZ is a Bluetooth stack implementation that complies with the Bluetooth special interest group specification. BlueZ is the official Bluetooth stack implementation for Linux [http://www.bluez.org]. BlueZ uses Bluetooth dongles to scan Bluetooth devices in range. BlueZ retrieves standard Bluetooth information from the devices and make them accessible to software running on the APs. We process this information using a Linux shell script. With this setup, each AP communicates with other nodes using Wi-Fi or Bluetooth. BlueZ is controlled to scan the wireless medium for Bluetooth devices in proximity.

We developed a Linux shell script that is capable of retrieving the BT address of nearby Bluetooth devices. The script gathers other information about the detected device including the time a device is detected. The gathered information is then relayed to car tracking server on the Internet using TCP/IP. The route from the AP that detects the Bluetooth device to the server is controlled by the WMN routing tables maintained at each AP and configured automatically by the OLSR protocol.

18.5.1 INTRAGRID USER SCENARIO

An example of a possible user scenario starts in a family room as the user "pairs" his handheld device to the home theater to consume a specific AV streaming service, which is discovered via SOWGA's SDR. The SOWGA Profile Manager creates a service profile that includes information pertaining to the service, the client's device capabilities, location, and possible user mobility information such as destination. An example service profile is shown in Figure 18.3. Potential compatibility problem is discovered as the home theater does not support encoding formats supported by the handheld device. In this case, the home theater establishes a connection with the office computer to perform the required encoding of the multimedia file. The computer streams the appropriately encoded file to the home theater. The home theater operates only as a repeater that forwards the data packets to the handheld device. Then, the service is started and the AV is streamed to the handheld device. SOWGA profile manager starts maintaining status information about the service (for example, number of seconds/frames streamed). In addition, the SOWGA context manager starts monitoring service triggers such as connection link quality, user location, and other vital context information.

Being part of the home wireless grid, the user device can use the collaboration of all available devices in delivering the service as defined in the service profile. In this scenario, the CPUs of different in-range devices can be shared to perform content tuning (for example, video coding) to adapt to limitations of thin mobile clients. While streaming, the user decides to go to the backyard. As the user moves, the link quality with the home theater degrades and improves with the office computer. The SOWGA context manager detects changes in the service levels and issues a trigger. The SOWGA roaming manager detects this trigger and starts pairing with the office computer by sending the computer the service profile, including service status information. A soft handoff takes

Service consumer mobile device

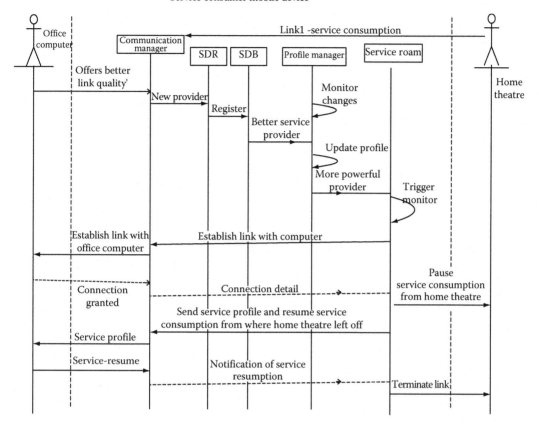

FIGURE 18.4 UML sequence diagram service consumption.

place, where a connection between the handheld device and the office computer is established, notifying the living room home theater to stream the service to the office computer. The office computer will in turn stream the service to the handheld device. Finally, the link between the home theater and the handheld device is broken. The user's device resumes receiving the video from the office computer. This roaming process is depicted in the UML sequence diagram of Figure 18.4.

The above roaming trigger is based on a predefined provisioning policy pertaining to the link quality between service consumer and service providers. The user can define other service provisioning policies. For example, location and destination coordinates can be used, such as that the user device can broadcast the service profile to all grid devices or only to selected devices based on its route during mobility. Another provisioning policy can specify the properties of preferred service providers to pair with. For instance, this policy can be configured to roam the AV stream from one service provider to another based on the available CPU, battery power, or any other vital device property that may affect the performance, service availability, and delivery. A third service provisioning policy can be configured to trigger the roaming sequence based on the credit charge of a service by different service providers. The SOWGA roaming manager can then roam the service to the most economical provider. These service provisioning policies can be defined by the user and get communicated by the local SOWGA to the wireless grid via the service profile. Such provisioning policies can enhance the user experience, service roaming, and the overall performance of the SOWGA architecture.

18.5.2   INTERGRID SERVICE MOBILITY MANAGEMENT SCENARIO

The second scenario implements intergrid service discovery and location-based service roaming management for mobile users. With the assistance of the gateway, the SDR protocol is able to discover services across multiple grids. The scenario involves four service providers, one gateway, and one service consumer device. The service providers and gateway devices are stationary, but the consumer device is mobile. Figure 18.5 illustrates this scenario.

To emulate the location of each device, friendly location names are used. For example, Provider1.1 (read provider 1 of WG one) is located at location 0, denoted as L0, in Figure 18.6. Unlike the stationary locations of providers and gateway, the location of the service consumer device is dynamic. Because the experiments were conducted using an emulator, the consumer's location changes were accomplished using a routine that changes the device's location based on a time-based formula. The Java ME location JSR 179 can be used to provide the location API. The boundaries of each WG were configured based on fixed locations. Figure 18.6 illustrates the two WGs. Wireless grid #1 extends from L0 to L4 and wireless grid #2 extends between L4 and L8. These fixed locations are programmed into the WG devices.

After both WGs are set up and discovery is performed, the scenario begins by the consumer device joining grid number one at location L0. Upon arrival, the consumer device advertises a discovery request. Provider 1.1 device, being the closest to the consumer device, responds with a copy of its SDB, which contains a current list of services and device information (for example, address and location). The consumer device then starts to download the 50 MB file from provider 1.1. As the consumer device moves from L0 to L1, it realizes, using the information in its SDB, that Provider 2.1 is in range and consequently initiates a roaming request from Provider 1.1 to Provider 2.1. Likewise, as the consumer device moves to other locations, it triggers a roaming sequence among near by providers.

This scenario demonstrates the use of location as a service roaming trigger. The consumer device uses its location and the location of other providers, as listed in the SDB, to perform the roaming. The SC device uses the service profile to preconfigure neighboring providers to efficiently deliver the service as the consumer devices enter their range. The experimental results and performance of the middleware main protocols are discussed next.

18.5.3   PERFORMANCE RESULTS

The performance of SOWGA is evaluated in both implementation scenarios above and using the following measures:

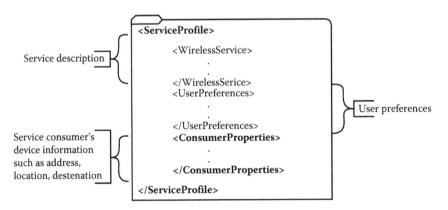

FIGURE 18.5   Example service profile.

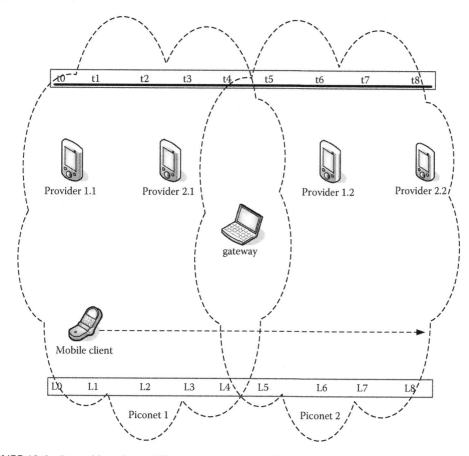

FIGURE 18.6 Intergrid service mobility management scenario.

- Discovery time, which is the time taken by a device to discover devices and services within one piconet. We refer to this also as grid setup time.
- Average time for discovering one service that just joined the grid.
- Service roaming time—the time it takes to roam a service from one provider to another.
- Discovery time and network over head. Time is measured for the service consumer device to discover devices and services within one WG. Network load is measured for this discovery.

The Bluetooth discovery performance has been enhanced with the SOWGA middleware layer due to the efficient design of the SDR protocol and the use of distributed SDB. However, due to the use of an expressive service description using the XML template, the initial discovery of a PAN, or the setup time of a WG is higher than the Bluetooth standard. Using SOWGA's SDR protocol, it took about 3.8 s to complete the discovery and set up of a wireless grid between seven devices. Whereas using Bluetooth SDP, it took about 2.3 s to complete full discovery of seven devices. The SDR's performance, however, significantly exceeds the Bluetooth SDP when compared in the scenario of a new device joining a WG. This scenario is important for mobile devices and SRP performance.

We observe that SDR response time is less than Bluetooth SDP response time. Unlike Bluetooth, in SDR, a new service advertises itself once it joins the grid. SDR discovers the new service only, and this makes the response time for SDRP constant, regardless of the number of devices in the grid. Because Bluetooth SDP does not support advertisements, Bluetooth SDP is required to rediscover all network services every time a new device or service joins the grid. The time required for discovering

TABLE 18.1
Service Roaming Time Performance

Test	Time Range (ms)	Average Time (ms)
Connection time	97 to 188	147.1
Session pause and resumption	25 to 75	43.8

a new service in Bluetooth SDP is proportional to the number of devices in the network, as shown above.

The SDR's performance significantly exceeds the Bluetooth SDP due to two design reasons. The first reason is due to the use of the XML-based SDB that is distributed on all grid devices. This enables the service consumer's SDR to save time discovering all devices one at a time. Instead the Consumer's SDR only discovers one device then it gets a copy of its up-to-date SDB containing discovered information of remaining grid devices. The second reason for the efficiency of the SDR is due to the use of the broadcast capability. This capability enables the announcement of new devices and services without the need to connect to every grid device to accomplish the registration.

Time performance of SRP to roam a service from one provider to another time is measured by the following time metrics: (1) connection time, which is the time to establish a connection with new provider; and (2) session resumption time, which is the time to pause session from old provider and resume it on new provider. The tests were repeated ten times and the time was recorded for each event. Table 18.1 shows the range and average time for the connection and session resumption.

The SRP's algorithm was found to provide sufficient session continuity support for Bluetooth-based WG. On average it took 147 ms to establish connection with a new provider. The average time for session pause and resumption was about 44 ms. Session pause and resumption include the time to send the session status to new provider, pause the session from old provider, and resume to new provider. Discovery time is not included in the above measurements.

18.6   CONCLUSION

This chapter presents a SOWGA to enable collaboration in wireless mesh networks. The architecture focuses on service discovery and brokerage and service roaming components. The service discovery and brokering protocols enable wireless devices to autonomously share their resources. The proposed context-aware SRP uses an XML-based service profile to efficiently manage services within a highly dynamic and ad-hoc wireless networks. SRP uses mobile user's geographic location, terminal's capabilities, network topology, and grid devices to provide efficient management of mobile users and services. As a proof-of-concept, the protocols are implemented for the IEEE 802.15 (Bluetooth) standards running on top a wireless mesh network deployment.

The protocols combine the advantages of the SOA and wireless grid technologies to provide the following two advantages: (1) allow automatic and transparent resource sharing through standard interfaces and resource discovery protocols regardless of the heterogeneities in device hardware, software capabilities, and location; and (2) handle the intermittent network connectivity nature of WMN through providing roaming functionality at the service layer and not the communication layer. Service roaming transparently manages connectivity so that grid devices are always best connected with services in the most efficient way.

The proposed service discovery protocol uses an easily extensible XML schema to describe services. This enables wireless devices to share a variety of services and not limit them, as with Bluetooth SDP, to predefined services. In addition, the XML-based service description allows more expressiveness in describing the functionality of a service and in how to actually consume it. It also allows introducing more capabilities to service sharing such as security and QoS. This is also not possible with the standard Bluetooth SDP. Moreover, SOWGA service discovery protocol supports service advertisement. This enables our protocol to do per-service advertisements and updates to other devices. Using Bluetooth SDP, all services on all devices must be announced to have an updated list of grid services.

The proposed protocols accommodate the highly dynamic ad-hoc contexts in which wireless grids operate. Each device uses a locally available service database that is automatically updated to account for dynamic situations such as a device leaving the grid. We also designed a service brokerage algorithm that enables the scheduling of resources among a number of wireless devices. The brokerage algorithm also enables the interactions between applications and resources. In addition, these middleware-based protocols transparently manage connectivity so that the grid user is always connected with best levels of services. As mobile users move from one location to another, grid devices collaborate together seamlessly to roam mobile users' services. The grid architecture provides consistent and efficient user experience for all services regardless of the user environment, network topology, or location.

Future work will focus on refining our approach along several dimensions. First, we plan to implement several applications that would demonstrate the utility of a wireless grid. Second, more sophisticated solutions based on P2P overlays will be investigated for service discovery and advertisement. This may eliminate the need to be constrained by the broadcasting capabilities of the underlying communication protocols. Third, brokering support will be enhanced to accommodate more expressiveness in specifying requirements. Currently, the broker can only perform simple name-value-based matching of resource requirements. In addition, a mechanism to build wireless grids that can span several heterogeneous wireless networks will be evaluated with the proposed SRP. Currently, our SRP protocol has been evaluated using Bluetooth scatternet. There is a need to evaluate the SRP design using heterogeneous wireless infrastructures that include Wi-Fi and Evolution-Data Only (EV-DO). Another extension to our current research is to investigate service composition within the SOWGA framework. Service composition mechanisms attempt to merge two or more services into a single composite service that combines and enhances the functions of the individual services being composed. A good service composition mechanism leverages each of the elements listed above to provide automatic composition of services without human intervention. This type of investigation is viewed as complementary extension to the SOWGA framework and will be added incrementally overtime.

REFERENCES

[1] A. Yamin, et al., Towards emerging context-aware, mobile and grid computing, *The International Journal of High Performance Computing Applications*, 17(2), 191–203, 2003.

[2] B. Hopkins and R. Antony, *Bluetooth for Java*, Apress, Barkeley, CA, 2003, ISBN: 1590590783.

[3] B. Hull, et al., CarTel: A distributed mobile sensor computing system, *Proceedings of the 4th international conference on Embedded Networked Sensor Systems (SenSys)*, Boulder, CO, Nov. 2006.

[4] C. Lee, et al., Konark: A system and protocols for device independent, peer-to-peer discovery and delivery of mobile services, *IEEE Transaction on Systems, Man and Cybernatics*, 33(6), 682–696, Nov. 2003.

[5] D. Raz, A. Juhola, J. Serrat-Fernandez, and A. Galis, *Fast and Efficient Context-Aware Services*, Wiley, London, U.K., p. 429, June 2006. ISBN:0-470-01668-X.

[6] F. Zhu, M.W. Mutka, and L. Ni, Service discovery in pervasive computing environments, *IEEE Pervasive Computing Magazine*, 4(4), 81–90, Dec. 2005.

[7] I.F. Akyildiza, et al., Wireless mesh networks: A survey, *IEEE Radio Communications*, 43(9), 429, Sep. 2005.

[8] J. Tourrilhes. L7-mobility: A framework for handling mobility at the application level, *Proceedings of the 15th IEEE International Symposium on Personal, Indor and Mobile Radio Communications*, PIMRC, Barcelona, Spain, 2004, ISBN: 0-7803-85233.

[9] K. Farkas and B. Plattner, Supporting real-time applications in mobile mesh networks, *Proceedings the MeshNets 2005 Workshop*, Budapest, Hungary, July 2005.

[10] L. McKnight, J. Howison, and S. Bradner, Wireless grids, *IEEE Internet Computing*, Aug. 2004.

[11] M. Bisignano, A. Calvagna, G. Di Modica, and O. Tomarch, Expeerience: A Jxta middleware for mobile ad-hoc netwoks, *Proceedings of the 3rd International Conference on Peer-to-Peer Computing (P2P2003)*, Linköping Sweden, 2003.

[12] M. Lee, Z. Jianliang, K. Young-Bae, and D. Shrestha, Emerging standards for wireless mesh technology, *IEE Wireless Communications*, 13(2), 56–63, April 2006.

[13] N. Mastronarde, D. Turaga, and M. Schaar, Collaborative resource exchanges for peer-to-peer video streaming over wireless mesh networks, *IEEE Journal on Selected Areas in Communications*, 25(1), 108–118, January 2007.

[14] S. Avancha, A. Joshi, and T. Finin, Enhanced service discovery in bluetooth, *IEEE Computer Magazine*, 35(6), 96–99, June 2002.

[15] V. Sacramento, et al., MoCA: A middleware for developing collaborative applications for mobile users sacramento, *IEEE Distributed Systems Online*, 5(10), 2004.

[16] H. Ahmed, M. Darieby, B.A. Hai, and Y. Morgan, A bluetooth and WiFi based MESH networks platform for traffic monitoring, *Proceedings of the National Academies Transportation Research Board 87th Annual Meeting, TRB-87*, Washington, DC, January 2008.

[17] B. Rashid and M. El-Darieby, Seamless service mobility for mobile devices communicating within wireless grids, *Proceeding of the IEEE 20th Canadian Concerence on Electrical and Computer Engineering (CCECE)*, Vancouver, British Columbia, Canada, 2007.

[18] M. El-Darieby and D. Krishnamurthy, A resource discovery and brokering mechanism for wireless grids *Proceedings of the 7th IFIP/ IEEE Symposium on Integrated Network Management (IM'07)*, Germany, 2007.

[19] J. Haerri, F. Filali, and C. Bonnet, Performance Comparison of AODV and OLSR in VANETs urban environments under realistic mobility patterns, in *Proceedings of 5th IFIP Mediterranean Ad-Hoc Networking Workshop*, 2006, Lipari, Italy.

[20] R. Sen, R. Handorean, G. Roman, and C. Gill, Service-oriented computing imperatives in ad hoc wireless settings, Chapter XII, *Service-Oriented Software System Engineering Challenges and Practices*, ISBN:1591404274, Idea Group Publishing, Hershey, PA, 2005.

[21] F. Zhu, M.W. Mutka, and L.M. Ni, Service discovery in Pervasive Computing environments, *IEEE Pervasive Computing*, 4(4), 81-90, December 2005.

19 Cooperation and Interference in Wireless Mesh Networks

Yuanzhu Peter Chen and Yong Wang

CONTENTS

A wireless mesh network is a multi-hop wireless network consisting of a large number of cooperative wireless nodes, which can be either stationary or mobile. The primary goal in wireless mesh networking is to achieve high system throughput in presence of numerous data traffic flows. However, because of the broadcasting nature of the wireless medium, the interference between simultaneous transmissions is a key factor affecting the throughput. Using all communication links simultaneously certainly does not guarantee high throughput of the entire network. This is because the interference between wireless links incurs unnecessary collisions and consequently retransmissions. Therefore, to achieve a certain level of throughput in mesh networks, the activities of these links should be arranged in a cooperative way such that a balance between interference and effective communication is attained. There have been significant research efforts to alleviate interference in wireless mesh networks and, thus, to improve the throughput. In this chapter, we discuss existing work aiming at high throughput while simultaneously incurring minimal interference in such networks. And this is an intriguing interplay among a multitude of factors. Throughout the chapter, we focus on unicasts (one-to-one communications). We first review some of the most accepted communication and interference models in the research of multi-hop wireless networking. We then discuss the three most important approaches to minimizing interference to support multiple flows simultaneously, that is, by routing, channel assignment, and scheduling. Afterward, we lead the readers to the not-so-conventional approach of reducing interference by mobility. We conclude the chapter by speculating on the difficulties in modeling wireless mesh networks and the interplay of different aspects of interference avoidance.

19.1 COMMUNICATION AND INTERFERENCE MODELS

When signals propagate in a wireless channel, their average strengths attenuate as a power function of the distance they travel, which is called large-scale path loss. In addition, the signal strengths can also vary significantly even by a short distance due to propagation mechanisms such as reflection, diffraction, and scattering. As a result, when a receiver detects a signal, it essentially receives a combination of multiple copies of the same original signal from different paths. These multiple copies are usually modulated by different amplitudes, phases, and even frequencies. For a signal to be reconstructed, its strength must be greater than all other adversary factors in the channel, that is, noise and interference, by a certain factor depending on the modulation scheme used by the signal. By Shannon-Hartley theorem, the capacity of a channel (in bps) is no more than

$$C = B \log \left(1 + \frac{S}{N} \right),$$

where
   B is the channel bandwidth (Hz)
   S/N is the signal-to-noise ratio

When there is interference, its strength can be included as part of the noise. Therefore, interference plays an extremely important role in wireless mesh networking, and the wireless channel must be effectively shared by different transceivers.

To hide the details of signal propagation in wireless channels, we need an abstract model to present to the upper layers to properly coordinate channel sharing. Ideally, such a model should be realistic enough to be useful and simple enough to be manageable. The communication or interference aspects of the network should be taken into account in the model. Thus, various trade-offs exist. A simple model of the communication capabilities of mesh networks can be represented as an undirected graph of nodes and edges. Nodes correspond to the hosts, and there is an edge between two nodes if these hosts are within range of each other. In contrast, modeling interference can be much more complicated and flexible. Various models have been proposed and adopted in the literature, including the conflict graph, interference graph, interference number, and the protocol and physical models. In practice, the relation between communication and interference is much more complex. For example, interference is not only a spatial concept directly related to distance but also a temporal one. In this section, we focus on the spatial aspect of interference in multi-hop wireless networks. We start with simple graph-based models for communication and interference. Then we move to the more realistic models that consider the strength ratio of the intended signal to interference and ambient noise. These models are used to derive the capacity upper bound in multi-hop wireless networks successfully. Essentially, the various techniques of interference-aware routing, channel assignment, and link scheduling are efforts to approach such an upper bound.

19.1.1 GRAPH-BASED MODELS

A straightforward approach to modeling a multi-hop wireless network is using graphs to represent certain geometric properties of the network. In a graph-based model, nodes are generally assumed to be in a two-dimensional (2-D) space, and their communication capabilities and interference potentials are determined by the distance between nodes. A disk connectivity graph is a geometric graph of the nodes in the 2-D plane. We use directed graph $G = (V, E)$ to denote the disk connectivity graph, where V corresponds to the nodes in the network and E contains a directed edge (u, v) if v is within the transmission range of u. Graph G is directed because the nodes can have different transmission ranges. As a special case, if all nodes have the same transmission range, the graph is also called a unit disk graph (UDG) [1] and it is undirected.

To incorporate interference among edges, Jain et al. [2] propose to use a conflict graph derived from the disk connectivity graph. The conflict graph $C = (V', E')$ is defined on all the links of the disk connectivity graph G, that is, $V' = E$. That is, each edge of G corresponds to a vertex in C. Let l_{ij} and l_{pq} be two vertices in C or two links in G equivalently, where i, j, p, and q are endpoints of the links. There is an undirected edge in C between l_{ij} and l_{pq} if $dist(i, q) \leq R_i$ or $dist(p, j) \leq R_p$, where $dist(\cdot)$ is a distance function. Here, a node u in the network has an interference range R_u, within which no other node can receive a packet successfully when u is transmitting. In other words, the conflict graph records that link l_{ij} and l_{pq} cannot be scheduled at the same time because either q is within i's interference range or j is within p's interference range. The degree of a vertex in the conflict graph quantizes how much interference a link is susceptible to. This is a "link-centric" interference model. Alternatively, an interference graph can also be defined on the nodes themselves. That is, the interference graph $G' = (V, E')$ has a vertex for each node in the network and has a directed edge (u, v) if v is within u's interference range. Because the interference range of a node is always assumed to be greater than its transmission range, we know that G is a spanning subgraph of G', that is, $G \subseteq G'$. Thus, the in-degree (out-degree, respectively) indicates how much interference a node may have as a receiver (transmitter, respectively). And this is a "node-centric" interference model, be it receiver-oriented or transmitter-oriented.

In research on topology control and power control, a goal is to minimize the graph-wise interference defined by any of the above quantities by tuning the transmission power of individual nodes so that certain graph-theoretic properties are satisfied, say connectivity and min-cut [3–7].

19.1.2 RELATIVE SIGNAL STRENGTH-BASED MODELS

In real network operation, the reception of a signal is determined by the ratio of the intended signal to the sum of ambient noise and interference, so the distance or power attenuation is not the only factor in modeling communication and interference in a wireless mesh network. Indeed, there is no such thing as transmission range or interference range in a real mesh network. Consider a transmitter and a receiver that are a short distance apart. The transmission can fail under strong noise or interference even though the separation distance is small. On the other hand, a transmitter may not garble an ongoing transmission when the transmission is taking place between two sufficiently close nodes even though the interfering transmitter is not far away. Therefore, a communication or interference model should consider the relative signal strengths to be more practical. Two such models are presented by Gupta and Kumar [8], called Protocol Model and Physical Model.

1. Protocol Model—Suppose all nodes in the network have the same transmission power. The transmission from node X_i to node X_j is successful if

$$|X_k - X_j| \geq |(1 + \Delta)|X_i - X_j|$$

for every other node X_k in the network that is transmitting simultaneously. The value of Δ can be understood as a guard zone factor around the receiver X_j to prevent any other node from interfering with the transmission.

2. Physical Model—Let $\{X_k | k \in \mathcal{T}\}$ be the set of nodes in the network that are transmitting at the same time. Let P_k denote the transmission power of node X_k. Then, the transmission from node X_i to X_j is successful if

$$\frac{\frac{P_k}{|X_i - X_j|^\alpha}}{N + \sum_{\substack{k \in \mathcal{T} \\ k \neq i}} \frac{P_k}{|X_k - X_j|^\alpha}} \geq \beta.$$

Here, the received signal strength after path loss must be stronger than the noise and interference combined by a given threshold β, that is, the minimum signal-to-interference-and-noise ratio. Typically, the path loss exponent is between 2 and 4.

There are a few differences between these two models, but the essential one is that the Protocol Model carries a "max" notion in that the interference is caused by the closest interfering node, whereas while the Physical Model carries a "sum" notion because the interference is accumulated from all interfering nodes plus the background noise. Such a difference makes the latter considerably harder to analyze mathematically. Using these two models, Gupta and Kumar are able to obtain the capacity limit of wireless mesh networks as sketched next.

19.1.3 CAPACITY OF MULTI-HOP WIRELESS NETWORKS

The capacity of a wireless network is a quantification of the data transportation capabilities of the network. This is collective of all the nodes in the network because of the broadcasting nature of wireless communications. Given a network of a certain physical dimension, say a 1000 m × 1000 m square, the capacity of the network is *how many data can be transported by how far in a unit of time*. This is an analogy of mechanical work and has a unit of bps × m. Gupta and Kumar [8] first explored the capacity of multi-hop wireless networks formally. There, the network dimension is normalized to a circle of unit area, that is, radius of $1/\sqrt{\pi}$. Assume that there are n nodes in the network sharing a channel of bandwidth W bps. In addition, each node is able to control its transmission power and, thus, the communication range. Gupta and Kumar show that the network capacity is no more than $O(W\sqrt{n})$ regardless of the placement of the nodes in the network, scheduling of transceivers, and assignment of transmission power. This leads to a per-node capacity of merely $O(W/\sqrt{n})$. The intuition is illustrated in Figure 19.1. Suppose node S wants to send a flow to node D. It enlists a number of intermediate notes as relays. The transmitters tune their communication ranges to d, indicated by the small circles. Thus, all nodes within the union of the communication ranges of the nodes on the path from S to D must be silent during the transportation of the flow. As a result, this flow

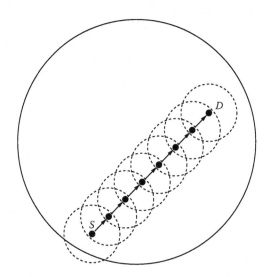

FIGURE 19.1 Capacity illustrated.

of a length $O(1)$ "cuts" through the network to provide a capacity of $O(W)$ bps×m. Consequently, there can be at most $O(\sqrt{n}/d)$ such flows active in parallel. Note that d cannot be arbitrarily small to have a connected network. This capacity bound of $O(W \times \sqrt{n})$ holds for both the physical and protocol models [8]. And such a limitation comes from the shared-channel and multi-hop nature of wireless mesh networks.

To increase the capacity, two approaches have been proposed in the literature.

1. Add relay-only nodes in the network [9–11]. Relay-only nodes use an out-of-band channel to provide long-haul capacities so that the shared wireless channel is used only to provide short-range, possibly multi-hop, transportation.
2. Introduce mobility among the nodes [12,13]. When nodes are moving, they can mechanically deliver messages in the network. As a result, a much smaller number of wireless transmissions are required to deliver a message.

The first approach is relatively easy to comprehend. We discuss the second approach later in Section 19.5.

19.2 INTERFERENCE REDUCTION VIA ROUTING

Data traffic flows in mesh networks are typically multi-hop. Among different route assignments to these flows, there can be a varying degree of colocation. To reduce the interference among the flows, the degree of co-location should be low ideally. However, this is at the cost of consuming more network resources that could be used potentially by other traffic flows otherwise. Therefore, the selection of routes in mesh networks must consider the interference among flows, and this should be done with high adaptiveness and at low costs. In this section, we first review some background issues concerning design of routing protocols in multi-hop wireless networks. We then reflect upon the rising awareness of minimizing interference in route calculation and packet forwarding.

19.2.1 BACKGROUND OF ROUTING IN AD HOC AND MESH NETWORKS

Transportation of packets from the source to destination is a core enabler of any multi-hop network. Consequently, routing has a central role in the research of ad hoc and mesh networking. Routing has two functionality components, that is, packet forwarding and route calculation. In packet forwarding, an intermediate node, after receiving a packet, must relay it to a neighbor based on the information contained in the packet or maintained by itself, or both. Route calculation is to prepare a node in the network, either as a packet source or forwarder, with the information that will be used to forward packets.

Apparently, design of a routing protocol can influence the network performance and software maintenance significantly. Several factors that must be considered when designing a routing protocol are listed below

- Forwarding strategy—When a packet is received by an intermediate node, should it contain all the information needed for forwarding? This decision determines the design of the other factors. Such a decision differentiates a source routing protocol from the rest.
- Global structure—Is there a global structure deduced and maintained, either completely stored at different nodes or distributed separately? Under certain assumptions, global knowledge is not needed even in a distributed fashion. In this case, greedy forwarding at individual nodes can attain end-to-end packet transportation. Position-based routing forms such a category of routing protocols.
- Complete global picture—Should a single node have a complete copy of the network topology or can the nodes in the network reconstruct the complete topology collectively? Conventionally, this aspect is a distinction between link-state and distance-vector routing.

- Information acquisition—The information needed by a source node to prescribe a path for a packet or by an intermediate node to forward the packet is collected from different parts of the network. The routing protocol can require each node to maintain (and help other nodes to maintain) fresh information at all times or can wait until when such information is needed by a source. These two strategies represent proactive (table-driven) and reactive (on-demand) routing protocols.
- Link metric—The network topology is typically represented as a weighted graph of nodes. It is compiled from the link parameters reported by relevant nodes. Traditionally, hop count has been used to find a shortest path in the network. Thus, "1" is essentially reported. However, it has been observed that more informative parameters facilitate designing a better routing protocol. These include measurements reflecting the link reliability, throughput, proximal interference, congestion level, and so on.
- Cross-layer—The networking module in a node is a complex software system. For clarity in functionality definition and ease in software design and maintenance, the module is usually divided into layers, predominantly according to the ISO-OSI reference model. The cost incurred by this is however paid by the loss of efficiency of the module. Using information from different layers for decision making at a certain layer has been proved effective and efficient. Such an approach is called cross-layer design. A balance of how transparent the different layers should be is imperative at an early stage of the design.

A great deal of efforts in routing protocol design have been exerted in the research of ad hoc and mesh networking. Most of the proposals have been covered in some excellent reviews over the years [14–18]. In addition, some more surveys focus on specialized routing protocols or issues, such as position-based routing [19], multipath routing [20], secure routing [21], and cross-layer design in routing [22]. Interested readers are referred to these articles for a comprehensive treatment.

19.2.2 INCREASING CONSCIOUSNESS OF INTERFERENCE AVOIDANCE

When designing a routing protocol for multi-hop wireless networks, the interference among close-by nodes has been considered more and more consciously. Here, we first reflect on the changes in selecting an appropriate routing metric. These interference- and load-aware routing metrics are the basis for designing routing protocols that are able to capture the characteristics of these networks. We review some of the most important and representative routing strategies in such endeavors subsequently.

19.2.2.1 Interference-Aware Metrics

It is being realized that minimum-hop is not the best metric for route selection. Instead, reliability and medium accessibility are more descriptive measures. Earlier explorations in this direction include ABR (associativity-based routing) [23] and SSR (signal-stability-based routing) [24]. In these proposals, a node broadcasts a periodic beacon as a probe. Each node evaluates the connectivity to a neighbor by counting the number of successful receptions of these beacons in recent history. SSR enhances the evaluation by measuring the signal strength experienced during reception of the probes. This idea is further refined by ETX (expected transmission count) [25]. In ETX, the bidirectional reliability between a pair of nodes is defined as

$$\text{ETX} = \frac{1}{p_\text{f} \times p_\text{r}},$$

where p_f and p_r denote the probability that a transmission is successful in the forward and reverse directions, respectively. Due to the link-layer reliability requirement, this measure reflects the expected amount of time needed to complete the two-way DATA/ACK handshake. Similar to ABR and SSA, measurements are taken from periodic broadcast of probes by the nodes. The reverse link probability p_r is recorded directly, and the forward link probability p_f is carried by the probes.

Because broadcast and unicast can be supported differently by the link layer, using a broadcast probe to evaluate a link usability for unicasts is not necessarily accurate. For example, in the IEEE 802.11 family, broadcasts are not reliable, that is, no ACK is required, whereas a unicast DATA frame should always be acknowledged by an ACK frame. In addition, a broadcast is always made at a basic data rate of the underlying physical layer protocol, but a unicast frame can be transmitted at a higher rate. Apparently, the difference in the data rates implies a difference in the BER (bit error rate) and thus in the reception success rates. In contract, per-hop round trip time (RTT) [26] is a measure based on periodic unicast frames. In the proposal, a node sends a probe to a neighbor, which, upon successful reception, immediately responds with a probe acknowledgment. The primary advantage of RTT is that the contention delay experienced by the probe and its acknowledgment is the same as that experienced by any unicast frame and acknowledgment. Because both the probe and its acknowledgment are placed on the transmission queue, RTT also measures the queuing delay on both sides. In addition, the frame loss due to bad channel conditions is also included in the measurement. Expected transmission time (ETT) [27] is a "bandwidth-adjusted ETX" in that

$$ETT = ETX \times \frac{S}{B},$$

where
  S denotes the size of the packet
  B denotes the bandwidth of the link

Such an extension recognizes the fact that different data rates have different effects in evaluating a link's data transportation capabilities in a mesh network.

19.2.2.2 Interference-Aware Routing

Different approaches can be taken when designing and operating a routing protocol that battles interference in ad hoc and mesh networks. This can be done locally at an intermediate forwarder, when prescribing routes for packets, or even off-line in a centralized fashion.

After an intermediate node has received a packet and attempts to forward to a downstream neighbor, it may have difficulties in doing so because the link layer experiences some temporary bad channel conditions or intensive contention for that particular downstream neighbor. Rather than dropping the packet, as is conventionally done, the forwarder can choose to send it to a different neighbor as long as it is certain that the alternative neighbor can forward the packet to the ultimate destination. Such local and temporary tweak of routes in packet forwarding is called route adaptation. It is usually done jointly by the link and network layers, thus a cross-layer design. Typically, the link layer has a multicast or anycast enhancement at each node so that a forwarder can test multiple neighbors to choose one as the next forwarder of a packet. This idea has been realized in several pieces of work [28–31].

Multipath routing [20] explicitly solicits multiple paths between a pair of source and destination. It originated from connection-oriented wired networks, such as PSTN and ATM networks, and was referred to as alternate path routing. Its initial goal was to reduce the call blocking probability in the network core. This idea was introduced in the research of multi-hop wireless networks later to increase the end-to-end communication reliability and to achieve better load balancing. Here, multiple routes are maintained and used during a single session. Depending on the independence requirement for these routes, they can be node-disjoint, link-disjoint, or nondisjoint. Apparently, the more independent these routes are from one another, the more reliable they are collectively, but the more difficult to discover and maintain. Examples of multipath routing in such networks include SMR [32] and AOMDV [30,33]. Multipath routing is not the most effective in avoiding interference

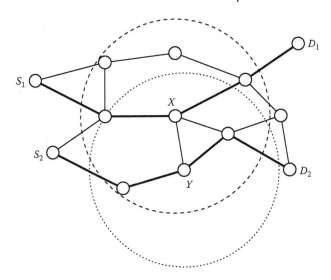

FIGURE 19.2   Interflow interference.

although it is an important piece of step stone. After all, even being node-disjoint does not imply noninterfering because the interference range is usually much larger than the communication range.

For global minimization of interference among multiple flows, we can use centralized mathematical programming. To understand interference at the network layer, we can divide interference into two types: interflow and intraflow.

Consider a network of nodes as in Figure 19.2, where the lines are the wireless links between nodes. Between two source/destination pairs $\{S_1, D_1\}$ and $\{S_2, D_2\}$, there are two flows, indicated by solid lines. Even though the routes taken by these flows are node-disjoint, there will still be interference between them. Consider nodes X and Y, which are within transmission range of each other. They cannot be activated at the same time because they must also be within interference range of each other, indicated by the dashed and dotted circles, respectively. In fact, because the interference range is considerably larger than the transmission range; even if X and Y were "2 hops away" in terms of transmission range, they may still be within interference range. This vital notion will be discussed more in detail shortly.

Even in a scenario of a single multi-hop flow, there is interference among the links comprising the route. Consider a linear topology of eight nodes as in Figure 19.3 and a flow from end node X to end node Y. In the network, the internodal distance is slightly less than the transmission range. Assume that the inference range is twice the transmission range, indicated by the circles. When node C is forwarding a packet to node D, nodes E and F must be silent to avoid interfering with D's reception. Similarly, for node C to receive the ACK from node D successfully, nodes A and B must be silent, too. That is, forwarding a packet from node C to D is susceptible to activities on all other links en route. With these in mind, we discuss what a centralized flow programming must cope with subsequently.

In a centralized route calculation, the formulation is typically based on a network flow model [34]. That is, the network is modeled with a graph with edge capacities, and a maximum combination (sum or λ-factor) of the given set of flows is pursued. The model must be enhanced to factor in the interference among and within flows as stated above. To do that, the Protocol or Physical Model (Section 19.1.2) or an approximate is used. In the sequel, we review some recent work to provide the readers an insight into how this is done.

Jain et al. [2] formulate the interference with a conflict graph (Section 19.1.1). The conflict graph derived from the connectivity graph of wireless nodes is a graph of the wireless links. In an

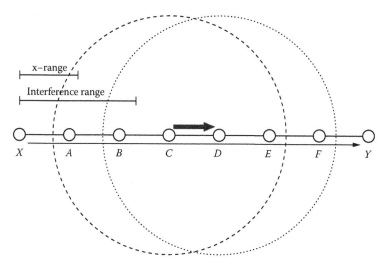

FIGURE 19.3 Intraflow interference.

undirected conflict graph, two vertices representing two links are adjacent if their activities interfere with each other. That is, their end points are within the interference range. In Ref. [2], the optimal routing problem is formulated with a multi-commodity flow problem augmented with constraints derived from the conflict graph. They show that finding the optimal throughput is NP-hard even for the single source/destination case. They are able to provide upper and lower bounds on the optimum. Note that the conflict graph is a simplification even for the protocol model because the interference here is determined by the absolute distance from the interferer; whereas in both the protocol and physical models discussed in the previous section, interference is relative to the transmitter–receiver (TR) separation distance. In this work, an attempt at using a physical model is made and shown to support similar upper and lower bounds. Again, the physical model is a simplification in that it does not consider the cumulative nature of SINR. In a later work, Kolar and Abu-Ghazaleh [35] extend to multiple flows using an interference model represented by a conflict matrix. Such a node matrix can represent an arbitrary interference relation between nodes. Again this is a protocol model not considering TR separation distance. Alicherry et al. [36] consider a joint channel assignment, scheduling, and routing. The routing component therein is an interesting extension of Jain et al. [2] in itself. To model the interference, they assume that the interference range is always q times greater than the communication range, where q is a constant. Using the geometric properties implied, the authors show that the aggregate throughput can be approximated within a constant factor. Careful readers must have noticed that, in all these efforts, only a simplified protocol model is used to mimic the interference in the network. Using more realistic models, being it protocol or physical, makes the flow constraint specification significantly more difficult, which indicates an avenue for further research.

19.3 INTERFERENCE REDUCTION VIA CHANNEL ASSIGNMENT

In current wireless communication technologies, a frequency band is usually divided into multiple nonoverlapping subcarriers, that is, channels, that can operate simultaneously. Even if two sufficiently close nodes in the network are transmitting at the same time, if they use different channels, they will not interfere with each other. Such spatial frequency reuse is an important technique to avoid interference and to support multiple concurrent transmissions in multi-hop wireless networks.

There has been a transition from using a single channel to multiple channels in the research on multi-hop wireless networking. In the earlier stage, it was assumed that a single channel is used in a network and that all nodes communicate using this single channel for technological and economic reasons. With a single channel in use, interference can be avoided either by sufficient space separation or time rotation. Later in the research community, it was accepted that a network can operate on multiple channels simultaneously. In fact, the IEEE 802.11 standard family, the most popular platform to implement wireless mesh networks, can in principle support multichannel architectures. In particular, IEEE 802.11b/g provides 3 nonoverlapping channels in the 2.4 GHz band, and IEEE 802.11a provides 12 nonoverlapping channels in the 5 GHz band. Therefore, as technologies mature, using multiple channels simultaneously in a wireless mesh network is believed to be an effective and practical approach to support an increased number of concurrent transmissions and thus to improve the throughput of the entire wireless network. In this section, we introduce two different types of multichannel architectures, depending on the number of interfaces that a node has. We then go forth to categorize the channel assignment solutions into two camps according to the time granularity of channel switching. The goal of these channel assignment algorithms is unquestionably to maximize the network throughput while minimizing interference.

19.3.1 SINGLE-NIC VS. MULTI-NIC

In a multichannel wireless mesh network, two node configurations are possible. One is that in which each wireless node has one network interface card (NIC), where the NIC needs to switch among the available wireless channels to fulfill different traffic demands. In the other configuration, each node has multiple NICs, each associated with one wireless channel, and can operate on these NICs at the same time. Typically, the number of NICs is assumed to be smaller than the number of available channels. We discuss these two configurations in the sequel. In either case, each NIC at a wireless node needs to choose a channel to operate on. By carefully assigning the available channels to the NICs, interference can be reduced and, thus, the throughput of the network will be improved. When assigning channels to NICs to minimize interference, another important factor that must be considered is the connectivity of the resulting topology. For example, if an NIC is assigned a channel that is not used by any other node in range, the NIC cannot communicate with the rest of the network even if there is no interference. Therefore, the channel assignment is also a trade-off between interference and connectivity.

Single-NIC multichannel wireless mesh networks are studied in Refs. [37–39]. The main motivation for this model is that most inexpensive wireless devices in practice are equipped with a single NIC. However, the IEEE 802.11 standards support multiple channels to enhance the network throughput. Therefore, to have the current widely deployed wireless hardware to take advantage of the multiple channels available, various algorithms have been proposed for single-NIC multichannel wireless mesh networks. They [37] propose a network layer approach, which works directly on top of the IEEE 802.11 MAC layer protocol. So and Vaidya present two heuristics to assign channels: one is to assign channels to nodes, and the other is to assign channels to traffic flows. The former separates route establishment and channel assignment and makes the solution to the two components less complicated. However, it may cause the so-called deafness problem, where the sender and receiver do not operate on the same channel. The latter considers route establishment and channel assignment together and, after a route is established, assigns all nodes on the same route a common channel. This makes the algorithm more involved but it can eliminate the deafness problem. Simulation results showed that employing multiple channels can greatly improve the network throughput compared with the single-channel architecture. The approaches taken by the authors in Refs. [38,39] are MAC layer solutions, which dynamically assign channels to the NICs in an on-demand fashion. We explore the details in the next section when we discuss the dynamic/static channel assignment.

With the development of wireless communication technologies, more and more multi-NIC wireless devices are deployed. Compared with single-NIC systems, integrating multiple NICs at each network node fits the multichannel architecture more naturally and can be used to further improve the network throughput. For a multi-NIC multichannel wireless network, multiple NICs at each node are tuned to different channels and can perform communications simultaneously. In this case, a node need not switch among channels to serve one at a time. This is particularly powerful for mesh networking because relaying nodes can transport traffic continually, mitigating the intra-flow interference problem (Section 19.2). Raniwala et al. [40] propose a centralized algorithm for channel assignment and routing in multi-NIC multichannel wireless networks. They use the graph-based communication and interference models (Section 19.1.1), where two nodes can communicate if they are within the transmission range of each other and they interfere with each other's transmission if they are within the interference range of each other, provided that the two nodes operate on a common channel. Because channel assignment and routing are closely related to each other, and the strategy of one may affect the other greatly, these two problems are investigated together in Ref. [40]. Their channel assignment consists of two phases: first, to determine through which interface a node communicates with each of its neighbors and, second, to assign a channel to each interface of a node. The second component of channel assignment interacts with the routing algorithm to refine the solution of each other in an iterated way until no improvement can be made. Their simulation results show that deploying multiple NICs at each network node can increase the network throughput significantly compared with the conventional single-channel network architecture. In a configuration of two NICs per node, the network throughput is increased by a factor of up to eight. In a later work, Raniwala and Chiueh [41] extend the centralized algorithm to a distributed one, which uses only local information from the $\lceil q + 1 \rceil$-hop neighborhood, where q is the ratio between the interference and transmission ranges. Compared with the single-channel network architecture, the simulation shows that the localized algorithm can effectively coordinate a network of nodes, each equipped with two NICs and can improve the network throughput by a factor of 6–7.

Most channel assignment algorithms focus on minimizing the interference of some sort to improve the throughput of the network. Nevertheless, the connectivity of the resulting topology is another metric to evaluate a channel assignment algorithm. Tang et al. [3] propose a channel assignment algorithm for multi-NIC multichannel wireless networks. Their study also uses the graph-based communication and interference models determined by the uniform transmission and interference ranges. Note that every channel assignment defines a resulting topology, where there exists an edge on channel γ between nodes u and v if u and v are within the transmission range of each other and if γ is assigned to one of the NICs in node u and one of the NICs in node v. In the resulting topology, they define the interference of a link as the number of other links that interfere with the link, that is, the degree of the corresponding vertex of the derived conflict graph (Section 19.1.1). Collectively, the interference of the resulting topology is defined as the maximum interference over all links. The proposed channel assignment algorithm in Ref. [3] strikes the balance between the interference and connectivity of the resulting topology and generates a network structure that is interference-minimum among all k-connected topologies for a given integer parameter k.

19.3.2 STATIC VS. DYNAMIC CHANNEL ASSIGNMENT

Assigning channels can be done statically or dynamically. For static channel assignment, once a channel assignment is calculated, it will not be changed unless the topology of the network or the traffic pattern has changed significantly. In contrast, dynamic channel assignment requires that NICs switch among the available channels on demand frequently. Compared with the static approach, dynamic channel assignment captures the traffic changes more precisely and, thus, can provide better solution accordingly. However, the delay and operational costs of channel switching should also be taken into account. In addition, dynamic channel assignment algorithms need to modify the MAC layer protocols, so they cannot be deployed directly atop the off-the-shelf wireless hardware currently available. Apparently, channel assignment can also be a hybrid of static and dynamic assignment. In hybrid channel assignment, some NICs dynamically switch channels to accommodate traffic changing, whereas other NICs are assigned channels statically.

Channel assignment algorithms can be static in the sense that the channel assignment will be updated only when there are significant traffic or topology changes in the network. The static channel assignment algorithms are usually network layer, or "global," solutions and do not need modification to the 802.11 MAC Standard. For example, the multi-NIC multichannel works discussed earlier in this section [3,40,41] are static channel assignment strategies.

Dynamic channel assignment requires that NICs be switched rapidly among channels. Such a fast switching mechanism can be challenging for node coordination. Dynamic channel assignment algorithms are usually MAC layer solutions [38,39]. The algorithm of So and Vaidya [38] maintains a preferable channel list (PCL) at each node to decide the order in selecting available channels. The PCL of a node records the use of channels inside the transmission range of the node and is divided into different levels of preference. To accomplish a transmission request, the sender and receiver negotiate on the best channel using the PCLs of both sides, where the best is defined as the least loaded channel. That is, the algorithm tends to balance the load among different channels to reduce the bandwidth waste due to the interference and subsequent retransmissions. The algorithm proposed by Bahl et al. [39] is a slotted contention-based protocol. Each node maintains a channel schedule containing the list of channels that the node plans to switch to in subsequent time slots and the time at which it plans to make switches. When a transmission to a particular destination using a certain channel fails, back-off is adopted to reduce the probability of the further transmission conflicts.

It is also possible to combine static and dynamic channel assignment to have hybrid approaches, for example, the hybrid channel assignment algorithms proposed in Refs. [42,43]. In a hybrid approach, some of the NICs in each node are constantly associated with certain preselected channels, and other NICs can switch among the remaining available channels. The fixed NICs are usually used to preserve the connectivity of the network topology and to exchange control information in the network. The switchable NICs are used for data transportation using channels selected dynamically. Furthermore, there are two possible settings for the NICs working on fixed channels. One is that the fixed channels are common to every node in the network [42]. Alternatively, each node has the liberty of choosing its own set of fixed channels [43]. These hybrid channel assignment algorithms are evaluated by simulations, and the results show that they achieve a balance between flexibility and stability.

19.4   INTERFERENCE REDUCTION VIA LINK SCHEDULING

If two nodes use the same channel and are not separated sufficiently from each other, their transmission and reception must be performed at different times. Conversely, even if two nodes are rather close and they use the same channel, there will not be interference between them if well scheduled. For routing or channel assignment, as discussed previously, the network resources are usually not sufficient to accommodate all traffic demands simultaneously in an interference-free fashion. In contrast, link scheduling uses the much more abundant resource of time to avoid interference as long as the delay and bandwidth requirements of the users or applications are met. The idea of link scheduling is to partition time into slots, each of which is used to accommodate a subset of the given link activity requests without interference. The objective of link scheduling is usually to minimize the makespan, that is, total number of time slots, to schedule a given set of transmissions. Here, interference can be modeled in different ways as introduced in Section 19.1, and this may affect the complexity of the scheduling problem significantly. Due to its similarity to the classic CPU scheduling problems, solutions to link scheduling can resemble those to CPU scheduling. Further complications, however, lie in the correlation among link activation due to interference.

Currently, the research on link scheduling in wireless mesh networks has been done at the following two levels. One is to focus on the link scheduling problem itself and assume the routes and channel assignment are given. The other is to study the link scheduling problem coupled with routing or channel assignment. The former is relatively easier to formulate, but the latter is more effective to maximize the network throughput. The reason is that routing, channel assignment, and link scheduling are not independent subproblems for minimizing interference and maximizing throughput of the network. Instead, they are closely correlated to one another, and the solution of one subproblem affects the decisions of the other subproblems to a great extent. Thus, to be more practical, these subproblems should be and often are considered together. In this section, we discuss link scheduling at these two levels.

19.4.1 STAND-ALONE LINK SCHEDULING

In a stand-alone link scheduling problem, a set of link activation requests are given, and the goal is to find the shortest makespan such that each request can be scheduled at least once under certain communication and interference models.

Mazumdar et al. [44] formulate the link scheduling problem as a k-hop matching problem in a UDG representing the network structure (Section 19.1.1). In graph theory, given a graph G, a matching is a subset of edges that do not share a common vertex. Here, this concept is generalized to "k-valid matching" (as termed in Ref. [45]), for any given constant k. In the generalization, the distance between two edges is defined as the number of hops between the two closest endpoints of these edges. Then a k-valid matching is a set of edges such that the distance between any two of them is at least k. Using a simplified interference model, where two nodes can interfere with each other if and only if they are within k hops in G, Mazumdar et al. reduce the link scheduling problem to k-valid match. They show that, if $k > 1$, the maximum k-valid matching problem is NP-complete for general graphs and cannot be approximated within a constant ratio. For UDGs, it permits a polynomial-time approximation scheme (PTAS). The weighted version of the maximum k-valid matching problem is also studied, where the weight of each edge depends on factors such as congestion cost, supported data rate, and queue length.

More realistic interference models have been used in link scheduling. In the study by Wang et al. [45], each node has a pair of fixed but different transmission and interference ranges. The interference model used is essentially the conflict graph [2] (Section 19.1.1). Using this model, the links can be scheduled within a constant approximation ratio to the optimum, and the authors are able to devise a contention-based distributed algorithm to do so. Moscibroda et al. [46] study the problem under the Physical Model (Section 19.1.2), which better reflects the fact that a successful reception of a transmission at a receiver depends on how strong the received signal strength is compared with the interference caused by other simultaneous transmissions and the noise level. Within each time slot of the network operation, a node is either idle or transmitting at a certain transmission power level, which should be large enough to reach the intended receiver. Therefore, a valid schedule of a set of transmissions is in fact a sequence of power assignments of the nodes, where each power assignment describes the transmission power levels of all nodes in a time slot, such that all the transmissions can be accomplished successfully under the Physical Model. The scheduling complexity of a given set of link activation requests is defined as the minimum makespan over all valid schedules in the network. It turns out that determining the scheduling complexity is NP-complete. The authors are able to devise an algorithm that computes a schedule of length $O(I_{\text{in}} \cdot \log^2 n)$, where n is the number of nodes in the network and I_{in} is a static interference measure obtained from the connectivity graph defined by the maximum transmission power. The work of Brar et al. [47] is similar to Ref. [46] published at about the same time. The major difference is that in Ref. [47] the link requests are weighted.

19.4.2 COUPLED WITH OTHER APPROACHES

The link scheduling problem can also be formulated with routing or channel assignment to better improve the network throughput. Here, we provide a few such examples.

1. Joint routing and scheduling—Kodialam and Nandagopal [48] consider a joint routing and scheduling problem. They use the graph-based communication model (Section 19.1.1) and a simplified interference model, where the only constraint is that each node can communicate

with at most one neighbor at any given time. In the communication model, each link can have a different but fixed data rate. The objective is to maximize the achievable fraction of a given set of flows specified by their sources and destinations. The authors develop an approximation algorithm with a performance ratio of $\frac{2}{3}$ compared with the upper bound obtained using a PTAS. Their experimental results show that the algorithm offers even higher throughput in practice.

2. Joint routing, channel assignment, and scheduling—A later work of Kodialam and Nandagopal [49] extends Ref. [48] by incorporating the channel assignment issue as well. In their heuristics, both static and dynamic channel assignment schemes are considered and combined with routing and scheduling. According to the simulation results, both the static and dynamic approaches perform fairly well in achieving high throughput and, especially, the dynamic approach has a performance close to the optimum on average.

The joint routing, channel assignment, and scheduling problem is also studied extensively by Alicherry et al. [36]. They use a graph-based communication and interference model, where the interference range is greater than the communication range by a fixed factor. It is assumed that each node has an aggregated traffic demand $l(u)$ from its associated users. The authors aim to formulate the maximum flow problem in multi-hop wireless networks in the presence of link interference, and they are able to maximize the fraction of a given vector of node traffic loads using mathematical programming. A relaxed version of the problem is formulated as a Linear Program and optimally solved. Then to get a feasible solution for the original problem, some adjustments on routing and channel assignment are made. Finally, an interference-free link schedule is computed. The algorithm is proved to achieve solutions that are only a constant factor away from the optimal ones. Furthermore, simulation results show that the average performance is much better than the theoretical worst-case bounds.

19.5 INTERFERENCE REDUCTION VIA MOBILITY

Mobility has been an adverse factor in wireless networking for most of the time. The shadowing and multipath fading of signal propagation can cause up to 40 dB of reception strength fluctuation in a very short period of time. In addition, mobility also creates new links and invalidate old ones when nodes change their relative positions. More recently, it has been realized that mobility can also be beneficial. The observation is that if a mobile node transmits only when it moves into the vicinity of the receiver, the interference caused by the transmission on other communication links will be decreased. The pioneering work of Grossglauser and Tse [13] shows that the capacity of a mobile ad hoc network can be increased dramatically using mobile relays. Apparently, this mechanism incurs a delay in data transportation, so a trade-off between delay and throughput should be sought.

Recall the results on the capacity of multi-hop wireless networks (Section 19.1.3). The major barrier to having a higher capacity in multi-hop wireless networks is that much of the traffic transmitted by a node is relayed traffic. If nodes are mobile, the situation can be fairly different. A source node can wait until it moves very close to the destination node before transmission. The Infostation Architecture [50] supports high-speed data transfer between a ground station and vehicles moving by. If nodes are not moving along roads but arbitrarily in a 2-D plane, the chance that they move very close is slim. For example, within a circle of radius $1/\sqrt{\pi}$, the situation that two nodes are within $O(1/\sqrt{n})$ lasts for approximately $1/n$ fraction of time. Forcing direct transmissions from source to destination in this case can cause an impractically long delay.

Grossglauser and Tse [13] then propose to use a number of relays that temporarily buffer packets until final delivery to the destination. When the number of relays is large, the chance that one copy of the packet is close to the destination comes sooner. Grossglauser and Tse show that it suffices to use a two-hop scheme for each packet. That is, the source only needs to broadcast a packet to its neighbors, which in turn transmit it to the destination when the time comes. Furthermore, this ensures that the per-node capacity in the network is constant as apposed to $O(1/\sqrt{n})$ as in static networks.

Bansal and Liu [12] show that a bounded delay can be achieved based on a slightly different set of assumptions. Here, the unit-area network circle has n static nodes and m mobile nodes. The sources and destinations can be both mobile and static while the relays are mobile. In the routing algorithm, a single copy of a packet is unicast through multiple hops of relays from source to destination. At each hop, the packet is forwarded to a node that is heading to the destination approximately, until it is close enough to the destination when a last delivery is made. With sufficiently large m of the order of n, there exists a constant $c > 0$ such that the per-node throughput can be $cW \times \min\{m, n\}/n \log^3 n$, and the delay can be bounded by $\frac{4}{\sqrt{\pi}} v$ from above, where v is the average velocity of the mobile nodes. More recent studies relax the requirements for the mobility models and obtain a more general relation among capacity, delay, and mobility [51,52]. With these studies, using mobility to avoid interference is becoming more practical from the original, relatively theoretic framework.

19.6   CONCLUSIONS AND OPEN ISSUES

Thus far, we have seen the research foci in multi-hop wireless networking from a relatively abstract point of view. We realize that, even under fairly simplified physical layer models for communication and interference, what we can achieve is not satisfactory. The reason is that the problem is far more complex than what our models can offer solutions for.

Building a model for upper layers of mesh networking is nontrivial, even just for individual wireless links. It is understood that the strength of a signal attenuates as it travels and that the propagation of wireless signals can be affected by reflection, diffraction, and scattering. In effect, what an antenna receives is a superposition of multiple distorted copies of the original signal mingled with additive white Gaussian noise (AWGN) and interference. Due to the stochastic nature of the above propagation mechanisms, such a superposition is random. In a relatively open area, the average received signal strength can be approximated as a power function of distance with an exponent $-4 < \alpha < -2$. Unfortunately, the variance of this received power can be orders of magnitude and the operation environments are typically not open. In a real wireless mesh network, where the environments are changing even if the transceivers are not moving, these random behaviors are inevitable. How much of these are captured by the models proposed in multi-hop wireless networking? And yet, how manageable are these models?

More difficulties come from the interaction among the activities in the network: from local interference among wireless links to global coupling of routes taken by different data traffic flows. It has been realized that load-sensitive routing in the Internet, where flows can change paths to adapt to the changing network conditions, is extremely difficult due to the convoluted correlations among the flows, queues, contention windows, and so on, and to the much coarser time granularity that a countermeasure can be made compared with the rate of these changes [53]. Remember that traffic in the Internet can be much more predictable at least in an average sense. The case for multi-hop wireless networking is even worse because of the more dynamic nature of the data flows. Furthermore, interference among wireless links can only add to the complexity. We are now able to better appreciate the simplicity and beauty of TCP where billions of autonomous machines form a fairly effective feedback system, literally the largest ever created by mankind. After all, all TCP cares for is the congestion window.

To achieve the capacity bound that a multi-hop wireless network can potentially achieve, we have discussed three techniques in the chapter, that is, routing, channel assignment, and link scheduling. Essentially, all these are to realize an effective sharing of the communication resources through interference reduction. Note that each of these approaches to interference reduction can be applied along with other ones at the same time although they typically work at two different layers of the protocol stack. The orthogonality and dependency among these factors allow different ways to combine and separate them. Note that, to realize a higher throughput in the network, there are other techniques that we are not able to cover in this chapter due to the limit of space. And they are power control and rate adaptation. In power control, a transmitter is able to tune its output power level to reach nodes at differing distances to provide sufficient SINR. Apparently, reaching farther nodes incurs stronger interference within the proximity of the transmitter. Alternatively, when the SINR is high, more sophisticated modulation schemes can be used to support higher data rates. Raising the transmission power is a way to obtain a high SINR at the receiver, but can the resultant interference be paid off by the shorter transmission time using a higher data rate? Interested readers are recommended to read an intriguing discussion of these extra factors by Kim et al. [54].

When studying wireless mesh networks, how closely do we want to resemble the real world?

REFERENCES

[1] Brent N. Clark, Charles J. Colbourn, and David S. Johnson. Unit disk graphs. *Discrete Mathematics*, 85(1–3):165–177, 1990.

[2] Kamal Jain, Jitendra Padhye, Venkat Padmanabhan, and Lili Qiu. Impact of interference on multi-hop wireless network performance. In *Proceedings of MobiCom*, pp. 66–80, San Diego, CA, September 2003.

[3] Jian Tang, Guoliang Xue, and Weiyi Zhang. Interference-aware topology control and QoS routing in multi-channel wireless mesh networks. In *Proceedings of MobiHoc*, pp. 68–77, Urbana-Champaign, IL, May 2005.

[4] Pascal von Rickenbach, Stefan Schmid, Roger Wattenhofer, and Aaron Zollinger. A robust interference model for wireless ad hoc networks. In *Proceedings IEEE WMAN*, vol. 13, p. 239.1, Denver, CO, April 2005.

[5] Martin Burkhart, Pascal von Rickenbach, Roger Wattenhofer, and Aaron Zollinger. Does topology control reduce interference? In *Proceedings MobiHoc*, pp. 9–19, Tokyo, Japan, May 2004.

[6] Kousha Moaveni-Nejad and Xiang-Yang Li. Low-interference topology control for wireless ad hoc networks. In *Proceedings of the Second IEEE Communications Society Conference on Sensor and Ad Hoc Communications and Networks (SECON)*, Santa Clara, CA, September 2005.

[7] Thomas Moscibroda and Roger Wattenhofer. Minimizing interference in ad hoc and sensor networks. In *Proceedings of DIALM-POMC*, pp. 24–33, Cologne, Germany, September 2005.

[8] Piyush Gupta and P. R. Kumar. The capacity of wireless networks. *IEEE Transactions on Information Theory*, 46(2):388–404, March 2000.

[9] Alexander Zemlianov and Gustavo de Veciana. Capacity of ad hoc wireless networks with infrastructure support. *IEEE Journal on Selected Areas in Communications*, 23(3):657–667, March 2005.

[10] Benyuan Liu, Zhen Liu, and Don Towsley. On the capacity of hybrid wireless networks. In *Proceedings of INFOCOM*, pp. 1543–1552, San Francisco, CA, March–April 2003.

[11] Ulas C. Kozat and Leandros Tassiulas. Throughput capacity of random ad hoc networks with infrastructure support. In *Proceedings of MobiCom*, pp. 134–146, San Diego, CA, September 2003.

[12] Nikhil Bansal and Zhen Liu. Capacity, delay and mobility in wireless ad-hoc networks. In *Proceedings of INFOCOM*, pp. 1553–1563, San Francisco, CA, March–April 2003.

[13] Matthias Grossglauser and David N. C. Tse. Mobility increases the capacity of ad-hoc wireless networks. In *Proceedings of INFOCOM*, pp. 1360–1369, Anchorage, AK, April 2001.

[14] S. Ramanathan and Martha Steenstrup. A survey of routing techniques for mobile communications networks. *Mobile Networks and Applications (MONET)*, 1(2):89–103, 1996.

[15] Josh Broch, David A. Maltz, David B. Johnson, Yih-Chun Hu, and Jorjeta Jetcheva. A performance comparison of multi-hop wireless ad hoc network routing protocols. In *Proceedings of the Fourth Annual ACM/IEEE International Conference on Mobile Computing and Networking*, ACM, Dallas, TX, October 1998.

[16] Elizabeth M. Royer and Chai-Keong Toh. A review of current routing protocols for ad-hoc mobile wireless networks. *IEEE Personal Communications*, 6(2):46–55, April 1999.

[17] Mehran Abolhasan, Tadeusz Wysocki, and Eryk Dutkiewicz. A review of routing protocols for mobile ad hoc networks. *Ad Hoc Networks*, 2(1):1–22, 2004.

[18] Changling Liu and Jorg Kaiser. A survey of mobile ad hoc network routing protocols. Technical Report TR-4, MINEMA, University of Magdeburg, October 2005.

[19] Martin Mauve, Jorg Widmer, and Hannes Hartenstein. A survey on position-based routing in mobile ad hoc networks. *IEEE Network*, 15(6):30–39, November–December 2001.

[20] Stephen Mueller, Rose P. Tsang, and Dipak Ghosal. Multipath routing in mobile ad hoc networks: Issues and challenges. In Maria Carla Calzarossa and Erol Gelenbe, editors, *Lecture Notes in Computer Science (LNCS 2965)*. Springer, Orlando, FL, 2004. (from MASCOTS tutorial, detailed information can be found at http://www.informatik.uni-trier.de/~ley/db/conf/mascots/mascotst2003.html)

[21] Yih-Chun Hu and Adrian Perrig. A survey of secure wireless ad hoc routing. *IEEE Security & Privacy*, 2(3):28–39, May–June 2004.

[22] Liang Qin and Thomas Kunz. Survey on mobile ad hoc network routing protocols and cross-layer design. Technical Report SCE-04-14, Carleton University, Ottawa, Ontario, Canada, August 2004.

[23] Chai-Keong Toh. A novel distributed routing protocol to support ad-hoc mobile computing. In *Proceedings of 15th IEEE Annual International Phoenix Conference on Computers and Communications*, 1996, 27–29 March, Phoenix, AZ, pp. 480–486, 1996.

[24] Rohit Dube, Cynthia D. Rais, Kuang-Yeh Wang, and Satish K. Tripathi. Signal stability based adaptive routing (SSA) for ad hoc mobile networks. *IEEE Personal Communications*, 4(1):36–45, February 1997.

[25] Douglas S. J. De Couto, Daniel Aguayo, John Bicket, and Robert Morris. A high-throughput path metric for multi-hop wireless routing. In *Proceedings of MobiCom*, pp. 134–146, San Diego, CA, September 2003.

[26] Atul Adya, Paramvir Bahl, Jitendra Padhye, Alec Wolman, and Lidong Zhou. A multi-radio unification protocol for IEEE 802.11 wireless networks. In *Proceedings of the First International Conference on Broadband Networks (BroadNets)*, pp. 344–354, San Jose, CA, October 2004.

[27] Richard Draves, Jitendra Padhye, and Brian Zill. Routing in multi-radio, multi-hop wireless mesh networks. In *Proceedings of MobiCom*, pp. 114–128, Philadelphia, PA, September 2004.

[28] Jian Zhang, Yuanzhu Peter Chen, and Ivan Marsic. MAC scheduling using channel state diversity for high-throughput IEEE 802.11 mesh networks. *IEEE Communications Magazine*, 46(11), 94–99, November 2007.

[29] Jing Ai, Alhussien A. Abouzeid, and Zhenzhen Ye. Cross-layer optimal decision policies for spatial diversity forwarding in wireless ad hoc networks. In *Third IEEE International Conference on Mobile Ad-hoc and Sensor Systems (MASS)*, Vancouver, British Columbia, Canada, October 2006.

[30] Shweta Jain and Samir R. Das. Exploiting path diversity in the link layer in wireless ad hoc networks. In *IEEE International Symposium on a World of Wireless Mobile and Multimedia Networks (WoWMoM)*, pp. 22–30, Taormina, Italy, June 2005.

[31] Vishnu Navda, Samrat Ganguly, and Samir Das. Interference-aware fast path adaptation in wireless mesh network. In *Proceedings of MobiCom (poster)*, Los Angeles, CA, September 2006.

[32] Sung-Ju Lee and Mario Gerla. Split multipath routing with maximally disjoint paths in ad hoc networks. In *IEEE International Conference on Communications (ICC)*, pp. 3201–3205, St. Petersburg, Russia, June 2001.

[33] Mahesh K. Marina and Samir R. Das. On-demand multipath distance vector routing in ad hoc networks. In *IEEE International Conference on Network Protocols (ICNP)*, pp. 14–23, Riverside, CA, November 2001.

[34] Ravindra K. Ahuja, Thomas L. Magnanti, and James B. Orlin. *Network Flows : Theory, Algorithms, and Applications*. Prentice Hall, Englewood Cliffs, NJ, 1993.

[35] Vinay Kolar and Nael B. Abu-Ghazaleh. A multi-commodity flow approach for globally aware routing in multi-hop wireless networks. In *Proceedings of the Fourth Annual IEEE International Conference on Pervasive Computing and Communications (PerCom)*, p. 10, Pisa, Italy, March 2006.

[36] Mansoor Alicherry, Randeep Bhatia, and Li Erran Li. Joint channel assignment and routing for throughput optimization in multiradio wireless mesh networks. *IEEE Journal on Selected Areas in Communications*, 24(11):1960–1971, November 2006.

[37] Jungmin So and Nitin Vaidya. A routing protocol for utilizing multiple channels in multi-hop wireless networks with a single transceiver. Technical Report, UIUC Technical Report, October 2004.

[38] Jungmin So and Nitin Vaidya. Multi-channel MAC for ad hoc networks: Handling multi-channel hidden terminals using a single transceiver. In *Proceedings of MobiHoc*, pp. 222–233, Roppongi, Japan, May 2004.

[39] Paramvir Bahl, Ranveer Chandra, and John Dunagan. SSCH: Slotted seeded channel hopping for capacity improvement in IEEE 802.11 ad-hoc wireless networks. In *Proceedings of MobiCom*, pp. 216–230, Philadelphia, PA, September 2004.

[40] Ashish Raniwala, Kartik Gopalan, and Tzicker Chiueh. Centralized channel assignment and routing algorithms for multi-channel wireless mesh networks. *ACM Mobile Computing and Communications Review (MC2R)*, 8(2):20–35, April 2004.

[41] Ashish Raniwala and Tzicker Chiueh. Architecture and algorithms for an IEEE 802.11-based multi-channel wireless mesh network. In *Proceedings of INFOCOM*, pp. 2223–2234, Miami, FL, March 2005.

[42] Shih-Lin Wu, Chih-Yu Lin, Yu-Chee Tseng, and Jang-Ping Sheu. A new multi-channel MAC protocol with on-demand channel assignment for multi-hop mobile ad hoc networks. In *Proceedings of International Symposium on Parallel Architectures, Algorithms, and Networks (I-SPAN)*, pp. 232–237, Dallas/Richardson, TX, December 2000.

[43] Pradeep Kyasanur and Nitin H. Vaidya. Routing and link-layer protocols for multi-channel multi-interface ad hoc wireless networks. *ACM Mobile Computing and Communications Review (MC2R)*, 10(1):31–43, January 2006.

[44] Gaurav Sharma, Ravi R. Mazumdar, and Ness B. Shroff. On the complexity of scheduling in wireless networks. In *Proceedings of MobiCom*, pp. 227–238, Los Angeles, CA, September 2006.

[45] Weizhao Wang, Yu Wang, Xiang-Yang Li, Wen-Zhan Song, and Ophir Frieder. Efficient interference-aware TDMA link scheduling for static wireless networks. In *Proceedings of MobiCom*, pp. 262–273, Los Angeles, CA, September 2006.

[46] Thomas Moscibroda, Roger Wattenhofer, and Aaron Zollinger. Topology control meets SINR: The scheduling complexity of arbitrary topologies. In *Proceedings of MobiHoc*, pp. 310–321, Florence, Italy, May 2006.

[47] Gurashish Brar, Douglas M. Blough, and Paolo Santi. Computationally efficient scheduling with the physical interference model for throughput improvement in wireless mesh networks. In *Proceedings of MobiCom*, pp. 2–13, Los Angeles, CA, September 2006.

[48] Murali Kodialam and Thyaga Nandagopal. Characterizing achievable rates in multi-hop wireless networks: The joint routing and scheduling problem. In *Proceedings of MobiCom*, pp. 42–54, San Diego, CA, September 2003.

[49] Murali Kodialam and Thyaga Nandagopal. Characterizing the capacity region in multi-radio multi-channel wireless mesh networks. In *Proceedings of MobiCom*, pp. 73–87, Cologne, Germany, August 2005.

[50] Richard H. Frenkiel, B. R. Badrinath, Joan Borras, and Roy D. Yates. The infostations challenge: Balancing cost and ubiquity in delivering wireless data. *IEEE Personal Communications*, 7(2):66–71, April 2000.

[51] Gaurav Sharma, Ravi Mazumdar, and Ness Shroff. Delay and capacity trade-offs in mobile ad hoc networks: A global perspective. In *Proceedings of INFOCOM*, pp. 1–12, Barcelona, Spain, April 2006.

[52] Michele Garetto, Paolo Giaccone, and Emilio Leonardi. On the capacity of ad hoc wireless networks under general node mobility. In *Proceedings of INFOCOM*, pp. 357–365, Anchorage, AK, May 2007.

[53] Eric Anderson and Thomas E. Anderson. On the stability of adaptive routing in the presence of congestion control. In *Proceedings of INFOCOM*, pp. 948–958, San Francisco, CA, March–April 2003.

[54] Tae-Suk Kim, Jennifer C. Hou, and Hyuk Lim. Improving spatial reuse through tuning transmit power, carrier sense threshold, and data rate in multi-hop wireless networks. In *Proceedings of MobiCom*, pp. 366–377, Los Angeles, CA, September 2006.

20 Cooperation in Wireless Sensor and Actor Networks

Erica Cecilia Ruiz-Ibarra
and Luis Armando Villasenor-Gonzalez

CONTENTS

In a wireless sensor and actor network (WSAN), several cooperation tasks must take place to comply with the specific requirements of the application, one of these cooperation tasks is related to the coordination process between the elements of the WSAN, namely the sensor and the actor nodes. The sensor node collects data from the environment and may report data-related events to the actors. As a result, the actors can proceed to execute a task in accordance with the event data received from the sensor nodes, this procedure requires the implementation of a coordination mechanism. In this chapter, a framework for the study and the analysis of the cooperation processes is presented and a novel taxonomy is proposed for the classification of the coordination mechanisms for WSAN. In addition, some of the most representative coordination mechanisms published in the literature are summarized followed by a classification and a comparative analysis using the proposed taxonomy. Finally, some open issues within the scope of cooperative services are stated to provide a background for future WSAN research activities.

20.1 INTRODUCTION

A WSAN is composed of sensor and actor nodes distributed in a geographic area of interest; the sensors are involved in monitoring the physical environment, while the actors are responsible for executing a task in accordance with the data collected and reported by the sensors during an event. To achieve a balanced performance, a WSAN architecture must implement an efficient cooperative communication strategy to allow the nodes to collaborate in the optimal assignment of resources and to execute tasks with the lowest possible delay. Such collaboration must take place by exchanging information and generating negotiated decisions while trying to extend the WSAN lifetime. As a result, wireless sensor networks (WSNs), as well as WSAN, are a fertile field for the study and experimentation of new cooperation models. This represents a challenge, especially in a WSAN environment, where sensor and actor nodes must coordinate their efforts in optimum ways.

The term *Cooperation* is derived from the Latin prefix *co-* (from the Latin preposition *cum*, meaning *with*) and the Latin term *operari* (meaning *work*); thus cooperation implies working together. From its etymological roots, cooperation may be considered the "action taken by two or more agents, working together to achieve a common and identical goal." Within the context of a WSAN, the coordination mechanisms promote the cooperation of multiple devices to achieve a specific goal; in consequence, WSANs are considered to be cooperative systems because their elements are capable of sensing, inferring, sharing information, and executing tasks to achieve a common goal. For example, in wildfire applications, the sensors cooperate by monitoring the physical environment and reporting event-related information, such as temperature and humidity. Once an event (e.g., fire) triggers an event-reporting process, the sensors proceed to transmit the event-related information toward the actor nodes; each actor then reacts and cooperates to take appropriate and timely actions. As part of the cooperation process, tasks must be delegated between neighboring entities to achieve optimum behavior. For example, when a source node does not have radio signal coverage with its intended destination node, the information can be relayed via intermediate nodes (i.e., multi-hop routing). Thus, the data transmission task is delegated to neighboring nodes, which will in turn delegate this task to other nodes, so the data can be delivered to the intended destination. The cooperation between these entities is not only related to the exchange of information, but may include other services, such as aggregation (i.e., data fusion), synchronization, localization, clustering, and power control, among others.

The main objective of this chapter is to provide a common framework for the study and analysis of cooperative mechanisms in a WSAN environment; a special interest is given to the coordination mechanisms and a novel taxonomy is proposed. The remainder of this chapter is organized as follows. Section 20.2 presents a description of the WSAN network architecture, with an emphasis on the cooperative nature of this type of system. Section 20.3 outlines a novel taxonomy for the study and the analysis of coordination mechanisms in a WSAN environment. Section 20.4 summarizes different coordination mechanisms presented in the literature. Section 20.5 provides a classification and a comparative analysis of the WSAN coordination mechanisms by employing the taxonomy introduced in Section 20.3. Finally, Section 20.6 presents the conclusions of this work, including a description of open issues within the scope of cooperative services in WSANs.

20.2 WSAN ARCHITECTURE

Recent advances in microelectronics and wireless technology have enabled the development of small-size devices, which are low-cost, limited in energy, and equipped with wireless communication capabilities. This has led to the development of WSNs, which are composed of hundreds of sensor nodes used to monitor multiple physical variables, such as temperature, humidity, sound, pressure, movement, vibrations, and so on [1]. A WSN can be deployed to support a large array of applications, which include environmental monitoring, inventory tracking, prediction of natural disasters (e.g., earthquakes, forest fires), home automation, traffic control, and military supervision in the battlefield [1]. However, there are some complex scenarios that require the cooperation between sensors and higher capability devices, such as actor nodes, to support the proper execution of specific tasks; WSAN have been proposed as an important extension of WSN [2]. A WSAN can be deployed for a great variety of applications, such as microclimate control in a building or a greenhouse, detection of biological, chemical, or nuclear attacks, automation of industrial processes, and control of ventilation systems and heating [3].

20.2.1 WSAN TOPOLOGY

A WSAN is composed of a large amount of sensor nodes and a few actors connected by wireless means; these devices cooperate among themselves to provide distributed sensing and to execute specific tasks [4]. Sensor and actor nodes can be spread in the field, while a sink node can be used to monitor the network and may be used to communicate with a task manager, as illustrated in Figure 20.1. In a WSAN the sensor nodes behave as passive elements, collecting information from the physical world, whereas the actors are active elements that make independent decisions and are capable of executing appropriate actions in accordance with the information collected; all these capabilities allow the user to monitor and to act when located in a remote location [5]. In a WSN scenario, it is usually assumed that sensor nodes cannot be locally configured or recharged while deployed in the field; therefore, these types of devices are required to be autonomous and energy efficient. The energy constraint of the sensor nodes imposes limitations on the size of the device; similarly, there is a reduction in resources such as memory, processing speed, computing power, and bandwidth [1], all this with the purpose of extending the lifetime of the sensor node. On the other hand, in a WSAN, the actor nodes are equipped with greater resources, such as increased computing capacity, powerful transmitters, and increased battery lifetime by means of rechargeable or replaceable power sources [5]. As a result, a WSAN can be seen as an ad hoc network composed of heterogeneous devices in which the sensor and the actor nodes have different capabilities.

Figure 20.2 shows a block diagram of the components of sensor and actor nodes in a WSAN architecture. The sensor nodes are equipped with a power unit, a communication subsystem, storage and processing subsystems, analog-to-digital converters (ADCs), and a sensing unit. The sensing unit is used to monitor the environmental variables, and the collected analog information is converted into digital data by the ADC. The data is then analyzed by the processor and the information is transmitted to the actors [5,6]. In the actor nodes the decision unit (i.e., controller) works as the entity that gathers the data provided by the sensor nodes and generates the proper commands for execution. The execution commands are translated into analog signals by the digital-to-analog converted (DAC), which results in specific actions by the actuation unit [5,6].

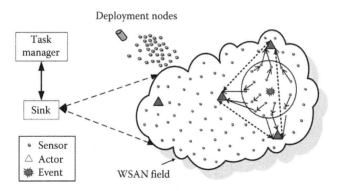

FIGURE 20.1 Physical architecture of a WSAN. (From Akyldiz, I.F. and Kasimoglu, I., *Ad Hoc Networks*, 2004. With permission.)

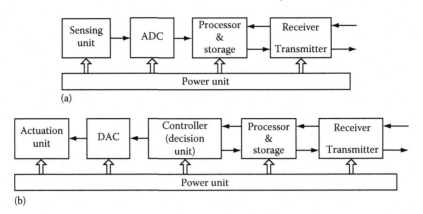

(a)

(b)

FIGURE 20.2 System components of (a) sensors and (b) actors. (From Akyldiz, I.F. and Kasimoglu, I., *Ad Hoc Networks*, 2004. With permission.)

20.2.2 WSAN ARCHITECTURE TYPES

A WSAN is a distributed system that can adapt and react to the environmental conditions that are reported by the collaborative effort of all the sensors and actors [7]. Two different types of architectures can be defined according to the way data is collected by the sensor nodes and are reported back to the actor nodes. These are defined as the automated and semiautomated architectures, illustrated in Figure 20.3.

In the automated architecture, data is collected by the sensors and are transmitted directly to the actor nodes, which efficiently coordinate to execute a specific task without collaboration from the sink. As a result, automated WSAN architectures are recommended for time-sensitive applications in which a fast reaction by the actors is a critical requirement. Another characteristic of automated WSAN architectures is the efficient consumption of energy in the network, since remote sensors

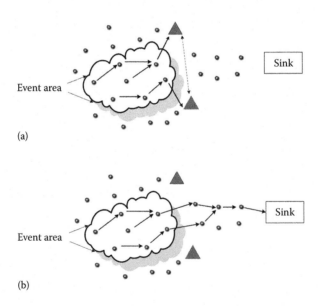

(a)

(b)

FIGURE 20.3 (a) Automated (b) semiautomated architecture. (From Akyldiz, I.F. and Kasimoglu, I., *Ad Hoc Networks*, 2004. With permission.)

(i.e., sensors located far from the event) do not get involved during the retransmission of data toward the actors; this results in an increased network lifetime. However, as a result of the direct communication between sensor and actor nodes, the automated WSAN architecture requires the implementation of an efficient coordination mechanism to support the collaboration between these devices [5]. In contrast, in the semiautomated architecture the sensor data is transmitted to a central controller (e.g., the sink), which processes the collected data and determines which actors must take action to execute a specific task; this is accomplished by transmitting a set of commands to the corresponding actors. The semiautomated architecture is similar to the one employed in WSN networks, so it is possible to use communication protocols developed for WSNs [5]. On the other hand, one of the main disadvantages of the semiautomated WSAN architecture is the centralized processing of the information at the sink, which may result in an increased delay during the execution of tasks by the actors. Consequently, the semiautomated WSAN architecture is not recommended for time-sensitive applications.

20.2.3   COORDINATION LEVELS IN A WSAN ARCHITECTURE

The communication process, in a WSN environment, mainly takes place from the sensor nodes to the sink; in contrast, the communication process in a WSAN environment can take place between the sensor and the actor nodes. Thus, a WSAN architecture requires the implementation of different coordination levels; these coordination levels are defined as

- Sensor–Sensor (SS). The SS coordination is employed to gather information from the physical world in an effective and energy efficient way.
- Sensor–Actor (SA). The SA coordination is employed to report new events and to transmit the characteristics of the event from the sensors to the actors [8]; in addition, the SA coordination may also be used over the downlink (i.e., from the actor toward the sensor) to inform the sensors to proceed with specific sensing tasks.
- Actor–Actor (AA). The AA coordination is required to generate the most appropriate way to execute a specific task while coordinating which actor nodes should respond within a certain area.

The objective of these mechanisms is to coordinate the actions between the sensor and the actor nodes, while making optimum use of the available resources and at the same time executing the required tasks within the time bound required by the application [9].

It is possible to define two types of cooperation strategies at the multiple coordination levels: horizontal and vertical cooperation. The horizontal cooperation takes place between devices of the same type. In a WSAN architecture, horizontal cooperation is accomplished at the SS and AA coordination levels. The cooperation between sensors involves multiple tasks, such as cluster formation, data fusion, and on–off radio synchronization, which contributes to improve the efficient use of network resources. In relation to the cooperation between actors, this may consist of exploiting their increased wireless communication capabilities to exchange information related to events. Vertical cooperation takes place between heterogeneous entities; such is the case of the SA coordination level. At this level, the sensors collaborate with the actors by providing them with reliable information related to events with the lowest possible delay; as a result, actors can make adequate decisions and be able to execute timely actions. The SA coordination must take place by implementing algorithms that are capable of providing orderly execution and synchronization of tasks, while avoiding any redundant efforts.

In relation to the task assignment process, this may require a single actor or it may require the cooperation of multiple actors [5]. In a WSAN, multiple actors can receive the information from the sensor nodes; this situation is defined as a multiple actor scenario, where each sensor independently selects an actor and transmits the collected data to this particular device. On the other hand, if only one actor receives the event information then it is defined as a single actor scenario. During the communication process, there are two possible scenarios, one involves the possibility of one-hop transmission, while the other involves the possibility of multi-hop transmission. It should be noted that one-hop transmission is usually inefficient in WSN scenarios given the large distance between the sensors and the sink; however, in a WSAN scenario this inefficiency might not be the case as the actor nodes can be located closer to the sensor nodes in the field [5].

The following section presents a coordination mechanism taxonomy. This taxonomy is proposed to provide a framework for the analysis and evaluation of coordination mechanisms used in WSAN architectures.

20.3   TAXONOMY FOR WSAN COORDINATION MECHANISMS

To the best of our knowledge, there is no specific coordination mechanism taxonomy for WSAN networks. There is some related work, such as Farinelli's [10] who presents a taxonomy for the coordination of multi-robot systems and is based on four levels: cooperation, knowledge, coordination, and organization. However, this taxonomy only considers cooperative systems consisting of robots and a coordination protocol-based/protocol-free decision-making process, but it does not include data and context sharing. Another related work by Salkhman et al. [11] presents a taxonomy for context-aware collaborative systems based on the commonalities of different context-aware systems that emphasized collaboration. The structure proposed by Salkhman is supported in three axes: goal, approaches, and means. However this taxonomy is too broad, as it can be applied to a great variety of systems, from small augmented artifacts to large-scale and highly distributed sensor/(actor) networks. As a result, the proposed structure does not provide the proper elements for the fine classification required in a WSAN coordination mechanism. A third work by Sameer et al. [12] develops a range of middle-ware services such as synchronization, localization, aggregation, and tracking to facilitate the coordination through self-organizing networked sensors.

A new taxonomy for WSAN coordination mechanisms is proposed in this section. The proposal is inspired from the work presented in Salkhman [11] and Sameer [12]. The taxonomy is divided into four sections: WSAN framework, collaborative procedures, performance criteria, and application requirements. Figure 20.4 shows a diagram of the proposed taxonomy.

WSAN framework. It represents the structure of a WSAN and includes the network architecture, the coordination levels, node mobility, and network density.

- *Network architecture.* It is related to the way information is sensed and reported to the actors, which may be automated or semiautomated.
- *Coordination levels.* It refers to the coordination levels employed by the coordination mechanism. These levels may be SS, SA, and AA.
- *Node mobility.* It is used to specify if the nodes in a WSAN (i.e., sensors, actors, or the sink) are mobile or fixed.
- *Network density.* It defines the ratio between the number of nodes spread throughout the field and the field dimensions; the network density may be classified as spread or dense.

Collaborative procedures. It refers to those collaborative procedures (or services) which are used to support the exchange of information by the coordination mechanism. These services include routing mechanism, synchronization, localization, aggregation, clustering, data cipher or encryption, power control, quality of service (QoS).

- *Routing mechanism.* It is related to the procedure implemented for selecting a route to transmit packets to a destination. The implementation of a specific type of routing protocol may be different at multiple coordination levels.

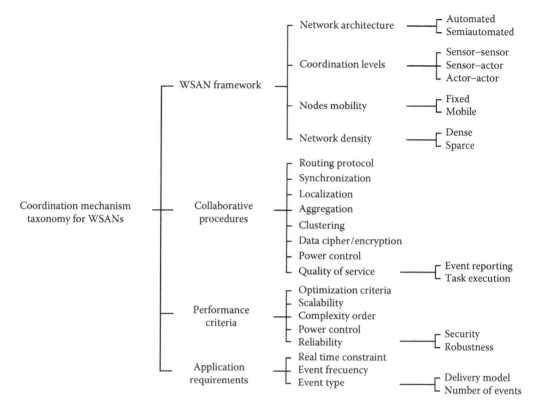

FIGURE 20.4 Coordination mechanism taxonomy for WSANs.

- *Synchronization.* It refers to the implementation of energy-efficient techniques required to associate time and location information with the sensed data; this is required to support collaborative processing.
- *Localization.* It is implemented to provide information regarding the geographical location of the sensor and actor nodes; this can be achieved by means of a global positioning system (GPS) receiver or through trilateration techniques.
- *Aggregation.* It has the objective of reducing the energy consumption and the control overhead associated with the transmission of information to a common destination. To achieve this goal, a single packet is created as a result of the fusion of data that is generated by multiple sources; this strategy reduces the traffic load in the network.
- *Clustering.* Clustering provides a hierarchy among nodes. The substructures that are collapsed in higher levels are called clusters, and there is at least one node in each cluster, which is denoted as the cluster head. This is usually an effective technique implemented to increase the network lifetime. Additional benefits of the clustering service are scalability, load balancing, and the possibility to implement a hierarchical routing strategy.
- *Data cipher or encryption.* It is related to the implementation of any cipher/encryption mechanism used to protect the integrity of data in the WSAN.
- *Power control.* It is related to the possibility of varying the transmission power; as a result, the radio signal coverage of a node can be modified.
- *Quality of service.* It is related to the mechanisms employed to provide guaranteed services in a WSAN architecture, in accordance with the application's requirements. In a WSAN scenario, it is possible to define two different functionalities to provide QoS support. One functionality is based on providing the required resource reservations to the nodes involved

in reporting an event at the SA coordination level, therefore providing differentiated services to the data being transmitted in the WSAN. A different functionality is to prioritize the execution of tasks by the actors in response to events in the WSAN. In consequence, the actors should be able to respond in accordance with the priority of the events reported by the sensor nodes.

Performance criteria. It is related to the criteria elements used to estimate the performance of the coordination mechanism, such as

- *Optimization criteria.* These criteria denote metrics on which the mechanism is based to reach the proposed objectives.
- *Scalability.* It is related to the capability of a network to maintain the performance level when the number of member nodes as well as the requirements on it increases.
- *Complexity order.* It provides a measurement of the computational complexity of the proposed algorithms and can be used to provide insight regarding an algorithm's usage of computational resources (usually running time or memory).
- *Communication complexity.* It is associated to the communication protocols where different measures, such us, total control messages, total message exchanges and synchronization rounds, are commonly used to evaluate the communication complexity.
- *Reliability.* It provides a measure of the level of security and robustness of a WSAN coordination mechanism. To classify a coordination mechanism as secure, it must implement additional functionalities to guarantee data integrity and to avoid access to the WSAN by intruders. On the other hand, robustness is related to the capability of the coordination mechanism to handle faults, while assuring data delivery from the sensors to the actors by means of acknowledgment or retransmission procedures.

Application requirements. The coordination mechanism must take into consideration the application requirements. Some of these requirements include real-time constraints, event frequency, and the possibility of supporting concurrent events.

- *Real-time constraints.* These are related to the amount of time required by the coordination mechanism to report events and execute a task. Depending on the application requirements, the coordination mechanism may be required to report events and execute tasks in real time with low delay; other applications may not require the imposition of real-time restrictions.
- *Event frequency.* It is related to the periodicity of events at which the system is capable of providing a proper response.
- *Event type.* The type of event can be categorized in terms of the delivery model and the number of events. According to the delivery model it is possible to identify four kinds of events: continuous, event-driven, observer-initiated and hybrid. In continuous type, the sensors will report their readings at a determined rate. In event-driven, sensors will inform to the application when certain events occur. In observer-initiated, sensors will reply to application requests. The hybrid approach implements a mixture of any of the three previous approaches. Regarding the number of events these can be classified as single, multiple or concurrent event types.

This section presents a proposal for a coordination mechanism taxonomy. This taxonomy provides a framework for the classification of coordination mechanisms designed for WSAN environments. The following section presents an overview of some of the most relevant coordination mechanisms published up to this date in the literature, including a description of some of their most important features.

20.4   COORDINATION MECHANISMS FOR WSAN

The sensor and actor nodes should be able to provide a proper response in accordance with the sensor-collected data and the application requirements. To achieve this goal, a compromise must be reached between the event-reporting and execution delay as well as the power consumption of the nodes in a WSAN. Consequently, the coordination mechanism must assign resources in optimum ways while providing a response in the shortest possible time.

20.4.1   COMMUNICATION AND COORDINATION IN WSAN

Melodia et al. [13,14] present a coordination mechanism for WSAN; the proposal presented in Ref. [14] is an extended version of the earlier work described in Ref. [13]. These proposals described a distributed coordination mechanism for WSAN, which implements an automated architecture; the proposed mechanism is organized in two-levels: SA and AA. The event area is segmented into variable-size clusters and assumes that both the sensor and actors have fixed locations (i.e., there is no mobility) and can sense/act in an area defined by their coverage range. This mechanism is suitable for scenarios where the events are not too frequent.

The proposed SA coordination model is based on an event-driven clustering paradigm where cluster formation is triggered by an event and clusters are created on the fly to optimally react to the event; in other words, only the event area is clustered and each cluster consists of sensors that send their data to the same actor, as is illustrated in Figure 20.5. In this way, the existing energy resources are better used, and the communication overhead to maintain clusters, before the event occurs, is eliminated. The optimal solution to the proposed event-driven clustering problem is by means of integer linear programming [15]. In addition, Melodia et al. [14] propose a distributed event-driven partitioning and routing protocol (DEPR) that includes an adaptive mechanism, which trades off energy consumption for delay when event data must be delivered to the actors within a predetermined latency bound. DEPR is a scalable consensus-free protocol for SA distributed coordination, as it is based on localized routing decisions. The protocol assumes that each sensor is aware of its location as well as the neighbor's and the actor's location. Furthermore, it is assumed that the network is synchronized (i.e., clock synchronization) by means of one of the existing synchronization protocols presented in the literature [16]. The SA coordination problem consists of establishing a data route from each sensor residing in the event area to the corresponding actors, assuring that the observed reliability (e.g., latency bound) r is above a predefined event threshold r_{th} (i.e., $r \geq r_{th}$) while minimizing the energy consumption associated with the data delivery route. When a sensor

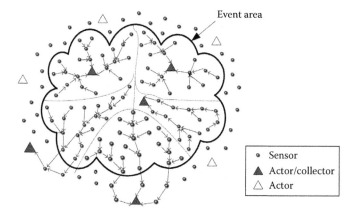

FIGURE 20.5   Event-driven partitioning with multiple actors. (From Melodia, T., Pompil, D., Gungor, V.C., and Akyildiz, I.F., *IEEE Trans. Mobile Comput.*, 2007. With permission.)

node receives data from at least two other nodes, it aggregates the received information using a data fusion strategy. In this way, all sensors in a cluster construct a data aggregation tree toward the selected actors. Each sensor implements a multistate protocol (i.e., DEPR), which alternates among four different states, namely the idle, the startup, the speed-up, and the aggregation state, as is described in Ref. [14]. The main objective of these state transitions is to reduce the number of hops when the reliability requirement is violated and to save energy when the reliability requirement is met.

The AA coordination model in Ref. [14] defines the problem as an optimization problem to share the action workload among different actors; this problem is formulated as a mixed integer nonlinear program (MINLP) [17]. Furthermore, the model proposes a distributed algorithm for the AA coordination problem, based on localized auctions among the actors. The collectors are those actors that receive the sensor readings; however, they may not be able to act on the whole area of the event nor be the best set of actors for that task in terms of action completion time or energy consumption. Therefore, the goal of AA coordination is to select the best actors to provide an appropriate action on the event area. The proposal in Ref. [14] considers two possible scenarios: an overlapping and a nonoverlapping area scenario. An overlapping area scenario results from the possibility of multiple actors reacting to the same event, whereas a nonoverlapping scenario considers only the possibility of a single actor reacting to the event. In an overlapping area the problem consists of selecting a subset of actors which maximizes the average residual energy of all actors involved in the action, while complying with the latency requirements. Whereas in a nonoverlapping area, the problem simplifies to selecting the power level for the actor that minimizes the energy consumption while satisfying the action completion bound. In the overlapping area scenario, a localized auction protocol is proposed as a solution for the AA coordination problem, which defines the behavior of actors participating in transactions as buyers and sellers. The aim of an auction is to select the best set of actors to perform a task on each overlapping area. The actors can assume different roles:

- *Seller:* The actor that receives event data and is responsible for that portion of the event area.
- *Auctioneer:* It is responsible for conducting the auction on a particular overlapping area.
- *Buyer:* The actor that can act on a specific overlapping area.

The auction takes place in each overlapping area. The bid of each actor that participates in the auction consists of the power level required to execute the action, the time needed to complete the action on the whole area, and the available energy of the actor. The aim is to maximize the total revenue of the team, constituted by the actors participating in the auction. A similar model is formulated by Gerkey et al. [18] who proposes the implementations of WSANs as a synthetic market system in which the nodes (sensors or actors) trade resources with each other to increase their own wealth. In this case the problem is modeled by sensors and actors which are considered to be competitive individuals that participate in an economy of task contracts.

20.4.2 A REAL-TIME COMMUNICATION FRAMEWORK FOR WSAN

Ngai et al. [19] propose a real-time communication framework for an automated architecture that minimizes the transmission delay from sensors to actors and provides a quick reaction by the actors. The approach is designed for event-driven applications in a self-organized network conformed by fixed sensors and mobile actors, which broadcast their location periodically. It is assumed that each sensor and actor knows its location by means of a GPS receiver [20] or that it can be derived by means of a localization technique [21]. The proposed framework is organized in multiple levels or steps: detection of the event, report of the event's gathered information, coordination among actors, and finally reaction to the event by the actors, as shown in Figure 20.6.

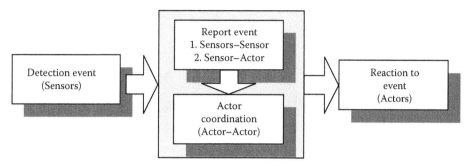

FIGURE 20.6 Workflow of the framework. (From Ngai, E.C.H., Lyu, M.R., and Liu, J., *IEEE Aerospace Conference*, 2006. With permission.)

The proposed mechanism supports the SS, SA, and AA coordination levels. At the first level (i.e., SS), the event is detected by a group of sensors located within the event area; they start the clustering process and aggregate data from their surrounding nodes. At the second level (i.e., SA), the event-gathered information is reported from the sensors to the actors by means of a real-time and distributed event-reporting algorithm. The event area is segmented into multiple map sections, and the corresponding data along with the map information are independently sent to the closest actors. This coordination mechanism (i.e., event-reporting algorithm) allows the sensor to transmit data to actors using a route with the minimum delay. In addition, important data is transmitted with a higher priority, thus ensuring that the actors can obtain a rough image of the event in the shortest possible time; as a result, the actors can start the coordination process without waiting for the complete event report. At the third level (i.e., AA), an efficient coordination algorithm between actors is proposed to let the actors share information regarding the event and allow them to react in a quick and opportune way. The actors combine the received maps and determine how many and which actors should be involved to take an action; then the selected actors move toward the event area and execute the appropriate tasks.

The proposed framework makes use of geographical-based routing protocols, as they scale up well and can easily adapt to any location change by the actors. In addition, it imposes a virtual grid structure on the network to simplify coordinate representation during the formation and combination of maps. To reduce the network traffic, the sensors aggregate event reports from the neighborhood. The first sensor n_1 detecting the event begins with the formation of maps. During an event, multiple nodes can start the formation of maps and begin to create clusters with their neighboring nodes. The clusters split the event area into map pieces, which are reported by sensor n_1 to the actor; any node located on the boundary between two maps stops forwarding the event-detection message and transmits its location coordinates, data value, and the event ID to the previous nodes. Each piece of the map can be represented by a tree structure with the sensor n_1 as the root, as illustrated in Figure 20.7. When a node receives the replies from its descendent nodes, it concatenates its own reply and forwards them to the previous hop. Nodes with even depth number h concatenate the reply with their own location coordinates and sensed data, whereas nodes with odd depth number h aggregate the data from their immediate descendants. Nodes with odd depth number calculate an average based on their own sensed data values and their descendants' (i.e., depth $h + 1$) sensed data values. Finally, the root sensor n_1 collects all the coordinates and sensed data from the nodes within their clusters, and it is responsible for reporting this event to its closest actor.

It should be noted that node n_1 divides the data into different layers according to their importance. The base layer corresponds to high priority data and contains the type of event, the time when the event is first detected, the location of the map, and the mean value of the collected data. The refinement layer data corresponds to additional information collected from the event as described in Ref. [19].

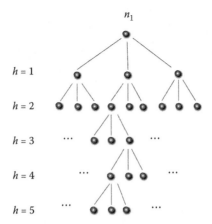

FIGURE 20.7   Tree representation of nodes on the map. (From Ngai, E.C.H., Lyu, M.R., and Liu, J., *IEEE Aerospace Conference*, 2006. With permission.)

In relation to the coordination process between actors, an actor can combine multiple maps if it receives several reports of the same event. Next, the actor can start the communication with other actors located close to the event area and exchange information to combine their maps and estimate the size of the event. The coordination between actors can begin before the arrival of the data in the refinement layer, thus allowing a faster response from actors. The localization of the mobile actors is updated and propagated to the sensors in the corresponding areas; for this purpose, the actor broadcasts its departure and arrival to the nearest sensors, which then proceed to determine a potential actor for a given cluster.

20.4.3   COORDINATION MECHANISM IN WSAN

Yuan et al. [22] propose a three-level coordination model (i.e., SS, SA, and AA coordination levels), as is illustrated in Figure 20.8. The proposal assumes the implementation of different routing protocols at each level. This model is based on a hierarchical geographical clustering paradigm and the clustering is achieved by segmenting the event area into fixed zones to optimally distribute the workload among different actors. In addition, it considers that the sensors are stationary and location-aware, whereas actors can be mobile. This model is suitable for application scenarios where multiple events occur simultaneously or with frequently occurring events.

At the SS coordination level, Ref. [22] proposes a clustering algorithm based on geographical adaptive fidelity (GAF), which is an energy-aware location-based routing algorithm, where each node uses its GPS-indicated location to associate itself with other nodes in the same grid. It should be noted that actors can be equipped with GPS but it is not necessary for each sensor; therefore, the localization techniques can use trilateration and receive signal strength indicator (RSSI) measurements to locate any sensor in the WSAN. Inside each grid, nodes collaborate with each other to switch between active and inactive states and to turn off unnecessary nodes in the network without affecting the level of routing fidelity. In addition, sensors choose a cluster head that aggregates the data received from each sensor associated with it; the cluster head is responsible for reporting the data to the appropriate actors. Unless an event occurs or a query command is received from the actors, most sensors will stay in a sleep mode.

Regarding the SA coordination level, the actors periodically broadcast their current location along with a timestamp. In this way the cluster head can achieve clock synchronization and maintain a routing table that includes nearby actors; this procedure helps to deal with the mobility of actors. The serial number of each grid can be used as the ID of each cluster head and can be shared among

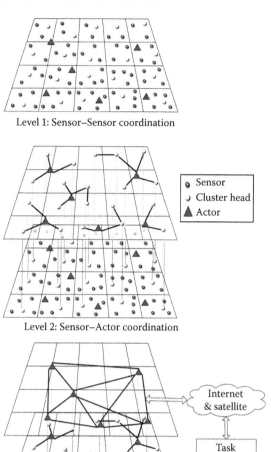

Level 1: Sensor–Sensor coordination

Sensor
Cluster head
Actor

Level 2: Sensor–Actor coordination

Internet
& satellite

Task
manager

Level 3: Actor–Actor coordination

FIGURE 20.8 Three-level coordination model. (From Yuan, H., Ma, H., and Liao, H., *First International Multi-Symposiums on Computer and Computational Sciences [IMSCCS'06]*, 2006. With permission.)

the sensors within the same cluster. All sensors in the event area coordinate with one another to select the appropriate actors according to different criteria, for example, the distance between the event area and the actor, the energy consumption of the sensors, or the acting ranges of the actors. Each cluster head is associated to one actor and must build a data aggregation tree toward the selected actor. All actors triggered by the same event build a second-level aggregation tree toward the actor in the center of the event area; this process is executed at the AA coordination level. When the cluster head is far from all the actors, the cluster head will relay the packets to the closest actor while coordinating with the nearest adjacent cluster-head toward the selected actor; this process is achieved by using geographical localization information.

At the AA coordination level, the actors process the data received from each associated cluster head and relay the information regarding the event to other actors. Depending on the characteristics of the event, one or more actors can execute one or more tasks. Two schemes are proposed to carry out the assignment of tasks among actors: action-first (AF) scheme and decision-first (DF) scheme. In the AF scheme, actors located close to the event area receive event information via SA coordination;

these actors immediately execute an action without any further negotiation with other actors. During the processing stage, each of these actors transmits the action information to other actors so that they can choose to join or retreat the action independently. To help control the number of actors that can potentially join the action, it is necessary to specify an action threshold level. If a certain actor cannot execute the task it may request help from other actors. In the DF scheme, all actors receiving the event information should coordinate strongly to execute the tasks. To avoid action competition among the available actors, a central controller is required. The event area is optimally split among different actors according to the event features and location-based grids. This arrangement helps to avoid the overlapping area problem while balancing the workload among different actors.

20.4.4   REAL-TIME COORDINATION AND ROUTING IN WSAN

Shah et al. [23] propose a real-time coordination and routing (RCR) framework for WSAN with automated and semiautomated architectures. The proposed RCR framework makes use of a hierarchical scheme, which is used to dynamically configure the sensors into clusters by applying the dynamic weighted clustering algorithm (DAWC). In addition, it provides real-time packet delivery by implementing the delay-constrained energy-aware routing (DEAR) protocol, which establishes a backbone network by integrating the forward-tracking and back-tracking mechanisms that provide all the possible routes toward a destination node (sink/actor). The path selection criteria is based on the packet delay as well as the balanced consumption of energy by the sensor nodes. In the presence of a sink, a centralized version of DEAR is implemented (C-DEAR) to coordinate actors through the sink. On the other hand, when there is no sink or the implementation ignores the presence of one, the coordination among sensors and actors is provided in a distributed way by D-DEAR.

The operation of DAWC consists of a cluster formation process and a delay budget estimation. The cluster formation process is based on a weighting equation that sets weights to different application parameters. When the sensors are not uniformly deployed in the sensor field, the density of the nodes can vary in different areas of the field. Hence choosing an optimal number of clusters K_{op} and the optimal size of clusters is an important design goal of DAWC; both are described in Ref. [23]. The first phase of DAWC is to form K_{op} clusters. During this cluster formation stage, each cluster head estimates the delay budget of each of its member nodes. The delay budget is used to identify an appropriate node to send delay-constrained data packets, as described in Ref. [23]. The cluster-head election procedure described in Ref. [23] is based on calculating a weight for each node, and then the cluster head is chosen from the node with the maximum weight. A weight threshold for the cluster head is defined to rotate the cluster-head responsibility among all the potential nodes. A cluster can include p-hop members, for $p \geq 1$, hence DAWC can adapt according to the dynamic topology of sensors and actors in the network. The approach assumes that the nodes are aware of their geographical location by means of a GPS receiver. The sink or actors can be located several hops from the source clusters; therefore, packets are forwarded through intermediate clusters. Some member nodes within a cluster can hear the members of neighboring clusters, and such nodes act as gateways. The cluster head keeps a record of all these gateways. Once the cluster formation is complete, each cluster gets the neighboring cluster list along with the gateways to reach them.

The DEAR algorithm is used to deliver packets from the source cluster to the target nodes (sink/actors) in a WSAN. This algorithm combines the forward-tracking and back-tracking approach to reduce the cost of path establishment. It establishes a distributed single path in which a cluster head selects the outgoing link with efficient energy consumption. An energy-efficient link does not merely mean the low-cost link, but a link that can satisfy the delay constraint and balances the energy consumption on all outgoing links [23].

Centralize DEAR (C-DEAR). This algorithm deals with a semiautomated architecture. The sink is stationary like the sensor nodes, and the path from the cluster head to the sink is built in a proactive way. The sink is the destination for all the source nodes. The source to sink path is divided into two phases: source to cluster head and cluster head to sink. The first phase is done during the cluster formation in a forward-tracking manner and the next phase from cluster head to sink uses back-tracking. To achieve this, Shah et al. [23] model the network as a connected directed graph $G = (V, E)$, where V represents the sensor nodes and E the set of directed edges; the algorithm visits the graph G and marks all the vertices (i.e., nodes). A mark is associated with the life of the node that is deleted when the node expires. A vertex can be marked if it has not been marked already or if the current path delay is less than the previously observed path delay. Once all the vertices are marked, a route is built toward the sink in a proactive way, and each node sets its hop distance to the sink, as described in Ref. [23].

Distributed DEAR (D-DEAR). The events detected by the sensors are directly routed to the actors without the intervention of the sink. D-DEAR decomposes the graph G into m number of subgraphs for each of the m mobile actors. The idea is similar to C-DEAR except that D-DEAR has m possible destinations. The marking process is triggered independently by all the m actors to build m representative routes. The marking criterion in D-DEAR is modified such that the vertices (i.e., node) accept the mark of an actor based on its Euclidian distance. The nearest one (i.e., actor) is the best candidate to mark the node. The algorithm proactively updates the route to reduce the chances of path failure. The selection of a particular link by the cluster head is based on the criteria required to balance the load in terms of delay and the energy of its member nodes. If the delay constraint can be satisfied through multiple links, then it selects the one where the minimum power available (PA) of any node is also larger than the minimum PA of a node in any other link [23].

20.4.5 COMMUNICATION ARCHITECTURE FOR MOBILE WSAN

In Ref. [24], Melodia et al. present a communication architecture for WSANs; this is an improved version of their previous work [13]. The proposal is based on an automated architecture, so the communication between sensors and actors is direct without the intervention of the sink. The communication architecture takes into account the SA and AA coordination levels. The proposed performance improvements consider the mobility of actors while trying to make an efficient consumption of energy and at the same time trying to stay within the bound of a reliability threshold as defined in Ref. [13]. To achieve these goals, a hybrid location management scheme is proposed to handle the mobility of actors with minimal energy expenditure from the sensor. Actors broadcast updates, limiting their scope based on Voronoi diagrams, whereas sensors predict the movements of actors based on Kalman filtering of previously received updates. This scheme reduces the energy consumption of sensors by avoiding over 75 percent of location updates with respect to existing location update algorithms. Also, an integrated routing/physical layer scheme for SA communication is developed based on geographical routing. The delay of the data-delivery process is controlled with this scheme based on power control, and the network congestion is reduced by forcing multiple actors to be recipients for traffic generated in the event area. Finally, the proposal also provides a new AA coordination model, which is used to coordinate the motion and the action of the participating actors based on the characteristics of multiple concurrent events. In particular, it selects the best actors to form a collaborative group to perform the required actions, based on the characteristics of the event. In addition, the coordination process is used to control the group motion toward the relevant area.

The coordination mechanism, described in Ref. [24], relies on a proactive location management approach, which is based on update messages sent by the mobile actors toward the sensors. The proposed location management approach tries to limit the location update broadcast in space and time. In the spatial domain, it limits the scope of actor-initiated location updates through Voronoi diagrams. In addition, they predict the actor movement by the AA coordination procedures. Every actor is responsible for transmitting location updates toward sensors in its Voronoi cell and regulates its power to limit interference beyond the farthest point in its Voronoi cell. With respect to flooding, the energy consumption for location updates is drastically reduced. In relation to the temporal domain, location updates are limited to actor positions that cannot be predicted at the sensor side. Location updates are triggered at the actors when the actual position of the actor is "considered to be far" from what can be predicted at the sensors, based on past measurements. Therefore, actors moving along predictable trajectories will need to update their position much less frequently than actors that follow temporally uncorrelated trajectories. A Kalman filter is used in both the sensor and the actor to reduce the exchange of messages during the location updates. The filter is used to estimate the position at the actors and to predict the position of the actors by the sensors. The results presented in Ref. [24] show that the proposed prediction procedure is effective even when complicated movement patterns are in place and shows good performance properties against noise.

The strategy used to control the delay during the data-delivery process depends on the traffic generated in the event area. For light traffic, an accurate connectivity model is proposed, which first derives the energy-efficient forwarding distance in the presence of a fast fading channel and then decreases the end-to-end delay by increasing the transmit power. In relation to the reliability, it is controlled by means of actor feedback messages. When an event occurs, all sensors start transmitting with the maximum forwarding range. Then, according to the actor feedback on the observed reliability, sensors may decrease their forwarding range until either the reliability is close to the required event reliability threshold r_{th}, or until the optimal forwarding range is reached. In case of high traffic, a new network congestion control mechanism is implemented at the network layer to force multiple actors to share the traffic generated in the event area. When an actor a_i detects very low reliability, it selects another actor to re-route the traffic from half of the sensors in its Voronoi cell to that actor. Each actor a_k is assigned by a_i a weight w_k, which measures its suitability to become a recipient for the traffic generated in the portion of the event area in which a_i is receiving data. The weight, w_k, is calculated as the weighted sum of three factors: congestion, directivity, and distance, according to [24]. A congested actor a_i selects the optimal actor a_{k*} with minimum weight w_{k*}. The procedure is applied recursively by actors that are still congested after splitting the traffic in two. The results show that in situations of low and moderate traffic, the end-to-end delay can be consistently decreased by increasing the forwarding range.

In relation to the AA coordination model, [24] formulate the multi-actor task allocation problem as a MINLP, which consists of selecting a team of actors and setting their velocity to optimally divide the action workload, so as to minimize the energy required to complete the action while respecting the action completion bound. According to the event features, each event is characterized by the event type, the event priority, the event area, the scope (the action area) and intensity, and the action completion bound. These characteristics are distributively reconstructed by the actors that receive sensor information and constitute inputs to the multi-actor task allocation problem. When all the available actors are forced to be part of a team, the action time can be reduced at the expense of energy consumption.

20.5   COMPARATIVE ANALYSIS

This section presents a comparison of the coordination mechanisms described in Section 20.4. This comparison is based on the coordination mechanism taxonomy introduced in Section 20.3. This taxonomy is divided into four sections, and Tables 20.1 through 20.4 show a summary of the comparative analysis for each of these. The coordination mechanism proposals are referenced using the following notation: the work presented by Melodia et al. in Ref. [14] is referenced as *A*; the framework proposed by Ngai et al. [19] is referenced as *B*; the coordination mechanism provided by Yuan et al. [22] is referenced as *C*; the proposal presented by Shah et al. [23] is referenced as *D*; and the architecture developed by Melodia et al. [24] is referenced as *E*.

WSAN framework. It represents the structure of a WSAN and includes the network architecture, the coordination levels, node mobility, and network density. Table 20.1 shows the comparative analysis for the WSAN framework.

- *Network architecture.* From Figure 20.1, it can be seen that all the coordination mechanisms described in Section 20.4 support the automated architecture as the direct communication between sensors and actors results in low latency. In proposal *D*, the authors present a

TABLE 20.1
WSAN Framework Comparison

	Coordination Mechanism Approach				
WSAN Framework	Melodia A [14]	Ngai B [19]	Yuan C [22]	Shah D [23]	Melodia-II E [24]
Network architecture					
Automated	Yes	Yes	Yes	Yes	Yes
Semiautomated	No	No	No	Yes	No
Coordination level					
SS	No	Yes	Yes	Yes	No
SA	Yes	Yes	Yes	Yes	Yes
AA	Yes	Yes	Yes	No	Yes
Node mobility					
Sensor	Fixed	Fixed	Fixed	Fixed	Fixed
Actor	Fixed	Mobile	Mobile	Mobile	Mobile
Sink	—	—	—	Fixed	—
Network density	Dense	Dense	Dense	Dense sparce	Dense

TABLE 20.2
Collaborative Procedures Comparison

	Coordination Mechanism Approach				
Collaborative Procedures	Melodia A [14]	Ngai B [19]	Yuan C [22]	Shah D [23]	Melodia-II E [24]
Routing support	Geographical	Geographical	GAF (SS) Ad-hoc (AA)	C-DEAR D-DEAR	Geographical
Synchronization	Yes	—	Yes	Yes	Probably
Localization	Yes	Yes	Yes	Yes	Yes
Aggregation	Yes	Yes	Yes	NS	NS
Clustering	Event-driven	Event-driven	Hierarchical Geographical	Dynamic Weighted	—
Data cipher/encryption	No	No	No	No	No
Power control	Yes	No	No	No	Yes
QoS					
Event reporting	Yes	No	No	No	Yes
Task execution	No	No	Yes	No	Yes

coordination mechanism and evaluate it using both the automated and semiautomated architectures. According to the authors, the automated architecture requires less configuration time as opposed to the semiautomated architecture; this implies that the packet delivery delay is not affected during the configuration process in an automated architecture. On the other hand, in the semiautomated architecture, the average packet delay increases with the number of sensor nodes in the network as the sink is the only one capable of providing coordination support for the cluster head and the actors.

• *Coordination levels.* The coordination levels supported by each of the proposed mechanisms differ. In *A*, the SS coordination functions do not take place, as the sensors proceed to associate with an actor immediately after the detection of an event; in this way the sensor

TABLE 20.3
Performance Criteria Comparison

Performance Criteria	Coordination Mechanism Approach				
	Melodia A [14]	Ngai B [19]	Yuan C [22]	Shah D [23]	Melodia-II E [24]
Optimization criteria	Energy Latency Loss packet	Energy Latency	Energy Latency	Energy Latency	Energy Latency Lost packet
Scalability	Yes	Yes	Yes	Yes	Yes
Communication and complexity order	—	—	—	—	—
Reliability					
Security	No	No	No	No	No
Robustness	No	No	No	No	No

TABLE 20.4
Application Requirements Comparison

Application Requirements	Coordination Mechanism Approach				
	Melodia A [14]	Ngai B [19]	Yuan C [22]	Shah D [23]	Melodia-II E [24]
Real-time constraint	Low latency	Low latency	Yes	Yes	Low latency
Event frecuency	Low	Low	High	High	High
Event type					
Delivery model	Ed	Ed	Ed	Ed	Ed
Concurrent events	No	No	Yes	Yes	Yes

Ed: Event-driven.

nodes form a cluster with the actor as the cluster head. In other words, the cluster formation and data aggregation process is supported between sensor and actor nodes and not between sensors. In *C*, the proposed coordination mechanism supports the implementation of the SS, SA, and AA coordination levels. In relation to proposal *E*, the coordination mechanism operates at the SS and AA coordination levels; it should be noted that proposal *E* considers mobile actors and was developed as an extension of proposal *A*. In some articles, the coordination level is described using a different description, as in proposals *B* and *D*. In *B* the architecture implements clustering and data aggregation techniques with sensors and the data corresponding to each cluster are reported to the closest actor. This is defined as SS and SA event notification, which, evidently, implies SS and SA coordination functions. In *D*, the sensors are grouped into clusters and the average packet delay is estimated via the DAWC protocol. All these tasks are related with the SS coordination.

- *Node mobility.* The proposals described in Section 20.4 assume that the sensor nodes are fixed (i.e., the sensor node does not move), and most of the proposals assume the actors to be mobile, with the exception of *A*, which assumes the actors to be fixed. However, the authors of proposal *A* have recently published an updated version where they assume the actor to be mobile. By considering mobile actors it is possible to consider a broader range of applications of a WSAN architecture at the expense of increased complexity, as the actor mobility requires an efficient coordination.

• *Network density.* In relation to the network density, all the proposals are capable of operating in a high-density topology; however, some of the proposals do not explicitly specify this capability, as in *B*. Nonetheless, it is assumed that *B* is scalable as it relies on cluster formation and data aggregation during the reporting of an event. Regarding proposal *D*, the authors explicitly state that it can manage dense as well as sparse deployment of nodes.

Collaborative procedures. It refers to collaborative procedures (or services), between the elements of a WSAN, used to support the exchange of information by the coordination mechanism. Table 20.2 shows the comparative analysis in this aspect.

• *Routing support.* Most of the proposed coordination mechanisms make use of geographical routing protocols, given their scalability and the fact that they can easily adapt to the location changes by the actor nodes [19]. In addition, it is possible to exploit the localization information of the nodes to route packets to the intended nodes in accordance with their localization within an event area. In *A*, *B*, and *E* the SA coordination is based on a geographical routing paradigm, whereas the proposal presented in *C* makes use of the GAF routing protocol in the first coordination level. The sensor nodes can use the geographical localization information to transmit the data to the cluster head, which will in turn forward the data to the closest actor. In addition, *C* proposes the use of a conventional ad hoc routing protocol in the last level (i.e., AA coordination level), given the reduced number of actor nodes and their higher performance capabilities. On the other hand, *D* proposes the implementation of energy and delay aware routing protocols, such as the DEAR protocol, to transmit the data to the sink (i.e., semiautomated scenario) or to the actors (i.e., automated scenario). In the semiautomated scenario, *D* proposes the implementation of a centralized DEAR (C-DEAR) protocol, whereas for automated architectures it implements a distributed DEAR (D-DEAR) protocol. These two implementations of the DEAR protocol are designed to comply with the end-to-end delay requirements for real-time applications. In proposal *E* the authors derive a simple yet optimal forwarding rule based on geographic position in the presence of Rayleigh fading channels. The geographical forwarding rules are optimal from the energy consumption standpoint.

• *Synchronization.* The synchronization technique is implemented only in proposals *A*, *C*, and *D*; proposals *B* and *E* do not make any reference to the requirement of a time synchronization mechanism. In *A*, it is assumed that the network is synchronized by means of the implementation of an existing synchronization protocol [16]. In *C* the actors periodically transmit their geographical coordinates along with a timestamp; this information allows the cluster head to synchronize with the actors. In *D*, the implementation assumes that the channel is asymmetric, and the authors state that a synchronization mechanism is not required. With respect to this last statement, it is assumed the authors refer to a link synchronization mechanism for communication and do not necessarily make reference to the requirement of a time synchronization mechanism. Nonetheless, a time synchronization mechanism is assumed to be implemented in proposal *D* as packet delay measurements are made by the receiver, which implies that packets are tagged with timestamp information. In relation to proposal *E*, it does not explicitly make a reference to the requirement of a synchronization technique; however, it is assumed that a time synchronization service is required, as the location estimation of the actors is made during specific time intervals.

• *Localization.* The localization service is considered to be implemented by all the proposals presented in Section 20.4. It is assumed that all the nodes are capable of knowing or determining their geographical localization; this can be implemented by means of a GPS receiver, through trilateration techniques or a similar approach. In proposal *E*, the authors make reference to a hybrid location management scheme to handle the mobility of actors with the minimal energy expenditure based on update messages sent by mobile actors to sensors. According to *E*, the proposed solution is appropriated for WSAN applications and overcomes the drawbacks of previously proposed localization services. Actors broadcast updates and limit their scope based on Voronoi diagrams, whereas sensors predict the movements of actors based on Kalman filtering of previously received updates. This scheme does not assume a particular localization technique for the actor (e.g., GPS, particle filter, etc.) to keep the model general.

- *Aggregation.* The data fusion service is implemented in *A*, *B*, and *C*, whereas *D* and *E* do not specify it. In *A*, data fusion is implemented only at the SA coordination level whenever a sensor receives information from at least two other sensors; the data is then relayed toward the actor node. In *B* the data fusion is implemented at the SS coordination level and further divided into different layers according to their importance. The aggregated data is then transmitted to the closet actor in the order of significance. In *C* data fusion is implemented at the three coordination levels. First, the cluster head performs data fusion on the data received from the member nodes (i.e., fusion at the first level). At the SA coordination level all the cluster heads associated to the same actor construct a data-aggregation tree toward the actor. Finally at the AA coordination level, all the actors activated by a common event construct a third data-aggregation tree toward the actor located at the center of the event area.

- *Clustering.* The cluster formation approach is implemented in proposals *A*, *B*, *C*, and *D*. In contrast, proposal *E* does not rely on a cluster formation mechanism. In relation to proposals *A* and *B*, they both implement the event-driven clustering paradigm. In event-driven clustering, the cluster formation process is triggered by an event and the clusters are created on the fly. Reactive cluster formation algorithms have the disadvantage that they consume valued time during the cluster formation as a result of an event occurrence. Hence, for real-time applications the event-based architecture is not suitable [23]. Proposal *C* is based on a hierarchical geographical clustering paradigm where the cluster formation is done by splitting the action area into smaller sections to create virtual grids; this strategy reduces the traffic load within each grid and makes an efficient use of resources. Finally, *D* proposes a clustering algorithm DAWC, which adapts to the dynamic topology of the networks. The procedure of cluster formation is not periodic, and it is based on a weighting equation, which sets weights to different parameters according to the application needs.

- *Data cipher and/or encryption.* None of the coordination mechanisms described in section 20.4 make use of a security mechanism, such as ciphering or encryption, to protect the integrity of data. As a result, there is an open opportunity area for the development of coordination mechanisms which may incorporate ciphering and or encryption techniques.

- *Power control.* A power control mechanism is implemented in proposals *A* and *E*. Proposal *A* employs the power control mechanism in the actors that can select their power among *L* different levels to perform the action. A higher power corresponds to a lower action completion time. On the other hand, proposal *E* employs the power control mechanism at the Sensor–Actor coordination level, where the sensors can increase the forwarding range with respect to the energy-efficient forwarding range in order to adjust the end-to-end delay when the traffic generated in the event is low or moderate. Furthermore, at the location management approach, the actors update their location to sensors in its Voronoi cell and regulate its power so as to limit interference beyond the farthest point in its Voronoi cell.

- *Quality of service.* In relation to QoS, it is possible to identify two different functionalities. One functionality is related to the support of different priorities during the transmission of data to report an event from sensors to actors. Another functionality is related to the prioritized assignment and execution of tasks by the actors which is related to the application requirements.
  - *Data transfer priorities (Event reporting).* Proposal *A*, *D* and *E* implement a data transfer priority scheme. Proposal *A* introduces a novel notion of "event reliability threshold,"

which is defined as the minimum latency required by the application to ensure that the observed reliability is above an event reliability threshold. To provide the required reliability with minimum energy expenditure, *A* proposes a DEPR protocol as explained in [14]. Proposal *D* relies on a clustering algorithm DAWC which estimates the delay budget for forwarding a packet from the cluster-heads. This algorithm guarantees the packet delivery delay to be within the given delay bound. The implemented DEAR routing protocol is capable of supporting real-time packet delivery, as the path selection criteria is based on the packet delay as well as the balance consumption of energy of sensors nodes [23]. Proposal *E*, similar to *A*, uses the concept of "reliability," but in this case the application "reliability" requirement is achieved by adjusting the end-to-end delay by means of a power control mechanism when the traffic generated in the event area is low, and by means of a actor-driven congestion control scheme in case of congestion. Proposals *C* and *B* are concerned with reducing the latency at the event reporting stage during the transmission of event related information from the sensors toward the actors, but do not implement a priority scheme to support different application requirements. For example, in proposal *B* during the event reporting stage, the sensor nodes first transmit a general event notification message and at a later time they proceed to transmit detailed event notification messages toward the actors. This is done with the purpose of letting the actors obtain a rough image of the event in the shortest possible time. Nonetheless, proposal *B* does not implement a priority scheme to provide any guarantees on the transmitted messages toward the actors.

- *Task assignment priorities (Task execution).* Proposals *C* and *E* make use of a task priority assignment technique. In proposal *C*, according to the characteristic of the event, one or more actors can be triggered to perform one or more task. At the AA coordination level, *C* proposes an AF scheme where the actors that receive event information perform an action without negotiating with further actors. The actors involved in a task assignment proceed to inform other actors by providing them with event information. In this way other actors make a decision to join or retreat the action independently. This decision process is based on evaluating different factors, such as the priority assigned to tasks according to the procedure defined in proposal *C*. In proposal *E*, the task assignment process is achieved by means of MINLP, where the event is characterized by a tuple that describes the event type, event priority, event area, scope and intensity, and action completion bound. The characteristics of an event allow the selection of a team of actors, and their velocity is defined to optimally divide the action workload. This helps minimize the energy required to complete the action while not exceeding the action completion bound. Furthermore, the AA coordination mechanism proposed by *E* includes an event preemption policy for multiactor task allocation for cases where resources are insufficient to accomplish a high-priority task.

Performance criteria. They are related to criteria elements used to estimate the performance of the coordination mechanism. Table 20.3 shows a summary of the comparative analysis for the performance criteria.

- *Optimization criteria.* All of the proposals make use of energy and latency as optimization metrics. In *A*, the maximum allowed latency (i.e., the delay between the detection of the event and the reception of this information by the actors) is defined as reliability. In addition to these parameters, proposals *A* and *E* measure the packet loss rate. Due to the nature of WSANs and their resource limitations, it is obvious that coordination mechanisms try to reduce the energy consumption of nodes while trying to reduce the delay related to the event reporting and the execution of tasks. In this way, it is possible to increase the network lifetime and at the same time guarantee an opportune response to reported events.

- *Scalability.* Proposals *A*, *B*, *C*, and *E* make use of geographical routing protocols, which are becoming the most promising scalable solutions for critically energy-constrained sensor networks. In addition, proposals *A*, *B*, *C*, and *D* make use of clustering schemes that promote scalability and an efficient use of energy in the network.
- *Communication and complexity order.* None of the proposals provide a clear analysis regarding the memory and computational resources required to support the proposed coordination mechanism. In relation to *D* and *E*, they provide a measure of the order of complexity for some aspects of the coordination mechanism. However, none of the proposals provide a complete analysis to help determine the complete order of complexity of the coordination mechanism as a whole. Proposal *D* describes that the marking process cost used in C-DEAR is $O(n)$ where n is the number of sensors. This can be considered as the actual cost of building a route from the source node to the sink node. In proposal *E*, the authors make reference only to the fact that the energy consumption in the proposed Location Management algorithm increases as a function of $O(n)$, assuming the actor is able to reach all sensors in its Voronoi cell in one hop. In relation to the communication complexity none of the proposals provide a clear analysis regarding the amount of control messages or message exchanges required to support the proposed coordination mechanism.
- *Reliability.* It is related to the security and the robustness of the coordination mechanism.
  - *Security.* With respect to security, none of the proposals make use of a procedure to guarantee data integrity or avoid network access to intruders.
  - *Robustness.* With respect to robustness, none of the proposals implement or propose a procedure to guarantee fault tolerance. Some proposals measure only the packet loss rate and show how the algorithms, implemented to reduce latency, also help to reduce the packet loss; however, no procedure is proposed to guarantee packet delivery through the use of acknowledgment messages or the retransmission of information.

Application requirements. The coordination mechanism must take into consideration the application requirements. Table 20.4 shows a summary of the comparative analysis of the application requirements.

- *Real-time constraints.* In general, the proposed coordination mechanisms support real-time applications, as they promote latency reduction during the event reporting process and the execution of tasks. In relation to proposals *A*, *B*, and *E*, they introduce a network configuration delay due to the event-based clustering approach they implement; nonetheless, these proposals try to reduce the event reporting latency. As a result, these proposals may not be suitable for real-time applications.
- *Frequency of events and event types.* With respect to the frequency of events that can be handled by the coordination mechanism, proposals *A* and *B* are not capable of providing an efficient operation under high-frequency event scenarios. This is also true for scenarios involving multiple events. This inefficiency is related to the limitations imposed by the event-driven clustering paradigm. The event-driven clustering approach wastes valuable time during the network configuration process, which is used to form the clusters and create the data aggregation trees; consequently, it is not possible to provide an opportune response to the reported events. In relation to proposals *C*, *D*, and *E*, they are able to operate in frequent-event scenarios as well as scenarios involving multiple simultaneous events. In relation to the type of events, all the proposals implement an event-driven approach where the sensors will inform to the application when certain events occur.

20.6   CONCLUSION AND OPEN ISSUES

This chapter presents a detailed overview of coordination mechanisms proposed in the literature for WSANs, including a description of the cooperative processes that take place between sensors and actors. In addition, a coordination mechanism taxonomy is proposed for the classification of these types of mechanisms. One of the main requirements of a WSAN is related to the support of real-time communication, which may be a vital requirement of the application. Furthermore, a well-designed WSAN architecture must provide efficient use of the network resources, such as energy consumption; consequently, coordination mechanisms proposed for WSANs must be able to respond to the events by executing timely actions while at the same time trying to increase the network lifetime.

From the comparative analysis presented in Section 20.5, it is clear that some functionalities described in the coordination mechanism taxonomy are not implemented. One of these functionalities is related to the support of QoS. Some coordination mechanisms do not provide the required support to provide prioritized execution of tasks by the actors. Similarly, some of the proposed coordination mechanisms do not provide the means to provide differentiated service to data and control messages transmitted in the WSAN.

Another issue is the reliability of the proposed architecture, which includes the system security and robustness. WSAN technology provides an opportunity for innovation, which involves unique challenges. For example, supporting multimedia applications and services imposes great challenges on the design of a coordination mechanism for WSANs. Fault-tolerant routing is a critical task for WSAN operating in dynamic environments. On the other hand, the rapid proliferation of wireless networks and mobile nodes has changed the landscape of network security. The task of providing security services for WSN is not trivial due to the resource constraints of the sensor nodes. As a result, there is a need for the development of new architecture and mechanisms to protect information in a WSAN. Furthermore, WSAN architectures are in their infancy stage, so there is an opportunity to design architectures that may consider the support of QoS, robustness, and security. In relation to QoS, fault-tolerant mechanisms, and security, there are multiple independent proposals, such as Refs. [25–27] among others; however, these proposals have not been considered within the scope of coordination mechanisms for WSANs.

To satisfy the application requirements of time and energy consumption, coordination mechanisms make use of clustering schemes, data aggregation, and some additional services, such as power control and localization. However, they do not employ a cross-layer design, which may be used to reduce the overhead in the network, thus allowing a reduction in energy consumption and latency. For example, the clustering, aggregation, and power control services involve several layers of the protocol stack, and an efficient solution is likely to require a cross-layer design. However, the schemes proposed so far often focus on few aspects, typically trying to merge routing and data aggregation techniques, but ignoring physical, MAC, and application layer issues. On the other hand, cooperative communication exhibits various forms at different protocol layers and introduces many opportunities for cross-layer design and optimization. From the perspective of the network, cooperative communications can benefit not only the involved nodes but the whole network in many different aspects, such as higher spatial diversity and throughput, adaptability to network conditions, reduced interference, lower transmitted power, and lower delay. The past few years have seen tremendous interest in cooperative communications, mostly at the physical layer. However, significant research challenges still exist [28].

In relation to the routing protocols, many of them are based on clustering, which directly allows aggregation of data at the cluster head. Such algorithms work well in relative static environments where the cluster structure remains unchanged for a sufficiently long time, but they may be fragile when used in more dynamic environments, such as WSAN with mobile actors. Often the cost required to maintain the hierarchical structure is substantial. Initial works address some of these problems, but further research efforts are required to keep a network functional under mobility [29].

Another aspect that needs to be addressed is the definition of standardized performance metrics to evaluate and compare the performance between different coordination mechanisms. There is no consensus in the community on how to design, implement, and evaluate the performance of a coordination mechanism. Finally, the problem of assigning prioritized tasks to the actors continues to be an open issue up to this date.

One of the main contributions of this work is the proposal of a coordination mechanism taxonomy for WSAN architectures. The taxonomy provides a framework suited to the specific requirements of a coordination mechanism designed for a WSAN environment. The proposed taxonomy is divided into four sections: WSAN framework, collaborative procedures, performance criteria, and application requirements. In addition to the taxonomy, some of the most representative coordination mechanisms published in the literature up to this date have been described and a comparative analysis has been provided. In general, the proposed mechanisms proceed to split the event area and perform a hierarchical coordination by employing localization information; this is done with the objective of selecting the proper nodes (i.e., sensors and actors) that will react in response to a specific event with the smallest possible response time. The proposed applications for each of the coordination mechanisms differ with respect to the frequency of the events. For example in A [14], which is one of the first published works, it is assumed that sensors and actors are fixed and the frequency of events is low. Later, WSAN-related research migrated to different scenarios where the frequency of events increased while considering the possibility of multiple-simultaneous events and the mobility of actors, as in proposal C [22]. In general, the proposed coordination mechanisms try to comply with the support of real-time response requirement along with an efficient use of energy in the WSAN. On a final note, none of the coordination mechanisms presented in Section 20.4 implement a mechanism that guarantees the security of the data nor system robustness.

ACKNOWLEDGMENT

This work was sponsored in part by the Mexican Council for Science and Technology (CONACYT) under project grant J48391-Y.

REFERENCES

[1] K. Romer and F. Mattern. The design space of wireless sensor networks. *IEEE Wireless Communications*, 11(6), 54–61, 2004.

[2] D. Vassis, G. Kormentzas, and C. Skianis. Performance evaluation of single and multi-channel actuator to actuator communication for wireless sensor actuator networks. *Ad Hoc Networks*, 4(4), 487–498, 2006.

[3] E.M. Petriu, N.D. Georganas, D.C. Petriu, D. Makrakis, and V.Z. Groza. Sensor-based information appliances. *IEEE Instrumentation and Measurement Magazine*, 3(4), 31–35, 2000.

[4] I.F. Akyildiz, W. Su, Y. Sankarasubramaniam, and E. Cayirci. A survey on sensor networks. *IEEE Communications Magazine*, 40(8), 102–114, 2002.

[5] I.F. Akyldiz and I. Kasimoglu. Wireless sensor and actuator networks: Research challenges. *Ad Hoc Networks*, 2(4), 351–367, 2004.

[6] D.V. Dinh, M.D. Vuong, H.P. Nguyen, and H.X. Nguyen. Wireless sensor actuator networks and routing performance analysis. *International Workshop on Wireless Ad-hoc Network*, 2005.

[7] E. Cayirci, T. Coplo, and O. Emiroglu. Power aware many to many routing in wireless sensor and actuator networks. *Proceeding 2nd European Workshop on Wireless Sensor Networks*, 2005.

[8] F. Hu, X. Cao, S. Kumar, and K. Sankar. Trustworthiness in wireless sensor and actuator networks: Toward low-complexity reability and security. *IEEE Global Telecommunications Conference (GLOBECOM)*, 2005.

[9] M. Daz, D. Garrido, L. Llopi, B. Rubio, and J.M. Troya. A component framework for wireless sensor and actuator network. *11th IEEE International Conference on Emerging Technologies and Factory Automation*, 2006.

[10] A. Farinelli, L. Iocchi, and D. Nardi, Multirobot systems: A classification focused on coordination. *IEEE Transactions on Systems, Man and Cybernetics*, Part B, 34(5), 2015–2028, 2004.

[11] A. Salkham, R. Cunningham, A. Senart, and V. Cahill. A taxonomy of collaborative context-aware systems. *Workshop on Ubiquitous Mobile Information and Collaboration Systems (UMICS'06)*, 2006.

[12] S. Sundresh, G. Agha, K. Mechitov, W.Y. Kim, and Y.M. Kwon. Coordination services for wireless sensor networks. *International Workshop on Advanced Sensors, Structural Health Monitoring and Smart Structures*, 2003.

[13] T. Melodia, D. Pompili, V.C. Gungor, and I.F. Akyildiz. A distributed coordination framework for wireless sensor and actuator networks. *Proceeding of the 6th ACM International Symposium on Mobile Ad Hoc Networking and Computing*, 2005.

[14] T. Melodia, D. Pompili, V.C. Gungor, and I.F. Akyildiz. Communication and coordination in wireless sensor and actor networks. *IEEE Transactions on Mobile Computing*, 6(10), 1116–1129, 2007.

[15] R.K. Ahuja, T.L. Magnanti, and J.B. Orlin. *Network Flows: Theory, Algorithms, and Applications*. Prentice Hall, Englewood Cliffs, NJ, Jan. 1993. ISBN-13: 9780136175490.

[16] B. Sundararaman, U. Buy, and A. Kshemkalyani. Clock synchronization for wireless sensor networks: A survey. *Ad-Hoc Networks*, Ad Hoc Networks, 3(3), 281–323, 2005.

[17] J. Czyzyk, M. Mesnier, and J. More. The NEOS server. *IEEE Journal on Computational Science and Engineering*, 5(3), 68–75, 1998.

[18] B.P. Gerkey and M.J. Matari. A market-based formulation of sensor-actuator network coordination. *AAAI Spring Symposium on Intelligent Embedded and Distributed Systems*, 2002.

[19] E.C.H. Ngai, M.R. Lyu, and J. Liu. A Real time communication framework for wireless sensor-actor networks. *IEEE Aerospace Conference*, 2006.

[20] J.G. McNeff. The global positioning system. *IEEE Transactions on Microwave Theory and Techniques*, 50(3), 645–652, 2002.

[21] L. Hu and D. Evans. Localization for mobile sensor networks. *Proceeding of the 10th International Conference on Mobile Computing and Networking ACM Mobicom*, 2004.

[22] H. Yuan, H. Ma, and H. Liao. Coordination mechanism in wireless sensor and actuator networks. *First International Multi-Symposiums on Computer and Computational Sciences (IMSCCS'06)*, 2006.

[23] G.A. Shah, M. Bozyigit, O.B. Akan, and B. Baykal. Real time coordination and routing in wireless sensor and actuator networks. *Proceding of the 6th International Conference on Next Generation Teletraffic and Wired/Wireless Advanced Networking (NEW2AN)*, 2006.

[24] T. Melodia, D. Pompili, and I. Akyldiz. A communication architecture for mobile wireless sensor and actor networks. *Proceeding of the IEEE International Conference on Sensor, Mesh and Ad Hoc Communications and Networks (SECON)*, 2006.

[25] F. Xia, Y.-C. Tian, Y. Li, and Y. Sun. Wireless sensor/actuator network design for mobile control applications. *Sensors*, 7(10), 2157–2173, 2007.

[26] X.-M. Huang, J. Deng, J. Ma, and Z. Wu. Fault tolerant routing for wireless sensor grid networks. *Sensors Applications Symposium Proceedings of the IEEE*, 2006.

[27] F. Hu, W. Siddiqui, and K. Sankar. Scalable security in wireless sensor and actuator networks (WSANs): Integration re-keying with routing. *Computer Networks: The International Journal of Computer and Telecommunications Networking*, 51(1), 285–308, 2007.

[28] P. Liu, Z. tao, Z. Lin, E. Erkip, and S. Panwar. Cooperative wireless communications: A cross-layer approach. *IEEE Wireless Communications*, 13(4), 84–92, 2006.

[29] E. Fasolo, M. Rossi, J. Widmer, and M. Zorzi. In network aggregation tecnhiques for wireless sensor networks: A survey, *IEEE Wireless Communications*, 14(2), 70–87, 2007.

21 Self-Healing Wireless Sensor Networks

Natalija Vlajic, Nelson Moniz, and Michael Portnoy

CONTENTS

Self-healing—the ability of a network to effectively combat coverage and routing holes and network disconnection—represents one of the most desired operational properties of large-scale wireless sensor networks (WSNs). In this chapter, we look at the advantages and inherent challenges of self-healing through the deployment of mobile nodes. We show that although bridging a routing hole by means of a mobile node provides unquestionable benefits in terms of hole confinement and reduced transmission delays, the deployment of the mobile is often hard to justify from the perspective of overall energy consumption. In particular, we prove that in case of circular- and rectangular-shaped holes, the deployment of a mobile bridge provides very questionable, if any, energy benefit for the network. Consequently, we demonstrate the need to jointly consider all of the relevant parameters (energy, transmission delay, and node failure), when deciding whether or where to deploy the mobile.

In the second part of the chapter we discuss optimal placement of a mobile node (OPlaMoN), a self-healing algorithm that determines the most optimal deployment location of a mobile bridge inside a routing hole of any arbitrary topology. The algorithm assumes distributed cooperative computation involving a minimum number of sensor nodes and as such is highly suited for energy and processing constrained large-scale WSNs.

The chapter is concluded with an overview of open research problems.

21.1 INTRODUCTION

21.1.1 LARGE-SCALE WIRELESS SENSOR NETWORKS

In recent years, WSNs have attracted the attention of the research community thanks to their tremendous potential in various application fields, including environment monitoring, security surveillance, disaster management, and combat operations [1–3]. Most of the these application scenarios assume a large-scale network, consisting of inexpensive, static, battery-powered nodes, randomly deployed in remote, hostile, or inaccessible areas (Figure 21.1). From the network operation point of view, most scenarios consider the use of a simple routing technique, such as geographic routing [4], over a multi-hop topology. (In a multi-hop network topology data generated by a "source" is passed from one node to another, until it eventually reaches the final destination—the so-called sink.)

FIGURE 21.1   Routing/coverage hole in a large-scale WSN.

Energy conservation is identified as the most critical issue in the design and operation of WSNs, due to its direct impact on the network efficacy and, more importantly, on the overall network lifetime. The most straightforward way to achieve effective energy conservation is

1. By minimizing the number of required transmissions among the network nodes, for example, through the use of node clustering and energy-efficient MAC protocols
2. By using optimized routing paths through the network

Unfortunately, the implementation of (2) turns out to be an especially great challenge in a number of application scenarios. Namely, as indicated earlier, a typical WSN setup assumes one or more of the following:

1. Sensor nodes are randomly scattered throughout the deployment field, for example, by being disseminated from a plane.
2. Deployment field is a region of irregular geographic composition, possibly comprising natural obstacles (lakes, cliffs, etc.) or man-made obstacles (buildings, overpasses, etc.).
3. Sensor nodes are prone to failure due to
   a. Component malfunctioning
   b. Battery depletion
   c. Environmental factors: extreme heat, flooding, freezing
   d. Man-caused factors: interference, accidental damage, explosion.

Based on the above, the WSN topology inevitably gets plagued by serious irregularities or areas completely void of functioning nodes, a.k.a. holes, as illustrated in Figure 21.2. Clearly, the existence of such topological anomalies has a major negative impact on both the network function (sensing of certain phenomenon) and operation (routing of data).

In the subsequent sections, we take a closer look at different types and characteristics of holes occurring in large-scale WSNs. We also discuss the most common ways of combating these holes.

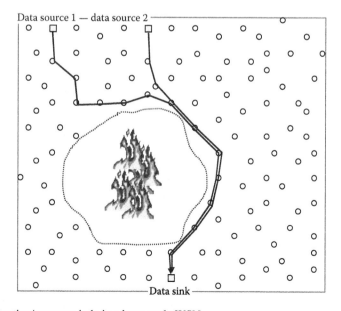

FIGURE 21.2 Routing/coverage hole in a large-scale WSN.

21.1.2 ROUTING HOLES IN LARGE-SCALE WSNs

The idea of holes in WSNs is one with numerous meanings [5]. For example, a routing hole comprises a group of nodes incapable of communicating with one another. A coverage hole comprises a group of nodes unable to sense their environment. An area completely void of nodes represents both a coverage hole and a routing hole. Other types of holes include jamming holes, sink holes, energy holes, and so on. Given the critical role of routing in large-scale WSNs, from the perspective of energy conservation and network operation, routing holes represent the most serious form of network anomaly. For that reason, routing holes are the major focus of our discussion.

The likelihood of encountering a routing hole in a large-scale WSN of random-node deployment and the average size and circumference of such holes have been investigated in Ref. [6]. The study shows that most routing holes naturally occurring in an obstacle-free deployment area can be mapped within ten hops and surpassed within four hops or less. The study presented in Ref. [7] takes a look at routing holes that get formed due to the presence of large physical obstacles, including buildings, overpasses, rivers, pools, and so on. In most cases, these holes can be modeled by using one of the standard geometric shapes: rectangle, circle, ring, or crescent. Also, the study notes that while some of these holes will appear and always be part of the network (e.g., holes due to natural obstacles such as rivers or lakes), other holes could possibly appear and disappear during the lifetime of the network (e.g., holes due to mobile man-made objects such as trains). Finally, in several works including Refs. [8–10], methods aimed at discovering routing holes have been proposed. The importance of such methods is unquestionable, as the effectiveness of any routing strategy will ultimately depend on the network's ability to identify the number and scale of its topological anomalies.

21.1.3 IMPACT OF ROUTING HOLES ON GEOGRAPHIC ROUTING

From the energy conservation point of view, finding an effective way to forward packets through a large-scale WSN is highly critical. It has been shown that traditional ad hoc routing protocols do not scale well when applied to the energy-constrained nodes of a WSN [4]. Specifically, the use of protocols such as Ad-Hoc On-Demand Distance Vector Routing (AODV) and Dynamic Source Routing (DSR) should be by all means avoided, because they require a considerable amount of information to be exchanged among the nodes.

In recent years, routing algorithms specially designed to meet the unique constraints of WSN systems have started emerging. Among the numerous proposed solutions, geographic routing and its variants [12–15] have been recognized as the most effective form of routing in large-scale WSNs. Namely, in the case of geographic routing, nodes have to know only the locations of their nearest neighbors. Consequently, packets are routed through the network by being forwarded to a neighbor that is physically closer (that is, closest) to the destination. The main advantages of geographic over other WSN routing strategies include

- Stateless, and therefore highly energy efficient, nature of routing
- Fast adaptability to network's topological changes
- Scalability

Unfortunately, when applied to a network with pronounced topological anomalies, geographic routing faces two major challenges. The first challenge is the so-called local minimum phenomenon, where none of a node's neighbors are closer to the destination than the forwarding node itself, as illustrated in Figure 21.3. The packets that arrive to such a node will get "stuck," unless one of the improved versions of geographic routing is employed (for example Refs. [9,10,12]). The second challenge of geographic routing, also faced by other routing strategies, is the so-called hole expansion phenomenon [10,11]. Namely, in the case of a routing hole and traffic streams as shown in Figure 21.3, the nodes along the perimeter of the hole suffer an increased energy burden, which ultimately leads to premature exhaustion of their batteries and further expansion of the hole. (Note that, in the absence of

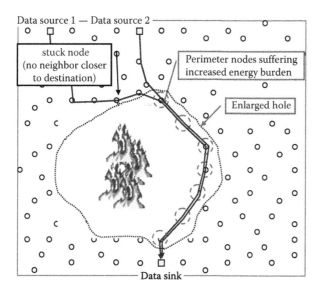

Data source 1 — Data source 2

stuck node
(no neighbor closer
to destination)

Perimeter nodes suffering
increased energy burden

Enlarged hole

Data sink

FIGURE 21.3 Local minimum and hole expansion phenomenon.

the routing hole in Figure 21.3, the optimal geographic-routing paths of the two traffic streams would be completely disjoint. Accordingly, the average energy consumption of sensor nodes involved in routing would be 50 percent lower, and their average lifetime 50 percent longer.) As most real-world environments abound in physical obstacles, successful deployment of WSNs in these environments asks for effective ways of dealing with their inherently occurring topological anomalies.

21.1.4 SELF-HEALING IN LARGE-SCALE WSNs

The techniques aimed at combating routing holes in a general class of wireless ad hoc network—one example being wireless sensor networks—are commonly referred to as self-healing techniques. Unfortunately, most of the known self-healing techniques [10,15–17] share a major common limitation—they focus on moving the routing function away from routing holes, for example, by finding alternate paths through the network, instead of attempting to directly combat the holes, for example, by minimizing their effects on routing or eliminating them completely. Note that although moving the routing function away from routing holes is an effective way of dealing with hole expansion phenomenon, it often results in a considerable increase in the overall routing delay and overall energy consumption.

A more direct and rather intuitive way of combating routing holes is the use of mobile nodes, as shown in Figure 21.4. (Examples of promising works on the use of mobile nodes for the purpose of combating coverage holes include [22–25].) From the perspective of network's routing function, the formation of a simple shortcut path involving a mobile bridge positioned in the interior of a routing hole provides two obvious advantages for the network: (1) reduced routing delay (two hops through the hole instead of multiple hops around the hole); and (2) reduced load on the boundary nodes, as they get relieved from routing operation. One should be aware, however, that from the perspective of overall energy consumption this approach to combating routing holes may be potentially disadvantageous, as multiple smaller hops are now substituted by one large hop. Generally, multi-hop short-range transmission is known to consume less energy than the energy required by one large-hop transmission [19].

In this chapter, we conduct a detailed investigation of the energy aspect of self-healing by means of mobile nodes in large-scale WSNs. By focusing on two particular types of routing holes—circular- and rectangular-shaped—we derive precise conditions under which the deployment of a mobile bridge can, or cannot, be considered justified. The obtained results indicate that in all cases of circular and one special case of rectangular holes (that is, square-shaped holes) the deployment of

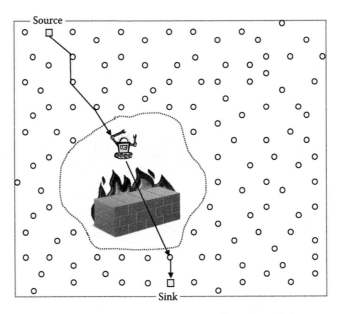

FIGURE 21.4   Self-healing by means of mobile nodes: one mobile node per hole.

the mobile comes at a very high energy cost, rendering it practically unjustifiable. On the other hand, the use of a mobile bridge in case of elongated rectangular holes is likely to result in energy savings and, accordingly, is more likely to be justified from the perspective of overall energy consumption.

The remainder of the chapter is organized as follows. The most important network assumptions are outlined in Section 21.2.1. The energy aspects of self-healing by means of a mobile node, in the case of circular and rectangular holes, are presented in Sections 21.2.2 and 21.2.3. In Section 21.3 an overview of a distributed cooperative algorithm for optimal placement of a mobile node (OPlaMoN) is provided. In Section 21.4, the findings from Section 21.2 are substantiated and the effectiveness of the algorithm from Section 21.3 is verified through simulation-based data. Section 21.5 summarizes the key findings and identifies a few related open research issues.

21.2   ENERGY AS MOBILE NODE DEPLOYMENT CRITERION

21.2.1   GENERAL NETWORK ASSUMPTIONS

- The radio range of individual sensor nodes is r [units].
- Each node is aware of its location.
- One single data source and one single respective data sink exist in the network.
- The network employs geographic routing to route data between the source and the sink.
- The optimal path between the source and the sink is intercepted by a routing anomaly—a routing hole.
- The node at which the source-to-sink traffic first encounters/touches the hole (that is, hole boundary) is denoted by $node(1)$, as shown Figure 21.5. The node at which the source-to-sink traffic exits the hole boundary is denoted by $node(n)$. Accordingly, a total of n boundary nodes, and $n - 1$ links, participate in the routing of the source-to-sink traffic.
- Sensor-network researchers generally agree that in a typical real-world sensor network, the deployment of only one or just a few mobile nodes may be feasible, given their considerably higher cost relative to the cost of static sensors. Hence, to "bridge" the hole, the use of only one mobile node is considered, as shown in Figure 21.6.
- When deployed, the mobile is placed at distance r [units] from one of the affected boundary nodes—we will denote this node with $node(k)$, $k \in \{1, .., n - 2\}$. (We do not consider

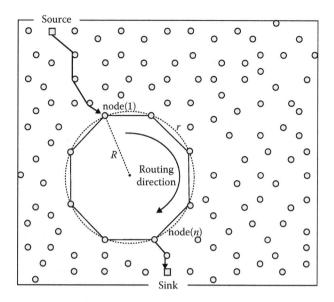

FIGURE 21.5 Circular shaped routing hole.

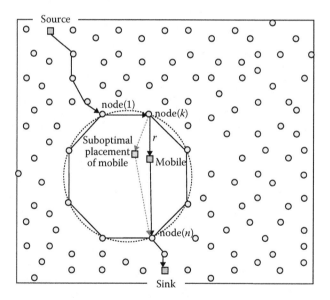

FIGURE 21.6 Circular hole bridged by a mobile node.

placing the mobile next to $node(n-1)$, because this node is the immediate neighbor of exit $node(n)$.) The distance of r [units] enables the mobile node to successfully receive a signal from $node(k)$.

- Upon its deployment, the mobile node passes the traffic directly across the hole, from $node(k)$ to exit $node(n)$.
- Finally, for all static nodes the minimum received-signal power required to reach the desired SNR ratio is $P_{received}$. Accordingly, by assuming a simple free-space path loss model and path loss gradient $\alpha = 2$, sensor nodes must emit signals with power $P_{received} \cdot r^2$, to ensure the radio range of r [units] [18].

In Sections 21.2.2 and 21.2.3 we focus on the energy aspects of self-healing by means of a mobile bridge in two particular types of routing holes likely occurring in real-world environments: circular- and rectangular-shaped [7]. Circular routing holes can be found in outdoor environments: countryside, battlefields, and so on, and they are typically caused by natural obstacles such as craters, large rocks, lakes, and ponds. Rectangular routing holes, on the other hand, can be found in urban environments, as they roughly correspond to buildings, large overpasses, and other man-made obstacles.

21.2.2 CIRCULAR-SHAPED HOLE

To evaluate the possible energy benefits of using a mobile node to bridge a circular-shaped routing hole, the assumptions from Section 21.2.1 are extended as follows:

- The routing hole, that is, area void of sensor nodes, matches the interior of a circle of radius R [units] (Figure 21.5).
- Accordingly, the sensor nodes on the boundary of the hole, and their respective radio links, form a convex polygon P of degree m (number of polygon sides = m). For the simplicity of the discussion, P is assumed to be a regular polygon with all sides of equal size r [units].

Clearly, under the above assumptions, as the size of the hole increases (R increases), more nodes will fit (that is, be found) on the boundary of the hole, ultimately increasing the degree of polygon P. In the general case of a regular polygon, the relationship between r, R, and m is given by the following formula [26]:

$$r = 2R \cdot \sin \frac{180[°]}{m} = 2R \cdot \sin \frac{\pi [\text{rad}]}{m} \tag{21.1}$$

In the worst-case routing hole scenario, in which the maximum possible number of boundary nodes is affected by a given source-to-sink traffic stream, as shown in Figure 21.6, $node(1)$, and $node(n)$ will happen to be on the opposite sides of polygon P. Consequently, assuming m is even, n and k will correspond to $n = \frac{m}{2} + 1$ and $k \in \{1, ..., \frac{m}{2} - 1\}$. (Note: the proceeding analysis and obtained results can be easily extended to an odd m.)

21.2.2.1 Routing around Hole

Now, let us begin our energy-related analysis by, first, considering the base case, involving no mobile node (Figure 21.5). The energy required to route a bit of source-to-sink traffic around the boundary of the hole is annotated by E_{1-n}. The respective expression is provided in Equation 21.2.

$$E_{1-n} = (n-1) \cdot P_{\text{received}} \cdot r^2 \tag{21.2}$$

By substituting Equation 21.1 in Equation 21.2,

$$E_{1-n} = (n-1) \cdot P_{\text{received}} \cdot (2R)^2 \cdot \left(\sin \frac{\pi}{m}\right)^2 \tag{21.3}$$

21.2.2.2 Routing through Mobile

Next, let us consider the case involving a mobile node, as shown in Figure 21.6. The energy required to route the traffic through the mobile placed next to $node(k)$, where $k \in \{1, ..., n-2\}$, and across the hole is annotated by $E_{1-k-mob(k)-n}$. The respective expression is provided in Equation 21.4, and its simplified version in Equation 21.5.

$$E_{1-k-mob(k)-n} = E_{1-k} + E_{k-mob(k)} + E_{mob(k)-n}$$
$$= (k-1) \cdot P_{\text{received}} \cdot r^2 + P_{\text{received}} \cdot r^2 + P_{\text{received}} \cdot d(mob(k), node(n))^2 \tag{21.4}$$

$$E_{1-k-mob(k)-n} = P_{received} \cdot \left(k \cdot r^2 + d^2_{mob(k)-n}\right) \tag{21.5}$$

In Equation 21.4, E_{1-k} represents the energy required to route a bit of traffic between $node(1)$ and $node(k)$, $E_{k-mob(k)}$ represents the energy required to route a bit of traffic between $node(k)$ and the mobile, and $E_{mob(k)-n}$ represents the energy required to route a bit of traffic between the mobile and the exit node ($node(n)$). $d(mob(k), node(n))$ represents the Euclidean distance between the mobile and $node(n)$. In Equation 21.5, $d_{mob(k)-n}$ is a short notation for $d(mob(k), node(n))$. The optimal placement of the mobile—the one that minimizes $d_{mob(k)-n}$, and ultimately minimizes $E_{1-k-mob(k)-n}$ (see Equation 21.5)—is along the line that passes through $node(k)$ and $node(n)$, as illustrated in Figure 21.6.

Under the optimal placement of the mobile node in the neighborhood of $node(k)$, as discussed above, $d_{mob(k)-n}$ can be expressed simply as

$$d_{mob(k)-n} = d_{k-n} - r \tag{21.6}$$

Consequently, Equation 21.5 can be rewritten into

$$E_{1-k-mob(k)-n} = P_{received} \cdot \left(k \cdot r^2 + (d_{k-n} - r)^2\right) \tag{21.7}$$

In Equation 21.7, $d(node(k), node(n)) = d_{k-n}$ is the Euclidean distance between $node(k)$ and $node(n)$. As such, d_{k-n} corresponds to the $(n-k)$th diagonal of polygon P and is given by [26]

$$d_{k-n} = 2R \cdot \sin \frac{\pi[rad] \cdot (n-k)}{m} \tag{21.8}$$

By substituting Equation 21.1 and Equation 21.8 in Equation 21.7, Equation 21.7 becomes

$$E_{1-k-mob(k)-n} = P_{received} \cdot (2R)^2 \cdot \left[k \cdot \left(\sin \frac{\pi}{m}\right)^2 + \left(\sin \frac{\pi \cdot (n-k)}{m} - \sin \frac{\pi}{n}\right)^2 \right] \tag{21.9}$$

21.2.2.3 Routing around Hole vs. Routing through Mobile

From the preceding discussion, one can simply conclude that the deployment of the mobile in a circular-shaped hole is energy justifiable as long as we can find a $k \in \{1, \dots, n-2\}$ that satisfies the following inequality:

$$E_{1-k-mob(k)-n} < E_{1-n} \tag{21.10}$$

Based on Equations 21.3 and 21.9, an equivalent form of Equation 21.10 is obtained:

$$\left(\sin \frac{\pi \cdot (n-k)}{m} - \sin \frac{\pi}{m}\right)^2 < (n-k-1) \cdot \left(\sin \frac{\pi}{m}\right)^2 \tag{21.11}$$

By substituting $(n-k)$ in Equation 21.11 with j ($j = n-1, n-2, \dots, 2$), and by annotating the left- and right-hand side of Equation 21.11 with $\mathbf{f_1}(j)$ and $\mathbf{f_2}(j)$, respectively,

$$\mathbf{f_1}(j) = \left(\sin \frac{j \cdot \pi}{m} - \sin \frac{\pi}{m}\right)^2 \tag{21.12}$$

$$\mathbf{f_2}(j) = (j-1) \cdot \left(\sin \frac{\pi}{m}\right)^2 \tag{21.13}$$

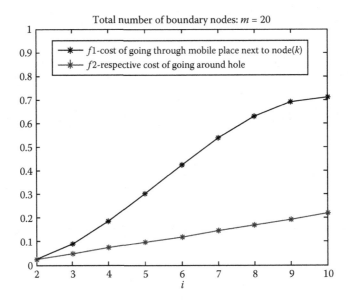

FIGURE 21.7 $\mathbf{f}_1(j)$ vs. $\mathbf{f}_2(j)$, for $m = 20$.

the initial statement can be rephrased into the use of a mobile-bridge node in a circular-shaped routing hole comprising a total of m boundary nodes is energy justifiable as long as

$$\exists\, \mathbf{j} \in \{2, \ldots, \mathbf{n} - 1\} \text{ such that } \mathbf{f}_1(j) < \mathbf{f}_2(j) \qquad (21.14)$$

Recall that in the worst-case scenario of source-to-sink traffic stream, $n = \frac{m}{2} + 1$, and $j \in \{m/2, \ldots, 2\}$.

Figure 21.7 serves as an illustrative example, because it shows the relationship between $\mathbf{f}_1(j)$ and $\mathbf{f}_2(j)$ for $m = 20$. (Similar graphs would be obtained for other possible values of m.) From these figures it is evident that the condition stated in Equation 21.14 never gets satisfied. This, consequently, leads to the following conclusion: the deployment of a mobile bridge in a circular-shaped routing hole will never be energy justifiable, regardless of the actual size of the hole or the number of boundary nodes actively involved in routing. Or, put another way, if using the total consumed energy as the sole criterion of efficiency, it will always be more favorable to route traffic around a circular-shaped hole (even if it means routing through tens or hundreds of static sensors!) than to bridge the hole by means of a single mobile node.

21.2.3 RECTANGULAR-SHAPED HOLE

In this section, we investigate the energy aspect of using a mobile node to bridge a rectangular-shaped routing hole. To do so, the assumptions from Section 21.2.1 will be extended as follows (Figure 21.8):

- Sensor nodes are organized into a grid of size $N \times N$. Each node can directly communicate with four of its nearest neighbors.
- The overall area affected by the routing hole can be modeled as a rectangle of dimensions $a \times b$, where a and b are two arbitrary real numbers ($a, b \in \Re$).

Based on the earlier analysis concerning circular-shaped holes, one can easily verify that Equation 21.2 can be (re)used to calculated the energy required to route a bit of source-to-sink traffic

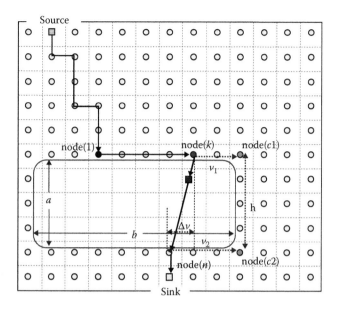

FIGURE 21.8 Rectangular hole bridged by a mobile node.

around a rectangular hole, while Equation 21.7 can be (re)used to calculate the energy required to route a bit of source-to-sink traffic through the given hole by means of a mobile, as shown in Figure 21.8. Also, as in the case of circular-shaped holes, we can argue that the deployment of the mobile node is energy justifiable as long as Equation 21.10 is satisfied. Finally, in the case of a rectangular hole, Equation 21.10 becomes Equation 21.15 and subsequently Equation 21.16.

$$(n-1) \cdot r^2 > k \cdot r^2 + (d_{k-n} - r)^2 \qquad (21.15)$$

$$(n - k - 1) \cdot r^2 > (d_{k-n} - r)^2 \qquad (21.16)$$

In the remainder of this section, by assuming completely arbitrary size and position of the routing hole with respect to the source-to-sink traffic, we investigate under which conditions any of the boundary nodes affected by the source-to-sink traffic will satisfy Equation 21.15, that is, Equation 21.16. To facilitate our analysis, we denote the nodes in the upper right and lower right corner of the observed routing hole, that is, respective rectangle, with $node(c1)$ and $node(c2)$ (see Figure 21.8). Furthermore, within the boundary nodes affected by the source-to-sink traffic ($node(1)$ to $node(n)$), the following three groups are identified and separately studied:

Group 1: $node(1)$ to $node(c1)$
Group 2: $node(c1)$ to $node(c2)$
Group 3: $node(c2)$ to $node(n)$

21.2.3.1 Mobile Placed Next to a Node of Group 1

In this section, the following auxiliary notation is employed (Figure 21.8):

v_1—Number of hops between $node(k)$ and $node(c1)$.
h—Number of hops between $node(c1)$ and $node(c2)$. Note that h is directly proportional to the dimension a of the routing hole, hence $h = \lceil a/r \rceil$. For the simplicity of our discussion, we will assume that $h = a/r$.

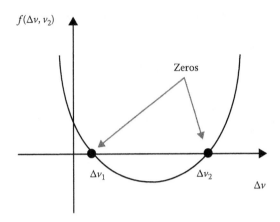

FIGURE 21.9 $f(\Delta v, v_2)$ function and its zeros assuming a fixed v_2.

v_2—Number of hops between $node(c2)$ and the exit node ($node(n)$).
(Based on their physical meaning, all three values $v_1, h, v_2 \in N$.)
$\Delta v = (v_1 - v_2)$—Defined in this way, Δv could be any integer number. Namely, if $\Delta v > 0$ the observed node ($node(k)$) is "above" and "left of" the exit node ($node(n)$), if $\Delta v = 0$ the observed node is "just above" the exit node, and if $\Delta v < 0$ the observed node is "above" and "right of" the exit node. The case presented in Figure 21.9 corresponds to a negative Δv.

For the nodes of Group 1, Equation 21.16 becomes

$$(v_1 - 1 + h + v_2) \cdot r^2 > (d_{k-n} - r)^2 \tag{21.17}$$

By substituting v_1 by $v_2 + \Delta v$, Equation 21.17 gets transformed into

$$(\Delta v + 2v_2 + h - 1) \cdot r^2 > (d_{k-n} - r)^2 \tag{21.18}$$

Given $d_{k-n} \geq r$, the following inequality will always hold

$$d_{k-n}^2 - r^2 = (d_{k-n} - r) \cdot (d_{k-n} + r) > (d_{k-n} - r)^2 \tag{21.19}$$

Based on Equation 21.19, $d_{k-n}^2 - r^2$ is an upper bound on $(d_{k-n} - r)^2$. Accordingly, given that $(d_{k-n} - r)^2$ is always strictly less than $d_{k-n}^2 - r^2$, Equation 21.19 will hold for as long as Equation 21.20 is satisfied:

$$(\Delta v + 2v_2 + h - 1) \cdot r^2 \geq d_{k-n}^2 - r^2 \tag{21.20}$$

Furthermore, for the nodes of Group 1, d_{k-n} can be represented as (see Figure 21.8)

$$d_{k-n}^2 = (\Delta v \cdot r)^2 + (h \cdot r)^2 \tag{21.21}$$

Hence, by employing Equation 21.21 in Equation 21.20, and then by dividing both sides of the inequality by r^2, Equation 21.20 becomes

$$\Delta v + 2v_2 + h - 1 \geq \Delta v^2 + (h^2 - 1) \tag{21.22}$$

Finally, Equation 21.22 can be transformed to Equation 21.23.

$$f(\Delta v, v_2) = \Delta v^2 - \Delta v - 2v_2 + (h - 1) \cdot h \leq 0 \tag{21.23}$$

Here, it is important to remind the reader that Equation 21.23 represents an alternative version of Equation 21.16, that is, Equation 21.10, adjusted to suit the nodes of Group 1.

To identify the conditions under which Equation 21.23 will be satisfied, let us first observe that $f(\Delta v, v_2)$ in Equation 21.23 is a linear function of v_2 and a convex function of Δv, (see Figure 21.9). The zeros of $f(\Delta v, v_2)$, with respect Δv, are

$$\Delta v_{1,2} = \frac{1 \pm \sqrt{1 + 8v_2 - 4(h-1)h}}{2} = \frac{1 \pm \sqrt{h(v_2)}}{2} \tag{21.24}$$

Based on the above, a condition necessary for Equation 21.23 to be satisfied is that $\Delta v_{1,2}$ (see Equation 21.24) exist and are "real." Otherwise, if $\Delta v_{1,2}$ did not exist and was real, $f(\Delta v, v_2)$ would lie entirely above x-axis, and the condition $f(\Delta v, v_2) \leq 0$ would never be satisfied!

Clearly, $\Delta v_{1,2}$ will exist and be real as long as the value under the square root $(h(v_2) = 1 + 8v_2 - 4(h-1)h)$ is positive, as indicated in Equation 21.25.

$$h(v_2) = 1 + 8v_2 - 4(h-1)h \geq 0 \tag{21.25}$$

Finally, Equation 21.25 will be satisfied provided the following holds:

$$v_2 \geq \frac{4h^2 - 4h - 1}{8} \tag{21.26}$$

Or, specifically,

$$v_2 = \text{ceil}\left(\frac{4h^2 - 4h - 1}{8}\right) + i, \ i = 1, 2, \ldots \tag{21.27}$$

21.2.3.2 Mobile Placed Next to a Node of Group 2 or Group 3

The energy justifiability analysis with respect to the nodes of Group 2 or 3 is presented in Appendices A and B. The results of this analysis show that, from the energy perspective, the deployment of the mobile "next to" a node of Group 2 or 3 will never be justifiable.

21.2.3.3 Important Theorems Arising from Sections 21.2.3.1 and 21.2.3.2

THEOREM 1

For a rectangular routing hole of a given h (where $h = a/r$), the deployment of the mobile is energy justifiable only if both of the following conditions C.1 and C.2 are satisfied (Figure 21.10):

C.1 The exit node (node(n)) lies at a specific hop distance v_2 from node(c2), where

$$\Delta v_2 \geq \frac{4h^2 - 4h - 1}{8} \tag{21.28}$$

C.2 The mobile is deployed at/near some node(k) that happens to be on the opposite side of the exit node and anywhere between $v_2 + \Delta v_1$ and $v_2 + \Delta v_2$ from node(c1), where

$$\Delta v_1 = \text{ceil}\left(\frac{1 - \sqrt{1 + 8v_2 - 4(h-1)h}}{2}\right) \tag{21.29}$$

$$\Delta v_2 = \text{floor}\left(\frac{1 + \sqrt{1 + 8v_2 - 4(h-1)h}}{2}\right) \tag{21.30}$$

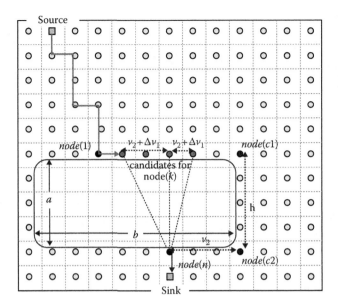

FIGURE 21.10 Topological conditions necessary for justifiability of mobile node deployment.

Proof:　To prove Theorem 1, let us consider the consequences of either C.1 or C.2 being violated.

- As shown in Section 21.2.3.1, any placement of the exit node at v_2 that contradicts C.1 will result in $f(\Delta v, v_2) > 0$ for the entire domain of Δv. This implies that Equation 21.23, that is, Equation 21.10, will never be satisfied. Accordingly, the deployment of the mobile will never be energy justifiable.
- The placement of the mobile at/near any boundary node other than the ones given by C.2 will again result in $f(\Delta v, v_2) > 0$, even if C.1 alone happens to be satisfied, and $f(\Delta v, v2)$ takes on negative values for some range of Δv. Specifically, for any $\Delta v < \Delta v_1$ or $\Delta v > \Delta v_2$ Equation 21.23 will not be satisfied given the convex nature of $f(\Delta v, v_2)$. Consequently, the deployment of the mobile will not be energy justifiable.

THEOREM 2

Let us consider a rectangular-shaped routing hole of size $S = a \times b$. If the deployment of a mobile node, aimed at bridging such a hole, is to be energy justifiable, then the hole's dimensions a and b must satisfy Equation 21.23,

$$b \neq k \cdot a \qquad (21.31)$$

where k is a small constant close to 1. (Note that Equation 21.31 implies that a and b must not be proportional to each other. In other words, the hole must not be of square shape.)

Proof:　First, in view of the discussion from Section 21.2.3.1, let us note that while one dimension of the hole (say dimension a) determines the value of h

$$h = \frac{a}{r} \qquad (21.32)$$

the other dimension (dimension b) determines the range of all possible positions v_2 of the exit node ($node(n)$):

$$1 \leq v_2 \leq \frac{b}{r} \tag{21.33}$$

To prove Theorem 2, let us assume Equation 21.31 is incorrect and a and b are directly proportional to each other ($b = k \cdot a$). By employing $b = k \cdot a$ and $a = h \cdot r$ (see Equation 21.32) into Equation 21.33, the following is obtained:

$$1 \leq v_2 \leq \frac{b}{r} = \frac{k \cdot a}{r} = \frac{k \cdot (r \cdot h)}{r} = k \cdot h \tag{21.34}$$

Clearly, v_2 given in Equation 21.34 does not satisfy condition C.1 of Theorem 1, except for very small values of h. Consequently, based on Theorem 1, the deployment of the mobile in a hole where $b = k \cdot a$ would never be energy justifiable. Thus, by using the concept of contrapositive, the correctness of Theorem 2 is proven. One can easily arrive at the same conclusion when reversing the roles of a and b: $h = b/r$ and $1 \leq v_2 \leq a/r$.

21.2.4 SUMMARY OF FINDINGS FROM SECTIONS 21.2.2 AND 21.2.3

By generalizing the findings of Sections 21.2.2 and 21.2.3, we deduce that in terms of the overall consumed energy, the deployment of a mobile bridge in any uniform-like shaped routing hole will not be justifiable, regardless of the hole's actual size or the number of its boundary nodes involved in routing. (By uniform-like shape, we mean a shape whose center of gravity is also its geometric center. Circle and square are two special cases of such holes.) Nevertheless, we argue that in such holes, especially the ones that affect large areas and a considerable number of boundary nodes, the deployment of a mobile should not be completely abandoned. Even though the path through the mobile may turn out to be more costly in terms of overall energy consumption, it is reasonable to expect that this path

- Offers lower transmission delay, by involving fewer nodes/hops
- Prevents further enlargement of the hole, by posing less demand on the boundary nodes (as discussed in Section 21.1.3)

For an illustration, see Figure 21.11. From the above, we further postulate: any algorithm aimed at finding the optimal position of the mobile, for the sake of bridging a routing hole, should not use energy as its sole criterion. Instead, the algorithm should incorporate transmission delay and hole-enlargement phenomenon as (equally) important parameters.

21.3 COOPERATIVE SELF-HEALING BY MEANS OF MOBILE NODES

21.3.1 INTRODUCTION

In this section, we present OPlaMoN, a self-healing algorithm that determines the most optimal deployment location of a mobile bridge inside a routing hole of any arbitrary topology. Building on the conclusions from the earlier sections, OPlaMoN takes into account all three of the critical network parameters: energy consumption, transmission delay, and node failure. Furthermore, OPlaMoN assumes distributed cooperative computation involving a minimum number of sensor nodes and as such is highly suited for energy and processing constrained large-scale WSNs.

The remainder of this section is organized as follows. In Section 21.3.2 we discuss the main features of distributed cooperative problem solving and the inherent advantages of using this form of

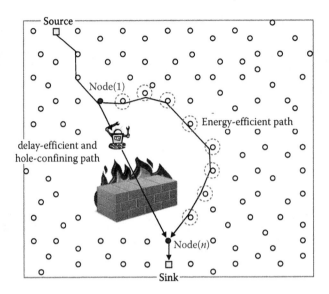

FIGURE 21.11 Energy-efficient path around hole vs. delay-efficient and hole-confining path through mobile.

problem solving in large-scale WSNs. In Section 21.3.3, we provide a detailed overview of OPlaMoN, together with a few illustrative examples of its operation.

21.3.2 DISTRIBUTED COOPERATIVE COMPUTATION

Distributed problem solving is a computational paradigm that refers the collective effort of multiple agents to solve a task at hand through combining of the agents' knowledge, information, and capabilities [20]. The main advantages of distributed over centralized problem solving (that is, computation) include increased reliability and fault tolerance, enhanced real-time response, lower communication costs, lower processing costs, reduced software complexity, and so on. [21]. Clearly, considering these properties, distributed problem solving is a form of computation much more suitable for large-scale WSNs than centralized problem solving.

From the implementation point of view, there are two main approaches to distributed problem solving:

1. Noncooperative. In this case, the problem is decomposed in such a way that each subtask can be performed completely by a single processing node, without the need for the node to see the intermediate states of processing at other nodes. Note, this also assumes that each processing node initially receives, and in the end returns, appropriate (often rather significant) portion of the overall problem-related information.
2. Cooperative. Here, on the other hand, each node performs useful processing using incomplete input data while simultaneously exchanging the intermediate results of its processing with other nodes to cooperatively construct a complete solution. In general, the amount of communication required to exchange these intermediate results is much less than the communication of raw data and processing results that are required if using the noncooperative method [21].

Given the fact that the cooperative approach has better chances of (further) minimizing the system's communication costs, this form of distributed problem solving emerges as the ultimate winner in the arena of large-scale WSNs.

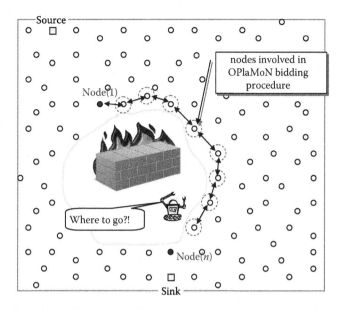

FIGURE 21.12   OPlaMoN: mobile node awaiting the highest bid.

21.3.3 OPLAMON

In this section, we take a look at OPlaMoN, a WSN self-healing algorithm aimed at finding the optimal location of a mobile bridge in a routing hole of any arbitrary topology. In its very essence, the algorithm assumes distributed cooperative computation among a very limited number of sensor nodes, with each node working on just a fraction of the overall problem-related information. As such, the algorithm poses minimal energy and processing-related requirements on the network.

In the context of the discussion from Section 21.2, the major steps of the algorithm include each affected boundary node ($node(k)$, $k = 1, \ldots, n$) calculates the so-called bidding value (see Section 21.3.1), which represents the overall gain that would be obtained if the mobile is to be placed next to this node (Figure 21.12). Through local exchange of bidding values, the nodes cooperatively identify the highest bid and the respective bidder (see Section 21.3.2). By deploying the mobile in the vicinity of the highest-bid node, the maximum network gain is ensured.

21.3.3.1 OPlaMoN: Calculation of Bidding Values

Each boundary $node(k), k = 1, \ldots, n$, calculates its bid value, $Bid(k)$, using the following weighted-sum formula:

$$Bid(k) = \begin{cases} -\infty, & \text{if battery level } < TH_{\text{critical}} \\ w_1 \cdot EG_k + w_2 \cdot DG_k \end{cases} \qquad (21.35)$$

In Equation 21.35, EG_k represents the relative energy gain that would be obtained if the mobile is to be placed next to $node(k)$ and is defined as

$$EG_k = \frac{E_{1-n} - E_{1-k-mob(k)-n}}{E_{1-n}} \qquad (21.36)$$

DG_k represents the relative delay gain (that is, reduction in transmission delay) that would be obtained if mobile is to be placed next to $node(k)$ and is defined as

$$DG_k = \frac{\text{\# hops around hole} - \text{\# hops through mobile}}{\text{\# hops around hole}} = \frac{(n-1) - (k-1+2)}{(n-1)} =$$

$$= 1 - \frac{k+1}{n-1} \qquad (21.37)$$

Note, to be able to calculate EG_k and DG_k, each node needs to know only the following:

- Its own location
- Its own hop distance from $node(1)$
- Location of exit $node(n)$
- Number of boundary nodes involved in the routing of the given source-to-sink traffic (n)

All of the above could be obtained (that is, calculated) in a single round of information exchange along the boundary of the routing hole.

Back in Equation 21.35, the weights w_1, $w_2 \in [0,1]$ determine the relative importance of EG_k and DG_k. They are meant to be adjustable and can be fine-tuned to each specific application, either by an expert or through an automated procedure. Clearly, in applications where the energy of the mobile is scarce, w_1 should be much greater than w_2. On the other hand, in time-critical applications, w_2 should be greater than w_1. Otherwise, $w_1 = w_2 = 0.5$ would be a reasonable choice.

21.3.3.2 OPlaMoN: Bidding Procedure

21.3.3.2.1 Initialization Phase
Using TENT rule [9], boundary nodes involved in the routing of source-to-sink traffic ($node(1)$ to $node(n)$) identify themselves. Subsequently, the exit node ($node(n)$) informs other boundary nodes ($node(1)$ to $node(n-1)$) of its location. The rest of the algorithm assumes that the mobile bridge is somewhere close to the boundary of the routing hole and within the radio range of at least one of the affected boundary nodes (Figure 21.13).

21.3.3.2.2 Steps Performed at Individual Nodes
Forward pass, from $node(1)$ **to** $node(n)$**:**

(1)   Calculate $Bid(k)$ according to Equation 21.35
(2.a) If $k = 1$,
 - $highestBid = Bid(1)$;
 - $higestBidderID = 1$;
 - Go to (4).
(2.b) Otherwise, wait for $Bid(k-1)$ from $node(k-1)$.
(3.a) After receiving bid-related data from $node(k-1)$, adjust own bid value:

$$Bid(k) = \begin{cases} -\infty, & \text{if } Bid(k-1) = -\infty \\ Bid(k), & \text{otherwise} \end{cases}$$

(3.b) Subsequently, calculate

$$highestBid = max\{highestBid, Bid(k)\}$$
$$highestBidder = index(max\{highestBid, Bid(k)\})$$

(4)   Send $[Bid(k), highestBid, highestBidderID]$ to $node(k+1)$.

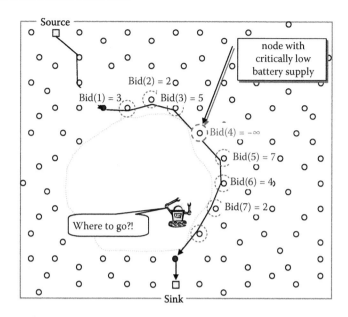

FIGURE 21.13 Boundary node with critically low battery supply.

Reverse pass, from *node*(*n*) **to** *node*(1):

(5.a) If $k = n$,
- *winningBid* = *highestBid*;
- *winningBidderID* = *highestBidderID*;
- Go to (6).

(5.b) Otherwise, wait for *highestBid* and *highestBidderID* from *node*(*n* + 1).

(6) Broadcast [*highestBid*, *highestBidderID*].

With Step (3.a), OPlaMoN ensures that the new "shortcut" path—path through the mobile—will never involve a node with critically low battery supply. As an illustration, let us look at the example of Figure 21.13. Although the initial bid value of *node*(5) is highest, placing the mobile bridge next to this node would result in traffic being routed to the mobile via *node*(4), which is already close to "dying" due to critically low battery levels. To avoid placing any further stress on *node*(4), OPlaMoN sets the bidding values of all downstream nodes (*node*(5) to *node*(7)) to infinity (∞) and, consequently, enables *node*(3) to win the bid. Also, with Steps (5) and (6), OPlaMoN ensures that the highest bid value will keep propagating back along the hole boundary, until it finally reaches the mobile node.

Final note: OPlaMoN assumes the involvement of only those boundary nodes that are actively involved in the routing of source-to-sink traffic (*node*(1) to *node*(*n*)). Each of the given nodes performs a minimum, that is, constant, number of computation and communication operations. This, consequently, yields excellent overall running time of $O(n)$.

21.4 SAMPLE SIMULATION RESULTS

To verify the work presented in earlier sections, a Qualnet*-based simulation framework has been developed. The framework assumes a virtual grid network layout, in which the overall deployment area is sectioned into a grid, and a single node is randomly placed in each cell of the formed grid [28].

---

* Qualnet is a network modeling tool by Scalable Network Technologies [27].

Sensor nodes have their communication range configured to allow them to reach all eight of their immediate neighbours.

In this section, the results of three selected simulation experiments are presented. The simulation setup features common to all three experiments include the following (see Figures 21.14 through 21.17)

- The deployed virtual grid comprises 50×50 cells.
- A routing hole is formed at the center of the network. Experiment 1, 2, and 3 assume a circular, square, and elongated shape hole, respectively.
- One single data source (represented with a larger-scale circle near point (100,100) in Figure 21.14) and one single data sink (represented with a larger-scale triangle near point [750,500] in Figure 21.14) exist in the network, at any point in time.

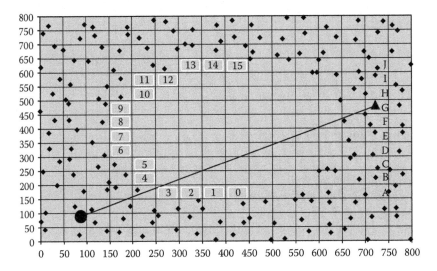

FIGURE 21.14 Self-healing in the presence of a circular-shaped hole: simulation setup.

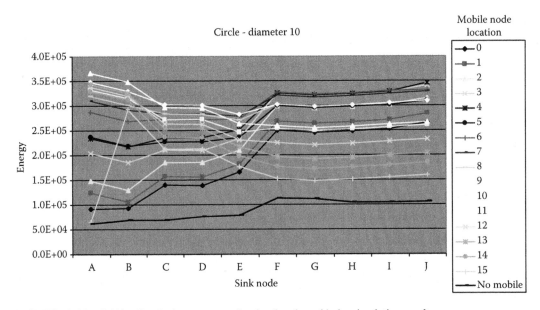

FIGURE 21.15 Self-healing in the presence of a circular-shaped hole: simulation results.

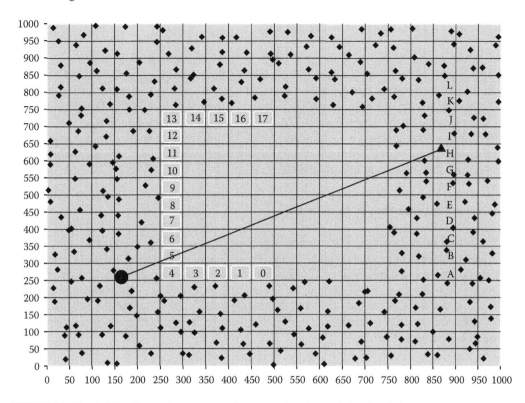

FIGURE 21.16   Self-healing in the presence of a rectangular-shaped hole: simulation setup.

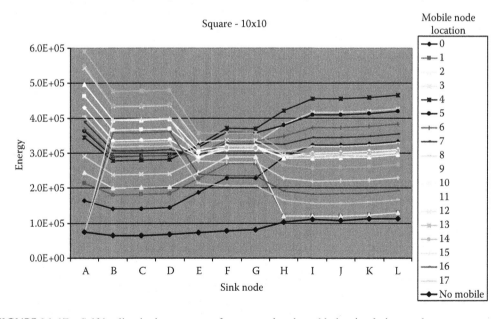

FIGURE 21.17   Self-healing in the presence of a rectangular-shaped hole: simulation results.

- In each simulation runs, the data sink occupies a different location. The investigated locations are marked with A, B, C, and so on.
- Within each simulation run, the energy aspect of deploying a mobile bridge at one of the cells marked 0, 1, 2, 3, and so on is investigated.

Experiment 1: Circular-Shaped Routing Hole

In this experiment, the operation of a WSN in the presence of a circular-shaped routing hole of diameter 10 [cell] is examined (Figure 21.14). Based on the obtained simulation results (Figure 21.15), it is evident that the energy required to route the sink-to-source traffic around the hole is considerably smaller than the energy required to route the given traffic through the mobile bridge, irrespective of the actual positions of the data sink and the mobile bridge. Hence, these results confirm our earlier hypothesis from Section 21.2.2: it is more favorable to route traffic around a circular-shaped hole—even if it means routing through tens, if not hundreds, of static sensors—then to bridge the hole by means of a single mobile node.

Experiment 2: Square-Shaped Routing Hole

The goal of this experiment is to examine the operation of a WSN in the presence of a 10×10 [cell^2] square-shaped routing hole (Figure 21.16). The obtained simulation results (see Figure 21.17) again show that the energy required to route the sink-to-source traffic around the hole is considerably smaller than the energy required to route the given traffic through the mobile bridge, irrespective of the actual positions of the data sink and the mobile bridge. Thus, our earlier hypothesis from Section 21.2.3 is once again confirmed: the deployment of a mobile bridge in square-shaped hole will never be energy justifiable.

Experiment 3: Routing Hole of "Elongated" Shape

In this experiment, the energy aspect of using a mobile bridge to self-heal a routing hole of an elongated shape is examined (Figure 21.18). The obtained simulation results indicate that in several deployment scenarios, the use of the mobile bridge results in energy savings for the network (Figure 21.19). These scenarios include the data sink placed at locations C, D, E, and F; and the mobile bridge placed at locations 4 or 5.

21.5 CONCLUSIONS AND OPEN RESEARCH ISSUES

In this chapter, the problem of self-healing and self-optimization in large-scale wireless sensor networks comprising a small number of mobile nodes has been studied. The obtained results have shown that, while bridging a routing hole by means of a mobile robot may seem intuitive, finding formal justification for the deployment of the mobile is not a trivial task. Namely, it has been proven that "energy"—the most commonly used measure of efficiency in the WSN literature—may fail to justify such a use of the mobile in a wide range of cases, including all uniformly (e.g., circle or square) shaped holes. Thus, the use of other parameters, including overall transmission delay and static-node failure, may have to be considered when deciding on whether, or where, to deploy the mobile. Based on the obtained theoretical results, and using the concepts of distributed cooperative computation, a new energy-efficient algorithm for the optimal placement of a mobile bridge has been proposed.

While this chapter has dealt with the deployment of a single mobile bridge in a routing hole affecting a single traffic stream, a number of related research problems remain to be examined, including

1. In some real-world applications more than one data source will exist and be required to report to the sink. This, consequently, will/could result in a situation as illustrated in Figure 21.20, where multiple data streams are affected by the same routing hole. Clearly, in such scenarios,

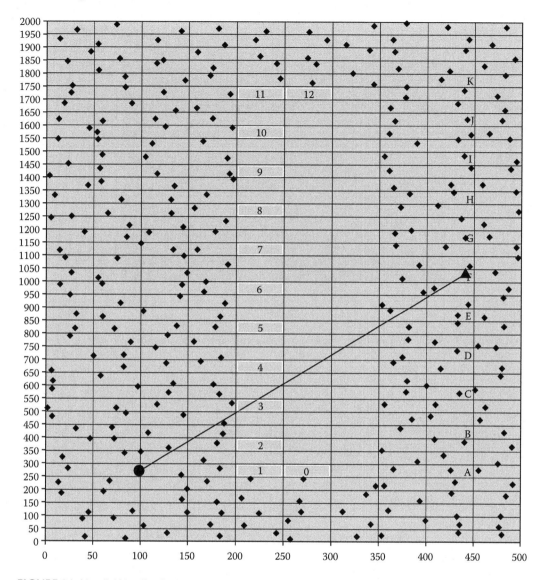

FIGURE 21.18 Self-healing in the presence of an elongated-shaped hole: simulation setup.

the problem of self-healing by means of a mobile bridge becomes even more complex, as the optimal location of the mobile will now also depend on the actual number, relative position, and intensity of the affected traffic streams.

2. In some real-world applications the utilization of more than one mobile bridge per single routing hole will be possible (Figure 21.21). In such scenarios, finding the most effective policies for the mobile nodes' deployment, for example, two in cascade vs. two in parallel, will be of paramount importance.

3. Finally, it is reasonable to expect that in some deployment environments the number of network anomalies will well exceed the number of available mobile nodes (Figure 21.22); consequently, some holes will have to be entirely left out from the healing process. In such application scenarios, a suitable hole-ranking criterion will be required to identify the holes most critical to heal. The devising of such a criterion will, clearly, be far from

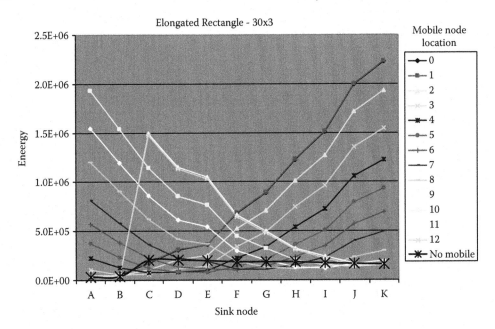

FIGURE 21.19 Self-healing in the presence of an elongated-shaped hole: simulation results.

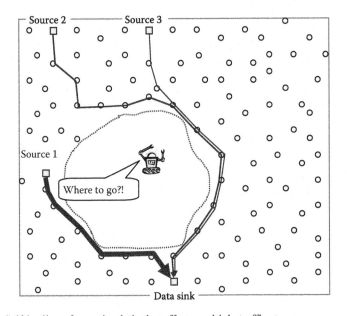

FIGURE 21.20 Self-healing of a routing hole that affects multiple traffic streams.

trivial, as it will have to take into account the size and position of each hole in the network, the relative position of holes with respect to each other,* their distance from the mobiles,† and so forth.

---

* By bridging one hole, the significance (that is, negative impact) of another hole might change, as the flow of traffic in the network gets redistributed.
† The actual locomotion of the mobile node consumes energy; hence, covering a nearby hole might ultimately be more energy efficient than covering a far-away hole.

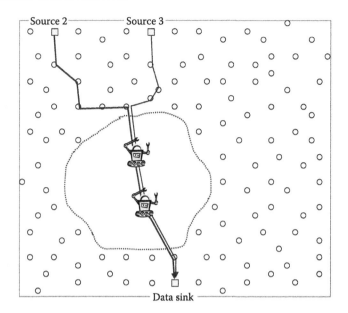

FIGURE 21.21 Self-healing by means of multiple mobile nodes.

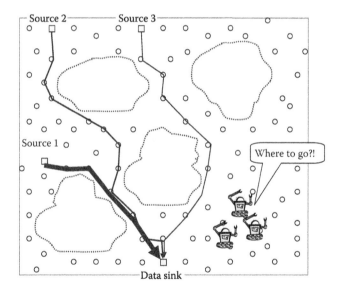

FIGURE 21.22 Self-healing involving multiple holes and multiple mobiles.

21.6 APPENDIX A: MOBILE BRIDGE PLACEMENT NEXT TO A GROUP 2 NODE

The analysis is based on the following auxiliary notation (Figure A.1):

z—number of hops between $node(k)$ and $node(c2)$.

v_2—number of hops between $node(c2)$ and the exit node $node(n)$.

Using this notation, Equation 21.15 becomes

$$(z - 1 + v_2) \cdot r^2 > (d_{k-n} - r)^2 \tag{A.1}$$

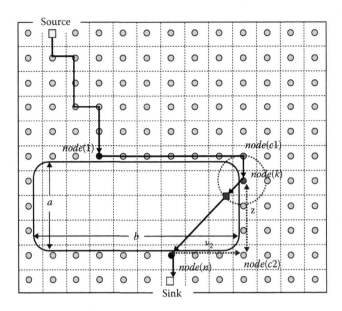

FIGURE A.1 Mobile deployed next to a node of Group 2.

Furthermore, by substituting the right-hand side of Equation A.1 with its upper bound as discussed in Section 21.2.3.1, Equation A.1 becomes

$$(z - 1 + v_2) \cdot r^2 \geq d_{k-n}^2 - r^2 \tag{A.2}$$

In the scenario of Figure 21.23, d_{k-n} can be represented as $d_{k-n}^2 = (v_2 \cdot r)^2 + (z \cdot r)^2$. Consequently, Equation A.2 gets transformed into Equation A.3, that is, Equation A.4

$$(z - 1 + v_2) \cdot r^2 \geq (v_2 \cdot r)^2 + (z \cdot r)^2 - r^2 \tag{A.3}$$

$$z + v_2 \geq v_2^2 + z^2 \tag{A.4}$$

Clearly, (given that z, v_2 can only take values in the range of positive integer numbers ≥ 1), Equation A.4 will be satisfied only in the extreme case: $z = 1$ and $v_2 = 1$. In other words, for the deployment of the mobile node to be justifiable, the exit node must be only one hop away from $node(c2)$, and the mobile must be placed next to $node(c2 - 1)$. Any other combination of z and v_2, from the range of allowable values, will not satisfy Equation A.5, that is, Equation A.1. Hence, we argue that placing the mobile next to the nodes of Group 2 is, generally, not energy justifiable.

21.7 APPENDIX B: MOBILE BRIDGE PLACEMENT NEXT TO A GROUP 3 NODE

The analysis is based on the following auxiliary notation (Figure B.1):
w—number of hops between $node(k)$ and $node(n)$; here, we assume $w > 1$, as $w = 1$ would lead to an absurd situation where the mobile is deployed next to the node, that is, just one hop away from the exit node.

Using this notation, Equation 21.8 becomes

$$(w - 1) \cdot r^2 > (d_{k-n} - r)^2 \tag{B.1}$$

In Equation B.1, d_{k-n} can be expressed simply as $d_{k-n} = w \cdot r$. Consequently, Equation B.1 becomes

$$(w - 1) \cdot r^2 > (w \cdot r - r)^2 = (w - 1)^2 \cdot r^2 \tag{B.2}$$

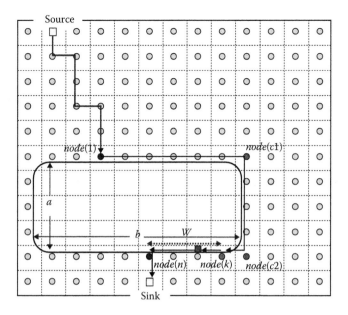

FIGURE B.1   Mobile deployed next to a node of Group 3.

Given that $w > 1$, both sides of Equation B.2 can be divided by $(w - 1)$, and Equation B.2 gets transformed into Equation B.3, that is, Equation B.4.

$$1 > (w - 1) \tag{B.3}$$

$$w < 2 \tag{B.4}$$

Clearly, the physical requirement $w > 1$, and $w < 2$ of Equation B.4, contradict each other. Hence, we conclude that Equation B.1 cannot be satisfied for any feasible value of w. In other words, for the nodes of Group 3, the deployment of the mobile node will never be justifiable.

REFERENCES

[1] J. K. Hart and K. Martinez, Environmental sensor networks: A revolution in the earth system science?, *Earth-Science Reviews*, 78(3–4), 177–191, 2006.
[2] N. Kurata, B. F. Spencer, and M. Ruiz-Sandoval, Risk monitoring of buildings with wireless sensor networks, *Journal of Structural Control and Health Monitoring*, 12, 315–327, 2005.
[3] BAA 07-46 LANdroids, Broad Agency Announcement, DARPA, 2007.
[4] J. N. Al-Karaki and A. E. Kamal, Routing techniques in wireless sensor networks: A survey, *IEEE Wireless Communications*, 11(6), 6–28, Dec. 2004.
[5] N. Ahmed, S. S. Kanhere, and S. Jha, The holes problem in wireless sensor networks: A survey, *ACM SIGMOBILE Mobile Computing and Communications Review*, 9(2), 4–18, 2005.
[6] M. Fayed and T. H. Mouftah, Characterizing the impact of routing holes on geographic routing, *Advanced Industrial Conference on Wireless Technologies (ICW'05)*, Montreal, Quebec, Canada, Aug. 2005.
[7] I. Chatzigiannakis, G. Mylonas, and S. Nikolestseas, Modeling and evaluation of the effect of obstacles on the performance of wireless sensor networks, *39th ACM/IEEE Simulation Symposium (ANSS)*, Los Alamitos, CA, 2006.
[8] S. Funke, Topological hole detection in wireless sensor networks and its applications, *3rd ACM/SIGMOBILE International Workshop on Foundations of Mobile Computing (DIAL-M-POMC)*, New York, Sep. 2005.

[9] Q. Fang, J. Gao, and L. Guibas, Locating and bypassing routing holes in sensor networks, *23rd Conference of the IEEE Communications Society (Infocom)*, pp. 2458–2468, Mar. 2004.

[10] K. Sha, J. Du, and W. Shi, WEAR: A balanced, fault-tolerant, energy aware routing protocol for wireless sensor networks, *International Journal of Sensor Networks*, 1(3/4), 156–168, Nov. 2006.

[11] Kranakis, E., Singh, H., and Urrutia, J., Compass routing on geometric networks, *11th Canadian Conference on Computational Geometry*, Vancouver, Canada, Aug. 1999.

[12] Karp, B. and Jung, H. T., GPSR: Greedy perimeter stateless routing for wireless networks, *ACM/IEEE MobiCom*, Boston, MA, Aug. 2000.

[13] P. Bose, P. Morin, I. Stojmenovic, and J. Urrutia, Routing with guaranteed delivery in ad hoc wireless networks, *Wireless Networks*, 7(6), 609–616, Kluwer Academic Publisher, Hingham, MA, 2001.

[14] A. Rao, S. Ratnasamy, C. Papadimitriou, S. Shenker, and I. Stoica, Geographic routing without location information, *MobiCom'03*, San Diego, CA, Sep. 2003.

[15] Q. Fang, J. Gao, L. Guibas, V. de Silva, and L. Zhang, GLIDER: Gradient landmark-based distributed routing for sensor networks, *24th Conference of the IEEE Communication Society (INFOCOM)*, pp. 339–350, Mar. 2005.

[16] X. Wu, G. Chen, and S. Das, Avoiding energy holes in wireless sensor networks with nonuniform node distribution, *IEEE Transactions on Parallel and Distributed Systems*, 19(5), 710–720, May 2008.

[17] W. Jia, T. Wang, G. Wang, and M. Guo, Hole avoiding in advance routing in wireless sensor networks, *IEEE Wireless Communications & Networking Conference*, Hong Kong, Mar. 2007.

[18] A. Howard, M. J. Mataric, and G. S. Sukhatme, Mobile sensor network deployment using potential fields: A distributed, scalable solution to the area coverage problem, *International Conference on Distributed Autonomous Robotic Systems (DARS 02)*, Fukuoka, Japan, 2002.

[19] J. Zhu and S. Papavassiliou, On the connectivity modeling and the tradeoffs between reliability and energy efficiency in large scale wireless sensor networks, *IEEE Wireless Communications and Networking Conference (WCNC 2003)*, pp. 1260–1265, Mar. 2003.

[20] V. Lesser and D. Corkill, Functionally accurate, cooperative distributed systems, *IEEE Transactions on Systems, Man, and Cybernetics*, SMC-11(1), 81–96, Jan. 1981.

[21] E. H. Durfee, Distributed problem solving and planning, In M. Luck, V. Marik, O. Stepankova, and R. Trappl (eds.), *Multiagent Systems and Applications*; Selected tutorial papers from the Ninth ECCAI Advanced Course (ACAI 2001) and AgentLink's Third European Agent Systems Summer School (EASSS 2001), pp. 118–149, *Springer-Verlag Lecture Notes in AI 2086*, Berlin, 2001.

[22] G. Wang, G. Cao, and T. La Porta, Proxy-based sensor deployment for mobile sensor networks, *IEEE International Conference on Mobile Adhoc and Sensor Systems (MASS'04)*, Fort Lauderdale, FL, Oct. 2004.

[23] N. Heo and P. K. Varshney, Energy-efficient deployment of intelligent mobile sensor networks, *IEEE Transactions on Systems, Man, and Cybernetics*, 35, 78–92, Jan. 2005.

[24] G. Wang, G. Cao, and T. La Porta, A bidding protocol for deploying mobile sensors, *IEEE International Conference on Network Protocols (ICNP)*, Atlanta, GA, Nov. 2003.

[25] M. Zhang, X. Du, and K. Nygard, Improving coverage performance in sensor networks by using mobile sensors, *IEEE Military Communications (MILCOM) 2005*, Atlantic City, NJ, Oct. 2005.

[26] B. Poonen and M. Rubinstein, The number of intersection points made by the diagonals of a regular polygon, *SIAM Journal on Discrete Mathematics*, 11, 135–156 1998.

[27] http://www.scalable-networks.com.

[28] D. Xai and N. Vlajic, Near-optimal node clustering in wireless sensor networks for environment monitoring, *21st Conference Advanced Networking and Applications (AINA 07)*, Niagara Falls, Ontario, Canada, May 2007.

Index

A

Access points (AP), 234
Action-first (AF) scheme, 461–462
Actor–actor (AA) coordination level, 461–462
Adaptive antenna, 326–327
Adaptive antenna array, 322, 324
Additive white Gaussian noise (AWGN)
  channel, 119
  model
    coding and relaying schemes, 15–16
    continuous channels, 15
    frequency division AWGN RC, 18–20
    nonorthogonal Gaussian CBC, 16–17
    orthogonal Gaussian CBC, 17–18
    unidirectional cooperation link, 15–16
  process, 286
Ad hoc networks
  average fairness, Shapley function, 295–297
  backbone node, 298–299
  boundary node, 291
  coalition game formation, 292–294
  inverse-cubic law, 298
  min–max fairness, nucleolus, 294–295, 298
  network connectivity *vs.* network size, 299–300
  packet-forwarding problem, 289–290
  repeated game theory, 290–291
  secondary network, 223–224
Admission control, 405
Alamouti scheme, 264, 266
All-IP architecture, 345
Analog-to-digital converters (ADCs), 451
Antenna array systems, 264–265
Application cooperation, wireless mesh networks
  collaboration, 414–415
  communication technology, 416–417
  high-definition audio/video (A/V) file, 415–416
  implementation and performance evaluation
    Bluetooth piconet, 423
    BlueZ, 424
    intergrid service mobility management scenario, 426–427
    intragrid user scenario, 424–425
    OpenWRT, 423–424
    performance measurement, 426–428
  intelligent transportation system (ITS), 416
  reference model, 414
  service management protocols
    resource broker (RB) component, 422
    service description, 420
    service discovery (SD) protocol, 420–422
    service-oriented wireless grid architecture (SOWGA), 419–420
    service roaming protocol (SRP), 422–423
  service-oriented architecture, 418–419
  wireless grid paradigm, 417–418

Authentication protocol phases, 400–402
Automated WSAN architecture, 452–453
Autonomous cooperation (AuCo), 353
Average energy gain, static user network, 155

B

Bayesian game, 204
Bessel function, 245
Binary phase shift keying (BPSK) modulation, 56, 303
Block fading environment, 274
Bluetooth piconet, 423
Bluetooth SDP, 427–428
BlueZ, 424
Boltzmann equation, 243
Byzantine agreement, 400–402

C

CacheData protocol, 375
CachePath protocol, 375
Cartesian coordinates, 241–242
CBTC, *see* Cone-based topology control approach
Centralized cooperation architecture, 350–351
Centralized power allocation
  power allocation algorithm
    multiuser channels, 127
    optimal power allocation, 127–129
    simulation, 130–131
  relay power allocation, optimization of, 125–127
  sum rate analysis, 123–125
Channel assignment
  nonoverlapping channel, 439–440
  single-*vs.* multi-network interface card (NIC), 440–441
  static *vs.* dynamic channel assignment, 441–442
Channel state information (CSI), 119
  cooperative cellular systems, 46–47
  error correction techniques, 83
  finite-SNR analysis, 87–88
  network strategies, 80–81
  OREL algorithms, 77
Channel state information at transmitter (CSIT), 140
Cluster cooperative (CC) protocol, 375
Coded cooperation, 288
Code division multiple access (CDMA), 120, 320
Coding/decoding schemes, 223
Cognitive relay, 288
  degrees of freedom and constraints, 215–216
  maximum throughput, 218
  packet acceptance control, 216
  scheduling probability, 217
  signaling protocols, 215
  spectrum sensing, 216
  traffic relaying, primary and secondary link, 214
Collaborative beamforming (CB), 300–302
Common power (COMPOW) protocol, 179–180

503